T0192123

The Handbook of Polyhydroxyalkanoates

The Handbook of Polyhydroxyalkanoates

Microbial Biosynthesis and Feedstocks

Edited by
Martin Koller

CRC Press
Taylor & Francis Group
Boca Raton London New York

CRC Press is an imprint of the
Taylor & Francis Group, an **informa** business

First edition published 2020
by CRC Press
6000 Broken Sound Parkway NW, Suite 300, Boca Raton, FL 33487-2742

and by CRC Press
4 Park Square, Milton Park, Abingdon, Oxon OX14 4RN

First issued in paperback 2023

© 2021 Taylor & Francis Group, LLC
First edition published by CRC Press 2021

CRC Press is an imprint of Taylor & Francis Group, an Informa business

No claim to original U.S. Government works

ISBN-13: 978-0-367-27559-4 (hbk)
ISBN-13: 978-0-367-54113-2 (pbk)
ISBN-13: 978-0-429-29661-1 (ebk)

DOI: 10.1201/9780429296611

Publisher's Note
The publisher has gone to great lengths to ensure the quality of this reprint but points out that some imperfections in the original copies may be apparent.

Typeset in Times
by Deanta Global Publishing Services, Chennai, India

Dedicated to the fond memory of our father, Josef Koller (1949–2019), who passed away during the creation of this book.

Contents

PART I Enzymology/Metabolism/Genome Aspects for Microbial PHA Biosynthesis

PART II   Feedstocks

Foreword I

Even in technologically advanced countries, relatively little of the massive amount of plastic produced each day is recycled. The vast majority of plastic placed in recycling bins is sent to landfill. We know about the harm to aquatic life and other detrimental effects of plastics that do not degrade in the natural environment. Polyhydroxyalkanoates (PHA) are not only biodegradable in most natural environments but are derived from renewable resources. Yet most people have never seen any products containing PHA because there are so few. The reason is the high cost of production. Without some sort of government subsidy, factors like substrate, fermentation, and separation costs prevent PHA from competing with conventional plastics. We need to produce better and cheaper PHA.

This collection of works is targeted primarily at providing researchers with state-of-the-art PHA technology required to address these challenges. Progress occurs more rapidly when the necessary information is available. Without adequate knowledge, we may stray down futile paths or draw incorrect conclusions. Years ago, we were given a bottle of toxic black smelly liquid called "nonvolatile residue" (NVR), the major waste stream from the manufacture of nylon 6'6' containing a mixture of ~100 different mono- and dicarboxylic acids. The objective was to use microorganisms to convert this waste material into something useful. After months of research, it was determined that NVR could be used to produce poly(3-hydroxybutyrate) (PHB) under chemostat and fed-batch conditions. This was wonderful, the conversion of a plastics manufacturing waste material into another plastic, PHB. Perfect, and who else could possibly think of such a weird idea? It turned out that researchers in Graz, Austria (the birthplace of this current collection of works) had arrived at exactly the same conclusion and that we were all wrong. At this point in time, the standard method for PHB analysis, also developed in Graz, was gas chromatography flame-ionization detection (GC-FID), but this method depends on the use of standards to establish the retention times of the target compounds. We ignored all of the other peaks since, to our knowledge, PHB was the only PHA in existence. Had modern rapid publication methods summarized the information available at that time, we would have been well aware that the use of NVR would lead to the formation of PHA copolymers.

Why should anyone care about the conversion of NVR into PHA? The first section in the current treatise deals with such issues. While it may be difficult for us in the Americas to comprehend, there is a shortage of farmland in much of the world. If one of the great attributes of PHA is that their production substrates are typically renewable, we must consider the effect of their large-scale production on food availability and avoid competition with agricultural materials destined for human nutrition. Speaking of competition with foodstuffs, there has been some effort to produce PHA in the plastids of various plants. Researchers trying to clone the genes for medium-chain-length PHA (*mcl*-PHA) synthesis into plants were stymied by a missing link. When the plant produces *mcl*-carboxylic acids by *de novo* synthesis, they are covalently linked to the acyl carrier protein (ACP), while they must be linked to coenzyme A (CoA) for PHA synthesis. Since some organisms can naturally produce *mcl*-PHA through *de novo* fatty acid synthesis, there must be an enzyme that can transfer the carboxylic acids to coenzyme A while still medium chain length. The discovery of *phaG* coding for the PhaG transacylase finally allowed progress to continue and demonstrates the importance of mapping key enzymes in a pathway and knowing the genes that code for them. Before developing a fermentation process for a specific PHA, it is imperative that one has some knowledge of key enzymes, their specificity, and how their activity is controlled. This knowledge is usually required when modeling the process as well. The first half of this Handbook deals with "Enzymology, Microbiology and Genetics," information that is, in combination with kinetic aspects dealt with in volume 2, required to model and or even plan a PHA production process.

Why do microorganisms accumulate PHA? Typical answers include that they serve as carbon and energy storage and/or that their production and degradation can be used by microbes to control their internal redox potential. Since there are different classes of PHA and many diverse organisms that synthesize them, there are likely specific reasons for different materials and microbes. Nevertheless, "stress" conditions are often cited as stimulating synthesis and stress-synthesis relationships are explored in several chapters of volume 2 dealing with "Environmental and Stress Factors." Stress conditions may also be used to prevent the growth of undesired microbes. Growing PHA-accumulating halophiles under osmotic conditions where few other microorganisms survive may eliminate the need for costly sterilization (one of the challenges facing the use of waste substrates that may resist sterilization) or perhaps finally allowing PHA to be produced in continuous culture over extended periods of time, greatly decreasing capital cost. Cost is presently limiting the widespread adoption of PHA as commodity plastics. The topics covered in the Handbook allow us to examine the latest approaches to better understand the processes that may allow economic PHA production, and the world should be a better place for it.

Bruce Ramsay and Juliana Ramsay

Foreword II

The Handbook of Polyhydroxyalkanoates comprises, in total, 42 chapters, which have been divided into three volumes. Volume 1 at hand focuses on various aspects of the production of polyhydroxyalkanoates (PHA) regarding suitable technical substrates, production organisms, and the cellular PHA biosynthesis machinery, whereas volumes 2 and 3 focus on the properties and downstream processes of PHA production although also substrates and mixed cultures for PHA production are subject of the second volume. A clear thematic cut between the topics of all volumes was not reached but is also difficult to reach. The three volumes are edited by Dr. Martin Koller from the University of Graz, who is very well known for his contributions to the field.

There are several chapters contributed to volume 1 by renowned scientists, which focus on the feedstocks that can be used for PHA production. A good overview and introduction are provided in Chapter 8, pledging for inexpensive and waste raw materials instead of the more expensive refined carbon sources. Other chapters review the use of long-chain fatty acids, vegetable oils, and byproducts, glycerol, CO_2, biogas, and syngas. Even the use of conventional plastics as a carbon source for PHA production is described. This means that conventional petrochemical plastics are converted into biodegradable plastics! One chapter describes how bacteria can be genetically engineered to produce PHA from inexpensive substrates. Another focus is the different organisms that can be used for PHA production. Phototrophic organisms like cyanobacteria will allow the production of PHA from CO_2 and sunlight. Haloarchaea and halophilic eubacteria will allow PHA production at high salt concentrations in the medium, which is highlighted in volume 2. Furthermore, PHA production by hydrogen-oxidizing bacteria, *Pseudomonas putida*, as well as *Burkholderia* and related bacteria is described in this Handbook. In addition, an *in silico* analysis of the genomes of PHA-synthesizing bacteria is provided in one chapter. Thermodynamic aspects and the mathematical modeling of PHA biosynthesis were also analyzed in two chapters at the beginning of volume 2. The relation of PHA metabolism and stress robustness, the analysis of the composition and structure of PHA granules, aspects of PHA degradation, and the control of PHA composition are subjects of other intriguing chapters.

We have to consider that intensive research on PHA has been conducted for more than 40 years. This research aimed at unraveling all biochemical factors that influence the biosynthesis and production of PHA. This aim has more or less been achieved. If one considers that about 40 years ago, not even the genes encoding the enzymes involved in PHA biosynthesis were known and only vague imaginations on the enzymes catalyzing the polymerization were present, we now know many details and much more than was expected. These details include the elements constituting the PHA granules and the regulation of PHA biosynthesis in the bacterial cells. This is fine.

The research on PHA was also aiming at finding conditions at which PHA could be commercially produced under economically reliable conditions. This has not at all been achieved. Despite some very short transient periods of appearance, no PHA production process was really established on a long term basis. Only poly(lactic acid) (PLA), which has a chemical structure related to that of bacterial PHA, has been established on the market but is produced by a quite different "semibiotechnologi-cal" process. However, it contributes only less than 0.1% of the total plastic market of more than 300 million tons per year. I see only one technically reliable PHA prod-uct, the copolyester of 3-hydroxybutyrate (3HB) and 3-hydroxyhexanoate (3HHx) produced by Kaneka in Japan, which is currently on the market. However, the share is very low: currently it is less than about 3% of PLA produced and less than about 0.003% of traditional plastics produced. To be honest, I do not see a significant com-mercial breakthrough coming in the near future, although biodegradable packaging materials are urgently required considering the persistent plastics and microplastic particles derived from them accumulating in our oceans, and although biodegradable plastics could be conveniently produced from renewable resources.

I see a couple of problems that will be difficult to overcome to make PHA com-mercially successful: (1) If they are produced from carbohydrates, maximally, only 33% of the carbon can be recovered as PHA. This is too little and is much less than the yield of conventional plastics from fossil resources – at least if produced from crude oil. This fact also explains why now other carbon sources than carbohydrates are intensively investigated as sources for PHA production. This is certainly a good way, and there are numerous examples shown in this volume. However, these carbon sources, in particular the gaseous or liquid C1-carbon sources, are often difficult substrates for microbial fermentations. (2) PHA are intracellular products. This lim-its the amount of PHA that can be produced per volume, which in turn requires also additional measures for the downstream processing. Remember, nearly all estab-lished biotechnological products – at least cheap bulk products – are obtained as extracellular products. There were reports on the so-called "extracellular" produc-tion of PHA or on the release of PHA from the cells after intracellular production. However, none of these reports could be transformed into a commercial process. (3) Conventional plastics are damned cheap, and the physicochemical and thermal properties of the materials produced from it are almost perfect! They have been optimized for packaging products and other applications for more than 70 years. The stable films that can be produced, for example, from poly(ethylene) are much, much thinner than films that can be currently manufactured from PHA if the stability is compared. And if one needs three times as much material to produce a PHA film or bottle than producing it from conventional plastics resources, it's again a matter of cost.

I have no doubt that niche applications and very specialized applications will be established for PHA in the future. If the costs do not matter that much, but the material properties in the widest sense are important, these applications will come probably even in large numbers. It is, however, difficult to give good advice on how to proceed in favor of real commercial production of bulk amounts of PHA as it is required for packaging materials. One should be aware that the PHA biosynthesis genes are available from many different bacteria and that they can be transferred and expressed (at least theoretically) into any other organism. In this regard, I had from the very beginning the expectation that the production of PHA in transgenic plants is a good idea. It was even shown at an early stage of this research that the PHA bio-synthesis genes of *Ralstonia eutropha* confer PHA biosynthesis to several plant species. People became very enthusiastic about this. Unfortunately, the processes were not successfully further developed. This was very disappointing because this would have allowed a direct production from CO_2 in the plants instead of a production from the carbohydrate that the natural plants are producing from CO_2 for bacterial fermentations. I never really understood why this problem could not have been overcome in a huge effort. It is certainly not trivial to get a complex organism like a plant to produce an unnatural compound as PHA is for plants, but it should be feasible.

It is good to have this volume 1 edited by Martin Koller to bring the entire PHA research further forward. This volume is very compact, and, in combination with volume 2 and 3, it may exert a new stimulus to PHA research. It summarizes current work and could be a good basis to think again about the hurdles of PHA production and how to overcome them.

Alexander Steinbüchel

About the Editor

Martin Koller was awarded his Ph.D. degree by Graz University of Technology, Austria, for his thesis on polyhydroxyalkanoate (PHA) production from dairy surplus streams, which was embedded into the EU-FP5 project WHEYPOL (Dairy industry waste as source for sustainable polymeric material production), supervised by Gerhart Braunegg, one of the most eminent PHA pioneers. As senior researcher, he worked on bio-mediated PHA production, encompassing the development of continuous and discontinuous fermentation processes, and novel downstream processing techniques for sustainable PHA recovery. His research focused on cost-efficient PHA production from surplus materials by eubacteria and haloarchaea and, to a minor extent, to the development of PHA for biomedical use.

He currently holds about 80 Web of Science listed articles often in high ranked scientific journals, authored twelve chapters in scientific books, edited three scientific books and five special issues on the PHA topic for diverse scientific journals, gave plenty of invited and plenary lectures at scientific conferences, and supports the editorial teams of several distinguished journals.

Moreover, Martin Koller coordinated the EU-FP7 project ANIMPOL (Biotechnological conversion of carbon containing wastes for eco-efficient production of high added value products), which, in close cooperation between academia and industry, investigated the conversion of the animal processing industry's waste streams toward structurally diversified PHA and follow-up products. In addition to PHA exploration, he was also active in microalgal research and biotechnological production of various marketable compounds from renewables by yeasts, chlorophyte, bacteria, archaea, fungi, and lactobacilli.

At the moment, Martin Koller is active as research manager, lecturer, and external supervisor for PHA-related projects.

Contributors

Francisca Acevedo
Department of Basic Sciences
Faculty of Medicine
and
Scientific and Technological
    Bioresource Nucleus
BIOREN
Universidad de La Frontera
Temuco, Chile

Daniela S Alvarez
Departamento de Química Biológica
Universidad de Buenos Aires
Buenos Aires, Argentina

Natalia Alvarez-Santullano
Molecular Microbiology and
    Environmental Biotechnology
    Laboratory
Department of Chemistry and and
    Center of Biotechnology Daniel
    Alkalay Lowitt
Universidad Técnica Federico Santa
    María
Valparaíso, Chile

Véronique Amstutz
University of Applied Sciences and Arts
    Western Switzerland – Valais-Wallis
Sion, Switzerland

Marina Basaglia
Department of Agronomy Food
    Natural Resources Animals and
    Environment (DAFNAE)
University of Padova
AGRIPOLIS
Legnaro, Italy

Neha Rani Bhagat
Defence Research and Development
    Organization (DRDO)
Chandigarh, India

Sergio Bordel
Department of Chemical Engineering
    and Environmental Technology
School of Industrial Engineering
University of Valladolid
Valladolid, Spain
and
Institute of Sustainable Processes (ISP)
Valladolid, Spain

Christopher J Brigham
Department of Interdisciplinary
    Engineering
Program of Biological Engineering
Wentworth Institute of Technology
Boston, MA, USA

Sergio Casella
Department of Agronomy Food
    Natural Resources Animals and
    Environment (DAFNAE)
University of Padova
AGRIPOLIS
Legnaro, Italy

Tiziano Cazzorla
Department of Agronomy Food
    Natural Resources Animals and
    Environment (DAFNAE)
University of Padova
AGRIPOLIS
Legnaro, Italy

Jiun Yee Chee
Ecobiomaterial Research Laboratory
School of Biological Sciences
Universiti Sains Malaysia
Minden, Penang, Malaysia

Nazim Cicek
Department of Biosystems Engineering
University of Manitoba
Winnipeg, MB, Canada

Chris Dartiailh
Department of Biosystems Engineering
University of Manitoba
Winnipeg, MB, Canada

Raúl Donoso
Programa Institucional de Fomento a la
    Investigación
Desarrollo e Innovación (PIDi)
Universidad Tecnológica Metropolitana
Santiago, Chile

Bernhard Drosg
University of Natural Resources and
    Life Sciences Vienna
Department IFA-Tulln
Tulln, Austria
and
BEST – Bioenergy and Sustainable
    Technologies GmbH
Graz, Austria

Lorenzo Favaro
Department of Agronomy Food
    Natural Resources Animals and
    Environment (DAFNAE)
University of Padova
AGRIPOLIS
Legnaro, Italy

Paulo Igor Firmino
Institute of Sustainable Processes (ISP)
Valladolid, Spain
and
Department of Hydraulic and
    Environmental Engineering
Federal University of Ceará
Fortaleza, Brazil

Ines Fritz
Department IFA-Tulln
University of Natural Resources and
    Life Sciences Vienna
Tulln, Austria

Jilagamazhi Fu
Department of Food Science and
    Bioengineering
Inner Mongolia University of
    Technology
Huhhot, Inner Mongolia, PR China

Gahlawat, Geeta
Defence Research and Development
    Organization (DRDO)
Chandigarh, India
and
Department of Microbiology
Panjab University
Chandigarh, India

Arup Giri
Defence Research and Development
    Organization (DRDO)
Chandigarh, India

Nils Hanik
University of Applied Sciences and Arts
    Western Switzerland – Valais-Wallis
Sion, Switzerland

Martin Koller
Institute of Chemistry Austria
NAWI Graz
Office of Research Management and
    Service
University of Graz, Austria
and
ARENA – Association for Resource
    Efficient and Sustainable
    Technologies
Graz, Austria

Preeti Kumari
Defence Research and Development
    Organization (DRDO)
Chandigarh, India

Manoj Lakshmanan
Ecobiomaterial Research Laboratory
School of Biological Sciences
Universiti Sains Malaysia
Minden, Penang, Malaysia
and
USM-RIKEN International Centre for
    Aging Science (URICAS)
School of Biological Sciences
Universiti Sains Malaysia
Minden, Penang, Malaysia

Raquel Lebrero
Department of Chemical Engineering
    and Environmental Technology
School of Industrial Engineering
University of Valladolid
Valladolid, Spain
and
Institute of Sustainable Processes (ISP)
Valladolid, Spain

David B Levin
Department of Biosystems
    Engineering
University of Manitoba
Winnipeg, MB Canada

Si Liu
UCD Earth Institute and School of
    Biomolecular and Biomedical
    Science
University College Dublin
Dublin, Ireland
and
BiOrbic – Bioeconomy Research Centre
University College Dublin
Dublin, Ireland

Juan Carlos López
Department of Environment, Bioenergy
    and Industrial Hygiene
Ainia, Paterna
Valencia, Spain

Beatriz Maestro
Microbial and Plant Biotechnology
    Department
Host-Parasite Interplay in
    Pneumococcal Infection
Centro de Investigaciones Biológicas
    Margarita Salas, CIB-CSIC
Madrid, Spain

Maria Tsampika Manoli
Polymer Biotechnology Group
Microbial and Plant Biotechnology
    Department
Centro de Investigaciones Biológicas
    Margarita Salas, CIB-CSIC
Madrid, Spain
and
Interdisciplinary Platform for
    Sustainable Plastics towards a
    Circular Economy
Spanish National Research Council
    (SusPlast-CSIC)
Madrid, Spain

Aranzazu Mato
Polymer Biotechnology Group
Microbial and Plant Biotechnology
    Department
Centro de Investigaciones
    Biológicas Margarita Salas,
    CIB-CSIC
Madrid, Spain
and
Interdisciplinary Platform for
    Sustainable Plastics towards a
    Circular Economy
Spanish National Research Council
    (SusPlast-CSIC)
Madrid, Spain

Katharina Meixner
University of Natural Resources and
    Life Sciences Vienna
Department IFA-Tulln
Tulln, Austria

Mariela P Mezzina
Departamento de Química Biológica
Universidad de Buenos Aires
Buenos Aires, Argentina

Maierwufu Mierzati
Department of Materials Science and
    Engineering
School of Materials and Chemical
    Technology
Tokyo Institute of Technology
Nagatsuta, Midori-ku
Yokohama, Japan

Nisha Mohanan
Department of Biosystems
    Engineering
University of Manitoba
Winnipeg, MB, Canada

Raúl Muñoz
Department of Chemical
    Engineering and Environmental
    Technology
School of Industrial Engineering
University of Valladolid
Valladolid, Spain
and
Institute of Sustainable
    Processes (ISP)
Valladolid, Spain

Tanja Narancic
UCD Earth Institute and School of
    Biomolecular and Biomedical
    Science
University College Dublin
Dublin, Ireland
and
BiOrbic – Bioeconomy Research
    Centre
University College Dublin
Dublin, Ireland

Rodrigo Navia
Doctoral Program in Sciences of
    Natural Resources and Scientific and
    Technological Bioresource Nucleus,
    BIOREN
and
Department of Chemical Engineering
    and Centre for Biotechnology and
    Bioengineering (CeBiB)
Faculty of Engineering and Sciences
Universidad de La Frontera
Temuco, Chile

Markus Neureiter
University of Natural Resources and
    Life Sciences Vienna
Department IFA-Tulln
Tulln, Austria

Juan Nogales
Interdisciplinary Platform for
    Sustainable Plastics towards a
    Circular Economy
Spanish National Research Council
    (SusPlast-CSIC)
Madrid, Spain
and
Department of Systems Biology
Centro Nacional de Biotecnología,
    CNB-CSIC
Madrid, Spain

Kevin E O'Connor
UCD Earth Institute and School of
    Biomolecular and Biomedical
    Science
University College Dublin
Dublin, Ireland
and
BiOrbic – Bioeconomy Research
    Centre
University College Dublin
Dublin, Ireland

Victor Pérez
Department of Chemical Engineering
    and Environmental Technology
School of Industrial Engineering
University of Valladolid
Valladolid, Spain
and
Institute of Sustainable Processes (ISP)
Valladolid, Spain

Danilo Perez-Pantoja
Programa Institucional de Fomento a la
    Investigación
Desarrollo e Innovación (PIDi)
Universidad Tecnológica Metropolitana
Santiago, Chile

M. Julia Pettinari
Departamento de Química Biológica
Universidad de Buenos Aires
Buenos Aires, Argentina

M. Auxiladora Prieto
Polymer Biotechnology Group
Microbial and Plant Biotechnology
    Department
Centro de Investigaciones Biológicas
    Margarita Salas, CIB-CSIC
Madrid, Spain
and
Interdisciplinary Platform for Sustainable
    Plastics towards a Circular Economy
Spanish National Research Council
    (SusPlast-CSIC)
Madrid, Spain

Bruce Ramsay
PolyFerm Canada
Harrowsmith, ON, Canada
and
Queen's University
Kingston, ON, Canada

Juliana Ramsay
Queen's University
Kingston, ON, Canada.

Sebastian L Riedel
Chair of Bioprocess
    Engineering
Institute of Biotechnology
Technische Universität Berlin
Berlin, Germany.

Yadira Rodríguez
Department of Chemical
    Engineering and Environmental
    Technology
School of Industrial
    Engineering
University of Valladolid
Valladolid, Spain
and
Institute of Sustainable Processes (ISP)
Valladolid, Spain

Jesús Miguel Sanz
Microbial and Plant Biotechnology
    Department
Host-Parasite Interplay in
    Pneumococcal Infection
Centro de Investigaciones Biológicas
    Margarita Salas, CIB-CSIC
Madrid, Spain

Michael Seeger
Molecular Microbiology and
    Environmental Biotechnology
    Laboratory
Department of Chemistry and Center
    of Biotechnology Daniel Alkalay
    Lowitt
Universidad Técnica Federico Santa
    María
Valparaíso, Chile

Mario Sepúlveda
Molecular Microbiology and
  Environmental Biotechnology
  Laboratory
Department of Chemistry and Center
  of Biotechnology Daniel Alkalay
  Lowitt
Universidad Técnica Federico Santa
  María
Valparaíso, Chile

Parveen KSharma
Department of Biosystems
  Engineering
University of Manitoba
Winnipeg, MB, Canada

John L Sorensen
Department of Chemistry
University of Manitoba
Winnipeg, MB, Canada

Alexander Steinbüchel
University of Münster
Münster, Germany

Kumar Sudesh
Ecobiomaterial Research
  Laboratory
School of Biological Sciences
Universiti Sains Malaysia
Minden, Penang, Malaysia
and
USM-RIKEN International
  Centre for Aging Science
  (URICAS)
School of Biological Sciences
Universiti Sains Malaysia
Minden, Penang, Malaysia

Natalia Tarazona
Polymer Biotechnology Group
Microbial and Plant Biotechnology
  Department
Centro de Investigaciones Biológicas
  Margarita Salas, CIB-CSIC
Madrid, Spain
and
Institute of Biomaterial Science and
  Berlin-Brandenburg Center for
  Regenerative Therapies
Helmholtz-Zentrum Geesthacht
Teltow, Germany.

Takeharu Tsuge
Department of Materials Science and
  Engineering
School of Materials and Chemical
  Technology
Tokyo Institute of Technology
Nagatsuta, Midori-ku, Yokohama, Japan

Camila Utsunomia
University of Applied Sciences and Arts
  Western Switzerland – Valais-Wallis
Sion, Switzerland

Ariel Vilchez
Doctoral Program in Sciences of
  Natural Resources and Scientific and
  Technological Bioresource Nucleus
BIOREN
Universidad de La Frontera
Temuco, Chile

Pamela Villegas
Programa Institucional de Fomento a la
  Investigación
Desarrollo e Innovación (PIDi)
Universidad Tecnológica Metropolitana
Santiago, Chile

Nick Weirckx
Institute of Bio- and Geosciences
IBG-1: Biotechnology and Bioeconomy
    Science Center (BioSC)
Forschungszentrum Jülich
Jülich, Germany

Idris Zainab-L
Ecobiomaterial Research Laboratory
School of Biological Sciences
Universiti Sains Malaysia
Minden, Penang, Malaysia
and
Department of Biochemistry,
Faculty of Science,
Bauchi State University Gadau.
Bauchi, Nigeria

Manfred Zinn
University of Applied Sciences and Arts
    Western Switzerland – Valais-Wallis
Sion, Switzerland

The Handbook of Polyhydroxyalkanoates, Volume 1: Introduction by the Editor

Nowadays, it is generally undisputed that we need alternatives for various fossil-resource based products such as plastics, which make our daily life indeed comfortable. Plastics, *per definitionem* a group of synthetic polymeric materials typically not produced by Mother Nature, are currently produced at increasing quantities, now in a magnitude of about 400 Mt per year. Such plastics, which are manufactured by well-established technologies, are used in innumerable fields of application, such as packaging materials, parts in the automotive industry, sports articles, biomedical devices, electronic parts, and many more. Despite their undoubted contribution to facilitating our all-day life, current plastic production is associated with essential shortcomings, such as the ongoing depletion of fossil resources, growing piles of waste consisting of non-degradable full-carbon-backbone plastics, microplastics accumulating in marine and other aquatic environments and also in food, and elevated CO_2 and toxin levels in the atmosphere generated by plastic incineration.

The last few decades have been dedicated to finding a way out of the fatal "Plastic Age" we live in today and to overcoming the above-mentioned evils. This goes in parallel with current political regulations in diverse global regions, such as the European Strategy for Plastics in a Circular Economy by the European Commission, or the forthcoming ban on disposable plastic items in megacities of PR China, which was announced just the other day (January 2020).

Switching from petrol-based plastics to bio-alternatives with plastic-like properties, which are based on renewable resources, and which can be subjected to biodegradation and composting, is regarded as one of these exit strategies. In this context, polyhydroxyalkanoates (PHA), microbial storage materials produced by numerous eubacterial and archaeal prokaryotes, are, to an increasing extent, considered auspicious candidates to replace traditional plastics in several market sectors, such as the packaging field, or even in sophisticated biomedical applications.

However, to make PHA competitive, they must cope with petrol-based plastics both in terms of quality and in economic aspects. Quality improvement of PHA-based materials is currently achieved by advanced microbial feeding strategies during the biosynthesis, by the generation of (nano)composites with diverse compatible and often cost-efficient (nano)filler materials, by blending with other suitable polymers, or development of novel (nano)composite materials. Importantly, the entire PHA production chain, encompassing the isolation of new robust production strains, improvement of the strains by means of genetic engineering, understanding the enzymatic machinery of PHA anabolism and catabolism, feedstock selection, fermentation technology, process engineering, and bioreactor design, and, last but not least, the downstream processing, needs to meet the criteria of sustainability. Hence, despite the myriad of premature praise articles found in the current literature, PHA and other "plastic-like" biopolymers cannot be regarded as the one and only panacea to solve the global plastic problem! The previously often-cited myth of biopolymers being intrinsically more sustainable than established petrochemical plastics nowadays has finally been abandoned by most serious scientists. This means that without considering and conceiving the entire life cycle of biopolymers like PHA and the products produced thereof, it is impossible to conclude *a priori* if they inherently outperform their petrochemical counterparts in terms of environmental benefit. This is only possible by using modern tools of cradle-to-grave life cycle assessment and holistic cleaner production studies.

Such economic, sustainability, and quality aspects are dealt with in the 42 chapters of this *Handbook of Polyhydroxyalkanoates*, which consists of carefully selected contributions by differently focused research groups, all of them belonging to the top global cohort regarding their individual PHA-related expertise. In general, the book consists of chapters each dedicated to one of the subsequently listed three major objectives:

a) How to better understand the mechanisms of PHA biosynthesis in scientific terms (genetics background, enzymology, metabolomics, "synthetic biology" approaches for engineering PHA production strains in a more effective way, etc.), and profiting from this understanding in order to enhance PHA biosynthesis in biotechnological terms and terms of PHA microstructure?
b) How to make smart materials based on PHA to be used for defined applications, both in the bulk and niche sector?
c) How to make PHA competitive for outperforming established petrol-based plastics on the industrial scale? What are the obstacles to market penetration of PHA?

In principle, these three major questions, especially (a) and (c), are treated by 15 contributions in the present volume 1 of *The Handbook of Polyhydroxyalkanoates*, which are dedicated to the subsequent central thrusts of PHA research.

ENZYMOLOGY/METABOLISM/GENOME ASPECTS FOR MICROBIAL PHA BIOSYNTHESIS

This broad topic covers the intracellular processes in PHA-accumulating microorganisms. As a core part of the entire book, it is covered by a total of seven chapters, which, of course, are also somehow related to substrate aspects described in the subsequent feedstock-focused chapters.

The first chapter in this section, provided by Maierwufu Mierzati and Takeharu Tsuge, describes the action of the enzymatic machinery involved in PHA biosynthesis by different microbial production strains. This encompasses a range of different biocatalysts, which finally provide the activated building blocks (monomers) that undergo polymerization by different PHA synthase enzymes; here, the focus is dedicated to hydroxyacyl-coenzyme A (HA-CoA) generation and subsequent polymerization of the hydroxyacyl moiety in HA-CoA as the two most essential elements in PHA biosynthesis. This chapter, for sure, is pivotal to understand both the interrelation between the synthetic mechanism of PHA formation and PHA's material properties.

A comprehensive chapter by Mariela P. Mezzina, Daniela S. Alvarez, and M. Julia Pettinari deals with physiological and metabolic aspects of PHA granules ("carbonosomes"), which constitute *de facto* organelles in prokaryotic microbes; these granules possess fascinating, complex attitudes regarding their composition and formation. The readers will learn that these "carbonosomes" are by far more than just simple "bioplastic inclusions in bacteria," and will obtain a deep insight into the broad range of PHA granule-associated proteins essential for *in vivo* PHA formation, such as synthases, polymerases, or phasins.

A genetically focused chapter by Parveen K. Sharma et al. from David Levin's team reviews the genomics and genetics of short-chain-length (*scl-*) PHA synthesis by *Cupriavidus necator* H16 (formerly *Ralstonia eutropha* H16) with the completely deciphered genome, PHA synthesis by recombinant *C. necator* strains, the genomics and genetics of medium-chain-length (*mcl-*) PHA synthesis by *Pseudomonas putida* and other *Pseudomonas* species, PHA synthesis by recombinant *Pseudomonas* species, and the genomics and genetics of PHA synthesis by *Halomonas* sp. and recombinant *Escherichia coli* strains. Special emphasis is dedicated to the different groups of PHA synthases found in individual PHA production strains. It is shown that genome analysis of PHA producers steadily identifies new genes; in the future, this knowledge should definitely be tapped for manipulating and advancing PHA production!

In the case of *mcl*-PHA, the enzymatic machinery and the metabolic pathways toward *mcl*-PHA differ considerably if compared to the events observed during *scl*-PHA biosynthesis; therefore, a specialized chapter by Maria Tsampika Manoli and other researchers associated with Auxiliadora Prieto is dedicated to the molecular basis of the PHA machinery in the best-known *mcl*-PHA producer, *Pseudomonas putida*, focusing on the involved genes, and diverse factors involved in the expression of the genes relevant for *mcl*-PHA biosynthesis.

Beyond that, PHA production by *Paraburkholderia* and *Burkholderia* species is reviewed in a separate chapter written by Natalia Alvarez-Santullano and colleagues from Michael Seeger's team. Here, the genes encoding enzymes and proteins involved in PHA metabolism by these powerful strains are presented, and the metabolic routes of PHA homo- and heteropolyester synthesis and the metabolism of substrates that are used by *Paraburkholderia* and *Burkholderia* to produce PHA are presented. Biotechnological applications, including biomedical uses of bacterial PHA produced by exactly these microbial species, are discussed.

The next highly specialized chapter by Lorenzo Favaro and colleagues from the team of Marina Basaglia and Sergio Casella summarizes recent relevant results dealing with PHA production from various organic byproducts by means of genetically engineered microbial strains. The most relevant and recent genomic tools for the genetic modification are initially described, with emphasis on hosts, genes, plasmids, promoters, and gene copy numbers. This chapter deals with two principal approaches in this direction, namely the engineering of highly efficient PHA producing microorganisms for their use of waste streams ("make the PHA producer convert an inexpensive substrate"), and the engineering of bacteria naturally able to use complex and inexpensive carbon sources, but unable to produce PHA ("make a converter of an inexpensive substrate accumulate PHA").

Because the composition of both scl-PHA and mcl-PHA on the molecular level and the exact microstructure of PHA (blocky structured PHA vs. random distribution of the monomers in heteropolyesters) are of major significance for the material properties and workability of PHA, an individual chapter by Camila Utsunomia, Nils Hanik, and Manfred Zinn covers the biosynthesis and sequence control of scl-PHA and mcl-PHA. This chapter presents the key elements to be considered in order to understand and fine-tune the microstructure and sequence-controlled molecular architecture of PHA copolyesters, including feeding regimes, genetic engineering of production strains, and artificial genetic networks.

FEEDSTOCKS

Eight chapters deal with the assessment of diverse feedstocks to be used as a carbon source for PHA production. Importantly, these feedstocks constitute carbonaceous (agro)industrial waste streams (lignocelluloses, waste glycerol, starchy waste, surplus whey, molasses, CO_2, CH_4, etc.) or their volatile follow-up products like syngas or biogas.

In this context, a comprehensive overview chapter by Sebastian Riedel and Christopher Brigham summarizes the current knowledge on PHA biosynthesis, starting from inexpensive waste feedstocks. Focus is dedicated to available industrial waste from agriculture and food processing as inexpensive feedstocks for PHA production, the types of polymers that are made of them, and the possibility of upscaling these processes to enable large-scale, low-cost PHA production.

The second chapter in this section, provided by Neha Rani Bhagat et al. from the team of Geeta Gahlawat addresses the fact that crude glycerol is a byproduct of many industrial processes, such as biodiesel production, and huge surplus amounts of it are released into the environment as waste, thereby necessitating the search for new methods of its utilization. These authors present the challenges, benefits, and drawbacks of PHA biosynthesis based on crude glycerol stemming from diverse, inexpensive resources by application of different microbial production strains and focus on the different types and properties of the generated PHA biopolyesters.

Another chapter by Manoj Lakshmanan and colleagues from the group of Kumar Sudesh discusses the use of vegetable oils and its byproducts, including oils without nutritional value, by various bacterial strains for PHA biosynthesis. This includes the production of PHA by both wild-type and genetically engineered bacterial strains. The potential application of these strain-substrate combinations for large-scale PHA production at low cost is also discussed in this chapter.

One specialized chapter in this feedstock section, provided by Chris Dartiailh and other associates of David Levin, reviews the synthesis of *mcl*-PHA using long-chain fatty acids (LCFAs) from different waste oils, highlighting the influence of selected substrates, carbon loading, and bioreactor systems on the yields of *mcl*-PHA in bacteria, the monomer composition, and the properties of synthesized biopolyesters. Emphasis is dedicated to functionalization and cross-linking of vinyl moieties present in obtained *mcl*-PHA to generate new biomaterials, and optimization of *mcl*-PHA production using LCFAs as feedstocks.

As a rather exotic, but currently emerging topic, even follow-up products of spent petrochemical plastics treated by chemo-biotechnological processes can be used as raw materials for microbial "bioplastic" production. Such "upcycling" of plastic waste to biodegradable polymers could indeed be part of the new circular economy paradigm. This chapter by Tanja Narancic and colleagues from Kevin E. O'Connor's team also provides a detailed, unprecedented analysis of the metabolic background of microbes being able to convert follow-up products of traditional spent plastics to PHA biopolyesters.

Among gaseous C1-substrates used for PHA biosynthesis, CO_2 from industrial effluent gases is of increasing interest and therefore is handled in a chapter comparing chemoheterotrophic with solar-based photoautotrophic PHA production. This chapter by Ines Fritz, Katharina Meixner, Markus Neureiter, and Bernhard Drosg provides an intriguing insight into the current state of autotrophic PHA biosynthesis by cyanobacteria, with a strong focus dedicated to the comparison of photoautotrophic and chemoheterotrophic PHA biosynthesis. Most of all, a frank discussion on the potential of CO_2-based PHA production for industrial implementation is provided.

In the context of gaseous substrates, the next chapter focuses on the use of the C1-compound methane for PHA production by type II methane-oxidizing α-proteobacteria. Here, it is shown by Yadira Rodríguez and colleagues from the research group of Raúl Muñoz how biopolyester production can be coupled to biogas generation based on anaerobic digestion of inexpensive organic waste materials. Importantly, it is shown in this chapter by techno-economic analysis that production costs of biogas-based PHA are competitive compared to established feedstocks for PHA production, provided that also advanced bioreactor facilities are available for efficient cultivation of microbes on gaseous substrate streams.

Finally, the last chapter written by Véronique Amstutz and Manfred Zinn covers the use of CO-rich syngas, which can be produced from various organic waste materials for PHA biosynthesis. Here, the pros and cons of tested carboxydobacterial production strains like *Rhodospirillum rubrum* in terms of substrate assimilation and PHA productivity are elucidated, and genetic engineering approaches to overcome specific metabolic hurdles are presented. In addition, the underlying processes of waste and biomass gasification and gas fermentation are elucidated, together with the metabolic pathways involved in the assimilation of syngas as carbon and energy sources.

Most importantly, all these feedstocks-to-PHA-related works aim to explore alternatives to commonly used first-generation feedstocks of value for human nutrition (pure sugars, edible starch, edible oils, etc.). This paradigm avoids the current "plate vs. plastic" dispute, thus making PHA production ethically clear, and contributes to the integration of PHA production into the principles of bioeconomy and circular economy. Further, these articles show how such alternative feedstocks have to be pre-treated in order to make them easily accessible for microbial conversion and to minimize the potential inhibitory effects on the production strains by appropriate upstream processing approaches.

Martin Koller

Part I

Enzymology/Metabolism/ Genome Aspects for Microbial PHA Biosynthesis

1 Monomer-Supplying Enzymes for Polyhydroxyalkanoate Biosynthesis

Maierwufu Mierzati and Takeharu Tsuge

CONTENTS

1.1 INTRODUCTION

Polyhydroxyalkanoates (PHA) are synthesized by some native bacterial strains, recombinant bacterial strains, and recombinant eukaryotes [1]. PHA are formed via the metabolic transformation of various carbon sources inside the microbial cell. In native PHA-producing bacteria, these polyesters are produced as intracellular carbon storage materials and energy sources. PHA are accumulated by bacteria as cytoplasmic inclusions; the number and size of each compound depend on the carboxyl group of the monomers that form an ester bond with the hydroxyl group of the adjacent monomers (see Figure 1.1) [2].

To date, more than 150 different PHA monomers have been identified [3] and categorized into short-chain-length (*scl*: ~C5), medium-chain-length (*mcl*: C6 to C14), and long-chain-length (*lcl*: C15~) 3-hydroxyalkanoates (3HAs) [4].

3

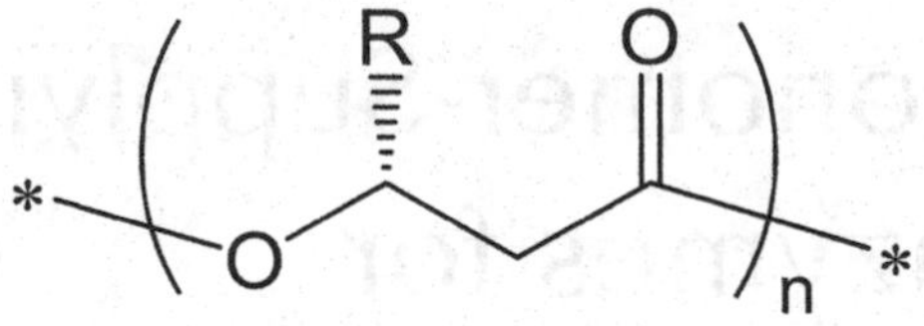

FIGURE 1.1 Chemical structure of typical PHA. The pendant group (R) stands for saturated or unsaturated side chains that consist of 1–13 carbons, in some cases it contains substituents. The most well-known PHA is P(3HB) where a methyl group is in the R position.

Among all of the biodegradable polymer materials, PHA are rather unique because the *in vivo* synthetic process of their manufacture is carried out inside a living organism, and they can be usually biodegraded in different environments and bio-systems compared with others. There are two major processes involved in the biosynthesis of PHA in microorganisms, generating the hydroxyacyl-coenzyme A (HA-CoA) from numerous metabolic pathways and polymerizing HA-CoAs into PHA. Commonly speaking, all metabolic pathways involved in PHA production follow these processes. PHA show a broad range of material properties from thermoplastics to rubber-like polymer depending on different monomeric compositions [5], which is the result of the polymerizing enzyme, PHA synthase (or polymerase), and the HA-CoA thioester precursors supplied to the enzyme based on the metabolic pathway and external carbon sources [6,7].

Because of the rich diversity in PHA biosynthesis metabolism and its broad range of applications as a material, modification via metabolic engineering may lead to promising applications in practical uses. PHA metabolic engineering is involved in controlling different factors related to the polymer's material characteristics, such as its monomeric composition, primary structure, and molecular weight of the polymer. In addition, material performance can be further enhanced by introducing genetically engineered PHA-related enzymes as reinforcement tools.

By engineering various PHA monomer-supplying enzymes via structure-based mutagenesis and function-based molecular evaluation, this gives us the possibility of maximizing the performance of each enzyme; PHA with optimizing properties can be obtained by collaboration of these enzymes with PHA synthase. In this chapter, we focus on the metabolic engineering of PHA-related monomer-supplying enzymes based on their structures and functions.

1.2 PHA BIOSYNTHESIS PATHWAYS AND RELATED ENZYMES

In microorganisms that spontaneously accumulate PHA, there are three main PHA biosynthetic pathways (see Figure 1.2). In pathway I, (*R*)-3-hydroxybutyrate [(*R*)-3HB] monomers are synthesized from two acetyl-CoA molecules. It is considered to be the most well-known and studied pathway and has been identified in a wide range of bacteria that accumulate *scl*-PHA, especially *Ralstonia eutropha* (currently designated as *Cupriavidus necator*) [8]. In *R. eutropha*, two acetyl-CoA molecules are condensed to generate acetoacetyl-CoA under the catalyzation of 3-ketothiolase (PhaA). The product of PhaA, acetoacetyl-CoA, is subsequently reduced to (*R*)-3HB-CoA by an NADH- or NADPH-dependent acetoacetyl-CoA reductase (PhaB).

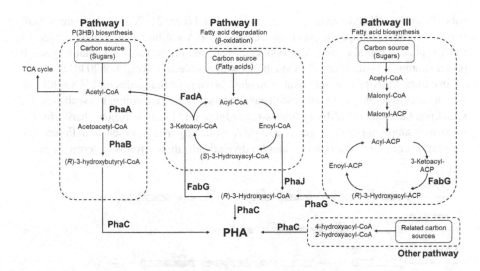

FIGURE 1.2 Metabolic pathways that supply various hydroxyalkanoates (HA) for PHA biosynthesis. PhaA, 3-ketothiolase; PhaB, NADPH- or NADH-dependent acetoacetyl-CoA reductase; PhaC, PHA synthase; PhaG, (*R*)-3-hydroxyacyl-ACP-CoA transferase; PhaJ, (*R*)-specific enoyl-CoA hydratase; FabG, NADPH-dependent 3-ketoacyl-ACP reductase; FabH, 3-ketoacyl-ACP synthase III; FadA, 3-ketoacyl-CoA thiolase.

Interestingly, only (*R*)-3HA-CoA isomers can be polymerized by PHA synthase (PhaC) if the hydroxyl group is on an asymmetric carbon.

Pathways II and III are mainly responsible for *mcl*-PHA production by utilizing intermediates from fatty acid degradation (β-oxidation) and fatty acid biosynthesis. As shown in Figure 1.2, various kinds of intermediate from each pathway are catalyzed into (*R*)-3HA-CoAs by monomer-supplying enzymes for PHA synthase, such as (*R*)-specific enoyl-CoA hydratase (PhaJ) and (*R*)-3HA-ACP-CoA transferase (PhaG) converting enoyl-CoA and (*R*)-3HA-ACP, respectively. However, 3-ketoacyl-ACP reductase (FabG) possesses the ability to provide a substrate for PHA biosynthesis in both pathways, reducing 3-ketoacyl-CoA to (*R*)-3HA-CoA and 3-ketoacyl-ACP to (*R*)-3HA-ACP.

PHA synthase (PhaC) is the most essential enzyme in PHA biosynthesis metabolism; it is responsible for the polymerization of HA units into PHA. HA-CoA is the main substrate for PhaC, and only *R*-enantiomer HA can be adopted, subsequently releasing CoA moiety.

To date, more than 60 different PHA synthases have been identified, and a total of 14 pathways have been reported to lead to PHA synthesis [9] since the first discovery of the PHA synthase operon in 1988 [10–12]. These natural PHA synthases have been categorized into four classes based on their primary structure, subunit composition, and substrate specificity for monomers with different chain lengths [13,14]. Class I PHA synthases consist of a single subunit (PhaC) with a molecular weight ranging from 60 to 70 kDa and form a homodimer. Class II PHA synthases also consist of single subunits, PhaC1 or PhaC2, and form a homodimer. Class III synthases consist of two heterodimers, PhaC and PhaE, where PhaC subunits are smaller than those of class I and II PhaC, with a molecular weight of about 40 kDa. The similarities in the amino acid sequences of PhaC subuntis in class III compared with those in class I and II are 21–28% [13], whereas PhaE subunits show no compelling similarity. Class IV PHA synthases are similar to class III PHA synthases and also have heterodimers, PhaC and PhaR. PhaR is a relatively small protein compared with other PHA synthases with a molecular weight of 20 kDa.

In natural PHA-producing organisms, *phaC* is the gene that codes for PHA synthase, which is usually clustered together with monomer-supplying enzyme genes such as *phaA*, *phaB*, *phaJ*, and other PHA biosynthesis-related genes [15]. A list of bacteria harboring the monomer-supplying enzyme genes, and PHA synthase genes is presented in Figure 1.3. The arrangement of *pha* operons can be divided into three groups in terms of *phaC*

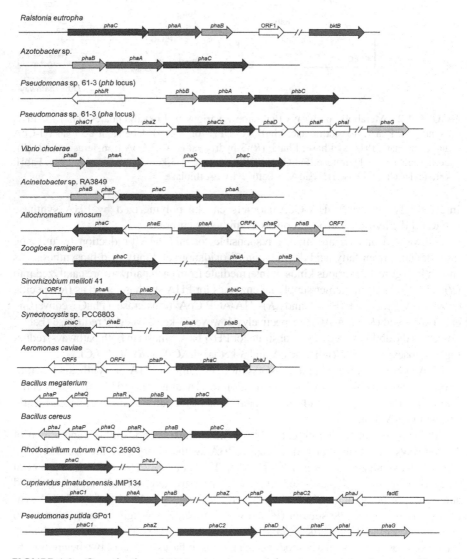

FIGURE 1.3 Organization of PHA monomer-supplying enzyme genes that are collocated with PHA synthase.

and consists of different monomer-supplying enzyme genes: (I) existence of the *phaC* gene by forming a cluster with *phaA* and/or *phaB* gene(s) (*R. eutropha* is a representative bacterium), (II) the *phaC* gene forming the same cluster with *phaJ* (*Aeromonas caviae* is a representative bacterium), and (III) the *phaC* gene and monomer-supplying genes are organized in a different array where the *phaC* gene exists alone in a separate locus or coexists with other related genes, such as a regulatory gene and PHA granule-associated protein (phasin) gene (*Pseudomonas putida* is a representative bacterium).

The monomer-supplying enzymes involved in PHA biosynthesis metabolisms are classified into three types as follows: (I) are responsible for (*R*)-specific reactions, such as PhaB, PhaJ, FabG, and LdhA, (II) are responsible for and capable of coenzyme A (CoA) addition or transfer, such as PhaG, FabH, HadA, Pct, and Bct, and (III) those responsible for the backbone synthesis for corresponding HA monomers, such as PhaA and BktB (see Table 1.1).

Lastly, there are proteins involved in the regulation of PHA biosynthesis by transcription and translation and also in forming and maintaining PHA granules inside of the bacteria body, such as PhaR, PhaF, and phasin (PhaI and PhaP).

1.3 MONOMER-SUPPLYING ENZYMES

Several monomer-supplying enzymes for PHA biosynthesis are presented in this section regarding their catalytic properties and substrate specificity. Furthermore, recent work on structural- and functional-based protein engineering of these subjects are thoroughly discussed (see Table 1.2).

1.3.1 3-KETOTHIOLASE (PHAA/BKTB)

Three-ketoacyl-CoA thiolase (acetyl-CoA acetyltransferase; EC 2.3.1.9) is an enzyme that catalyzes the thiolytic cleavage of 3-oxoacyl-CoA into acyl-CoA (shortened by two carbon atoms and generates acetyl-CoA), which happens to be the final step of fatty acid β-oxidation. In contrast, the reverse reaction is catalyzed by PhaA to form acetoacetyl-CoA via the condensation of two acetyl-CoA molecules in the first step of poly(3-hydroxybutyrate) [P(3HB); a.k.a. PHB] biosynthesis. Therefore, thiolases can function either as degradative in the β-oxidation pathway of fatty acids or biosynthetic agents. Additionally, PhaA has a narrow substrate specificity in the range of C3–C5 monomers chain length [46]; therefore, PhaA is specialized for *scl*-PHA biosynthesis. There were two 3-ketothiolases found in *R. eutropha*, encoded as PhaA and BktB, that can act in the biosynthetic pathway of PHA synthesis with the difference in substrate specificity [47]. All of the bacterial PHA biosynthetic 3-ketothiolases are classified as homotetramers [48–52], and the molecular weights of PhaA and BktB are 44 and 46 kDa, respectively [16,17]. BktB is proven to have substrate specificity mainly in longer monomers (C4–C10) compared with PhaA [53] and is also considered responsible for synthesizing 3-hydroxyvalerate (3HV) monomers in poly(3-hydroxybutyrate-*co*-3-hydroxyvalerate) [P(3HB-*co*-3HV)] biosynthesis [17].

The 3-ketothiolase involved in PHA biosynthesis has been broadly identified and studied in a wide range of microorganisms. The crystal structure of PHA biosynthetic

TABLE 1.1

Related Enzymes and Proteins in PHA Biosynthesis Metabolism

Enzyme or protein name	Classification[a]	Abbreviation	Function and role in PHA biosynthesis	Ref.
Genes originally located in the pha operon				
3-Ketothiolase	B	PhaA/BktB	Monomer-supplying enzyme	[12]
Acetoacetyl-CoA reductase	R	PhaB	Monomer-supplying enzyme	[12]
PHA synthase	–	PhaC	Catalytic subunit of class I, II, III, IV PHA synthase	[12]
Transcriptional regulator protein	–	PhaD	*Mcl*-PHA synthesis regulator	[16]
PHA synthase subunit	–	PhaE	Subunit of class III PHA synthase	[17]
Mcl-PHA GAP1	–	PhaF	*Mcl*-PHA granule associated protein	[18]
(*R*)-3-Hydroxyacyl-ACP-CoA transferase	C	PhaG	Monomer-supplying enzyme	[19]
Mcl-PHA GAP2	–	PhaI	*Mcl*-PHA granule associated protein	[18]
(*R*)-specific enoyl-CoA hydratase	R	PhaJ	Monomer-supplying enzyme	[20]
Scl-PHA GAP	–	PhaP	*Scl*-PHA-granule associated protein	[21]
Regulator protein or PHA synthase	–	PhaR	DNA binding regulator or subunits for class IV PHA synthase	[22]
PHA oligomer hydrolase	–	PhaY	PHA oligomer degradation	[23]
PHA depolymerase	–	PhaZ	PHA depolymerization	[24]
Genes outside of the pha operon				
3-Ketoacyl-ACP reductase	R	FabG	Monomer-supplying enzyme	[25]
3-Ketoacyl-ACP synthase III	C	FabH	Monomer-supplying enzyme	[26,27]
Lactate dehydrogenase	R	LdhA	Monomer-supplying enzyme	[28]
Lactate-CoA transferase	C	HadA	Monomer-supplying enzyme	[28]
Propionyl-CoA transferase	C	Pct	Monomer-supplying enzyme	[29]
Butyryl-CoA transferases	C	Bct	Monomer-supplying enzyme	[30]

[a] Classification of monomer-supplying enzymes: B, backbone synthesis; C, CoA addition or transfer; R, *R*-specific reaction.

TABLE 1.2
Monomer-Supplying Enzymes' Kinetic Parameters

Enzyme	Organism	Mutation position	Mutation function	Substrate	K_m [μM]	V_{max} [U mg^{-1}]	k_{cat} [s^{-1}]	Ref.
PhaA	Z. ramigera	–	–	Acetyl-CoA	330			[31]
		–	–	Acetoacetyl-CoA (thiolysis)	24			[31]
		C89S	Active-site residue identification	Acetyl-CoA	1100		0.03	[32]
		C378S	Active-site residue identification	Acetyl-CoA	11		0.05	[33]
	R. eutropha	–	–	Acetyl-CoA	1100			[34]
BktB	R. eutropha	–	–	Acetoacetyl-CoA (thiolysis)	11.58	1.5	102.18	[35]
PhaB	Z. ramigera	–	–	Acetoacetyl-CoA	2		300	[36]
		–	–	3-Ketovaleryl-CoA	2		124	[36]
		–	–	3-Ketohexanoyl-CoA	10		9	[36]
	R. eutropha	–	–	Acetoacetyl-CoA	5.7		102	[37]
		Q47L	Enhanced enzyme activity	Acetoacetyl-CoA	15.9		249	[37]
		T173S	Enhanced enzyme activity	Acetoacetyl-CoA	13.6		361	[37]
PhaJ	A. caviae	–	–	Crotonyl-CoA (C$_4$)	24		1922	[38]
		–	–	Hexenoyl-CoA (C$_6$)	40		294	[38]
		–	–	Octenoyl-CoA (C$_8$)	42		0.58	[38]
		–	–	Decenoyl-CoA (C$_{10}$)	42		0.65	[38]
		–	–	Dodecenoyl-CoA (C$_{12}$)	43		0.15	[38]
		L65A	Enhanced enzyme activity	Crotonyl-CoA (C$_4$)	33		899	[38]

(Continued)

TABLE 1.2 (CONTINUED)
Monomer-Supplying Enzymes' Kinetic Parameters

Enzyme	Organism	Mutation position	Mutation function	Substrate	K_m [μM]	V_{max} [U mg⁻¹]	k_{cat} [s⁻¹]	Ref.
		—	Enhanced enzyme activity	Hexenoyl-CoA (C_6)	18		89	[38]
		—	Enhanced enzyme activity	Octenoyl-CoA (C_8)	21		27	[38]
		—	Enhanced enzyme activity	Decenoyl-CoA (C_{10})	27		18	[38]
		—	Enhanced enzyme activity	Dodecenoyl-CoA (C_{12})	34		1.7	[38]
		V130G	Enhanced enzyme activity	Crotonyl-CoA (C_4)	154		3141	[38]
		—	Enhanced enzyme activity	Hexenoyl-CoA (C_6)	102		402	[38]
		—	Enhanced enzyme activity	Octenoyl-CoA (C_8)	76		405	[38]
		—	Enhanced enzyme activity	Decenoyl-CoA (C_{10})	13		108	[38]
		—	Enhanced enzyme activity	Dodecenoyl-CoA (C_{12})	5		16	[38]
	R. rubrum	—		Crotonyl-CoA (C_4)	21.4	2850	728	[39]
		—		Pentenoyl-CoA (C_5)	9.0	798	204	[39]
		—		Hexenoyl-CoA (C_6)	9.1	986	252	[39]
PhaJ1	P. aeruginosa	—		Crotonyl-CoA (C_4)	14	3600	995	[40]
		—		Hexenoyl-CoA (C_6)	43	3300	920	[40]
		—		Octenoyl-CoA (C_8)	55	8.3	2.3	[40]

(Continued)

TABLE 1.2 (CONTINUED)
Monomer-Supplying Enzymes' Kinetic Parameters

Enzyme	Organism	Mutation position	Mutation function	Substrate	K_m [μM]	V_{max} [U mg⁻¹]	k_{cat} [s⁻¹]	Ref.
PhaJ4	P. aeruginosa	V72I	Substrate specificity change					[41]
		–		Hexenoyl-CoA (C$_6$)	198	4100	1121	[40]
		–		Octenoyl-CoA (C$_8$)	118	3400	948	[40]
		–		Decenoyl-CoA (C$_{10}$)	37	1100	315	[40]
	P. putida	I72V	Substrate specificity change					[41]
FabG	Synechocystis sp. PCC 6803	P151V/F	Enhanced enzyme activity					[42]
		F188Y	Enhanced enzyme activity					[42]
FabH	E. coli	–		Malonyl-ACP	20		8	[43]
		–		Acetyl-CoA	60		7.4	[43]
		F87C/I/S/T	Substrate specificity change					[27]
	S. aureus	–		Malonyl-ACP	19		261	[44]
		–		Acetyl-CoA	373		431	[44]
Pct	C. propionicum	A243T	Enhanced enzyme activity					[45]
		V193A	Enhanced enzyme activity					[45]

A   B

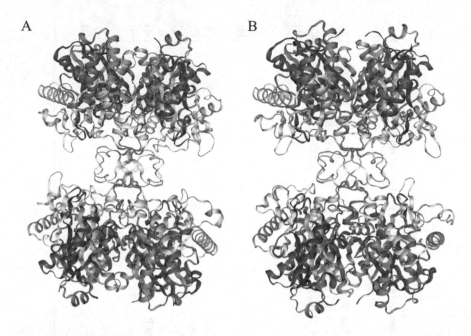

FIGURE 1.4   Crystal structure of 3-ketothiolases (PhaA) from *R. eutropha* H16, (A) PhaA$_{Re}$ (pdb: 4O99, 4O9A, 4O9C) [56]; (B) BktB$_{Re}$ (pdb: 4NZS, 4W61) [35].

3-ketothiolase from *Zoogloea ramigera* (PhaA$_{Zr}$) has been identified by Modis et al. [54,55], which indicates a two-step "ping pong" reaction system and the functional catalytic residues, His348 (activation of Cys89), Cys89 (covalent acyl-CoA intermediate formation), and Cys378 (substrate activation). Afterward, the crystal structures of PhaA and BktB from *R. eutropha* H16 were determined (see Figure 1.4) [56,35]. The overall structure of PhaA$_{Re}$ and BktB$_{Re}$ are identical to PhaA from *Z. ramigera*, where PhaA$_{Re}$ consists of two distinct domains, a core and a loop domain; BktB$_{Re}$ however, consists of three distinctive domains, two cores and a loop domain. Instead of Cys89 and Cys378 residues in PhaA$_{Zr}$, two catalytic residues, Cys88 and Cys379 from PhaA$_{Re}$, are located in similar locations. Furthermore, the His349 residue is located near Cys88 residue and functionalized as a general base. In the case of BktB$_{Re}$, His350, Cys90, and Cys380, residues are located at the corresponding sites similar to PhaA$_{Zr}$.

BktB$_{Re}$ is involved in producing 3-ketovaleryl-CoA, which is an intermediate for synthesizing the 3HV monomer, while PhaA$_{Re}$ is less capable of supplying 3-ketovaleryl-CoA. The difference in the substrate specificity of PhaA$_{Re}$ and BktB$_{Re}$ can be further explained by comparing their crystal structures. The most significant distinction between the two thiolases is related to the binding pocket of the covalently bonded acetyl or propionyl group. Val87 residue is located near the Cys88 covalent catalytic residue in PhaA$_{Re}$, while the corresponding Leu88 residue is located near the Cys90 covalent catalytic residue in BktB$_{Re}$. This leaves a larger space around covalent catalytic residues in BktB$_{Re}$ and thus results in substrate specificity for both acetyl and propionyl groups. As for substrate binding, stabilizing the adenosine diphosphate (ADP) moiety of CoA with a unique hydrophilic His219, Arg221, and

Asp228 residue has been confirmed by site-directed mutagenesis approaches. First, the H219F and H219A mutants exhibited 74% and 24% activity, respectively, compared with their parent enzyme. The R221K and R221A mutants exhibited 67% and 38% activity, respectively, compared with their parent enzyme. This result suggests that the His219 residue functions as a supplementary substrate-binding pocket rather than a stabilizing substrate, where Arg221 supports the stabilization of a substrate through hydrogen bond formation with an adenine moiety of CoA due to arginine's high capability of forming hydrogen bonds. However, the D228A mutant showed a 150% enzyme activity compared with its parent enzyme.

1.3.2 ACETOACETYL-COA REDUCTASE (PHAB)

NADPH- or NADH-dependent acetoacetyl-CoA reductase (PhaB) (EC 1.1.1.36) is the enzyme responsible for the second step in pathway I (see Figure 1.2) for P(3HB) biosynthesis; it reduces the 3-ketone group of acetoacetyl-CoA to synthesize (R)-3-hydroxybutyryl-CoA, with the oxidation of NADH or NADPH as the cofactor. Most of the acetoacetyl-CoA reductases are considered to be NADPH-dependent enzymes, such as PhaB from *Z. ramigera* and *R. eutropha* [8,46,57]. There are also other bacteria that require NADH for the reduction of acetoacetyl-CoA in the P(3HB) biosynthesis pathway, such as *Allochromatium vinosum* and *Azotobacter viinerandi* [8,57].

The crystal structure of PhaB from *R. eutropha* H16 (PhaB$_{Re}$) was extensively studied to fully understand the detailed molecular mechanisms of acetoacetyl-CoA reductase. It indicates that PhaB$_{Re}$ is categorized in a short-chain dehydrogenase/reductase family and exists as a tetramer (see Figure 1.5) [36,37,58]. The monomeric structure

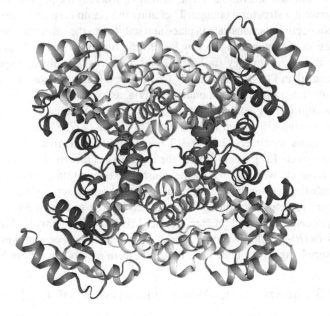

FIGURE 1.5 Crystal structure of acetoacetyl-CoA reductase (PhaB) from *R. eutropha* H16, PhaB$_{Re}$ (pdb: 3VZP, 4N5L).

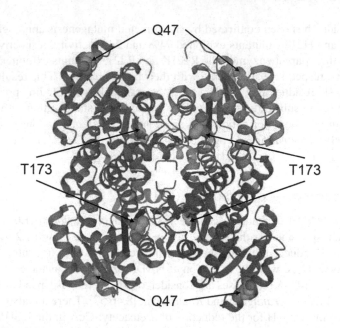

FIGURE 1.6   The mutation positions of Q47L and T173S in PhaB$_{Re}$ from *R. eutropha* H16.

of PhaB$_{Re}$ contains a Rossmann fold and a clamp domain. While under substrate binding, the Rossmann fold had the same conformation, whereas the position of the clamp domain underwent a structural change. Therefore, the clamp domain is assumed to be involved in connecting the substrate in place and stabilizing the substrate conformation.

The reduction of acetoacetyl-CoA by PhaB is highly important in PHA accumulation because it provides energy for P(3HB) biosynthesis. The substrate specificity of PhaB is broader than PhaA; it can supply C6 and even longer monomers based on an analysis of substrate range in recombinant bacteria, as revealed by *in vivo* studies [59].

A protein engineering study on PhaB$_{Re}$ by means of directed evolution revealed two mutants bearing Gln47Leu (Q47L) and Thr173Ser (T173S) substitutions. They showed k_{cat} values that were 2.4- and 3.5-fold higher than that of the wild-type enzyme, respectively [37]. According to the crystal structure analysis, the flexibility around the active site was affected by the constructive mutations, which makes them valuable in substrate recognition, thus resulting in enhanced activity (see Figure 1.6).

The primary structure of PhaB$_{Re}$ is considered similar to that of *Escherichia coli* NADPH-dependent 3-ketoacyl-ACP reductase (FabG), which is involved in fatty acid biosynthesis for (*R*)-3-hydroxyacyl-ACP production. FabG has been proposed to act as a monomer-supplying enzyme for PHA biosynthesis in recombinant *E. coli* [60,25].

1.3.3   (*R*)-3-HYDROXYACYL-CoA-ACP THIOESTERASE (PHAG)

PhaG was first reported as a 3-hydroxyacyl-ACP-CoA transferase in *P. putida*, possibly functioning as a 3-hydroxyacyl-ACP thioesterase (EC:2.4.1.-). To date, there are three different metabolic pathways found in *P. putida* that generate 3-hydroxyacyl-CoA [19,61,62]; they are the (I) β-oxidation pathway, (II) fatty acid *de novo* biosynthesis pathway, and (III) the chain elongation reaction pathway. The molecular weight of PhaG from *P. putida* is 34 kDa; unfortunately, detailed information on the quaternary structure of this protein has not yet been revealed. This suggests that an (*R*)-3HA-CoA ligase is necessary to introduce *mcl*-3HA units into the polymer chain through the fatty acid biosynthesis pathway considering PhaG only produces free (*R*)-3-hydroxyalkanoic acids rather than (*R*)-3HA-CoA. Wang et al. [63] reported an acyl-CoA ligase from *P. putida* (PP_0763) catalyzed on (*R*)-3-hydroxyacids derived from fatty acid biosynthesis via (*R*)-3-hydroxyacyl-ACP thioesterase. By co-expressing a thioesterase, CoA ligase, and PHA polymerase in recombinant *E. coli*, up to 7.3 g/L of *mcl*-PHA can be obtained from glucose as a sole carbon source [64].

1.3.4 (*R*)-SPECIFIC ENOYL-COA HYDRATASE (PHAJ)

(*R*)-Specific enoyl-CoA hydratase (PhaJ) (EC:4.2.1.119) catalyzes the stereoselective (*R*-specific) hydration of *trans*-2-enoyl-CoA and converts it into (*R*)-3HA-CoA for PHA accumulation. In other words, PhaJs conduct β-oxidation products into the PHA biosynthesis pathway. PhaJ from *A. caviae* (PhaJ$_{Ac}$) is the first (*R*)-hydratase that has been identified in the PHA biosynthesis pathway and is well studied [65]. PhaJ$_{Ac}$ exhibits activity toward *scl trans*-2-enoyl-CoA substrates consisting of 4–6 carbons [20]. The molecular size of the PhaJ$_{Ac}$ monomer is 14 kDa; it is the smallest among the (*R*)-hydratase family. Currently, various types of PhaJs from PHA-producing microorganisms have been identified and studied, specifically, those that fall into the category of an *scl* (C4–C6) enoyl-CoA-specific type and an *mcl* (C8–C14) enoyl-CoA-specific type. *Pseudomonas aeruginosa* is known to have both types of *R*-hydratase that work as a monomer-supplying enzyme in PHA biosynthesis. In the genome of *P. aeruginosa* DSM1707, four types of PhaJ genes (*phaJ1*$_{Pa}$ to *phaJ4*$_{Pa}$) have been identified, and their substrate specificities have been studied [66,40]. Among these *phaJ* genes, only *phaJ1*$_{Pa}$ showed relatively narrow substrate specificity toward enoyl-CoAs consisting of 4–6 carbons atoms, whereas *phaJ2*$_{Pa}$ to *phaJ4*$_{Pa}$ showed broader substrate specificity toward enoyl-CoAs consisting of 6–12 carbons atoms. The *phaJ* genes were also found in other *Pseudomonas* species as well [67– 69]. A *phaJ* gene involved in PHA biosynthesis found in strain *P. putida* encodes as *phaJ1*$_{Pp}$. Additionally, this *phaJ1*$_{Pp}$ shares 67% similarity in its amino acid sequence with that of *phaJ1*$_{Pa}$, and yet *phaJ1*$_{Pp}$ has a broader substrate specificity than *phaJ1*$_{Pa}$. Recently, the difference in the substrate specificities of two PhaJ1 enzymes was investigated by site-directed mutagenesis [41]. The replacement of valine in position 72 to isoleucine for PhaJ1$_{Pp}$ increased the preference for enoyl-CoA with *scl*s; however, the replacement of isoleucine in position 72 to valine for PhaJ1$_{Pa}$ increased the preference for enoyl-CoA with *lcl*s. In other words, this mutation has narrowed and broadened the substrate specificity of PhaJ1$_{Pp}$ and PhaJ1$_{Pa}$. By comparing the crystal structures of these two PhaJ1 enzymes, the only difference is the amino acid at position 72. Therefore, the bulkiness of the amino acid at position 72 became the main factor in determining the chain length specificity of PhaJ1.

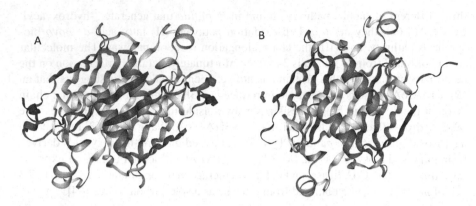

FIGURE 1.7 Crystal structure of (*R*)-specific enoyl-CoA hydratase (PhaJ). (A) PhaJ1$_{Pa}$ (pdb: 5CPG) [41]; (B) PhaJ$_{Ac}$ (pdb: 1IQ6) [40].

The crystal structure of (*R*)-specific enoyl-CoA hydratases PhaJ1$_{Pa}$ and PhaJ$_{Ac}$ are shown in Figure 1.7.

A similar site-directed mutagenesis study was carried out on PhaJ$_{Ac}$ [38]. Amino acid residues Ser62, Leu65, and Val130 in PhaJ$_{Ac}$ were responsible for the width and depth of the acyl-chain-binding pocket. By introducing amino acid substitutes Leu65Ala, Leu65Gly, and Val130Gly, the mutant PhaJ$_{Ac}$ showed a major increase in specificity toward *lcl* substrates than the wild type. This was the first protein engineering study that reported the successful modification of a PHA monomer-supplying enzyme based on rational design.

There were three enzymes found in strain *R. eutropha* H16 that were highly homologous to the *mcl*-specific (*R*)-specific enoyl-CoA hydratase PhaJ4$_{Pa}$, encoded PhaJ4a$_{Re}$ to PhaJ4c$_{Re}$. The expression of these PhaJs in recombinant *R. eutropha* producing poly[(*R*)-3-hydroxybutyrate-*co*-(*R*)-3-hydroxyhexanoate] [P(3HB-*co*-3HHx)] from soybean oil increased the 3HHx monomer fraction. Specifically, PhaJ4a$_{Re}$ and PhaJ4b$_{Re}$ showed a ten-fold higher catalytic performance than PhaJ4c$_{Re}$ [70]. Similarly, five PhaJs in *Haloferax mediterranei*, named PhaJ1 to PhaJ5, were identified, and several show high catalytic efficiencies toward poly(3-hydroxybutyrate-*co*-3-hydroxyvalerate) (PHBV) biosynthesis. In these PhaJs, PhaJ1 is the only PhaJ associated with PHA granules and has contributed to PHA mobilization [71]. A recent study showed that PhaJ is also present in PHA-accumulating bacteria *Bacillus cereus* [72]. PhaJ from *B. cereus* YB-4 (PhaJ$_{YB4}$) expressed in recombinant *E. coli* seems to have substrate specificity toward *scl* enoyl-CoA based on the composition of the PHA produced. *B. cereus* is also a strain with a *phaJ* gene involved in the *pha* cluster but not *B. megaterium*.

From the perspective of PHA accumulation, a *phaJ* gene from *Rhodospirillum rubrum* was bioengineered to overexpress in its host strain by a factor of at least five, a mere increase in PHA production was observed [73]. This result is associated with those with overexpression of PhaJ in *A. hydrophila* [74] and *A. caviae* [75]. PhaJ links the reaction between fatty acid biosynthesis and PHA biosynthesis. Yet, it is physically unachievable for microorganisms to set up a futile cycle between fatty acid biosynthesis and support the related fatty acid into PHA biosynthesis as an intermediate. This also indicated that PhaJ plays a less important role as a rate-limiting factor of PHA production.

1.3.5 3-KETOACYL-ACP REDUCTASE (FABG)

As mentioned before, 3-ketoacyl-ACP reductase (FabG) (EC 1.1.1.100), which is responsible for the catalysis of the NADPH-dependent reduction of 3-ketoacyl-ACP to (R)-3-hydroxyacyl-ACP, is a key enzyme in fatty acid biosynthesis pathway in bacteria, plants, and algae [76,77]. In the presence of a fatty acid, FabG can capture the 3-ketoacyl-CoA intermediate in the β-oxidation pathway and produce (R)-3-hydroxyacyl-CoA, which is the substrate for PHA synthase. FabG requires NADPH as a driving factor [78,79]. FabG from *Brassica napus* has a molecular weight of about 25.5 kDa and exists in the form of a tetramer in solution [80]. It was reported that FabG plays an essential role in an *E. coli* strain [81], and overexpression of FabG resulted in a two- to three-fold increase in fatty acid production [82]. Regarding PHA accumulation, it has been reported that coexpression of *E. coli* FabG and *P. aeruginosa* PhaC leads to the production of *mcl*-PHA from recombinant *E. coli* when cultivated with fatty acids, such as decanoate or dodecanoate [60,25,83].

Recently, detailed studies on both FabG and PhaB from cyanobacteria *Synechocystis* sp. PCC6803 have been carried out (referred to as $FabG_{sp}$ and $PhaB_{sp}$, respectively) [42,84]. The high similarity in the structure of these enzymes implies that $FabG_{sp}$ might be involved in P(3HB) biosynthesis as well. More importantly, $FabG_{sp}$ can act as a controlling agent in carbon flux distribution between the fatty acid and P(3HB) in *Synechocystis* sp. PCC6803. Based on the structural study, the most obvious difference between $FabG_{sp}$ and $PhaB_{sp}$ occurred in the region between amino acid residues 190 to 200, where $FabG_{sp}$ shows a flexible region and $PhaB_{sp}$ exhibits an α-helix shape. This region is considered to be a substrate-binding lid when compared with FabG from *E. coli* [85]. The flexible region of $FabG_{sp}$ may be suitable for larger molecules, whereas the α-helix region of $PhaB_{sp}$ accounts for narrower substrate selectivity only for acetoacetyl-CoA. Following this result, Liu et al. [42] designed a high-performance FabG on acyl-CoA with mutations P151V and P151F. By suppling (R)-3-hydroxyacyl-CoAs, it has the potential to serve as a critical enzyme to convert carbon flow from fatty acid synthesis to PHA.

1.3.6 3-KETOACYL-ACP SYNTHASE III (FABH)

A 3-ketoacyl-ACP synthase III, FabH (EC 2.3.1.180), is a member of the 3-ketoacyl synthase family of enzymes [86]. It works not only as a PHA monomer-supplying enzyme, but it is also active in other metabolic pathways. The original catalytic metabolism is the condensation of malonyl-ACP and acyl-CoA units to form aceto-acetyl-ACP and CO_2 in fatty acid biosynthesis, which can be converted into monomer-supplying units recognizable by PHA synthase enzymes [27,60,87].

With the overexpression of FabH in recombinant *E. coli*, P(3HB) homopolymer can be produced under the presence of glucose [87]. Based on this, researchers started to develop an engineered FabH mutant to produce PHA copolymers. FabH from different microorganisms exhibits wide substrate specificity even though they share a rather similar primary amino acid sequence [88– 90]. For example, FabH from *E. coli* and *Mycobacterium tuberculosis* have specificity for a substrate of 2–4 carbons in length and 10–16 carbons in length *in vitro*, respectively [89]. From the comparison of the primary amino acid sequence and crystal structure of *E. coli* FabH and *M. tuberculosis* FabH, a possible difference in the substrate specificity can be observed. There is an amino acid, Phe, at position 87 in the *E. coli* FabH protein, and Thr in the *M. tuberculosis* FabH, which are the binding sites of the substrate with different carbon chain lengths; thus, the modification on the codon encoding amino acid 87 of the FabH protein from *E. coli* has been established in order to bind fatty acid chains with a length longer than four carbons.

Overexpression of FabH with mutations to amino acid 87 of the binding pocket supplied *mcl*-PHA monomer in recombinant *E. coli* but low PHA yield when compared with the parent enzyme [27]. This drawback in the production of PHA has been overcome by adding the His244Ala mutation to these binding pocket mutants, which resulted in more than a six-fold PHA yield increase. Engineering of FabH with an improved PHA monomer supply is also a promising approach for PHA copolymer production.

1.4 MONOMER-SUPPLYING PATHWAYS AND ENZYMES INVOLVED

Since the first discovery of other monomers beside 3HB from active sludge in 1974 [91], more and more sophisticated monomer-supplying pathways have been comprehensively developed; PHA with different monomer compositions exhibited a wide range of material properties and possibilities for application. Value-added monomers, such as 3-hydroxypropionic (3HP), 3-hydroxyvalerate (3HV), and 4-hydroxybutyrate (4HB) along with 2-hydroxyalkanoates (2HAs), have been extensively studied; modified supply pathways in microorganisms have been successfully designed via gene/protein engineering and reacted in excellent manners in regard of increased production of PHA and improved material properties. Here we present monomer-supplying pathways for the above-mentioned monomers supported by inexpensive carbon sources and key monomer-supplying enzymes involved in all these pathways.

1.4.1 3HB-Supply Pathway

As mentioned previously in this chapter, the most common biosynthetic pathway of P(3HB) starting from acetyl-CoA consists of three enzymatic reactions activated by 3-ketothiolase (PhaA), NADPH-dependent acetoacetyl-CoA reductase (PhaB), and PHA synthase (PhaC) [1,57]. At first, two acetyl-CoA moieties are condensed by PhaA to acetoacetyl-CoA. The product then goes through a reduction by PhaB, which produces an (*R*)-isomer of 3HB-CoA. With the catalyzation of PhaC, P(3HB) polymer can then be synthesized from (*R*)-3HB-CoA.

Recently, Matsumoto et al. [92] reported a new P(3HB) biosynthetic pathway by utilizing an acetoacetyl-coenzyme A synthase (AACS) in *E. coli* and *Corynebacterium glutamicum* (see Figure 1.8). AACS was identified from terpenoid-producing *Streptomyces* sp. strain CL190, and it catalyzes a single condensation of acetyl-CoA

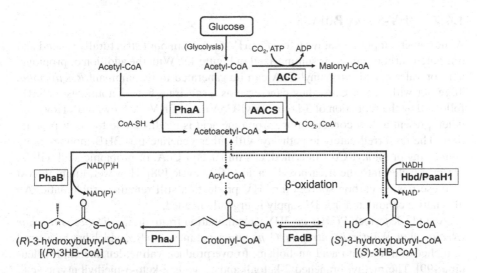

FIGURE 1.8 Stereoselective 3HB-CoA monomer-supplying pathway; the dotted lines indicate the forward direction of fatty acid β-oxidation. PhaA, 3-ketothiolase; PhaB, NADPH- or NADH-dependent acetoacetyl-CoA reductase; PhaJ, (*R*)-specific enoyl-CoA hydratase; ACC, acetyl-CoA carboxylase; AACS, acetoacetyl-CoA synthase; Hbd, (*S*)-3-hydroxybutyryl-CoA dehydrogenase from *C. acetobutylicum* ATCC 824; PaaH1, (*S*)-3-hydroxybutyryl-CoA dehydrogenase from *R. eutropha*; FadB, β-oxidation multienzyme that exhibits *S*-hydratase and 3-hydroxyacyl-CoA dehydrogenase activities.

and malonyl-CoA to generate acetoacetyl-CoA, different from the acetoacetyl-CoA thiolysis of PhaA; acetoacetyl-CoA synthesis provides a promising application in the heterologous production of P(3HB).

Acetoacetyl-CoA can be reduced to (*S*)-3HB-CoA by (*S*)-3HB-CoA dehydrogenase (PaaH1) from *R. eutropha*, or (*S*)-3HB-CoA dehydrogenase (Hbd) from *C. acetobutylicum* ATCC 824, which is catalyzed in the reverse β-oxidation pathway. Interestingly, even though they both catalyze acetoacetyl-CoA in order to produce 3HB-CoA, the final products exhibit different chirality, (*R*)-form and (*S*)-form, respectively [93,94]. Additionally, the monomer epimerization from (*S*)-3HB-CoA to (*R*)-3HB-CoA via a crotonyl-CoA intermediate was proposed in *R. rubrum* [95]. It indicates forward and reverse reactions catalyzed by two different hydratases: PhaJ, specific for the *R*-enantiomer [(*R*)-specific enoyl-CoA hydratase, *R*-hydratase], and NADH-dependent acetoacetyl-CoA reductase, which is function specific for the *S*-enantiomer [(*S*)-specific enoyl-CoA hydratase, *S*-hydratase]. To date, the only other bacterium confirmed to possess such an (*R*)-3HB-CoA supply pathway is the methanol-utilizing bacterium *Methylobacterium rhodesenium* [96]. Furthermore, Sato et al. [97] demonstrated a modified hydratase-involved pathway by expressing *R*-hydratase from *A. caviae* (PhaJ$_{Ac}$) in recombinant *E. coli* strains CAG18497 and LS5218, which can provide a β-oxidation multienzyme (FadB$_{Ec}$) that exhibits *S*-hydratase and 3-hydroxyacyl-CoA dehydrogenase activities. The P(3HB) that accumulated in this pathway exhibits high molecular weight up to 1.1 to 1.2 × 10^6 and 1.8 to 2.1 × 10^6 for M_n and M_w, respectively.

1.4.2   3HV-Supply Pathway

A monomer-supplying pathway for (R)-3HV-CoA from both structurally related and unrelated carbon sources is summarized in Figure 1.9. With the addition of propionic acid or valeric acid, propionyl-CoA can be generated in recombinant *R. eutropha*. Together with acetyl-CoA, they can serve as a substrate for 3-ketothiolase (BktB), followed by the reduction of 3-ketovaleryl-CoA to (R)-3HV-CoA by PhaB. However, when present at low concentration, propionic acid is not favorable for 3HV production. The preferred catalytic pathway will either generate first 3HB monomers by converting propionyl-CoA intermediates into acetyl-CoA, or propionic acid will be directly fueled into the tricarboxylic acid (TCA) cycle [98]. However, the high cost of these related carbon sources for 3HV production still remains problematic. An alternative candidate for 3HV supply is urgently needed.

A study described P(3HB-*co*-3HV) accumulation from an *Alcaligenes eutrophus* (*R. eutropha, Wautersia eutropha, C. necator*) mutant strain R3, which has an altered branched-chain amino acid anabolism, to overproduce valine, leucine, and isoleucine [99]. The methyl-branched 2-ketoalkanoate acids 2-keto-3-methylbutyric acid, 2-keto-4-methylvaleric acid, and 2-keto-3-methylvaleric acid from valine, leucine, and isoleucine, respectively, are considered to be overproduced when corresponding

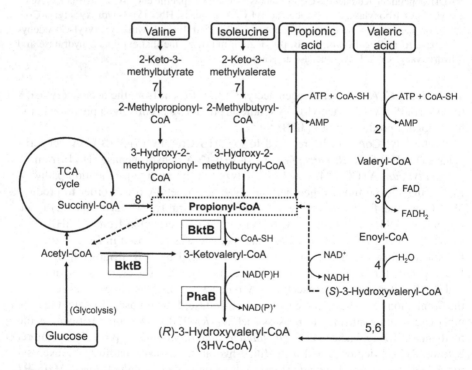

FIGURE 1.9   3HV-CoA monomer-supplying pathway from various carbon sources. BktB, 3-ketothiolase; PhaB, NADPH- or NADH-dependent acetoacetyl-CoA reductase. 1. Propionyl-CoA synthase; 2. Acyl-CoA synthase; 3. Acyl-CoA dehydrogenase; 4. Enoyl-CoA hydratase; 5. 3-hydroxyacyl-CoA dehydrogenase; 6. 3-ketoacyl-CoA thiolase; 7. Branched-chain 2-ketoacid dehydrogenase complex; 8. Sleeping beauty mutase gene complex.

amino acids are not synthesized under ammonium deficiency conditions, or other nutrients limit the growth of the cells. These generated 2-ketoalkanoates subsequently undergo decarboxylation to CoA thioesters by branched-chain 2-keto-acid dehydrogenase complex and are further degraded into propionyl-CoA. This 3HV-CoA supply system uses an amino acid degradation pathway and indicates that the central intermediate propionyl-CoA for 3HV accumulation can be successfully generated without the addition of a structurally related carbon source. In addition, Srirangan et al. [100] recently developed a novel 3HV supply pathway directly from unrelated carbon sources, such as glucose and glycerol, in recombinant *E. coli*. In *E. coli* strains, there is a dormant but potentially propionate-producing pathway named the Sleeping beauty mutase (Sbm) pathway. The Sbm operon with four related genes (*sbm-ygfD-ygfG-ygfH*) is responsible for canalization of the succinyl-CoA from the TCA cycle to the 3HV producing pathway intermediate, propionyl-CoA, which is further condensed into 3HV-CoA.

1.4.3 3HP-Supply Pathway

A biosynthetic pathway for poly[(*R*)-3-hydroxybutyrate-*co*-3-hydroxypropionate] P(3HB-*co*-3HP) has been constructed in recombinant *R. eutropha* from structurally unrelated carbon sources [101]. In this engineered pathway, 3-hydroxypropionyl-CoA (3HP-CoA) was generated by introducing two enzymes found in the 3-hydroxypropionate cycle from green non-sulfur bacterium, *Chloroflexus aurantiacus*, and malonyl-CoA reductase and the 3HP-CoA synthetase domain of propionyl-CoA synthase, namely, Acs$_{Ca}$ and Mcr$_{Ca}$, respectively [102,103] (see Figure 1.10). Furthermore, P(3HP) successfully gathered from glycerol has been reported by Andreessen et al. [104]. The genes for glycerol dehydratase (*dhaB1/dhaB2*) from *Clostridium butyricum* and propionaldehyde dehydrogenase (*pduP*) from *Salmonella enterica* serovar Typhimurium LT2 was introduced into recombinant *E. coli*, and the intermediate 3HP-CoA for 3HP accumulation was generated from an unrelated carbon source.

1.4.4 4HB-Supply Pathway

In a biosynthetic pathway in recombinant *E. coli* designed by Valentin et al. [105], the production of 4-hydroxybutyryl-CoA (4HB-CoA) was successfully facilitated by introducing genes from the 4HB-CoA supply pathway where succinyl-CoA is an intermediate from the bacterium *C. kluyveri* [106]. These genes were succinic semialdehyde dehydrogenase (*sucD*), 4-hydroxybutyrate dehydrogenase (*4hbD*), and 4-hydroxybutyryl-CoA: CoA transferase (*orfZ*). The 4HB-CoA monomer is solely generated in the presence of a glucose carbon source where succinyl-CoA is simply adapted from the TCA cycle without the initiation of a 4HB precursor (see Figure 1.11).

1.4.5 Other Monomer-Supply-Related Enzymes

Besides related and unrelated carbon sources such as sugar and fatty acids as potential monomer-supplying resources for PHA biosynthesis, amino acids are also

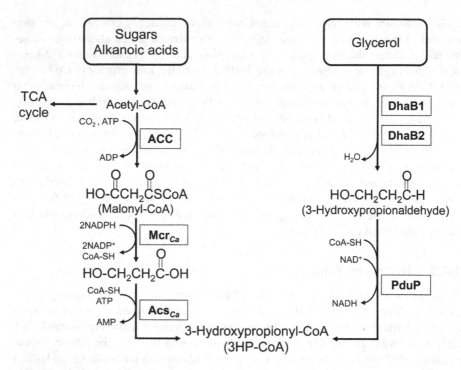

FIGURE 1.10 3HP-CoA monomer-supplying pathway from various carbon sources. ACC, acetyl-CoA carboxylase; Mcr_{Ca}, 3HP-CoA synthetase domain of propionyl-CoA synthase; Acs_{Ca}, malonyl-CoA reductase; DhaB1 and DhaB2, glycerol dehydratase; PduP, propionaldehyde dehydrogenase.

precursors for introducing various side-chain structures into polymer if appropriate enzymes are utilized to convert amino acids into 2HA-CoA. Then the converted substrate can be catalyzed by a lactate-polymerizing enzyme (LPE) (see Figure 1.12).

LPE is an engineered class II PHA synthase derived from *Pseudomonas* sp. 61-3 by the introduction of several amino acid mutations, S325T and Q481K (named $PhaC1_{Ps}STQK$), with substrate specificity for 2HA-CoA. The copolymerization of 2HAs such as lactate (2-hydroxypropionate; 2HP), glycolate (2-hydroxyacetate), 2-hydroxybutyrate (2HB), 2-hydroxy-4-methylvalerate (2H4MV), and 2-hydroxy-3-phenylpropionate (2H3PhP) has been reported [107–110].

In an anaerobic bacterium *C. difficile* (currently designated *Peptoclostridium difficile*), there is a special pathway for amino acid degradation from leucine to 4-methylvalerate (4MV) [28,111,112]. At first, the deamination of leucine to 2-keto-4-methylvalerate by branched-chain amino acid aminotransferase (IlvE) and aromatic amino acid aminotransferase (TyrB), followed by reduction to (R)-2-hydroxy-4-methylvalerate by 2H4MV dehydrogenase (LdhA). Subsequently, 2H4MV is converted to 2H4MV-CoA by 2H4MV-CoA transferase (HadA), and then dehydrated to 4-methyl-2-pentenoyl-CoA (4M2PE-CoA) by 2H4MV-CoA dehydratase (HadBC) with the help of its activator (HadI). Finally, 4M2PE-CoA is converted to 4MV by two additional enzymatic reactions. In this metabolic pathway, because

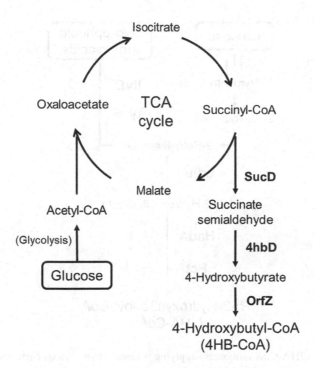

FIGURE 1.11 4HB-CoA monomer-supplying pathway using glucose as carbon source. SucD, succinic semialdehyde dehydrogenase; 4hbD, 4-hydroxybutyrate dehydrogenase; OrfZ, 4-hydroxybutyryl-CoA: CoA transferase.

2H4MV-CoA is the substrate for LPE and the intermediate of 2H4MV biosynthesis, the enzyme responsible for 2H4MV-CoA supply is considered to be a key monomer-supplying enzyme for biosynthesis of PHA containing 2HA units.

Studies have reported PHA copolymer biosynthesis by utilizing the enzymes involved in this pathway where related and unrelated amino acids and sugars have been used as a carbon source [110,113–115]. Recently, Mizuno et al. [110] demonstrated the biosynthesis of a 3HB-based copolymer containing a 2H4MV unit and other 2HA units when cultivated with glucose and corresponding amino acids in the presence of LdhA and HadA in recombinant *E. coli*. Other than PHA containing a 2H4MV unit when leucine is used as a precursor as expected, 2-hydroxy-3-methylbutyrate (2H3MB), 2-hydroxy-3-methylvalerate (2H3MV), and 2-hydroxy-3-phenylpropionate (2H3PhP) from related amino acids with similar backbones such as valine, isoleucine, and phenylalanine, respectively, were also successfully produced. This indicates that LdhA and HadA have broad substrate specificity, not only for the reduction of leucine but also for several other amino acids and HAs. In particular, HadA was able to catalyze aromatic substrates such as 2H2PhLA, 4HPhLA, 2H3PhP, and 2H4PhB; aliphatic substrates such as 2H4MV, glycolate, LA (2HP), 2HB, 3HB, 4HB, 5HV, and 6HHx; and also propionate and butyrate [116].

When compared with other D-2-hydroxyalkanoate or (*R*)-2-hydroxyalkanoate (2HA) dehydrogenases, (*R*)-2H4MV dehydrogenase LdhA shows high identities in

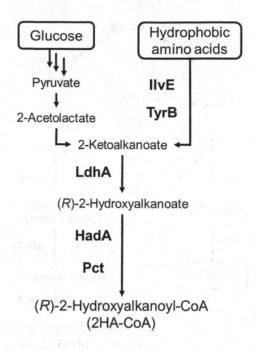

FIGURE 1.12 2HA-CoA monomer-supplying pathway from various carbon sources. IlvE, branched-chain amino acid aminotransferase; TyrB, aromatic amino acid aminotransferase; LdhA, lactate dehydrogenase; HadA, lactate-CoA transferase; Pct, propionyl-CoA transferase.

its amino acid sequence. For example, 38% when compared with D-2-hydroxyisocaproate dehydrogenase (d-HicDH) from *Lactobacillus casei*, 36% when compared with (R)-2-hydroxyglutarate dehydrogenase (HGDH) from *Acidaminococcus fermentans*, and 45% and 40% when compared with D-lactate dehydrogenases from *L. pentosus* and *L. bulgaricus*, respectively. These enzymes belong to the D-lactate dehydrogenase (D-LDH) family, which has substrate stereospecificity different from counterpart L-lactate dehydrogenase. Moreover, the crystal structures of D-LDH from *L. pentosus* [117] and *L. bulgaricus* [118], D-HicDH from *L. casei* [119], and HGDH from *A. fermentans* [120] identified key catalytic amino acid residues, especially E265 and H297, that act as general acids to protonate at the substrate carbonyl oxygen, whereas R236 connects the substrate carboxylate anion by a salt bridge. The 2HA dehydrogenase seems to have rich substrate specificity among various microorganisms' fermentation of amino acids through their (R)-2-hydroxyacyl acids, such as D-lactate dehydrogenase in alanine fermentation by *C. propionicum* [121], (R)-2-hydroxyglutarate dehydrogenase in glutamate fermentation by *A. fermentans* and *Fusobacterium nucleatum* [120], and (R)-3-phenyllactate dehydrogenase in phenylalanine fermentation by *C. sporogenes* [122].

In recent years, the biosynthesis of poly(lactate) (PLA) or lactate acid (LA)-based polyesters has been extensively studied. As mentioned previously, the utilization of LPE with key monomer-supplying enzymes involved in the PLA biosynthesis of metabolic pathways indicates a promising future in large scale PLA bio-production.

Propionyl-CoA transferases (Pct) can catalyze LA into LA-CoA by using propionyl-CoA or acetyl-CoA as a CoA supplier. Pct was first found in the alanine fermentation pathways of *C. propionicum* and later identified in other bacteria such as *Megasphaera elsdenii, Bacteroides ruminicola*, and *C. homopropionicum* [29]. Among all these Pcts, those from *C. propionicum* and *M. elsdenii* have been used widely not only for LA and 2HA production, but also for other HAs including glycolate (GA), 3-hydroxypropionate (3HP), 4-hydroxybutyrate (4HB), 2-hydroxybutyrate (2HB), 5-hydroxyvalerate (5HV), and 6-hydroxyhexanoate (6HHx) [45,107,123–125]. Unfortunately, when expressed in *E. coli* at a high expression level, Pct from *C. propionicum* (Pct_{Cp}) showed an inhibitory effect on cell growth [29]. Two Pct mutants were subsequently constructed in order to reduce the effect on cell growth, namely, $Pct532_{Cp}$ (A243T, and one silent nucleotide mutation of A1200G) and $Pct540_{Cp}$ (V193A, and four silent nucleotide mutations of T78C, T669C, A1125G, and T1158C), which resulted in an increased polymer content and LA fraction in P(3HB-*co*-LA) [45].

1.5 CONCLUSIONS AND OUTLOOK

Protein engineering has evolved from plain identification of catalytic amino acid residue of the enzyme to direct evolution and generate a high-function mutant. The main focus of direct evolution and systematic enzyme engineering strategies is to develop a potential host strain for PHA biosynthesis that effectively produces polyesters. In recent years, gene and protein engineering have been carried out extensively in numerous natural and unnatural PHA-producing microorganisms; novel biosynthetic pathways introducing a wide range of monomers supplied to PHA accumulation metabolism have been established. For example, 2HAs such as lactate, glycolate, and 2H4MV monomers have been introduced into the PHA family. They have until now not been able to naturally accumulate in PHA-producing bacteria, which lack such monomer-supplying pathways and enzymes for providing associated 2HA-CoAs. Metabolic system engineering, along with artificial gene and protein mutation, will allow us to further extend the PHA monomer spectrum.

Although we are still struggling with the problems in commercializing PHA production, designing novel pathways and suppling various value-added monomers might not stand as a game-changing solution, but generally speaking, the material properties of polyesters are extremely dependent on the type and composition of those monomers that compose the polymer. Hence it is essential to design constructive metabolic PHA synthetic systems of tailor-made polymers with preferred application value. It is safe to say that the power of rational enzyme engineering and direct evolution can accelerate the rapid advance in PHA development, alongside promoting a thriving and sustainable ecosystem.

REFERENCES

1. Sudesh K, Abe H, Doi Y. Synthesis, structure and properties of polyhydroxyalkanoates: Biological polyesters. *Prog Polym Sci.* 2000;25(10):1503–1555. doi:10.1016/S0079-6700(00)00035-6.

2. Philip S, Keshavarz T, Roy I. Polyhydroxyalkanoates: Biodegradable polymers with a range of applications. *J Chem Technol Biotechnol.* 2007;82(3):233–247. doi:10.1002/jctb.1667.

3. Amaro TMMM, Rosa D, Comi G, Iacumin L. Prospects for the use of whey for polyhydroxyalkanoate (PHA) production. *Front Microbiol.* 2019;10(MAY):992. doi:10.3389/fmicb.2019.00992.

4. Steinbüchel A, Valentin HE. Diversity of bacterial polyhydroxyalkanoic acids. *FEMS Microbiol Lett.* 1995;128(3):219–228. doi:10.1016/0378-1097(95)00125-O.

5. Doi Y. *Microbial Polyesters.* VCH, 1990. https://books.google.co.jp/books?id=f0FR AAAAMAAJ.

6. Steinbüchel A, Hustede E, Liebergesell M, Pieper U, Timm A, Valentin H. Molecular basis for biosynthesis and accumulation of polyhydroxyalkanoic acids in bacteria [published correction appears in FEMS Microbiol Rev. 1993 Apr;10(3–4):347–50]. *FEMS Microbiol Rev.* 1992;9(2–4):217–230. doi:10.1111/j.1574-6968.1992.tb05841.x.

7. Aldor IS, Keasling JD. Process design for microbial plastic factories: Metabolic engineering of polyhydroxyalkanoates. *Curr Opin Biotechnol.* 2003;14(5):475–483. doi:10.1016/j.copbio.2003.09.002.

8. Anderson AJ, Dawes EA. Occurrence, metabolism, metabolic role, and industrial uses of bacterial polyhydroxyalkanoates. *Microbiol Rev.* 1990;54(4):450–472. https://www.ncbi.nlm.nih.gov/pubmed/2087222.

9. Meng DC, Shen R, Yao H, Chen JC, Wu Q, Chen GQ. Engineering the diversity of polyesters. *Curr Opin Biotechnol.* 2014;29(1):24–33. doi:10.1016/j.copbio.2014.02.013.

10. Slater SC, Voige WH, Dennis DE. Cloning and expression in *Escherichia coli* of the *Alcaligenes eutrophus* H16 poly-β-hydroxybutyrate biosynthetic pathway. *J Bacteriol.* 1988;170(10):4431–4436. doi:10.1128/jb.170.10.4431-4436.1988.

11. Schubert P, Steinbüchel A, Schlegel HG. Cloning of the *Alcaligenes eutrophus* genes for synthesis of poly-beta-hydroxybutyric acid (PHB) and synthesis of PHB in *Escherichia coli. J Bacteriol.* 1988;170(12):5837–5847. doi:10.1128/jb.170.12.5837-5847.1988.

12. Peoples OP, Sinskey AJ. Poly-β-hydroxybutyrate (PHB) biosynthesis in *Alcaligenes eutrophus* H16. Identification and characterization of the PHB polymerase gene (*phbC*). *J Biol Chem.* 1989;264(26):15298–15303.

13. Rehm BHA. Polyester synthases: Natural catalysts for plastics. *Biochem J.* 2003;376(1):15–33. doi:10.1042/BJ20031254.

14. Pötter M, Steinbüchel A. Poly(3-hydroxybutyrate) granule-associated proteins: Impacts on poly(3-hydroxybutyrate) synthesis and degradation. *Biomacromolecules.* 2005;6(2):552–560. doi:10.1021/bm049401n.

15. Rehm BHA, Steinbüchel A. Biochemical and genetic analysis of PHA synthases and other proteins required for PHA synthesis. *Int J Biol Macromol.* 1999;25(1–3):3–19. doi:10.1016/S0141-8130(99)00010-0.

16. Huisman GW, Wonink E, Meima R, Kazemier B, Terpstra P, Witholt B. Metabolism of poly(3-hydroxyalkanoates) (PHAs) by *Pseudomonas oleovorans*: Identification and sequences of genes and function of the encoded proteins in the synthesis and degradation of PHA. *J Biol Chem.* 1991;266(4):2191–2198.

17. Liebergesell M, Sonomoto K, Madkour M, Mayer F, Steinbüchel A. Purification and characterization of the poly(hydroxyalkanoic acid) synthase from *Chromatium vinosum* and localization of the enzyme at the surface of poly(hydroxyalkanoic acid) granules. *Eur J Biochem.* 1994;226(1):71–80. doi:10.1111/j.1432-1033.1994.tb20027.x.

18. Prieto MA, Bühler B, Jung K, Witholt B, Kessler B. PhaF, a polyhydroxyalkanoate-granule-associated protein of *Pseudomonas oleovorans* GPo1 involved in the regulatory expression system for pha genes. *J Bacteriol.* 1999;181(3):858–868. https://www.ncbi.nlm.nih.gov/pubmed/9922249.

19. Huijberts GNM, De Rijk TC, De Waard P, Eggink G. 13C nuclear magnetic resonance studies of *Pseudomonas putida* fatty acid metabolic routes involved in poly(3-hydroxyalkanoate) synthesis. *J Bacteriol*. 1994;176(6):1661–1666. doi:10.1128/jb.176.6.1661-1666.1994.

20. Fukui T, Shiomi N, Doi Y. Expression and characterization of (*R*)-specific enoyl coenzyme A hydratase involved in polyhydroxyalkanoate biosynthesis by *Aeromonas caviae*. *J Bacteriol*. 1998;180(3):667–673. https://www.ncbi.nlm.nih.gov/pubmed/9457873.

21. Wieczorek R, Pries A, Steinbuchel A, Mayer F. Analysis of a 24-kilodalton protein associated with the polyhydroxyalkanoic acid granules in *Alcaligenes eutrophus*. *J Bacteriol*. 1995;177(9):2425–2435. doi:10.1128/jb.177.9.2425-2435.1995.

22. McCool GJ, Cannon MC. Polyhydroxyalkanoate inclusion body-associated proteins and coding region in *Bacillus megaterium*. *J Bacteriol*. 1999;181(2):585–592. http://jb.asm.org/content/181/2/585.abstract.

23. Kobayashi T, Uchino K, Abe T, Yamazaki Y, Saito T. Novel intracellular 3-hydroxybutyrate-oligomer hydrolase in *Wautersia eutropha* H16. *J Bacteriol*. 2005;187(15):5129–5135. doi:10.1128/JB.187.15.5129-5135.2005.

24. Kawaguchi Y, Doi Y. Kinetics and mechanism of synthesis and degradation of poly(3-hydroxybutyrate) in *Alcaligenes eutrophus*. *Macromolecules*. 1992;25(9):2324–2329. doi:10.1021/ma00035a007.

25. Ren Q, Sierro N, Witholt B, Kessler B. FabG, an NADPH-dependent 3-ketoacyl reductase of *Pseudomonas aeruginosa*, provides precursors for medium-chain-length poly-3-hydroxyalkanoate biosynthesis in *Escherichia coli*. *J Bacteriol*. 2000;182(10):2978–2981. doi:10.1128/JB.182.10.2978-2981.2000.

26. Lee S, Jeon E, Yun HS, Lee J. Improvement of fatty acid biosynthesis by engineered recombinant *Escherichia coli*. *Biotechnol Bioprocess Eng*. 2011;16(4):706–713. doi:10.1007/s12257-011-0034-6.

27. Nomura CT, Taguchi K, Taguchi S, Doi Y. Coexpression of genetically engineered 3-ketoacyl-ACP synthase III (fabH) and polyhydroxyalkanoate synthase (*phaC*) genes leads to short-chain-length-medium-chain-length polyhydroxyalkanoate copolymer production from glucose in *Escherichia coli* JM109. *Appl Environ Microbiol*. 2004;70(2):999–1007. doi:10.1128/AEM.70.2.999-1007.2004.

28. Kim J, Hetzel M, Boiangiu CD, Buckel W. Dehydration of (*R*)-2-hydroxyacyl-CoA to enoyl-CoA in the fermentation of α-amino acids by anaerobic bacteria. *FEMS Microbiol Rev*. 2004;28(4):455–468. doi:10.1016/j.femsre.2004.03.001.

29. Selmer T, Willanzheimer A, Hetzel M. Propionate CoA-transferase from *Clostridium propionicum*: Cloning of the gene and identification of glutamate 324 at the active site. *Eur J Biochem*. 2002;269(1):372–380. doi:10.1046/j.0014-2956.2001.02659.x.

30. David Y, Joo JC, Yang JE, Oh YH, Lee SY, Park SJ. Biosynthesis of 2-hydroxyacid-containing polyhydroxyalkanoates by employing butyryl-CoA transferases in metabolically engineered *Escherichia coli*. *Biotechnol J*. 2017;12(11):1700116. doi:10.1002/biot.201700116.

31. Davis JT, Moore RN, Imperiali B, et al. Biosynthetic thiolase from *Zooglea ramigera*. I. Preliminary characterization and analysis of proton transfer reaction. *J Biol Chem*. 1987;262(1):82–89.

32. Thompson S, Mayerl F, Peoples OP, Masamune S, Sinskey AJ, Walsh CT. Mechanistic studies on beta-ketoacyl thiolase from *Zoogloea ramigera*: Identification of the active-site nucleophile as Cys89, its mutation to Ser89, and kinetic and thermodynamic characterization of wild-type and mutant enzymes. *Biochemistry*. 1989;28(14):5735–5742. doi:10.1021/bi00440a006.

33. Williams SF, Palmer MA, Peoples OP, Walsh CT, Sinskey AJ, Masamune S. Biosynthetic thiolase from *Zoogloea ramigera*. Mutagenesis of the putative active-site base Cys-378 to Ser-378 changes the partitioning of the acetyl S-enzyme intermediate. *J Biol Chem*. 1992;267(23):16041–16043.

34. Steinbuchel A, Schlegel HG. Physiology and molecular genetics of poly(beta-hydroxy-alkanoic acid) synthesis in *Alcaligenes eutrophus*. *Mol Microbiol*. 1991;5(3):535–542. doi:10.1111/j.1365-2958.1991.tb00725.x.

35. Kim EJ, Son HF, Kim S, Ahn JW, Kim KJ. Crystal structure and biochemical characterization of beta-keto thiolase B from polyhydroxyalkanoate-producing bacterium *Ralstonia eutropha* H16. *Biochem Biophys Res Commun*. 2014;444(3):365–369. doi:10.1016/j.bbrc.2014.01.055.

36. Ploux O, Masamune S, Walsh CT. The NADPH-linked acetoacetyl-CoA reductase from *Zoogloea ramigera* characterization and mechanistic studies of the cloned enzyme over-produced in *Escherichia coli*. *Eur J Biochem*. 1988;174(1):177–182. doi:10.1111/j.1432-1033.1988.tb14079.x.

37. Matsumoto K, Tanaka Y, Watanabe T, et al. Directed evolution and structural analysis of nadph-dependent acetoacetyl coenzyme A(acetoacetyl-CoA) reductase from *Ralstonia eutropha* reveals two mutations responsible for enhanced kinetics. *Appl Environ Microbiol*. 2013;79(19):6134–6139. doi:10.1128/AEM.01768-13.

38. Tsuge T, Hisano T, Taguchi S, Doi Y. Alteration of chain length substrate specificity of *Aeromonas caviae* R-enantiomer-specific enoyl-coenzyme A hydratase through site-directed mutagenesis. *Appl Environ Microbiol*. 2003;69(8):4830–4836. doi:10.1128/AEM.69.8.4830-4836.2003.

39. Reiser SE, Mitsky TA, Gruys KJ. Characterization and cloning of an (R)-specific trans-2,3-enoylacyl-CoA hydratase from *Rhodospirillum rubrum* and use of this enzyme for PHA production in *Escherichia coli*. *Appl Microbiol Biotechnol*. 2000;53(2):209–218. doi:10.1007/s002530050010.

40. Tsuge T, Taguchi K, Taguchi S, Doi Y. Molecular characterization and properties of (R)-specific enoyl-CoA hydratases from *Pseudomonas aeruginosa*: Metabolic tools for synthesis of polyhydroxyalkanoates via fatty acid β-oxidation. *Int J Biol Macromol*. 2003;31(4–5):195–205. doi:10.1016/S0141-8130(02)00082-X.

41. Tsuge T, Sato S, Hiroe A, et al. Contribution of the distal pocket residue to the Acyl-chain-length specificity of (R)-specific enoyl-coenzyme a hydratases from *Pseudomonas* spp. *Appl Environ Microbiol*. 2015;81(23):8076–8083. doi:10.1128/AEM.02412-15.

42. Liu Y, Feng Y, Cao X, Li X, Xue S. Structure-directed construction of a high-performance version of the enzyme FabG from the photosynthetic microorganism *Synechocystis* sp. PCC 6803. *FEBS Lett*. 2015;589(20):3052–3057. doi:10.1016/j.febslet.2015.09.001.

43. Marcella AM, Barb AW. A rapid fluorometric assay for the S-malonyltransacylase FabD and other sulfhydryl utilizing enzymes. *J Biol Methods*. 2016;3(4):e53. doi:10.14440/jbm.2016.144.

44. Qiu X, Choudhry AE, Janson CA, et al. Crystal structure and substrate specificity of the beta-ketoacyl-acyl carrier protein synthase III (FabH) from *Staphylococcus aureus*. *Protein Sci*. 2005;14(8):2087–2094. doi:10.1110/ps.051501605.

45. Yang TH, Kim TW, Kang HO, et al. Biosynthesis of polylactic acid and its copolymers using evolved propionate CoA transferase and PHA synthase. *Biotechnol Bioeng*. 2010;105(1):150–160. doi:10.1002/bit.22547.

46. Haywood GW, Anderson AJ, Chu L, Dawes EA. Characterization of two 3-ketothiolases possessing differing substrate specificities in the polyhydroxyalkanoate synthesizing organism *Alcaligenes eutrophus*. *FEMS Microbiol Lett*. 1988;52(1–2):91–96. doi:10.1111/j.1574-6968.1988.tb02577.x.

47. Slater S, Houmiel KL, Tran M, et al. Multiple β-ketothiolases mediate poly(β-hydroxyalkanoate) copolymer synthesis in *Ralstonia eutropha*. *J Bacteriol*. 1998;180(8):1979–1987. https://www.ncbi.nlm.nih.gov/pubmed/9555876.

48. Pantazaki AA, Ioannou AK, Kyriakidis DA. A thermostable β-ketothiolase of polyhydroxyalkanoates (PHAs) in *Thermus thermophilus*: Purification and biochemical properties. *Mol Cell Biochem*. 2005;269(1):27–36. doi:10.1007/s11010-005-2992-5.

49. Kim SA, Copeland L. Acetyl coenzyme A acetyltransferase of *Rhizobium* sp. (Cicer) strain CC 1192. *Appl Environ Microbiol.* 1997;63(9):3432–3437. https://www.ncbi.nlm .nih.gov/pubmed/16535684.

50. Taroncher-Oldenburg G, Nishina K, Stephanopoulos G. Identification and analysis of the polyhydroxyalkanoate-specific β-ketothiolase and acetoacetyl coenzyme a reductase genes in the cyanobacterium *Synechocystis* sp. strain PCC6803. *Appl Environ Microbiol.* 2000;66(10):4440–4448. doi:10.1128/AEM.66.10.4440-4448.2000.

51. Steinbüchel A, Schlegel HG. Physiology and molecular genetics of poly(β-hydroxyalkanoic acid) synthesis in *Alcaligenes eutrophus. Mol Microbiol.* 1991;5(3):535–542. doi:10.1111/j.1365-2958.1991.tb00725.x.

52. Nishimura T, Saito T, Tomita K. Purification and properties of β-ketothiolase from *Zoogloea ramigera. Arch Microbiol.* 1978;116(1):21–27. doi:10.1007/BF00408729.

53. Bonk BM. *Novel Applications and Methods for the Computer-Aided Understanding and Design of Enzyme Activity.* Thesis, 2018.

54. Modis Y, Wierenga RK. A biosynthetic thiolase in complex with a reaction intermediate: The crystal structure provides new insights into the catalytic mechanism. *Structure.* 1999;7(10):1279–1290. doi:10.1016/S0969-2126(00)80061-1.

55. Modis Y, Wierenga RK. Crystallographic analysis of the reaction pathway of *Zoogloea ramigera* biosynthetic thiolase. *J Mol Biol.* 2000;297(5):1171–1182. doi:10.1006/jmbi.2000.3638.

56. Kim EJ, Kim KJ. Crystal structure and biochemical characterization of PhaA from *Ralstonia eutropha*, a polyhydroxyalkanoate-producing bacterium. *Biochem Biophys Res Commun.* 2014;452(1):124–129. doi:10.1016/j.bbrc.2014.08.074.

57. Madison LL, Huisman GW. Metabolic engineering of poly(3-hydroxyalkanoates): From DNA to plastic. *Microbiol Mol Biol Rev.* 1999;63(1):21–53. http://www.ncbi.nlm. nih.gov/pubmed/10066830%0Ahttp://www.pubmedcentral.nih.gov/articlerender.fcgi? artid=PMC98956.

58. Kim J, Chang JH, Kim EJ, Kim KJ. Crystal structure of (*R*)-3-hydroxybutyryl-CoA dehydrogenase PhaB from *Ralstonia eutropha. Biochem Biophys Res Commun.* 2014;443(3):783–788. doi:10.1016/j.bbrc.2013.10.150.

59. Dennis D, McCoy M, Stangl A, Valentin HE, Wu Z. Formation of poly(3-hydroxybutyrate-*co*-3-hydroxyhexanoate) by PHA synthase from *Ralstonia eutropha. J Biotechnol.* 1998;64(2–3):177–186. doi:10.1016/S0168-1656(98)00110-2.

60. Taguchi K, Aoyagi Y, Matsusaki H, Fukui T, Doi Y. Co-expression of 3-ketoacyl-ACP reductase and polyhydroxyalkanoate synthase genes induces PHA production in *Escherichia coli* HB101 strain. *FEMS Microbiol Lett.* 1999;176(1):183–190. doi:10.1016/S0378-1097(99)00215-3.

61. Kessler B, Kraak MN, Ren Q, Klinke S, Prieto M, Witholt B. *Enzymology and Molecular Genetics of PHAmcl Biosynthesis.* Biochem Princ Mech Biosynth Degrad Polym Wiley-VCH, Weinheim, Germany, 1998, 48–56.

62. Rehm BHA, Krüger N, Steinbüchel A. A new metabolic link between fatty acid de novo synthesis and polyhydroxyalkanoic acid synthesis. The *phaG* gene from *Pseudomonas putida* KT2440 encodes a 3-hydroxyacyl-acyl carrier protein-coenzyme a transferase. *J Biol Chem.* 1998;273(37):24044–24051. doi:10.1074/jbc.273.37.24044.

63. Wang Q, Tappel RC, Zhu C, Nomura CT. Development of a new strategy for production of medium-chain-length polyhydroxyalkanoates by recombinant *Escherichia coli* via inexpensive non-fatty acid feedstocks. *Appl Environ Microbiol.* 2012;78(2):519–527. doi:10.1128/AEM.07020-11.

64. Tappel RC, Pan W, Bergey NS, et al. Engineering *Escherichia coli* for improved production of short-chain-length-*co*-medium-chain-length poly[(*R*)-3-hydroxyalkanoate] (SCL-co-MCL PHA) copolymers from renewable nonfatty acid feedstocks. *ACS Sustain Chem Eng.* 2014;2(7):1879–1887. doi:10.1021/sc500217p.

65. Fukui T, Doi Y. Cloning and analysis of the poly(3-hydroxybutyrate-*co*-3-hydroxyhexanoate) biosynthesis genes of *Aeromonas caviae*. *J Bacteriol*. 1997;179(15):4821–4830. doi:10.1128/jb.179.15.4821-4830.1997.

66. Tsuge T, Fukui T, Matsusaki H, et al. Molecular cloning of two (*R*)-specific enoyl-CoA hydratase genes from *Pseudomonas aeruginosa* and their use for polyhydroxyalkanoate synthesis. *FEMS Microbiol Lett*. 2000;184(2):193–198. doi:10.1016/S0378-1097(00)00046-X.

67. Fiedler S, Steinbüchel A, Rehm BH. The role of the fatty acid β-oxidation multienzyme complex from *Pseudomonas oleovorans* in polyhydroxyalkanoate biosynthesis: Molecular characterization of the *fadBA* operon from *P. oleovorans* and of the enoyl-CoA hydratase genes phaJ from *P. oleovorans* and *Pseudomonas putida*. *Arch Microbiol*. 2002;178(2):149–160. doi:10.1007/s00203-002-0444-0.

68. Wang Q, Nomura CT. Monitoring differences in gene expression levels and polyhydroxyalkanoate (PHA) production in *Pseudomonas putida* KT2440 grown on different carbon sources. *J Biosci Bioeng*. 2010;110(6):653–659. doi:10.1016/j.jbiosc.2010.08.001.

69. Sato S, Kanazawa H, Tsuge T. Expression and characterization of (*R*)-specific enoyl coenzyme A hydratases making a channeling route to polyhydroxyalkanoate biosynthesis in *Pseudomonas putida*. *Appl Microbiol Biotechnol*. 2011;90(3):951–959. doi:10.1007/s00253-011-3150-5.

70. Kawashima Y, Cheng W, Mifune J, Orita I, Nakamura S, Fukui T. Characterization and functional analyses of *R*-specific enoyl coenzyme a hydratases in polyhydroxyalkanoate-producing *Ralstonia eutropha*. *Appl Environ Microbiol*. 2012;78(2):493–502. doi:10.1128/AEM.06937-11.

71. Liu G, Cai S, Hou J, et al. Enoyl-CoA hydratase mediates polyhydroxyalkanoate mobilization in *Haloferax mediterranei*. *Sci Rep*. 2016;6(1):24015. doi:10.1038/srep24015.

72. Kihara T, Hiroe A, Ishii-Hyakutake M, Mizuno K, Tsuge T. *Bacillus cereus*-type polyhydroxyalkanoate biosynthetic gene cluster contains *R*-specific enoyl-CoA hydratase gene. *Biosci Biotechnol Biochem*. 2017;81(8):1627–1635. doi:10.1080/09168451.2017.1325314.

73. Jin H, Nikolau BJ. Evaluating PHA productivity of bioengineered *Rhodosprillum rubrum*. *PLoS One*. 2014;9(5):1–8. doi:10.1371/journal.pone.0096621.

74. Han J, Qiu YZ, Liu DC, Chen GQ. Engineered *Aeromonas hydrophila* for enhanced production of poly(3-hydroxybutyrate-*co*-3-hydroxyhexanoate) with alterable monomers composition. *FEMS Microbiol Lett*. 2004;239(1):195–201. doi:10.1016/j.femsle.2004.08.044.

75. Fukui T, Kichise T, Iwata T, Doi Y. Characterization of 13 kDa granule-associated protein in *Aeromonas caviae* and biosynthesis of polyhydroxyalkanoates with altered molar composition by recombinant bacteria. *Biomacromolecules*. 2001;2(1):148–153. doi:10.1021/bm0056052.

76. Beld J, Lee DJ, Burkart MD. Fatty acid biosynthesis revisited: Structure elucidation and metabolic engineering. *Mol Biosyst*. 2015;11(1):38–59. doi:10.1039/c4mb00443d.

77. Lai CY, Cronan JE. Isolation and characterization of β-ketoacyl-acyl carrier protein reductase (*fabG*) mutants of *Eschelichia coli* and *Salmonella enterica* serovar typhimurium. *J Bacteriol*. 2004;186(6):1869–1878. doi:10.1128/JB.186.6.1869-1878.2004.

78. Alberts AW, Majerus PW, Talamo B, Vagelos PR. Acyl-carrier protein. II. Intermediary reactions of fatty acid synthesis. *Biochemistry*. 1964;3(10):1563–1571. doi:10.1021/bi00898a030.

79. Toomey RE, Wakil SJ. Studies on the mechanism of fatty acid synthesis XV. Preparation and general properties of β-ketoacyl acyl carrier protein reductase from *Escherichia coli*. *Biochim Biophys Acta (BBA)/Lipids Lipid Metab*. 1966;116(2):189–197. doi:10.1016/0005-2760(66)90001-4.

80. Sheldon PS, Kekwick RGO, Smith CG, Sidebottom C, Slabas AR. 3-Oxoacyl-[ACP] reductase from oilseed rape (*Brassica napus*). *Biochim Biophys Acta (BBA)/Protein Struct Mol.* 1992;1120(2):151–159. doi:10.1016/0167-4838(92)90263-D.

81. Zhang Y, Cronan JE. Transcriptional analysis of essential genes of the *Escherichia coli* fatty acid biosynthesis gene cluster by functional replacement with the analogous *Salmonella typhimurium* gene cluster. *J Bacteriol.* 1998;180(13):3295–3303. https://www.ncbi.nlm.nih.gov/pubmed/9642179.

82. Jeon E, Lee S, Lee S, Han SO, Yoon YJ, Lee J. Improved production of long-chain fatty acid in *Escherichia coli* by an engineering elongation cycle during fatty acid synthesis (FAS) through genetic manipulation. *J Microbiol Biotechnol.* 2012;22(7):990–999. doi:10.4014/jmb.1112.12057.

83. Jun Choi Y, Hwan Park J, Yong Kim T, Yup Lee S. Metabolic engineering of *Escherichia coli* for the production of 1-propanol. *Metab Eng.* 2012;14(5):477–486. doi:10.1016/j.ymben.2012.07.006.

84. Zhang H, Liu Y, Yao C, Cao X, Tian J, Xue S. FabG can function as PhaB for poly-3-hydroxybutyrate biosynthesis in photosynthetic cyanobacteria *Synechocystis* sp. PCC 6803. *Bioengineered.* 2017;8(6):707–715. doi:10.1080/21655979.2017.1317574.

85. Price AC, Zhang YM, Rock CO, White SW. Cofactor-induced conformational rearrangements establish a catalytically competent active site and a proton relay conduit in FabG. *Structure.* 2004;12(3):417–428. doi:10.1016/j.str.2004.02.008.

86. Davies C, Heath RJ, White SW, Rock CO. The 1.8 Å crystal structure and active-site architecture of β-ketoacyl-acyl carrier protein synthase III (FabH) from *Escherichia coli*. *Structure.* 2000;8(2):185–195. doi:10.1016/S0969-2126(00)00094-0.

87. Taguchi K, Aoyagi Y, Matsusaki H, Fukui T, Doi Y. Over-expression of 3-ketoacyl-ACP synthase III or malonyl-CoA-ACP transacylase gene induces monomer supply for polyhydroxybutyrate production in *Escherichia coli* HB101. *Biotechnol Lett.* 1999;21(7):579–584. doi:10.1023/A:1005572526080.

88. Choi KH, Heath RJ, Rock CO. β-ketoacyl-acyl carrier protein synthase III (FabH) is a determining factor in branched-chain fatty acid biosynthesis. *J Bacteriol.* 2000;182(2):365–370. doi:10.1128/JB.182.2.365-370.2000.

89. Choi KH, Kremer L, Besra GS, Rock CO. Identification and substrate specificity of β-ketoacyl (acyl carrier protein) synthase III (mtFabH) from *mycobacterium tuberculosis*. *J Biol Chem.* 2000;275(36):28201–28207. doi:10.1074/jbc.M003241200.

90. Khandekar SS, Gentry DR, Van Aller GS, et al. Identification, substrate specificity, and inhibition of the *Streptococcus pneumoniae* β-ketoacyl-acyl carrier protein synthase III (FabH). *J Biol Chem.* 2001;276(32):30024–30030. doi:10.1074/jbc.M101769200.

91. Wallen LL, Rohwedder WK. Poly-β-hydroxyalkanoate from activated sludge. *Environ Sci Technol.* 1974;8(6):576–579. doi:10.1021/es60091a007.

92. Matsumoto K, Yamada M, Leong CR, Jo SJ, Kuzuyama T, Taguchi S. A new pathway for poly(3-hydroxybutyrate) production in *Escherichia coli* and *Corynebacterium glutamicum* by functional expression of a new acetoacetyl-coenzyme A synthase. *Biosci Biotechnol Biochem.* 2011;75(2):364–366. doi:10.1271/bbb.100682.

93. Kim J, Chang JH, Kim KJ. Crystal structure and biochemical properties of the (S)-3-hydroxybutyryl-CoA dehydrogenase PaaH1 from *Ralstonia eutropha*. *Biochem Biophys Res Commun.* 2014;448(2):163–168. doi:10.1016/j.bbrc.2014.04.101.

94. Tseng HC, Martin CH, Nielsen DR, Prather KLJ. Metabolic engineering of *Escherichia coli* for enhanced production of (R)- and (S)-3-hydroxybutyrate. *Appl Environ Microbiol.* 2009;75(10):3137–3145. doi:10.1128/AEM.02667-08.

95. Moskowitz GJ, Merrick JM. Metabolism of poly-β-hydroxybutyrate. II. Enzymic synthesis of D-(–)-β-hydroxybutyryl coenzyme A by an enoyl hydrase from *Rhodospirillum rubrum*. *Biochemistry.* 1969;8(7):2748–2755. doi:10.1021/bi00835a009.

96. Mothes G, Babel W. *Methylobacterium rhodesianum* MB 126 possesses two stereo-specific crotonyl-CoA hydratases. *Can J Microbiol.* 1995;41(13):68–72. doi:10.1139/m95-170.

97. Sato S, Nomura CT, Abe H, Doi Y, Tsuge T. Poly[(*R*)-3-hydroxybutyrate] formation in *Escherichia coli* from glucose through an enoyl-CoA hydratase-mediated pathway. *J Biosci Bioeng.* 2007;103(1):38–44. doi:10.1263/jbb.103.38.

98. Bhubalan K, Lee WH, Loo CY, et al. Controlled biosynthesis and characterization of poly(3-hydroxybutyrate-*co*-3-hydroxyvalerate-*co*-3-hydroxyhexanoate) from mixtures of palm kernel oil and 3HV-precursors. *Polym Degrad Stab.* 2008;93(1):17–23. doi:10.1016/j.polymdegradstab.2007.11.004.

99. Steinbüchel A, Pieper U. Production of a copolyester of 3-hydroxybutyric acid and 3-hydroxyvaleric acid from single unrelated carbon sources by a mutant of *Alcaligenes eutrophus. Appl Microbiol Biotechnol.* 1992;37(1):1–6.

100. Srirangan K, Liu X, Tran TT, Charles TC, Moo-Young M, Chou CP. Engineering of *Escherichia coli* for direct and modulated biosynthesis of poly(3-hydroxybutyrate-*co*-3-hydroxyvalerate) copolymer using unrelated carbon sources. *Sci Rep.* 2016;6:36470. doi:10.1038/srep36470.

101. Fukui T, Suzuki M, Tsuge T, Nakamura S. Microbial synthesis of poly((*R*)-3-hydroxybutyrate-*co*-3-hydroxypropionate) from unrelated carbon sources by engineered *Cupriavidus necator. Biomacromolecules.* 2009;10(4):700–706. doi:10.1021/bm801391j.

102. Herter S, Farfsing J, Gad'On N, et al. Autotrophic CO_2 fixation by *Chloroflexus aurantiacus*: Study of glyoxylate formation and assimilation via the 3-hydroxypropionate cycle. *J Bacteriol.* 2001;183(14):4305–4316. doi:10.1128/JB.183.14.4305-4316.2001.

103. Herter S, Fuchs G, Bacher A, Eisenreich W. A bicyclic autotrophic CO_2 fixation pathway in *Chloroflexus aurantiacus. J Biol Chem.* 2002;277(23):20277–20283. doi:10.1074/jbc.M201030200.

104. Andreessen B, Lange AB, Robenek H, Steinbüchel A. Conversion of glycerol to poly(3-hydroxypropionate) in recombinant *Escherichia coli. Appl Environ Microbiol.* 2010;76(2):622–626. doi:10.1128/AEM.02097-09.

105. Valentin HE, Dennis D. Production of poly(3-hydroxybutyrate-*co*-4-hydroxybutyrate) in recombinant *Escherichia coli* grown on glucose. *J Biotechnol.* 1997;58(1):33–38. doi:10.1016/S0168-1656(97)00127-2.

106. Söhling B, Gottschalk G. Molecular analysis of the anaerobic succinate degradation pathway in *Clostridium kluyveri. J Bacteriol.* 1996;178(3):871–880. doi:10.1128/jb.178.3.871-880.1996.

107. Taguchi S, Yamada M, Matsumoto K, et al. A microbial factory for lactate-based polyesters using a lactate-polymerizing enzyme. *Proc Natl Acad Sci USA.* 2008;105(45):17323–17327. doi:10.1073/pnas.0805653105.

108. Matsumoto K, Ishiyama A, Sakai K, Shiba T, Taguchi S. Biosynthesis of glycolate-based polyesters containing medium-chain-length 3-hydroxyalkanoates in recombinant *Escherichia coli* expressing engineered polyhydroxyalkanoate synthase. *J Biotechnol.* 2011;156(3):214–217. doi:10.1016/j.jbiotec.2011.07.040.

109. Matsumoto K, Shiba T, Hiraide Y, Taguchi S. Incorporation of glycolate units promotes hydrolytic degradation in flexible poly(glycolate-*co*-3-hydroxybutyrate) synthesized by engineered *Escherichia coli. ACS Biomater Sci Eng.* 2017;3(12):3058–3063. doi:10.1021/acsbiomaterials.6b00194.

110. Mizuno S, Enda Y, Saika A, Hiroe A, Tsuge T. Biosynthesis of polyhydroxyalkanoates containing 2-hydroxy-4-methylvalerate and 2-hydroxy-3-phenylpropionate units from a related or unrelated carbon source. *J Biosci Bioeng.* 2018;125(3):295–300. doi:10.1016/j.jbiosc.2017.10.010.

111. Kim J, Darley D, Buckel W. 2-Hydroxyisocaproyl-CoA dehydratase and its activator from *Clostridium difficile*. *FEBS J*. 2005;272(2):550–561. doi:10.1111/j.1742-4658.2004.04498.x.

112. Kim J, Darley D, Selmer T, Buckel W. Characterization of (*R*)-2-hydroxyisocaproate dehydrogenase and a family III coenzyme A transferase involved in reduction of L-leucine to isocaproate by *Clostridium difficile*. *Appl Environ Microbiol*. 2006;72(9):6062–6069. doi:10.1128/AEM.00772-06.

113. Saika A, Watanabe Y, Sudesh K, Abe H, Tsuge T. Enhanced incorporation of 3-hydroxy-4-methylvalerate unit into biosynthetic polyhydroxyalkanoate using leucine as a precursor. *AMB Express*. 2011;1(1):1–8. doi:10.1186/2191-0855-1-6.

114. Saika A, Watanabe Y, Sudesh K, Tsuge T. Biosynthesis of poly(3-hydroxybutyrate-*co*-3-hydroxy-4-methylvalerate) by recombinant *Escherichia coli* expressing leucine metabolism-related enzymes derived from *Clostridium difficile*. *J Biosci Bioeng*. 2014;117(6):670–675. doi:10.1016/j.jbiosc.2013.12.006.

115. Saika A, Ushimaru K, Mizuno S, Tsuge T. Genome-based analysis and gene dosage studies provide new insight into 3-hydroxy-4-methylvalerate biosynthesis in *Ralstonia eutropha*. Zhulin IB, ed. *J Bacteriol*. 2015;197(8):1350–1359. doi:10.1128/JB.02474-14.

116. Yang JE, Park SJ, Kim WJ, et al. One-step fermentative production of aromatic polyesters from glucose by metabolically engineered *Escherichia coli* strains. *Nat Commun*. 2018;9(1):79. doi:10.1038/s41467-017-02498-w.

117. Stoll VS, Kimber MS, Pai EF. Insights into substrate binding by D-2-ketoacid dehydrogenases from the structure of *Lactobacillus pentosus* D-lactate dehydrogenase. *Structure*. 1996;4(4):437–447. doi:10.1016/S0969-2126(96)00049-4.

118. Razeto A, Kochhar S, Hottinger H, Dauter M, Wilson KS, Lamzin VS. Domain closure, substrate specificity and catalysis of D-lactate dehydrogenase from *Lactobacillus bulgaricus*. *J Mol Biol*. 2002;318(1):109–119. doi:10.1016/S0022-2836(02)00086-4.

119. Dengler U, Niefind K, Kieß M, Schomburg D. Crystal structure of a ternary complex of D-2-hydroxy-isocaproate dehydrogenase from *Lactobacillus casei*, NAD+ and 2-oxoisocaproate at 1.9 Å resolution. *J Mol Biol*. 1997;267(3):640–660. doi:10.1006/jmbi.1996.0864.

120. Martins BM, Macedo-Ribeiro S, Bresser J, Buckel W, Messerschmidt A. Structural basis for stereo-specific catalysis in NAD⁺-dependent (*R*)-2-hydroxyglutarate dehydrogenase from *Acidaminococcus fermentans*. *FEBS J*. 2005;272(1):269–281. doi:10.1111/j.1432-1033.2004.04417.x.

121. Schweiger G, Buckel W. On the dehydration of (*R*)-lactate in the fermentation of alanine to propionate by *Clostridium propionicum*. *FEBS Lett*. 1984;171(1):79–84. doi:10.1016/0014-5793(84)80463-9.

122. Dickert S, Pierik AJ, Linder D, Buckel W. The involvement of coenzyme a esters in the dehydration of (*R*)-phenyllactate to (*E*)-cinnamate by *Clostridium sporogens*. *Eur J Biochem*. 2000;267(12):3874–3884. doi:10.1046/j.1432-1327.2000.01427.x.

123. Park SJ, Lee TW, Lim SC, et al. Biosynthesis of polyhydroxyalkanoates containing 2-hydroxybutyrate from unrelated carbon source by metabolically engineered *Escherichia coli*. *Appl Microbiol Biotechnol*. 2012;93(1):273–283. doi:10.1007/s00253-011-3530-x.

124. Park SJ, Kim TW, Kim MK, Lee SY, Lim SC. Advanced bacterial polyhydroxyalkanoates: Towards a versatile and sustainable platform for unnatural tailor-made polyesters. *Biotechnol Adv*. 2012;30(6):1196–1206. doi:10.1016/j.biotechadv.2011.11.007.

125. Choi SY, Park SJ, Kim WJ, et al. One-step fermentative production of poly(lactate-*co*-glycolate) from carbohydrates in *Escherichia coli*. *Nat Biotechnol*. 2016;34(4):435–440. doi:10.1038/nbt.3485.

2 PHA Granule-Associated Proteins and Their Diverse Functions

Mariela P. Mezzina, Daniela S. Alvarez, and M. Julia Pettinari

CONTENTS

2.1 INTRODUCTION

Polyhydroxyalkanoates (PHA) accumulate in intracellular inclusion bodies called granules that can take up to 90% of cell volume [1] and are surrounded by a layer whose composition has been the subject of study for many years. The first experimental evidence of the presence of this layer was provided in 1964 by Lundgren et al., who described it in granules from *Bacillus* species as "discrete membrane-like structures which encase the granules" [2]. A later study published in 1968 shed light on the chemical composition of the granules, as it revealed that these inclusions were not only composed of polymer, which accounted for approximately 98% of the granule but also contained 2% proteins and a small number of lipids on their surface [3]. Over the years, numerous studies revealed that PHA granules are surrounded by an organized protein layer composed of several different granule-associated proteins (GAPs) [4–9].

GAPs include some proteins with enzymatic functions, such as polymerases and depolymerases, and also other proteins, known as phasins, that were initially proposed to act solely as a barrier to protect cellular components from interacting with the hydrophobic polymer but have in recent years been shown to possess a remarkable variety of different functions.

The presence of biosynthetic, catabolic, structural, and regulatory proteins in the granule surface indicates that PHA granules are not just cytoplasmic polymer inclusions but organized and complex organelle-like subcellular structures [4] that play an active role in carbon and electron flows, contributing to cellular homeostasis [10].

GAPs affect granule formation and PHA synthesis and degradation, and their role in these processes differs according to different models proposed for granule formation.

2.2 GRANULE ASSEMBLY MODELS

Although the pathways for PHA biosynthesis have been elucidated, and the enzymes involved have been identified, there is not a universally accepted mechanism for granule formation. This process has been mostly studied in the model organism, *Cupriavidus necator* (previously known as *Ralstonia eutropha*), but also in several other bacteria using a variety of microscopy techniques, giving rise to several hypotheses.

1) The first theoretical model to explain granule formation was the micelle model, based on immunocytochemical studies performed using *C. necator* that revealed that the PHA synthase was located on the surface of the PHA granules [6]. This model proposed that granules begin to form randomly in the cytoplasm when nascent PHA polymer chains remain covalently bound to PHA synthase molecules. As a result of their hydrophobic nature, polymer chains are oriented to the interior of the aggregates, forming micelle-like structures. As PHA micelles grow, other GAPs bind to their surface, contributing to the formation of the granule.

2) The budding model was proposed as a result of the experimental observation of a membrane-like material, which varied in thickness, surrounding the surface of PHA granules from different bacterial species [11]. According to this model, PHA synthase binds to the inner face of the plasma membrane, and the nascent PHA chain is synthesized into the hydrophobic environment of the bilayer. When the polymer chains reach a certain size, they bud from the membrane into the cytoplasm, where other GAPs bind to the PHA granules. However, it was demonstrated that no phospholipids are bound to PHA granules *in vivo* in representatives of *alpha-*, *beta-*, and *gammaproteobacteria* [12].

3) A third model, the scaffold model, was proposed based on transmission electron microscopy observations of *C. necator* cells in which nascent granules were observed to arise from the center of the cell. A scaffold of unknown nature, visualized as a dark-stained mediation element, was proposed to provide sites for the PHA synthase to initiate granule formation [13]. In *Pseudomonas putida*, the association of the PHA granules to the nucleoid, located in the center of the cells, was observed to be mediated by the binding of the GAP PhaF [14], while in *C. necator*, this interaction seems to be mediated by PhaM [15].

Although PHA granule formation was studied in different bacteria through a variety of techniques, there is still no conclusive model that can account for all the experimental observations. It is possible that there could be different mechanisms of PHA granule formation in different bacteria, and there is growing evidence that proteins other than the PHA synthase have a central role in this process [14–16].

2.3 GAPS WITH ENZYMATIC ACTIVITY: PHA SYNTHASES AND DEPOLYMERASES

PHA are synthesized and degraded in a highly regulated cycle that affects carbon and reduces power availability for many bacterial processes [17]. Studies performed by Merrick and Doudoroff in 1961 using PHA granules from *Bacillus megaterium* and *Rhodospirillum rubrum* revealed that the enzymes involved in the synthesis and degradation of PHA are located in the granule surface [18]. The presence of the enzymes responsible for synthesis and degradation of the polymer in the surface of the granules facilitates the turnover of PHA, as synthesis and degradation processes are simultaneously active.

In carbon excess and nutrient limitation conditions, the extra carbon substrate is stored as PHA. The PHA synthase or polymerase (PhaC) is the key enzyme in the biosynthesis of polyhydroxyalkanoates, as it catalyzes the polymerization of coenzyme A (CoA) thioesters of hydroxyalkanoic acids provided by different metabolic routes into PHA with the concomitant release of CoA [19]. Under carbon starvation conditions, PHA depolymerase (PhaZ) degrades PHA and releases 3-hydroxyalkanoic acid monomers, which after activation to 3-hydroxyacyl-CoAs, can be oxidized by the β-oxidation pathway. PhaZ is also attached to the granules and is produced under both PHA production and PHA mobilization conditions [20].

PHA synthases are classified into four groups depending on their subunit composition and substrate specificity. Class I PHA synthases consist of only one type of subunit and use monomers of three to five carbon atoms as substrates, producing short-chain-length PHA [21]. They are represented by the best-known synthase, PhaC, from *C. necator*. Class II synthases, mostly synthases from *Pseudomonas*, preferentially use monomers of six to fourteen carbon atoms and produce medium-chain-length PHA [19]. Class III synthases, like the one present in *Allochromatium vinosum*, are composed of two different subunits, PhaC and PhaE, and have the same substrate specificity as class I synthases [22]. Class IV synthases are present in *Bacillus* species and are also composed of two different subunits: PhaC, the catalytic subunit, and PhaR. These synthases are capable of polymerizing mainly short-chain-length monomer units, such as 3-hydroxybutyrate (C4) and 3-hydroxyvalerate (C5), but can polymerize some unusual monomers as minor components [23].

2.4 NON-ENZYMATIC GAPS: TRANSCRIPTIONAL REGULATORS AND PHASINS

Apart from the proteins that catalyze the synthesis and degradation of PHA, granules contain several other GAPs, including proteins with regulatory activities and a variety of small proteins with different functions, collectively known as phasins, the most abundant components of the protein layer surrounding the granules.

In many bacteria, PHA synthesis is transcriptionally regulated by PhaR, a protein that binds to PHA granules extensively studied in *C. necator*. In 2002, two different research groups discovered that in this microorganism, PhaR acts as a transcriptional repressor that binds to three different targets: (i) the promoter region of *phaP1* (that encodes for the most abundant phasin), (ii) its own promoter region, and (iii) the granule surface [7,24]. In the absence of poly(3-hydroxybutyrate) (PHB) granules, PhaR binds to a regulatory sequence located upstream of *phaP1*, repressing PhaP1 synthesis. As PHB accumulates, PhaR is titrated by binding to the granule surface, thus decreasing its cytoplasmic concentration and relieving *phaP1* repression. As PhaP1 concentration rises, PhaR is displaced from the surface of the granules, causing the PhaR cytoplasmic concentration to rise. As free PhaR binds to the regulatory sequence of *phaP1* and to its own regulatory sequence, the expression of both genes is repressed [7,24].

Many homologs of *phaR* have been described in other PHB-producing species near PHB-synthesis genes [7], but in some cases, it has different effects. A very similar regulator named PhbR was identified in *Azotobacter vinelandii*, but a mutant strain lacking *phbR* showed lower levels of expression of genes associated with polymer synthesis compared with the wild type strain [25], indicating that this regulator, unlike PhaR of *C. necator*, could be an activator of PHB synthesis. Furthermore, PhbR not only activates PHB synthesis in *A. vinelandii* (inducing *phbB* transcription) but its own transcription as well [26].

Analysis of possible regulators of PHA synthesis in *Pseudomonas* identified two proteins that affected PHA accumulation. In 1999, Prieto et al. showed that insertional inactivation of *phaF*, which encodes a GAP with a C-terminal domain similar to histones, reduced PHA levels and that PhaF inhibits *phaC* expression in non-PHA accumulating (glucose-containing) medium [27]. A year later, a transcriptional regulator that also affects PHA synthesis in *P. oleovorans* but does not bind to PHA granules was described and named PhaD. Interestingly, *phaD* mutants were also unable to produce PhaI, one of the main GAPs in this microorganism [28]. Later studies performed on mutant strains derived from *P. putida* KT2442 showed that, in the presence of some carbon sources, PhaD acts as a positive regulator, activating the transcription of its operon (that also includes genes that code for the PHA polymerases and depolymerase) and also of *phaI* and *phaF*, the main phasins of *P. oleovorans*, that are part of an operon [29].

Phasins were first described in 1994 by Pieper-Fürst et al., who identified a low molecular weight protein, GA14, associated with PHA granules in *Rhodococcus ruber*. Since these proteins form a layer at the surface of PHA granules, they were named phasins (PhaP) in analogy to oleosins, which form a layer at the surface of triacylglycerol inclusions in oilseed plants [8]. Early studies performed on *C. necator* revealed that Tn5 insertions affecting a gene coding for a 24KDa protein had reduced PHA synthesis [30]. This protein was very abundant in the granule surface and considered to have a similar role as GA14 from *R. ruber*, so it was also called a phasin [31].

Analysis of other PHA-producing bacteria revealed that all of them presented phasins as the major protein components of PHA granules. These amphiphilic low molecular weight proteins play an important structural function, forming an interphase between the hydrophobic content of PHA granules and the hydrophilic cytoplasm content. It has been proposed that phasins prevent coalescence of granules by forming a protein layer that separates the hydrophobic PHA from the hydrophilic cytoplasmic content [31]. Phasins have been described as promoting bacterial growth and PHA synthesis and affect the number, size, and distribution of the granules [32,33].

2.5 FUNCTIONAL DIVERSITY OF PHASINS

Based mainly on observations of the phasins of *C. necator* and *Pseudomonas*, the first studies that analyzed the role of phasins proposed three mechanisms to explain the enhancing effect of these proteins on PHA synthesis.

First, binding of phasins to the surface of PHA granules increases the surface/volume rate of the granules, stabilizing them and preventing their coalescence, leading to an increase in PHA synthesis. This is supported by several studies that showed that mutants unable to produce PhaP have fewer and larger granules compared with strains that produce PhaP [8,31].

A second interesting observation is the interaction of phasins with PHA synthases, activating polymer synthesis. Early studies showed that PhaC had lower activity in *phaP* mutants in *C. necator* [31]. Additionally, PhaP from this organism was shown to increase the *in vitro* activity of PhaC from *Aeromonas caviae* and also of class II PHA synthases PhaC1 and PhaC2 from *Pseudomonas aeruginosa* [34]. A similar observation was reported in *Synechocystis* sp. strain PCC 6803, in which a mutant that does not produce phasin showed reduced activity of the PHB synthase [35]. Another phasin from *C. necator*, PhaM, was also described as the physiological activator of the PHB synthase in this microorganism [36].

A third mechanism proposes that phasins promote polymer synthesis indirectly by preventing other unspecific proteins from binding to the granules [31,37]. Experiments performed on recombinant PHA-producing *Escherichia coli* showed that, in the absence of PhaP, polymer synthesis was diminished, and large amounts of the heat-shock protein HspA were synthesized and bound to the granule's surface. HspA was observed to mimic the role of the phasin, preventing coalescence of granules into one single granule, but with limited effectiveness, as polymer content was lower in the absence of PhaP [38].

As phasin studies continued, many other functions of these remarkable proteins have been described [10]. ApdA from *R. rubrum* was observed to activate PHB granules isolated from different species, including recombinant *E. coli* so that the granules could be hydrolyzed by soluble *R. rubrum* PHB depolymerase *in vitro* [39], and Mms16 (later called ApdA) from *Magnetospirillum gryphiswaldense* was also observed to bind to PHB granules *in vivo* and to activate the PHB depolymerase of *R. rubrum in vitro* [40]. PhaP from *C. necator* was also reported to affect PHB degradation, although the mechanism is not clear. The effect could be direct, through interaction with the PHB depolymerase, or indirect, by providing a PHB depolymerase access to the surface of the PHB granules [41].

Phasins have been observed to control PHA synthesis through a variety of mechanisms, including both positive and negative transcriptional regulation. On the one hand, PhaF acts as a negative transcriptional regulator of PhaC in *P. putida*, as the disruption of *phaF* was observed to lead to higher expression levels of PhaC [27]. On the other hand, overexpression of *phaP* in *A. hydrophila* was observed to increase the expression of *phaC*, indicating that it acts as a positive regulator [42].

Some phasins, such as PhaM from *C. necator* and PhaF from *P. putida*, play a key role in intracellular granule localization, as these GAPs have DNA-binding motifs that allow interaction with the bacterial nucleoid. PhaM and PhaF are also involved in the equal distribution of accumulated PHB granules to daughter cells during cell division [14,15,43]. In addition, PhaM interacts with the synthase forming an active PhaM-PhaC initiation complex and also increases the specific activity of purified PHB synthase *in vitro* [36].

Phasins are not restricted solely to interaction with PHA and PHA-related proteins and their genes. In recombinant PHB-producing *E. coli*, the expression of PhaP from *Azotobacter* sp. FA8 relieved the stress produced by PHB accumulation, increasing growth and polymer accumulation [33], and reduced the expression of stress-related genes, such as *ibpA* and *dnaK*. Surprisingly, the protective effect was also observed in a non-PHB-synthesizing *E. coli* strain, in which the expression of the phasin increased growth and resistance to both heat shock and superoxide stress by paraquat [44]. Since this microorganism does not have the capacity to produce PHB, the protective effect of the phasin could not be attributed to the protection of the cells against the negative effects of polymer accumulation, but it reflected a more general protective role. Further characterization of PhaP from *Azotobacter* sp. FA8 revealed that it protected citrate synthase from thermal aggregation *in vitro* and facilitated its refolding after chemical denaturation, revealing chaperone activity. This protein also has chaperone activity *in vivo*, as it was observed to bind to inclusion bodies formed by an overexpressed protein, reducing their number and size [45]. Considering these studies, many diverse functions can be added to the traditional roles assigned to phasins (see Figure 2.1):

   i) Activators of polymerases, enhancing polymer synthesis
   ii) Activators of depolymerases, which favors polymer utilization
   iii) Transcriptional regulators
   iv) Control of intracellular localization of PHA granules and equal distribution of PHA granules to daughter cells during cell division
   v) Stress-relieving effects
   vi) *In vivo* and *in vitro* chaperone activities

2.6 WHAT MAKES A PHASIN A PHASIN?

Several types of phasin families have been distinguished based on the presence of different protein motifs that were defined by comparing a great number of phasins found in bacteria belonging to many taxonomic groups. The Pfam database (http://pfam.xfam.org/) considers four phasin-related families, each containing a characteristic domain. The largest family (PF09361) groups sequences found in bacteria that belong to *alpha-*, *beta-*, and *gammaproteobacteria*, including the most studied phasin, PhaP1 from *C. necator*. Phasins found in *Bacillus* species comprise

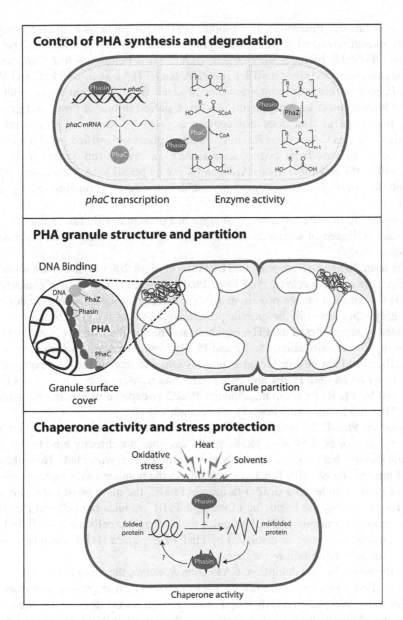

FIGURE 2.1 Functional diversity of phasins. *Top panel*: Some phasins regulate the transcription of PHA genes, while others activate PhaC or PhaZ activity. *Central panel*: Phasins are the major components of the PHA granule's surface, and some can bind DNA and/or participate in granule partition. *Bottom panel*: Phasins can have stress protection properties, probably by protecting protein from misfolding due to stress conditions. Phasins bind to misfolded proteins and could help them regain a folded state.

a separate family (PF09602), and another (PF09650) contains a diverse group of mostly uncharacterized proteins belonging to different *Proteobacteria*. The last family (PF05597) includes all characterized phasins belonging to *Pseudomonas* that accumulate medium-chain-length PHA (*mcl*-PHA), such as PhaF and PhaI from *P. putida*, but also contains other proteins from different *Proteobacteria* [46]. Although most known phasins can be classified in one of these four groups, there is a handful that does not contain any of the identified phasin-related domains, such as GA14 from *R. ruber*, the first phasin identified [8]. This is also the case of phasins from cyanobacteria, such as PhaP from *Synechocystis* sp. PCC 6803 [35] and those from Archaea [47]; it is possible that the discovery of new phasins in related microorganisms could, in the future, constitute new phasin families.

Considering that the different phasins studied present a wide functional diversity and that the degree of sequence similarity among some of them is very low, what defines a phasin?

The term "*sensu stricto*" was used by Pötter et al. in 2005 to define the so-called "true" phasins of *C. necator*, PhaP1 and PhaP3, based on a number of characteristics: (i) they present phasin protein motifs, (ii) direct binding to PHA granules, (iii) are highly represented in the granule protein layer [48], and (iv) their expression is regulated by repressors of the PHA metabolism (PhaR in the case of *C. necator*) [49]. Following this classification, PhaP2 and PhaP4 from *C. necator*, which are similar to PhaP1, would not be considered as phasins *sensu stricto*, as they are expressed at much lower levels than PhaP1 and even PhaP3, and transcription of both genes is not repressed by PhaR. In addition, although PhaP2 is capable of binding to artificial poly(3HB), it was not found bound to the granules *in vivo* [49].

However, PhaM, the activator of PhaC in *C. necator*, would also not fit into this category, as it does not have a phasin motif and does not directly bind to the PHB granule surface, but attaches indirectly to it via interaction with PhaC. Nevertheless, PhaM presents phasin-like functions as it affects the number and surface-to-volume ratio of PHB granules in a similar fashion as PhaP1, the main phasin of *C. necator* [15]. PhaM ensures the formation of multiple PHB granules per cell and the equal distribution of accumulated PHB granules to the daughter cells during cell division [15], functions that are also displayed by PhaF of *P. putida* [14] that can be considered a phasin by definition *sensu stricto*.

Furthermore, by this definition, GA14 from *R. ruber*, the first granule-associated protein called a phasin, would not be a *sensu stricto* or true phasin, as it does not present conserved phasin motifs [46]. For this reason, GA14 does not belong to any of the four phasin families, although (i) it is indeed present in the PHA granule surface, (ii) its gene is located in the PHA cluster of *R. ruber*, and (iii) its expression is strongly dependent on PHA synthesis [8].

In view of the conflicts that arise in the classification of these proteins, it would perhaps be better to keep the original idea and call all GAPs without a catalytic function phasins. As knowledge about the different structures, interactions, and physiological functions of these proteins increases, it might be convenient to introduce some kind of additional classification categories (e.g., different phasin classes) as has been applied in the case of the PHA synthases.

2.7 BIOTECHNOLOGICAL APPLICATIONS OF GAPS

Some of the properties of GAPs have been exploited for biotechnological applications. The capability to bind to the PHA granules has been used to develop systems for protein purification and drug display using either synthases or phasins.

PhaC catalyzes PHA synthesis and remains covalently attached to the surface of the PHA granules. Transcriptional fusions of PHA synthase to target proteins have been extensively used for the self-assembly of PHA particles displaying such proteins. Peters and Rehm used this strategy to immobilize β-galactosidase at the PHA granule surface, opening the way to the application of this technology in the recycling of biocatalysts [50]. Since then, a great number of enzymes for diverse applications have been immobilized on PHA beads through PhaC fusions, exhibiting enhanced functionality and stability [51–57].

PhaC fusions have also been described for the engineering of PHA granules, called polyester beads, displaying foreign antigens and are used in various biomedical applications, such as particulate vaccines. In one of the first studies that describe fusions to PhaC, PHA inclusions displaying two *Mycobacterium tuberculosis* antigens were developed and successfully used for mouse vaccination [58]. Over the past few years, many studies explored the capacity of engineered PHA beads to serve as diagnostic tools and vaccines against tuberculosis [59–63], *Streptococcus pneumoniae* [64,65], *Neisseria meningitidis* [66], *P. aeruginosa* [67], and Hepatitis C virus [68,69].

PHA synthases have also been used to develop protein purification methods through the separation of the granule-bound protein from cell components. Two different approaches have been proposed that allow the recovery of the target protein: the use of a self-cleavable intein [70] or a self-cleaving module from a modified sortase A from *Staphylococcus aureus* [71]. Additionally, a specific method for immunoglobulin G (IgG) purification was developed, fusing the PHA synthase to the IgG-binding ZZ domain of protein A from *S. aureus*. The PHA granules displaying a ZZ domain were able to efficiently purify IgG from human serum and mouse hybridoma culture supernatants [72,73].

Most of these methods have been developed in *E. coli* and *Lactococcus lactis*, but there are a few examples that use other bacteria. Grage et al. manipulated *B. megaterium* to produce functionalized PHB granules using plasmid expressed PhaC protein fusions [74]. Lee et al. engineered the human pathogen *P. aeruginosa* for the production of PHA inclusions displaying its own antigens by harnessing the inherent PHA production system of the bacteria [67]. A similar approach was used with a recombinant PHB-producing variant of the non-pathogenic *M. smegmatis* to develop a particulate vaccine displaying mycobacterial antigens [63].

The affinity of PhaP for PHA granules has also been used to purify recombinant proteins through simple methods that enable separation of the polymer granules from the cell medium and cellular components or to fuse antigens to PHA nanoparticles. The main phasin from *C. necator* has been used to construct fusions to several different proteins in PHB-synthesizing *E. coli*, allowing easy purification of the granule-bound recombinant protein after cell lysis. One study proposed the use of an intein to achieve a self-cleaving fusion, thus enabling the recovery of the protein of interest devoid of PhaP [75]. Polypeptide or protein-ligand fusions to the same phasin were also used as a means to direct PHA nanoparticles to several kinds of human target cells by receptor recognition both *in vitro* and in an *in vivo* mouse model [76] and to construct PHA particles that display PhaP-fused antigens that can be used to bind antibodies to enhance fluorescence-activated cell sorting [77]. A different phasin, PhaF from *P. putida*, was used in similar applications. Proteins fused to the N-terminal domain of this phasin, named BioF, could be produced in *P. putida* and co-purified with PHA granules, maintaining their enzymatic activity after release [78]. An interesting variant of this system involves the utilization of the PHA granules as substrates for the immobilization of the fused protein, as exemplified by the use of the granule-bound Cry toxin for environmental insect control [79].

Other applications are related to the intrinsic properties of phasins, independent of their capability to bind to PHA granules. Since phasins are amphipathic proteins, they can be used as biosurfactants. This was studied using PhaP from *A. hydrophila* that could efficiently emulsify petrochemical and vegetable oils and diesel *in vitro* and could be used as a biocompatible emulsifier for the cosmetic and pharmaceutical industry [80]. Lastly, the discovery of the chaperone-like properties observed for some phasins, such as PhaP from *Azotobacter* sp. FA8, open the way for a number of potential applications ranging from the enhancement of recombinant protein expression to the production of many different biotechnologically relevant products [10]. Expression of this phasin in *E. coli* enhanced tolerance to ethanol, butanol, and 1,3-propanediol and increased the production of ethanol and 1,3-propanediol in recombinant strains [81]. Additionally, expression of this protein in strains containing multiple mutations to optimize the synthesis of 1,3-propanediol in *E. coli* doubled the production of this compound in bioreactor cultures, showing that PhaP could boost the production of the diol even further [82].

2.8   CONCLUSIONS AND OUTLOOK

Since the discovery of PHA inclusions in bacteria, they have been related to fitness and survival, mainly as carbon and reducing power reservoirs. Later studies showed that PHA synthesis and degradation are dynamic processes that involve many proteins, among which GAPs have a major role. These proteins are involved in PHA turnover, but also in the coordination of granule synthesis with cell division. In the last few decades, many studies have revealed the remarkable functional diversity of GAPs, including their role in cellular homeostasis through the regulation of PHA metabolism, and also chaperone-like properties that could further potentiate the fitness enhancement properties of PHA. The diverse capabilities of different GAPs have been used for a variety of applications, and recent studies in areas such as nanotechnology and the production of biotechnologically relevant products suggest an even broader range of new potential applications.

REFERENCES

1. Madison LL, Huisman GW. Metabolic engineering of poly (3-hydroxyalkanoates): from DNA to plastic. *Microbiol Mol Biol Rev.* 1999; 63(1): 21–53.

2. Lundgren DG, Pfister RM, Merrick JM. Structure of poly-beta-hydroxybutyric acid granules. *J Gen Microbiol.* 1964; 34(3): 441–446.
3. Griebel R, Smith Z, Merrick JM. Metabolism of poly-β-hydroxybutyrate. I. Purification, composition, and properties of native poly-β-hydroxybutyrate granules from *Bacillus megaterium*. *Biochemistry.* 1968; 7(10): 3676–3681.
4. Jendrossek D. Polyhydroxyalkanoate granules are complex subcellular organelles (carbonosomes). *J Bacteriol.* 2009; 191(10): 3195–3202.
5. Grage K, Jahns AC, Parlane N, et al. Bacterial polyhydroxyalkanoate granules: biogenesis, structure, and potential use as nano-/micro-beads in biotechnological and biomedical applications. *Biomacromolecules.* 2009; 10(4): 660–669.
6. Gerngross TU, Reilly P, Joanne S, et al. Immunocytochemical analysis of poly-beta-hydroxybutyrate (PHB) synthase in *Alcaligenes eutrophus* H16: localization of the synthase enzyme at the surface of PHB granules. *J Bacteriol.* 1993; 175(16): 5289–5293.
7. Pötter M, Madkour MH, Mayer F, et al. Regulation of phasin expression and polyhydroxylkanoate (PHA) granule formation in *Ralstonia eutropha* H16. *Microbiology.* 2002; 148(8): 2413–2426.
8. Pieper-Fürst U, Madkour MH, Mayer F, et al. Purification and characterization of a 14-kilodalton protein that is bound to the surface of polyhydroxyalkanoic acid granules in *Rhodococcus ruber*. *J Bacteriol.* 1994; 176(14): 4328–4337.
9. Pfeiffer D, Jendrossek D. Localization of poly(3-Hydroxybutyrate) (PHB) granule-associated proteins during PHB granule formation and identification of two new phasins, PhaP6 and PhaP7, in *Ralstonia eutropha* H16. *J Bacteriol.* 2012; 194(21): 5909–5921.
10. Mezzina MP, Pettinari MJ. Phasins, multifaceted polyhydroxyalkanoate granule-associated proteins. *Appl Environ Microbiol.* 2016; 82(17): 5060–5067.
11. Stubbe JA, Tian J. Polyhydroxyalkanoate (PHA) homeostasis: the role of the PHA synthase. *Nat Prod Rep.* 2003; 20(5): 445–457.
12. Bresan S, Sznajder A, Hauf W, et al. Polyhydroxyalkanoate (PHA) granules have no phospholipids. *Sci Rep.* 2016; 6: 26612.
13. Tian J, Sinskey AJ, Stubbe J. Kinetic studies of polyhydroxybutyrate granule formation in *Wautersia eutropha* H16 by transmission electron microscopy. *J Bacteriol.* 2005; 187(11): 3814–3824.
14. Galán B, Dinjaski N, Maestro B, et al. Nucleoid-associated PhaF phasin drives intracellular location and segregation of polyhydroxyalkanoate granules in *Pseudomonas putida* KT2442. *Mol Microbiol.* 2011; 79(2): 402–418.
15. Pfeiffer D, Wahl A, Jendrossek D. Identification of a multifunctional protein, PhaM, that determines number, surface to volume ratio, subcellular localization and distribution to daughter cells of poly(3-hydroxybutyrate), PHB, granules in *Ralstonia eutropha* H16. *Mol Microbiol.* 2011; 82(4): 936–951.
16. Beeby M, Cho M, Stubbe J, et al. Growth and localization of polyhydroxybutyrate granules in *Ralstonia eutropha*. *J Bacteriol.* 2012; 194(5): 1092–1099.
17. López NI, Pettinari MJ, Nikel PI, et al. Polyhydroxyalkanoates: much more than biodegradable plastics. *Adv Appl Microbiol.* 2015; 93: 73–106.
18. Merrick JM, Doudoroff M. Enzymatic synthesis of poly-beta-hydroxybutyric acid in bacteria. *Nature.* 1961; 189: 890–892.
19. Rehm BHA. Polyester synthases: natural catalysts for plastics. *Biochem J.* 2003; 376(1): 15–33.
20. De Eugenio LI, Escapa IF, Morales V, et al. The turnover of medium-chain-length polyhydroxyalkanoates in *Pseudomonas putida* KT2442 and the fundamental role of PhaZ depolymerase for the metabolic balance. *Environ Microbiol.* 2010; 12(1): 207–221.
21. Qi Q, Rehm BHA. Polyhydroxybutyrate biosynthesis in *Caulobacter crescentus*: molecular characterization of the polyhydroxybutyrate synthase. *Microbiology.* 2001; 147(12): 3353–3358.

22. Taguchi S, Doi Y. Evolution of Polyhydroxyalkanoate (PHA) production system by "enzyme evolution": successful case studies of directed evolution. *Macromol Biosci.* 2004; 4(3): 145–156.

23. Tsuge T, Hyakutake M, Mizuno K. Class IV polyhydroxyalkanoate (PHA) synthases and PHA-producing *Bacillus. Appl Microbiol Biotechnol.* 2015; 99(15): 6231–6240.

24. Maehara A, Taguchi S, Nishiyama T, et al. A repressor protein, PhaR, regulates polyhydroxyalkanoate (PHA) synthesis via its direct interaction with PHA. *J Bacteriol.* 2002; 184(14): 3992–4002.

25. Peralta-Gil M, Segura D, Guzmán J, et al. Expression of the *Azotobacter vinelandii* poly-β-hydroxybutyrate biosynthetic *phbBAC* operon is driven by two overlapping promoters and is dependent on the transcriptional activator PhbR. *J Bacteriol.* 2002; 184(20): 5672–5677.

26. Hernandez-Eligio A, Castellanos M, Moreno S, et al. Transcriptional activation of the *Azotobacter vinelandii* polyhydroxybutyrate biosynthetic genes *phbBAC* by PhbR and RpoS. *Microbiology.* 2011; 157(11): 3014–3023.

27. Prieto MA, Bühler B, Jung K, et al. PhaF, a polyhydroxyalkanoate-granule-associated protein of *Pseudomonas oleovorans* GPo1 involved in the regulatory expression system for *pha* genes. *J Bacteriol.* 1999; 181(3): 858–868.

28. Klinke S, de Roo G, Witholt B, et al. Role of *phaD* in accumulation of poly (3-hydroxyalkanoates) in *Pseudomonas oleovorans. Appl Environ Microbiol.* 2000; 66(9): 3705–3710.

29. De Eugenio LI, Galán B, Escapa IF, et al. The PhaD regulator controls the simultaneous expression of the *pha* genes involved in polyhydroxyalkanoate metabolism and turnover in *Pseudomonas putida* KT2442. *Environ Microbiol.* 2010; 12(6): 1591–1603.

30. Schubert P, Steinbüchel A, Schlegel HG. Cloning of the *Alcaligenes eutrophus* genes for synthesis of poly-beta-hydroxybutyric acid (PHB) and synthesis of PHB in *Escherichia coli. J Bacteriol.* 1988; 170(12): 5837–5847.

31. Wieczorek R, Pries A, Steinbüchel A, et al. Analysis of a 24-kilodalton protein associated with the polyhydroxyalkanoic acid granules in *Alcaligenes eutrophus. J Bacteriol.* 1995; 177(9): 2425–2435.

32. York GM, Junker BH, Stubbe J, et al. Accumulation of the PhaP phasin of *Ralstonia eutropha* is dependent on production of polyhydroxybutyrate in cells. *J Bacteriol.* 2001; 183(14): 4217–4226.

33. de Almeida A, Nikel PI, Giordano AM, et al. Effects of granule-associated protein PhaP on glycerol-dependent growth and polymer production in poly(3-hydroxybutyrate)-producing *Escherichia coli. Appl Environ Microbiol.* 2007; 73(24): 7912–7916.

34. Ushimaru K, Motoda Y, Numata K, et al. Phasin proteins activate *Aeromonas caviae* polyhydroxyalkanoate (PHA) synthase but not *Ralstonia eutropha* PHA synthase. *Appl Environ Microbiol.* 2014; 80(9): 2867–2873.

35. Hauf W, Watzer B, Roos N, et al. Photoautotrophic polyhydroxybutyrate granule formation is regulated by cyanobacterial phasin PhaP in *Synechocystis* sp. strain PCC 6803. *Appl Environ Microbiol.* 2015; 81(13): 4411–4422.

36. Pfeiffer D, Jendrossek D. PhaM is the physiological activator of poly(3-hydroxybutyrate) (PHB) synthase (PhaC1) in *Ralstonia eutropha. Appl Environ Microbiol.* 2014; 80(2): 555–563.

37. Maehara A, Ueda S, Nakano H, et al. Analyses of a polyhydroxyalkanoic acid granule-associated 16-kilodalton protein and its putative regulator in the *pha* locus of *Paracoccus denitrificans. J Bacteriol.* 1999; 181(9): 2914–2921.

38. Tessmer N, König S, Malkus U, et al. Heat-shock protein HspA mimics the function of phasins *sensu stricto* in recombinant strains of *Escherichia coli* accumulating polythioesters or polyhydroxyalkanoates. *Microbiology.* 2007; 153(2): 366–374.

39. Handrick R, Reinhardt S, Schultheiss D, et al. Unraveling the function of the *Rhodospirillum rubrum* activator of polyhydroxybutyrate (PHB) degradation: the activator is a PHB-granule-bound protein (phasin). *J Bacteriol*. 2004; 186(8): 2466–2475.

40. Schultheiss D, Handrick R, Jendrossek D, et al. The presumptive magnetosome protein Mms16 is a poly(3-hydroxybutyrate) granule-bound protein (phasin) in *Magnetospirillum gryphiswaldense*. *J Bacteriol*. 2005; 187(7): 2416–2425.

41. Kuchta K, Chi L, Fuchs H, et al. Studies on the influence of phasins on accumulation and degradation of PHB and nanostructure of PHB granules in *Ralstonia eutropha* H16. *Biomacromolecules*. 2007; 8(2): 657–662.

42. Tian SJ, Lai WJ, Zheng Z, et al. Effect of over-expression of phasin gene from *Aeromonas hydrophila* on biosynthesis of copolyesters of 3-hydroxybutyrate and 3-hydroxyhexanoate. *FEMS Microbiol Lett*. 2005; 244(1): 19–25.

43. Maestro B, Galán B, Alfonso C, et al. A new family of intrinsically disordered proteins: structural characterization of the major phasin PhaF from *Pseudomonas putida* KT2440. *PLoS One*. 2013; 8(2): e56904.

44. de Almeida A, Catone M V., Rhodius VA, et al. Unexpected stress-reducing effect of PhaP, a poly(3-hydroxybutyrate) granule-associated protein, in *Escherichia coli*. *Appl Environ Microbiol*. 2011; 77(18): 6622–6629.

45. Mezzina MP, Wetzler DE, de Almeida A, et al. A phasin with extra talents: a polyhydroxyalkanoate granule-associated protein has chaperone activity. *Environ Microbiol*. 2015; 17(5): 1765–1776.

46. Mezzina MP, Wetzler DE, Catone MV, et al. A phasin with many faces: structural insights on PhaP from *Azotobacter* sp. FA8. *PLoS One*. 2014; 9(7): e103012.

47. Cai S, Cai L, Liu H, et al. Identification of the haloarchaeal phasin (PhaP) that functions in polyhydroxyalkanoate accumulation and granule formation in *Haloferax mediterranei*. *Appl Environ Microbiol*. 2012; 78(6): 1946–1952.

48. Pötter M, Steinbüchel A. Poly(3-hydroxybutyrate) granule-associated proteins: impacts on poly(3-hydroxybutyrate) synthesis and degradation. *Biomacromolecules*. 2005; 6(2): 552–560.

49. Pötter M, Müller H, Steinbüchel A. Influence of homologous phasins (PhaP) on PHA accumulation and regulation of their expression by the transcriptional repressor PhaR in *Ralstonia eutropha* H16. *Microbiology*. 2005; 151(3): 825–833.

50. Peters V, Rehm BHA. In vivo enzyme immobilization by use of engineered polyhydroxyalkanoate synthase. *Appl Environ Microbiol*. 2006; 72(3): 1777–1783.

51. Blatchford PA, Scott C, French N, et al. Immobilization of organophosphohydrolase OpdA from *Agrobacterium radiobacter* by overproduction at the surface of polyester inclusions inside engineered *Escherichia coli*. *Biotechnol Bioeng*. 2012; 109(5): 1101–1108.

52. Hooks DO, Blatchford PA, Rehm BHA. Bioengineering of bacterial polymer inclusions catalyzing the synthesis of N-acetylneuraminic acid. *Appl Environ Microbiol*. 2013; 79(9): 3116–3121.

53. Jahns AC, Rehm BHA. Immobilization of active lipase B from *Candida antarctica* on the surface of polyhydroxyalkanoate inclusions. *Biotechnol Lett*. 2015; 37(4): 831–835.

54. Ran G, Tan D, Dai WE, et al. Immobilization of alkaline polygalacturonate lyase from *Bacillus subtilis* on the surface of bacterial polyhydroxyalkanoate nano-granules. *Appl Microbiol Biotechnol*. 2017; 101(8): 3247–3258.

55. Tan D, Zhao JP, Ran GQ, et al. Highly efficient biocatalytic synthesis of L-DOPA using in situ immobilized *Verrucomicrobium spinosum* tyrosinase on polyhydroxyalkanoate nano-granules. *Appl Microbiol Biotechnol*. 2019; 103(14): 5663–5678.

56. Ran G, Tan D, Zhao J, et al. Functionalized polyhydroxyalkanoate nano-beads as a stable biocatalyst for cost-effective production of the rare sugar D-allulose. *Bioresour Technol*. 2019; 289: 121673.

57. Wong JX, Rehm BHA. Design of modular polyhydroxyalkanoate scaffolds for protein immobilization by directed ligation. *Biomacromolecules*. 2018; 19(10): 4098–4112.
58. Parlane NA, Wedlock DN, Buddle BM, et al. Bacterial polyester inclusions engineered to display vaccine candidate antigens for use as a novel class of safe and efficient vaccine delivery agents. *Appl Environ Microbiol*. 2009; 75(24): 7739–7744.
59. Chen S, Parlane NA, Lee J, et al. New skin test for detection of bovine tuberculosis on the basis of antigen-displaying polyester inclusions produced by recombinant *Escherichia coli*. *Appl Environ Microbiol*. 2014; 80(8): 2526–2535.
60. Parlane NA, Chen S, Jones GJ, et al. Display of antigens on polyester inclusions lowers the antigen concentration required for a bovine tuberculosis skin test. *Clin Vaccine Immunol*. 2016; 23(1): 19–26.
61. Chen S, Sandford S, Kirman JR, et al. Innovative antigen carrier system for the development of tuberculosis vaccines. FASEB *J*. 2019; 33(6): 7505–7518.
62. Rubio Reyes P, Parlane NA, Wedlock DN, et al. Immunogencity of antigens from *Mycobacterium tuberculosis* self-assembled as particulate vaccines. *Int J Med Microbiol*. 2016; 306(8): 624–632.
63. Lee JW, Parlane NA, Rehm BHA, et al. Engineering mycobacteria for the production of selfassembling biopolyesters displaying mycobacterial antigens for use as a tuberculosis vaccine. *Appl Environ Microbiol*. 2017; 83(5): e02289–e02316.
64. González-Miró M, Radecker AM, Rodríguez-Noda LM, et al. Design and biological assembly of polyester beads displaying pneumococcal antigens as particulate vaccine. *ACS Biomater Sci Eng*. 2018; 4(9): 3413–3424.
65. González-Miro M, Rodríguez-Noda L, Fariñas-Medina M, et al. Self-assembled particulate PsaA as vaccine against *Streptococcus pneumoniae* infection. *Heliyon*. 2017; 3(4): e00291.
66. González-Miró M, Rodríguez-Noda LM, Fariñas-Medina M, et al. Bioengineered polyester beads co-displaying protein and carbohydrate-based antigens induce protective immunity against bacterial infection. *Sci Rep*. 2018; 8(1): 1888.
67. Lee JW, Parlane NA, Wedlock DN, et al. Bioengineering a bacterial pathogen to assemble its own particulate vaccine capable of inducing cellular immunity. *Sci Rep*. 2017; 7: 41607.
68. Parlane NA, Grage K, Lee JW, et al. Production of a particulate hepatitis C vaccine candidate by an engineered *Lactococcus lactis* strain. *Appl Environ Microbiol*. 2011; 77(24): 8516–8522.
69. Martínez-Donato G, Piniella B, Aguilar D, et al. Protective T cell and antibody immune responses against hepatitis C virus achieved using a biopolyester-bead-based vaccine delivery system. *Clin Vaccine Immunol*. 2016; 23(4): 370–378.
70. Du J, Rehm BHA. Purification of target proteins from intracellular inclusions mediated by intein cleavable polyhydroxyalkanoate synthase fusions. *Microb Cell Fact*. 2017; 16(1): 184.
71. Hay ID, Du J, Reyes PR, et al. In vivo polyester immobilized sortase for tagless protein purification. *Microb Cell Fact*. 2015; 14(1): 190.
72. Lewis JG, Rehm BHA. ZZ polyester beads: an efficient and simple method for purifying IgG from mouse hybridoma supernatants. *J Immunol Methods*. 2009; 346(1–2): 71–74.
73. Brockelbank JA, Peters V, Rehm BHA. Recombinant *Escherichia coli* strain produces a ZZ domain displaying biopolyester granules suitable for immunoglobulin G purification. *Appl Environ Microbiol*. 2006; 72(11): 7394–7397.
74. Grage K, McDermott P, Rehm BHA. Engineering *Bacillus megaterium* for production of functional intracellular materials. *Microb Cell Fact*. 2017; 16(1): 211.

75. Banki MR, Gerngross TU, Wood DW. Novel and economical purification of recombinant proteins: Intein-mediated protein purification using in vivo polyhydroxybutyrate (PHB) matrix association. *Protein Sci.* 2005; 14: 1387–1395.

76. Yao YC, Zhan XY, Zhang J, et al. A specific drug targeting system based on polyhydroxyalkanoate granule binding protein PhaP fused with targeted cell ligands. *Biomaterials.* 2008; 29(36): 4823–4830.

77. Bäckström BT, Brockelbank JA, Rehm BHA. Recombinant *Escherichia coli* produces tailor-made biopolyester granules for applications in fluorescence activated cell sorting: functional display of the mouse interleukin-2 and myelin oligodendrocyte glycoprotein. *BMC Biotechnol.* 2007; 7(1): 3.

78. Moldes C, García P, García JL, et al. In vivo immobilization of fusion proteins on bioplastics by the novel tag BioF. *Appl Environ Microbiol.* 2004; 70(6): 3205–3212.

79. Moldes C, Farinós GP, De Eugenio LI, et al. New tool for spreading proteins to the environment: Cry1Ab toxin immobilized to bioplastics. *Appl Microbiol Biotechnol.* 2006; 72(1): 88–93.

80. Wei DX, Chen CB, Fang G, et al. Application of polyhydroxyalkanoate binding protein PhaP as a bio-surfactant. *Appl Microbiol Biotechnol.* 2011; 91(4): 1037–1047.

81. Mezzina MP, Alvarez DS, Egoburo DE, et al. A new player in the biorefineries field: phasin PhaP enhances tolerance to solvents and boosts ethanol and 1,3-propanediol synthesis in *Escherichia coli. Appl Environ Microbiol.* 2017; 83(14): e00662–e00717.

82. Egoburo DE, Díaz Peña R, Alvarez DS, et al. Microbial cell factories *à la Carte*: elimination of global regulators Cra and ArcA generates metabolic backgrounds suitable for the synthesis of bioproducts in *Escherichia coli. Appl Environ Microbiol.* 2018; 84(19): e01337–e01418.

3 Genomics of PHA Synthesizing Bacteria

*Parveen K. Sharma, Jilagamazhi Fu,
Nisha Mohanan, and David B. Levin*

CONTENTS

3.1 INTRODUCTION

Polyhydroxyalkanoates (PHA) are a family of bacterial polyesters synthesized as intracellular energy storage molecules that allow cells to survive in conditions of nutritional stress [1]. To date, over 150 types of PHA monomers have been identified that endow PHA polymers with a wide range of properties that may be tailored for various potential applications [2]. The PHA monomers can be generally divided into two groups according to the number of carbon atoms in their side-chains. Short-chain-length polyhydroxyalkanoates (*scl*-PHA) contain 3–5 carbon atoms, for example, polyhydroxybutyrate [PHB a.k.a. P(3HB)] consists of subunits with four carbon atoms. Medium-chain-length PHA (*mcl*-PHA) contain 3-hydroxyalkanoate subunits that contain 6–14 carbon atoms and are more structurally diverse than *scl*-PHA [3].

A wide range of eubacteria, archaea, and even eukaryotes, such as some yeast species, are known to store PHA as a carbon and energy source, but the majority of microorganisms that synthesize PHA are found in eubacteria. Attempts have been made to correlate various bacteria taxonomies with their genome content associated with PHA biosynthesis, although the machinery for PHA biosynthesis has various degrees of similarity. In this chapter, we review the literature on different bacteria that synthesize PHA polymers, with special emphasis on the genomics of the metabolic pathways used for PHA synthesis, and the genes and gene products involved.

3.2 *SCL*-PHA PRODUCING BACTERIA

Cupriavidus necator H16, previously known as *Alcaligenes eutrophus* and *Ralstonia eutropha* H16, is the most well-studied *scl*-PHA producing bacterium [4]. *C. necator* is well known for its ability to synthesize large quantities of intracellular PHB, which accumulates as much as 72% to 76% of the cell dry mass (CDM) when cultured in media containing soybean oil [5]. *C. necator* can also synthesize and accumulate up to 75 wt.-% of poly(3-hydroxybutyrate-*co*-3-hydroxyvalerate) [P(3HB-*co*-3HV) a.k.a. PHBV] in CDM copolymers when grown in cultures containing glucose and propionate [6]. *Alcaligenes latus*, unlike *C. necator*, does not require nutrient limitation to stimulate the synthesis of PHB from various sugars. However, increased PHB accumulation was observed under nitrogen-limiting conditions, resulting in an increase in the accumulation of PHB from 50% to 78% CDM [7].

Halomonas strains were isolated from deep-sea hydrothermal sites characterized by high pressure, high temperature gradients, and high concentrations in toxic elements (sulfides, heavy metals). Their ability to grow under conditions of high salt concentration and high pH has opened commercial applications in biotechnology. The presence of 3-hydroxyalkanoic acids in sediments recovered from the deep-sea hydrothermal vents suggests the synthesis of PHA by bacteria associated with these deep-sea vents [8].

A number of *Halomonas* spp. have been characterized for PHA production [9,10]. PHA production in cells of *Halomonas* spp. allows them to survive under high osmotic pressure. PHA production appears to be a universal phenomenon among halophiles and protects bacteria from the harmful effects of high salt concentration [11,12]. Genome sequences of 120 *Halomonas* species are available on the Integrated Microbial Genome (IMG) platform. Recently, *Halomonas* sp. SF2003 has been studied in detail for the presence of genes typically involved in PHA biosynthesis, such as *phaA*, *phaB*, and *phaC*, which has enabled preliminary analysis of their organization and characteristics [13].

Paracoccus species also show great potential for accumulating PHA. These Gram-negative methylotrophic bacteria are well known for possessing denitrification abilities and thus are considered very useful in the biotreatment of wastewater [14]. *Paracoccus denitrificans* and *Paracoccus pantotrophus* are the two most widely studied bacteria that can metabolize various carbon sources, including glycerol, methanol, n-pentanol, and carbon dioxide (CO_2) to PHA [15,16]. They display both autotrophic as well as heterotrophic physiology to produce PHA on simple substrates, as well as mixtures of other compounds that are mostly present in industrial effluent, wastewater, and ligno-cellulosic biomass hydrolysates [17,18]. Interestingly, PHA production in these bacteria has been observed to be both growth-associated as well as growth-limited and associated with carotenoid production [19]. However, large-scale production of PHA using *Paracoccus* spp. has not yet been demonstrated. The complete genome sequences of 59 *Paracoccus* species are available on IMG but have not been studied in detail.

3.3 *MCL*-PHA PRODUCING BACTERIA

To date, *Pseudomonas* species have been extensively investigated for their ability to synthesize *mcl*-PHA; the species tested include *P. aeruginosa*, *P. entomophila*, *P. fluorescens*, *P. oleovorance*, *P. putida*, *P. mendocina*, and *P. stutzeri* (and many more). *Pseudomonas* sp. can use substrates unrelated to the 3-hydroxyalkanoate structure of PHA polymer side-chains (glucose and glycerol), as well as substrates related to PHA side-chain structures (fatty acids, oils) for PHA production.

The carbon source used to grow the bacterial species that synthesize *mcl*-PHA has a significant effect on the subunit composition of the polymer produced (see Table 3.1). *Pseudomonas* species can use a wide range of carbon sources for synthesizing *mcl*-PHA producing versatile, functional groups in their side-chains, such as branched alkyl groups [35], halogens [36], epoxy moieties [37], alkyl esters [38], unsaturated aliphatic groups [39], and aromatic groups [40]. These functional groups can greatly change the physical properties of the *mcl*-PHA polymers, imparting excellent modifications for different applications.

Species such as *P. chlororaphis*, *P. fluorescens*, and *P. fulva* synthesize *mcl*-PHA with "uncommon" monomer compositions, and polymers synthesized by these bacteria may contain subunits with significant amounts of unsaturated or longer carbon chain length side-chains. *P. mendocina* CH50 is the only wild-type strain producing C8 homo-polymer from octanoate that has been reported to date [41]. A number of reviews on the genome organization and regulation of PHA synthesis are available for *Pseudomonas* sp. [42–45].

3.4 *SCL-CO-MCL*-COPOLYMER PRODUCERS

Mcl-PHA polymers containing medium length 3-hydroxyalkanoate side-chains are very amorphous, with low to no crystallinity, which makes them tacky at room temperature and unsuitable for structural or fiber applications. *Scl*-PHA

TABLE 3.1

PHA Production and Monomer Composition of Polymers Produced by *Pseudomonas* Species

Strain	Substrate	Concentration (g/L)	C/N ratio	PHA (wt.-%)	C6	C8	C10	C12	C12:1	C14	Ref.
P. putida LS46	Glucose	10	22.4	22	2.2	15.7	76.1	5.6	–	–	[19]
	Octanoic acid	3.2	11.7	56	6.9	89.3	3.4	–	–	–	
	Decanoic acid	2.8	10.7	34	4.6	47.5	46.6	1.3	–	–	
	Biodiesel glycerol			16.3	2.78	37.4	62.3	0.9	–	0.5	
P. putida KT2440	Sodium octanoate	12	39	52	18.8	81.2	–	–	–	–	[20]
	Glucose	18.5	41	25	5	11.8	70.3	4.7	9.4	0.8	[21]
	Glycerol	3.7	40	19	–	–	–	–	–	–	
P. entomophila L48	Dodecanoic acid	15	–	50.4	6.4	44.5	38.6	10.6	–	–	[22]
P. putida CA3	Phenylacetic acid	2	24	28	–	–	–	–	–	–	[23]
P. aeruginosa PAO1	Sodium gluconate	15	48	20	14	21	61	4	–	–	[24]
P. aeruginosa L2-1	Waste fryer oil	2	–	43	–	37.5	42.1	13.2	2.2	2.1	[25]
P. aeruginosa IFO3924	Palm oil	7	27	36	4	38	43	12	–	–	[26]
P. aeruginosa P14	Decanoic acid	5.2	30	16.9	8.2	55	36.8	–	–	–	[27]
P. mendocina ymp	Gluconate	15	54	13	3	15	68	13.8	–	–	[28]
	Octanoic acid	10	65	30	9.1	90.9	–	–	–	–	[29]
P. fulva TY16	Glucose	10	39	27	2.1	17.7	60.1	8.5	11.7	–	[30]
	Gluconic acid	10	36	34	2.4	20.5	63.5	7.1	6.5	–	
	Octanoic acid	4	26	55	9.6	79.1	11.3	3.7	–	–	

(Continued)

TABLE 3.1 (CONTINUED)
PHA Production and Monomer Composition of Polymers Produced by *Pseudomonas* Species

Strain	Substrate	Concentration (g/L)	C/N ratio	PHA (wt.-%)	C6	C8	C10	C12	C12:1	C14	Ref.
P. chlororaphis PA23-63-1	Glucose	20	11.2	30.56	5.7	46.7	36.3	12.4	–	1.6	[31]
	Octanoic acid	3.2	11.7	32.5	10.0	83.2	3.6		–	–	
	Glycerol bottom	20	–	24.37	7.2	40.1	36.2	12.8	2.2	0.4	
	Glycerol bottom + Valeric acid#	20	–	30.8	(27) 5.2	33.8	24.0	6.2	2.2		This study
	Glycerol bottom + Hexanoic acid#	20	–	36.5	(7.4) 38.3	26.3	20.9	5.4	–	1.6	
P. fluorescens GK13	Glucose	15	58	38	1.6	9.4	68.5	19.3	–		[32]
	Octanoic acid	10	65	31	4.2	95.8	–	–	–	–	
P. fluorescens BM07	Gluconate	12.6	42	8.2	–	5.2	35.5	19.3	32.5	7.4	[33]
	Octanoate	5.8	32	23.3	11.8	84.4	1.8	0.4	0.4	0.4	
P. stutzeri 13167	Glucose	10	45	58	2.4	21.3	63.2	4.2	6.1	1	[34]
	Soybean oil	10	–	63	–	8.2	63.4	2.6	–	–	

3.2 g/L hexanoic acid, 5.0 g/L valeric acid.

consists of 100% 3-hydroxybutanoate side-chains and has high tensile strength, but low elasticity, which makes them very brittle. *Scl*-PHA polymers consisting of poly(3-hydroxybutyrate-*co*-3-hydroxyvalerate) (PHBV) have slightly lower tensile strength, but greater elasticity, and *scl-co-mcl* copolymers composed of mostly C4 monomer and small amounts of C6 monomer have been shown to have properties most similar to polypropylene, with the added advantage that they are biodegradable [46,47]. The PHA synthases derived from copolymer producing strains, such as *Pseudomonas* sp. 61-3 and *Thiocapsa pfennigii*, have significant amino acid sequence identities and many highly conserved amino acid residues with the class I–IV PHA synthases [48,49].

3.5   GENOMICS OF *MCL*-PHA PRODUCING BACTERIA

Pseudomonas putida KT2440 is the most extensively studied *mcl*-PHA synthesizing bacterium. Omic studies of *P. putida* KT2440 have led to the development of new strategies for PHA production. In our laboratory (Department of Biosystems Engineering, University of Manitoba), we have focused on *P. putida* LS46, a recently isolated strain that is closely related to *P. putida* KT2440. We have compared the genome of *P. putida* LS46, which was isolated specifically for its ability to produce *mcl*-PHA, with 13 other *mcl*-PHA producing *Pseudomonas* species, which were isolated from diverse environments and not for their ability to produce PHA (see Table 3.2).

Some of the *P. putida* strains (KT2440, BIRD-1, as well as *P. fluorescens* Pf0-1), were isolated from rhizosphere soil, while *P. chlororaphis* PA23, *P. stutzeri* A1501,

TABLE 3.2

Pseudomonas **Genomes Used in this Study**

Genome name	Isolated from	Genome size (Mb)	ANI with *P. putida* ls46 (%)	Gene count	GC ratio	Horizontally transferred gene count
P. putida KT2440	Rhizosphere soil	6.18	97.34	5481	0.62	173
P. fluorescens SBW25	Leaf surface	7.14	78.67	6492	0.6	40
P. fluorescens Pf0-1	Agricultural soil	6.43	78.80	5857	0.61	256
P. entomophila L48	Fruit fly	5.88	85.41	5293	0.64	369
P. putida F1	Polluted creek	5.95	98.38	5422	0.62	133
P. aeruginosa PAO1	Infected wound	6.05	77.63	6401	0.66	2
P. mendocina ymp	Sediment surface	5.07	78.30	4730	0.65	231
P. chlororaphis PA23	Soybean roots	7.12	79.88	6264	0.65	45
P. monteilii SB3078	Contaminated soil	5.86	90.00	5525	0.63	2
P. putida BIRD-1	Rhizosphere soil	5.73	97.18	5046	0.62	112
P. putida LS46	Wastewater	5.87	100	5346	0.62	41
P. stutzeri A1501	Rice roots	4.56	76.91	4237	0.64	133
P. pseudoalcaligenes NBRC 14167	Sinus drainage	4.68	78.15	4713	0.62	68

and *P. fluorescens* SBW25 were isolated from plant roots or leaf surfaces. *P. putida* LS46, *P. putida* F1, *P. monteilii* SB3078, and *P. mendocina* ymp were isolated from polluted environments. *P. aeruginosa* PAO1 and *P. pseudoalcaligenes* NBRC14167 were isolated from wounds and human sinus drainage. Originally, none of these strains were isolated as PHA producers, but later it was observed the all these strains have the genetic machinery for *mcl*-PHA synthesis, and some of these strains were developed as industrial strains for PHA production. The genome sizes of these bacteria range from 4.56 to 7.14 Mb, with gene counts of 4,237 to 6,401. These bacteria showed average nucleotide identities of 76.91% to 98.38%.

3.5.1 Evolutionary Relationship among Pseudomonas Species

The most widely used method to study evolutionary relationships among bacteria is 16S rRNA gene sequence analysis. Protein-encoding genes are known to provide higher levels of taxonomic resolution than non-protein-encoding genes, like 16S rRNA [50]. Therefore, a protein-coding gene, *cpn60* (also known as *hsp60* or GroEL), was used for phylogenetic analysis of *Pseudomonas* species and strains [51]. Sequences were aligned using ClustalW, and a neighbor-joining tree was constructed using MEGA7 [52,53]. Phylogenetic relationships based on Cpn60 protein amino acid sequences separated these 14 strains into 4 clades. *P. putida* LS46 formed a clade along with ten other strains. In this clade, BIRD-1, F1, KT2440, *P. monteilii* SB3078, and *P. entomophila* L48 formed a subclade, which was separate from the clade consisting of *P. aeruginosa* PAO1, *P. stutzeri* A1501, *P. mendocina* ymp, *P. oleovorans*, and *P. pseudoalcaligenes* NBRC14167. *P. fluorescens* SBW25 and *P. fluorescens* Pf01 were different from all the other strains and related in a separate subclade. *P. chlororaphis* PA23 was different from all other *P. chlororaphis* strains and formed an out-group clade in the phylogenetic tree based on Cpn60 (Figure 3.1).

3.5.2 Shared Genes among Pseudomonas Strains

P. putida strains have 68,811 genes, of which 44,900 (65.25%) were present among the 14 strains analyzed. A large number of the genes (87.2% to 90.5%) were conserved among the 14 *P. putida* strains analyzed (see Table 3.3). All *P. putida* strains are closely related and have highly conserved fatty acid metabolism and PHA synthesis genes. *P. entomophila* L48 and *P. monteilii* SB3078 are closely related to *P. putida* strains and share 76% and 84% of the *P. putida* genes, respectively. In contrast, the 14 *P. putida* strains shared only 58.0% and 59.7% of the conserved genes with *P. stutzeri* and *P. pseudoalcaligenes*.

3.5.3 The PHA Synthesis Operon in Pseudomonas Species

The enzymes involved in *mcl*-PHA polymer synthesis are encoded by six genes that are organized into a PHA synthesis operon, which is identical in all the *Pseudomonas* strains analyzed. The six genes are *phaC1, phaZ, phaC2, phaD, phaF*, and *phaI*. The *mcl*-PHA biosynthesis cluster forms two putative transcriptional units: *phaC1-phaZ-phaC2-phaD* and *phaF-phaI*, which are under the regulation of a transcriptional

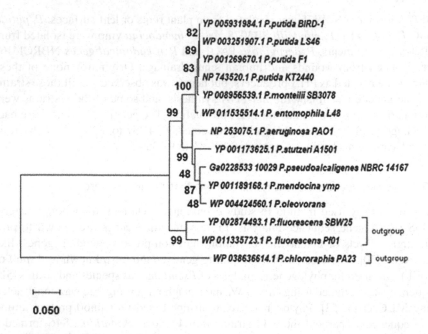

FIGURE 3.1 Phylogenetic tree depicting the relationships among *Pseudomonas* species. The tree is based on Cpn60 amino acid sequences, which were aligned by ClustalW; a neighbor-joining tree was generated using MEGA7 program. The tree is drawn to scale, with branch lengths in the same unit as those of the evolutionary distance used to infer the phylogenetic tree. The evolutionary distances were calculated using the JTT (Jones, Taylor, Thornton) method. The bootstrap method was used as a test of the phylogeny with 500 replications. Bootstrap values are indicated at the nodes.

regulator, PhaD [54]. The regulator was thought to be activated by an intermediate from the fatty acid β-oxidation pathway, resulting in higher transcriptional levels of *mcl*-PHA synthesis genes when the cell is grown -on fatty acid substrates (such as octanoic acid) versus glucose (Figure 3.3). The *phaZ* encodes a PHA depolymerase, which hydrolyzes the PHA monomers when they are required as a carbon source and are catabolized via a central metabolism for growth [55,56]. *PhaF* and *phaI* encode phasin proteins involved in PHA granule structure and size [57].

3.5.4 *Mcl*-PHA SYNTHESIS GENES

Pseudomonas species code for class II PHA synthases and the PHA synthesis operon of most *Pseudomonas* species have two PHA synthases: PhaC1 and PhaC2. These are highly conserved among *Pseudomonas* species, with up to 90% amino acid sequence identity, but within *P. putida*, the *phaC1* and *phaC2* genes share only 71% nucleotide identity and 55% amino acid sequence identity. PHA synthases catalyze the stereoselective conversion of (*R*)-3-hydroxyacyl-CoA substrates to PHA with the concomitant release of CoA [58]. Recognition of the (*R*)-3-hydroxyacyl-CoA intermediate is highly dependent on the substrate specificity of the PHA synthase itself.

TABLE 3.3
Shared Genes in Different PHA Producing *Pseudomonas* Strains

Strain	LS46	KT2440	BIRD-1	F1	PAO-1	SBW25	Pf0-1	L48	ymp	SB3078	PA23	A1501	NBRC14167
LS46	100												
KT2440	87.2	100											
BIRD-1	88.5	86.0	100										
F1	90.5	87.0	89.9	100									
PAO-1	63.6	66.7	64.5	63.7	100								
SBW25	64.1	62.7	64.4	64.6	61.2	100							
Pf0-1	70.3	68.7	71.3	71.1	66.8	71.2	100						
L48	76.5	74.5	77.5	76.2	64.0	65.2	73.4	100					
ymp	62.4	61.3	63.4	62.3	67.1	57.5	62.2	61.4	100				
SB3078	83.9	82.7	83.5	85.4	63.8	63.9	70.5	74.3	61.5	100			
PA23	67.8	66.5	67.9	68.1	68.1	70.4	77.2	70.1	60.2	67.9	100		
A1501	59.7	56.5	58.6	58.0	60.0	51.3	56.2	55.7	66.9	57.3	54.8	100	
NBRC14167	58.4	57.7	59.1	58.7	60.2	53.2	57.3	56.7	75.2	59.0	55.5	64.7	100

Mcl-PHA synthases are preferentially active toward CoA thioesters of various *mcl*-3-HA comprising 6–14 carbon atoms.

PHA synthases have different structures and subunit compositions among class I, II, III, and IV synthases. Class I and class II PHA synthases have only one subunit: PhaC for class I; PhaC1 and/or PhaC2 for class II. Class III and class IV PHA synthases are heterodimers: PhaC and PhaE for class III; PhaC and PhaR for class IV [59,60]. However, polymerases within and between PHA synthase classes are highly conserved in all four classes containing six conserved blocks, with eight conserved amino acid residues, and a lipase box [GX (S/C) XG]. With respect to specific catalytic activity and substrate specificity, the conserved Cys-319, Asp-480, and His-508 of the class I PHA synthases, and the conserved Asp-452 and His-453 of the class II PHA synthases, are crucial catalytic residues [61].

3.5.5 EVOLUTIONARY RELATIONSHIP AMONG PHAC1 AND PHAC2 PROTEINS

The PhaC1 and PhaC2 of different *Pseudomonas* species formed different clusters in neighbor-joining trees. The PhaC1 and PhaC2 in *Pseudomonas* strains were highly conserved among *Pseudomonas* species (Figure 3.2). However, *P. stutzeri* A1501 had only one PhaC, while *P. pseudoalcaligenes* NBRC 14167 encodes three PhaC proteins. PhaC from *P. stutzeri* A1501 and *P. pseudoalcaligenes* formed an outgroup. PhaC1 and PhaC2 of *P. aeruginosa* PAO1 formed a separate group, while the rest of the PhaC1 and PhaC2 proteins are in another group. *P. putida* PhaC1 proteins are closely related to *P. entomophila* and *P. monteilii*, while *P. fluorescens* PhaC1 and PhaC2 are clustered with *P. chlororaphis*, *P. mendocina*, and *P. pseudoalcaligenes*. The PhaC1 proteins have a significant effect on the subunit composition of PHA polymers [34,48].

3.6 THE GENOMICS OF *MCL*-PHA METABOLISM

3.6.1 *MCL*-PHA SYNTHESIS VIA THE FATTY ACIDS DE NOVO SYNTHESIS PATHWAY

Carbon metabolism pathways are highly conserved among 14 *Pseudomonas* genomes, with 75% to 100% amino acid sequence homology in glycerol metabolism enzymes. The fatty acid *de novo* route for *mcl*-PHA biosynthesis of *P. putida* refers to *mcl*-PHA synthesis from sugars, glycerol, and other carbon sources that are metabolized by the fatty acid *de novo* synthesis pathway to provide the (*R*)-3-hydroxyacyl-ACP (acyl carrier protein) intermediates that are polymerized to form *mcl*-PHA (see Figure 3.3) [62].

The mechanisms of initial carbon uptake and metabolism (such as glucose vs. glycerol) are virtually identical in *Pseudomonas* species. A glycerol uptake cluster is specifically required for glycerol metabolism in *Pseudomonas* [63], while the transportation of glucose into the cytoplasm requires an ABC (ATP binding cassette)-type glucose transporter. Glycerol uptake clusters and ABC transporters for glucose are present in all *Pseudomonas* species. Simultaneously, enzymes, such as glucose dehydrogenase and gluconate dehydrogenase, carry out oxidation reactions in the periplasmic space, converting glucose into gluconate and

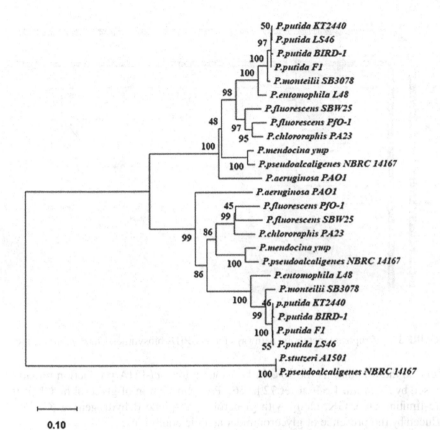

FIGURE 3.2 Phylogenetic tree depicting the relationship among PHA synthase proteins (PhaC1 and PhaC2) of *Pseudomonas* species. Amino acid sequences were aligned by ClustalW, and a neighbor-joining tree was generated using the MEGA7 program. The tree is drawn to scale, with branch lengths in the same unit as those of the evolutionary distances used to infer the phylogenetic tree. The evolutionary distances were calculated using the JTT method. The bootstrap method was used as a test of the phylogeny with 500 replications. Bootstrap values are indicated at the nodes.

keto-gluconate, respectively [64]. These are the enzymes of the Entner–Duodoroff (ED) pathway.

Glycerol uptake in *Pseudomonas* species is mediated by glycerol "uptake facilitators," which are integral membrane proteins (*glp*F, locus tag PPUTLS46_022196 in the *P. putida* LS46 genome), catalyzing the rapid equilibration of glycerol concentration gradients across the cytoplasmic membrane [65]. Glycerol is converted to glycerol-3-phosphate (G3P) by phosphorylation of glycerol by the ATP-dependent glycerol kinase (*glp*K, locus tag PPUTLS46_022201 in *P. putida* LS46), followed by the dehydrogenation of G3P into dihydroxyacetone phosphate (DHAP) by three glyceraldehyde-3-phosphate dehydrogenase (GPDHs: locus tags PPUTLS46_022211, PPUTLS46_005991, and PPUTLS46_012690 in *P. putida* LS46). Glycerol is not a preferential carbon source for PHA production, and about two-thirds of the glycerol

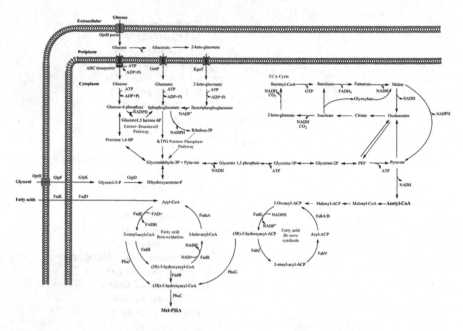

FIGURE 3.3   Proposed metabolic pathways for *mcl*-PHA biosynthesis in *P. putida* LS46.

added to the medium as the sole carbon source for *mcl*-PHA production remained unused by *P. putida* LS46 after 72 h [66]. Phosphorylation of glycerol by GK is the rate-limiting step. GK, along with glycerol-3-phosphate dehydrogenase (GlpD), is induced by the presence of glycerol under aerobic conditions.

Mcl-PHA biosynthesis by *P. putida* is regulated at the substrate transportation step. A repressor protein (coded by *glp*R) located in the glycerol uptake cluster was knocked out to reduce the lag phase of *P. putida* KT2440 grown on glycerol [67]. Biodiesel-derived waste glycerol, despite the presence of various impurities, was as good a carbon source as pure glycerol for PHA production. Heavy metal contamination in biodiesel glycerol led to the induction of transcripts for Cu, Cd, Ni, and Hg stress response genes, and proteomic analyses detected a number of proteins encoded by genes scattered across the *P. putida* LS46 genome, which were putatively involved in sensing, transporting, and the efflux of (heavy) metal ions, such as iron, cobalt/zinc/cadmium, and nickel [68].

In contrast, the glucose uptake pathway has been modified by knocking-out glucose dehydrogenase in *P. putida* KT2440, leading to increased *mcl*-PHA production from glucose [22]. Glucose is metabolized by the ED pathway because the Embden–Meyerhof–Parnas (EMP) pathway is not functional in *P. putida* due to the absence of the key glycolytic enzyme, 6-phosphofructo-1-kinase (Pfk). The biomass yield could be maximized by constructing a functional EMP pathway in *P. putida*. Sánchez et al. [69] designed two modules to construct a functional EMP pathway in *P. putida* KT2440. In engineered *P. putida*, 95% of the pyruvate was generated by the engineered EMP pathway, compared with 93% pyruvate generated by the ED pathway in wild-type *P. putida*. The authors did not test *mcl*-PHA production in the engineered strain but reported an increase in the carotenoid yield due to the production of higher pyruvate concentrations.

During *de novo* fatty acid synthesis in *P. putida* LS46, acetyl-CoA was used by the pathway enzymes to generate 3-hydroxyacyl-ACPs, the key precursor molecule for *mcl*-PHA biosynthesis. "Omics" data has suggested a few gene products were up-regulated during active *mcl*-PHA synthesis at the steps that provide the key pathway intermediates for polymer synthesis. The protein levels of a putative ketoacyl-ACP reductase (*fabG*, encoded by PPUTLS46_023353 in *P. putida* LS46), is one of eight homologs identified in the *P. putida* LS46 genome that provide various 3-hydroxyacyl-ACP intermediates. FabG was highly up-regulated in the stationary phase of waste glycerol cultures when there was active *mcl*-PHA synthesis [62]. RNA sequencing (RNAseq) analysis, however, indicated down-regulation of the *fabG* gene at the transcriptional level. Two isoforms of 3-hydroxyacyl-ACP dehydratases (FabA and FabZ) identified in the genome of *P. putida* LS46 carry out dehydration reactions to produce *trans*-2-acyl-ACP, and FabA also carries out an isomerization reaction leading to the biosynthesis of unsaturated fatty acids [70], which could also be a critical point for unsaturated *mcl*-PHA production from fatty acid *de novo* synthesis.

In the next step, intermediates are converted into acetyl-CoA via partial glycolysis through the ED pathway, which is then fluxed to fatty acid biosynthesis (see Figure 3.3). (*R*)-3-hydroxyalkanoates-acyl carrier proteins are used as the intermediate for *mcl*-PHA biosynthesis. However, (*R*)-3-hydroxyalkanoates-acyl carrier proteins must be converted into their CoA derivatives, which is carried out by PhaG, a 3-hydroxyacyl-ACP-CoA transacylase. PhaG (also known as an *mcl*-PHA monomer supplying protein) is identified biochemically as a unique protein that is required for *mcl*-PHA synthesis from fatty acid *de novo* synthesis in *Pseudomonas*. Its expression was highly up-regulated in *P. putida* grown in either glycerol or glucose cultures under *mcl*-PHA producing conditions [71].

3.6.2 MCL-PHA SYNTHESIS VIA THE FATTY ACID β-OXIDATION PATHWAY

The fatty acid β-oxidation pathway is used for *mcl*-PHA biosynthesis when fatty acids are used as a sole carbon source. The key precursor for *mcl*-PHA synthase, (*R*)-3-hydroxyalkanoate-CoA, is derived from the conversion of trans-2-enoyl-CoA, the intermediate of fatty acid β-oxidation. The key enzyme that carries out this reaction is an (*R*)-specific enoyl-CoA hydratase coded by *phaJ* in *Pseudomonas* species. There are four *phaJ* homologs identified in *P. aeruginosa*, and their gene products expressed in recombinant *Escherichia coli* were demonstrated to provide monomer for *mcl*-PHA biosynthesis from fatty acids [72]. When expressed in *E. coli*, PhaJ1 of *P. aeruginosa* showed substrate specificity toward enoyl-CoAs with acyl chain lengths from C4 to C6 (*scl* to *mcl*), while PhaJ2, PhaJ3, and PhaJ4 exhibited substrate specificities toward enoyl-CoAs with acyl chain lengths from C6 to C12 (*mcl*). Only two *phaJ* genes (*phaJ1* and *phaJ4*, based on the sequence similarity to those of *P. aeruginosa*) are present in *P. putida*, and the product of the *phaJ4* ortholog was shown to act as the primary *mcl*-PHA monomer supplying enzyme in *P. putida* [73]. Another putative monomer supplying enzyme for mcl-PHA synthesis from fatty acid β-oxidation is a 3-ketoacyl-CoA reductase coded by fabG. This enzyme is an NADPH-dependent 3-ketoacyl reductase and identified in P. aeruginosa as an mcl-PHA monomer supplying protein by reducing mcl-3-ketoacyl-CoAs to mcl-3-hydroxyacyl-CoA from fatty acid β-oxidation [74]. Yet, no evidence has shown that the protein (FadG) played a role as a monomer supplying enzyme for mcl-PHA biosynthesis from fatty acid β-oxidation in P. putida.

Pseudomonas spp. are known to produce other secondary metabolites like rhamnolipids and antimicrobial substances (phenazines). These secondary metabolites have common precursors and compete with each other for the partitioning of carbon substrates and reducing power. A defect in phenazine or rhamnolipid production led to higher PHA production in P. chlororaphis PA23 and P. aeruginosa [28,32]. Furthermore, the regulatory pathways for phenazine production and rhamnolipid production are very similar and regulated by gacA/gacS, rpoS stationary phase sigma, and stringent response (relA/spoT) [75].

3.7 MCL-PHA SYNTHESIS FROM VEGETABLE OILS AND FATS

Genome analyses of 13 Pseudomonas species indicated the presence of genes encoding triacylglycerol lipase/esterase (COG 1075), acetyl esterase (COG0657), and lipase chaperone (COG5380) proteins. Many Pseudomonas species, like P. putida KT2440, P. putida LS46, P. entomophila L48, P. fluorescens Pf01, and P. corrugata 388, can use long-chain fatty acids (LCFAs) derived from vegetable oils as sole carbon sources, but are unable to use vegetable oils directly for growth and PHA production because they lack genes encoding lipase/esterase enzymes that can be secreted extracellularly. The extracellular lipase/esterase enzymes cleave the ester bonds between the LCFAs and glycerol, making the LCFAs available for metabolism via the β-oxidation pathway. Evidence in support of this statement was provided by experiments in which recombinants of P. corrugata 388 were constructed by cloning and expressing a gene encoding a secreted lipase enzyme, which enabled the recombinant strains to grow on animal fat (lard) and synthesize PHA [76]. Plasmid-containing Pseudomonas lipase genes were procured from the American Type Culture Collection.

In contrast, a number of Pseudomonas species, including P. chlororaphis PA23, P. stutzeri A1501, P. aeruginosa PAO1, P. mendocina ymp, P. monteilii SB3078, and P. pseudoalcaligenes NBRC14167, can use vegetable oils directly as a carbon source for growth and PHA synthesis [32]. Analyses of the genomes of P. stutzeri A1501, P. aeruginosa PAO1, and P. mendocina ymp revealed the presence of genes encoding triacylglycerol lipases with signal peptides, which suggests that they are secreted and function extracellularly. These three species also encode lipase chaperone proteins, which ensure proper folding of the lipase proteins that are excreted from the cell [32]. The lipase chaperone gene is absent from the genomes of P. putida KT2440, P. putida LS46, P. putida BIRD-1, and P. putida F1. Strains that can use vegetable oils directly for growth and PHA synthesis, such as P. chlororaphis PA23, P. monteilii SB3078, and P. pseudoalcaligenes NBRC14167, also encode genes for a protease/lipase (EYO4_16115) ABC transporter.

3.8 GENOME ANALYSIS OF *HALOMONAS* SPECIES

As indicated in the Introduction, *Halomonas* strains isolated from deep-sea hydrothermal sites have been shown to produce PHA polymers. Three major classes of PHA were identified in *Halomonas* species: poly(3-hydroxybutyrate) (PHB), poly(3-hydroxyvalerate) (PHV), and poly(3-hydroxybutyrate-*co*-3-hydroxyvalerate) (PHBV) [77]. Under non-sterile conditions, *H. elongata, H. halophila, H. nitroreducens*, and *H. boliviensis* produced 40% to 69% PHA in the presence of 6% to 10% salinity [78,79]. *Halomonas* species can accumulate PHA up to 80% of CDM [80]. In addition to PHA production, *H. smyrnensis* AAD6T and other *Halomonas* strains can also produce extracellular polymeric substances (EPS) and levan [81–83]. Halophiles with osmo-adaptations and metabolic diversity have not yet been exploited for industrial applications to produce bioplastics, EPS, and/or halophilic enzymes.

A large number (120) of *Halomonas* sp. genome sequences (draft, permanent, and finished) are available on the IMG platform (https://img.jgi.doe.gov). Complete genome sequences of *Halomonas* sp. SF2003 and *H. smyrnensis* AAD6T have been recently published [13,81]. We selected 15 genomes of *Halomonas* species for further study. The genome size of *Halomonas* species varies from 2.87 to 5.33 Mb, with gene contents between 2,636 to 4,807. The *H. halocynthiae* DSM 14573 genome is 2.87 Mb, while the *H. titanicae* BH1 genome is 5.33 Mb (see Table 3.4). The % GC ratio is in the range of 0.52–0.62. A maximum of 1,002 genes was horizontally transferred into *H. anticariensis* DSM 16096. The 15 genomes of *Halomonas* spp. had up to 77.4% average nucleotide identity with *Halomonas anticariensis* SDSM16096. Analyses of genome size, gene counts, and % average nucleotide identity (ANI) indicate that there is wide diversity among the genomes of *Halomonas* species.

3.8.1 CARBON METABOLISM IN HALOMONAS SPECIES

The *H. smyrnensis* AAD6T genome has genes for glycolysis via the EMP and ED pathways, as well as for gluconeogenesis, the pentose phosphate pathway, and all *de novo* amino acid biosynthesis pathways. Genes for the complete tricarboxylic acid (TCA) cycle and glyoxylate shunt were also present. However, the absence of phosphofructokinases (PFKs) indicated the absence of a functional EMP pathway [81]. *Halomonas* species can use a wide variety of carbon substrates for PHA production. *Halomonas elongate* 2FF was reported to produce PHB from sucrose, glucose, galactose, pyruvic acid, and acetic acid. *H. campisalis* and *H. boliviensis* used xylose to produce PHB [82–86]. *H. halophila* produced PHA in high salt conditions (20–100 g/L NaCl) and was able to use chemical-grade sugars like glucose, fructose, sucrose, cellubiose, mannose, rhamnose, and arabinose, as well as carbohydrates derived from complex compounds, such as hydrolyzed cheese whey, hydrolyzed lignocellulose, and molasses [87], as carbon sources for growth and PHA production.

The TCA cycle provides energy and intermediates for the synthesis of many important biological compounds. *H. bluephagenesis* TD01 is unable to synthesize PHBV because the precursor propionyl-CoA is not synthesized. A recombinant

TABLE 3.4

Halomonas Genomes Used in the Present Study

Genome name	Genome size	Gene count	% GC	16S rRNA count	Horizontally transferred genes	% ANI
H. smyrnensis AAD6	3561919	3326	0.68	1	311	76.10
H. meridiana R1t3	3507875	3525	0.57	8	130	72.39
H. anticariensis DSM 16096	5019539	4807	0.59	1	1002	74.79
H. illicicola DSM 19980	3960292	3746	0.63	5	413	77.40
H. lutea DSM 23508	4533090	4368	0.59	1	617	100.00
H. zincidurans B6	3554760	3292	0.64	2	209	78.34
H. elongata DSM 2581	4061296	3556	0.64	4	743	75.55
H. stevensii S18214	3693745	3523	0.6	4	154	73.62
H. campaniensis LS21	4074048	3631	0.53	6	76	70.74
H. zhejiangensis DSM 21076	4060520	3739	0.55	3	330	71.75
H. titanicae BH1	5339792	2908	0.55	1	138	71.86
H. halocynthiae DSM 14573	2878444	2773	0.54	2	316	71.07
H. alkaliantarctica FS-N4	3797897	3484	0.52	1	125	70.63
H. boliviensis LC1	4136366	3914	0.55	4	245	71.79
H. halodenitrificans DSM 735	3466026	3256	0.64	1	119	74.99

strain with increased metabolic flux toward PHBV synthesis, *H. bluephagenesis* TD08AB, was constructed by deleting two genes in the TCA cycle: the succinate dehydrogenase gene, *sdhE*, and the isocitrate lyase gene, *icl*. The addition of α-ketoglutarate, citrate, or succinate to the culture media did not increase the cell mass production or the total PHBV content of the cells. It did, however, increase the intracellular concentrations of acetyl-CoA and propionyl-CoA, which are precursors for PHB and PHBV, and increased the molar ratio of 3-hydroxyvalerate in the PHBV to 90% [88].

The general pathway for synthesis of *scl*-PHA is very similar in both *Halomonas* and *C. necator*. There are three main steps: (i) formation of 3-ketoacyl-CoA from acyl-CoA and acetyl-CoA, a reaction catalyzed by 3-ketothiolase (PhaA); (ii) conversion of 3-ketoacyl-CoA to (*R*)-3-hydroxyacyl-CoA by NADP-dependent 3-keto-acyl-CoA reductase (PhaB); and (iii) polymerization by stereoselective conversion of (*R*)-3-hydroxyacyl-CoA to *scl*-PHA polymers by PHA synthase (PhaC). *Halomonas* spp. carries the genes for supplying fatty acyl-CoA precursors for PHA synthesis as well as the gene for degradation of stored PHA.

3.8.2 EVOLUTIONARY RELATIONSHIP AMONG HALOMONAS

The phylogenetic relationships of 15 *Halomonas* genomes (permanent or finished) were determined using Cpn60 amino acid sequences. This analysis indicated a wide diversity among *Halomonas* species (see Figure 3.4). The 15 strains were clustered in two clades: nine strains were present in clade I and six strains were present in clade II. Clade I was further split into three subclades, and clade II was split into two subclades. Diversity among *Halomonas* strains was identified using 16S rRNA genes and *Halomonas* spp. SF2003 and *Halomonas halodurans* are closely related to *Chromohalobacter salarius* and *Cobetia* species [13].

3.8.3 PHA SYNTHASE IN HALOMONAS SPECIES

Genome analyses of the 15 *Halomonas* strains indicated wide variation in PhaC proteins (see Figure 3.5). Encoding of multiple *phaC* genes in *Halomonas* species is very common among the genomes examined. The genomes of three *Halomonas* spp. each encoded more than one *phaC* gene (see Figure 3.5). *H. boliviensis* LC1 and *H. anticariensis* DSM16096 encoded genes for two PHA synthases. *H. illicicola* DSM 19980 is unique and encodes genes for five PHA synthases. Two PHA synthases of *H. illicicola* are identical and present in the same cluster, while three other

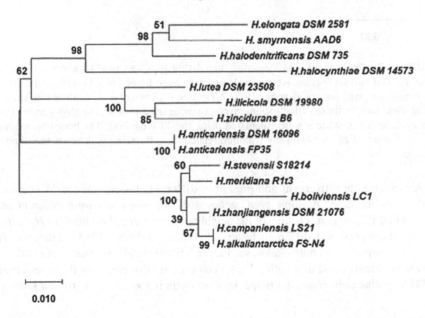

FIGURE 3.4 Phylogenetic tree depicting the relationships among *Halomonas* species. The tree is based on Cpn60 amino acid sequences, which were aligned by ClustalW, and a neighbor-joining tree was generated using MEGA7. The tree is drawn to scale, with branch lengths in the same unit as those of the evolutionary distance used to infer the phylogenetic tree. The evolutionary distances were calculated using the JTT method. The bootstrap method was used as a test of the phylogeny with 500 replications. Bootstrap values are indicated at the nodes.

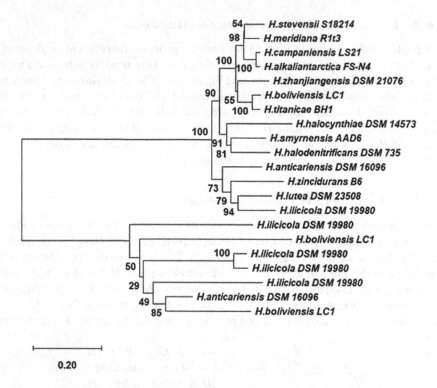

0.20

FIGURE 3.5 Phylogenetic tree depicting the relationships among PHA synthase proteins (PhaC) of *Halomonas* species. Amino acid sequences were aligned by ClustalW, and a neighbor-joining tree was generated using MEGA7. The tree is drawn to scale, with branch lengths in the same unit as those of the evolutionary distances used to infer the phylogenetic tree. The evolutionary distances were calculated using the JTT method. The bootstrap method was used as a test of the phylogeny with 500 replications. Bootstrap values are indicated at the nodes.

PHA synthases are divergent and present in different clusters. Of the 15 genomes studied, 14 contained multiple *phaC* genes with homologs that were different and present in different gene clusters of the neighbor-joining tree. Two *phaC* of *H. boliviensis* LC1 are present in different clusters. Homologs of the five PHA synthases of *H. illicicola* are present in both clades (see Figure 3.5). Two PHA synthases of *H. illicicola* are identical, while three other PHA synthases are diverse. How these variations in PHA synthases in *Halomonas* species control their specificity is still not known.

3.8.4　PHA Synthesis in Halomonas Species

Halomonas species can metabolize glucose, glycerol, sucrose, maltose, volatile fatty acids (VFAs), lignocellulosic waste (sawdust, corn stover), and industrial effluent to produce PHA (see Table 3.5). Batch and fed-batch, or one-step or two-step processes have been tried for PHA production. Accumulation of PHA was between 23% and 88% of cell weight under different conditions, with 10% to 60% salt concentrations.

TABLE 3.5
PHA Production by *Halomonas* Species

Strain	NaCl [%]	Carbon source	CDM [g/L]	PHB [%]	Condition	Ref.
H. elongata 2FF	10.0	Glucose (1%)	2.5	40.0	Batch	[78]
H. halophila	10.0	Glucose (2%)	3.2	48.0	Batch	[87]
CCM 3662	60.0	Glucose (2%)	5.1	72.0		
H. boliviensis	4.5	Butyrate & Acetate 0.8%	2.0	88.0	Batch	[89]
Halomonas sp. KM-1	1.94	Glycerol (10%)	0.67	24.5	Batch	[90]
H. venusta KT832796	1.5	Glucose	37.9	88.1	Fed-batch	[91]
H. boliviensis	–	Glucose + VFA	44.0	79.5		[92]
Halomonas TD01	1.94	Glucose	83.0	78.0		[93]
Halomonas KM-1	1.94	Glucose	72.1	83.0	Batch	[90]
Halomonas sp. SF2003	1.1	Glucose + VA$ 10 mol%	1.4	64.0	Two steps	[94]
			2.0	(15)*	Two steps	
		Glucose + VA 46 mol%		56.0 (27)		
		#IE	4.2	31.0	One step	
		IE + VA	6.2	23 (35)	One step	
H. smyrnensis AAD6	13.72	Glucose	1.34	45.8	Batch	[95]
	13.72	Sucrose (5%)	0.5	26.9	Batch	
H. campaniensis LS21	2.62	Glucose	8.0	75.0	Fed-batch	[96]
H. nitroreducens JCCO25.8	25% @	Glucose	–	33.0	Batch	[97]
H. campisalis	4.5	Maltose	1.3	81.0	Batch	[98,99]
MCM B-1027	3.0	Banana peel extract (1%)	0.7	21.5	Batch	

$ Valeric acid, *3hydroxyvalerate %, # IE industrial effluent, @ seawater.

H. boliviensis, *H. venusta*, *Halomonas* KM-1, and *Halomonas* TD-1 were the best producers and accumulated PHA to 80% to 88% CDM [90–92]. The addition of propionate, pyruvate, or valerate to the culture medium led to the production of copolymers of PHB and PHV with 15% to 35% 3-hydroxyvalerate (3HV) [100].

A number of biotechnology techniques have been used to enhance PHA production in halophilic bacteria. These include increasing substrate specificity, modifying the β-oxidation pathway, and constructing new metabolic pathways [101,102]. An open and continuous process using seawater with mixed substrates containing cellulose, starch, lipids, and proteins was used to produce PHB by recombinant *H. companiensis* LS21. It produced 70% PHB after 65 days in the presence of 27 g/L NaCl [80]. Deletion of phasin genes like *phbP1* and *phaP2* in *H. bluephagenesis* resulted in greater PHB granule size. Further deletion of *minC* and *minD* genes, which regulate cell fission, increased the PHB granule size to 10 μm [103].

3.9 GENOME ANALYSIS OF *PARACOCCUS* SPECIES

The genus *Paracoccus* (alphaproteobacteria) currently comprises 40 recognized and validly named species. *Paracoccus* spp. are Gram-negative methylotrophs and are known to inhabit a wide range of environments, like contaminated soil, rhizospheric soil of leguminous plants, and marine sediment. Some *Paracoccus* spp. are opportunistic human pathogens. *Paracoccus* spp. have applications in bioremediation due to their ability to carry out denitrification and degrade toxic compounds, such as herbicides (chlorpyrifos), acetone, dichloromethane, formamide, N,N-dimethylformamide (DMF), and methylamine [104,105]. The metabolic flexibility of *Paracoccus* is based on its ability to use a variety of nitrogen sources, including nitrate, nitrite, nitrous oxide, and nitric oxide, and C1-compounds, like methane, methanol, methylated amines, halogenated methanes, and methylated sulfur species, as alternative electron acceptors. In addition, *Paracoccus* spp. have a great ability to accumulate PHA polymers using glycerol, methanol, n-pentanol, and CO_2 as carbon sources. PHA is synthesized under autotrophic as well as heterotrophic growth conditions. Microbial mats are excellent starting materials for isolating PHA producing bacteria, and a number of *Paracoccus* species have been isolated as PHA producers from marine mats [106]. *Paracoccus* sp. LL1 has been reported to produce carotenoids along with PHA production [19].

In the present study, we selected 22 *Paracoccus* genomes from the IMG platform, which were draft, permanent draft, or finished genomes (Table 3.5). These bacteria were isolated from different environments and geographic locations. The genome size of the *Paracoccus* species analyzed ranged between 3.03 and 5.62 Mb, and the gene contents ranged between 3,258 and 5,522. The genomes of *Paracoccus* spp. have multiple replicons of chromosomes as well as plasmids. *Paracoccus denitrificans* PD1222 is composed of two chromosomes (ChI = 2.9 Mb and ChII = 1.7 Mb), plus a single mega-plasmid (Plasmid 1) of 653 kb [107]. *P. aminophilus* JCM 7686 carries a single circular chromosome plus eight plasmids (pAMI1 to pAMI8), ranging in size from 5.6 to approximately 440 kb. Plasmid pAMI2 carries genes for DMF degradation [108]. The genome of another strain of *P. aminophilus*, JCM 7685, is composed of four circular DNA molecules [109]. *P. yeei* is an opportunistic human pathogen, which has a circular chromosome and eight extrachromosomal replicons [109]. *P. yeei* has PHA production machinery like other *Paracoccus* strains. The other genomes, like *P. versutus* DSM 532 and *P. bengalensis* DSM 17099, have genome sizes comparable to those of *P. aminophilus* JCM 7686 and *Paracoccus denitrificans* PD1222, which are finished genomes.

Eighteen plasmids have been identified and sequenced in different *Paracoccus* species. It has been speculated that the bacteria of genus *Paracoccus* usually carry at least one mega-plasmid of more than 100 kb [110]. Finished genomes have reported one or more plasmids, but draft or permanent draft genomes have not identified plasmids in *Paracoccus* species. It is possible that draft genomes, when finished, will also show other replicons and plasmids. Some of these plasmids are associated with carotenoid production [111]. *P. denitrificans* DSM 15148, *P. denitrificans* DSM 415, *P. versutus* DSM 582, *P. bengalensis* DSM 17099, and *P. aminophilus* JCM 7686 genomes each have four 16S rRNA genes, while the genomes of other species have fewer 16S rRNA genes. *Paracoccus* species acquired up to 21% of their genes from other bacteria by horizontal gene transfer (see Table 3.6). The average nucleotide identity is 76.95% to 78.52%.

3.9.1 EVOLUTIONARY RELATIONSHIP AMONG PARACOCCUS

The evolutionary relationship among the 22 *Paracoccus* isolates, belonging to 17 species, was studied using Cpn60 proteins. Generally, most eubacteria encode a single copy of the *cpn*60 gene. However, all of the *Paracoccus* species studied have multiple copies of the *cpn*60 gene, all of which are identical. The genome of *P. pantotrophus* J46 encoded four copies of the *cpn*60 gene, while *P. nitrificans* DSM 415 encoded three *cpn*60 genes. The genomes of *P. aminophilus* JCM 7686, *P. versutus* DSM 582, and *P. bengalensis* contain two copies of the *cpn*60 gene (see Figure 3.6).

TABLE 3.6
Paracoccus **Genomes Used in the Present Study**

Genome name	Genome size	Gene count	% GC	16S rRNA count	% ANI	Horizontally transferred gene count
P. sanguinis DSM 29303	3588279	3457	0.71	1	76.95	139
P. alkenifer DSM 11593	3191900	3099	0.67	3	77.36	323
P. halophilus CGMCC 1.6117	4008709	3912	0.65	2	77.44	335
P. versutus DSM 582	5627664	5522	0.68	4	78.52	487
P. denitrificans DSM 15418	4767709	4677	0.67	4	78.20	156
P. saliphilus DSM 18447	4570660	4442	0.61	1	77.77	399
P. chinensis CGMCC 1.7655	3632870	3608	0.68	1	77.09	390
P. bengalensis DSM 17099	4989582	5009	0.67	4	78.43	188
P. aminovorans DSM 8537	3946204	3832	0.68	2	78.17	191
P. isoporae DSM 22220	3523473	3411	0.66	1	75.94	179
P. denitrificans PD1222	5236194	5158	0.67	3	78.04	976
P. solventivorans DSM 6637	3377602	3258	0.69	1	77.22	147
P. contaminans RKI 16-01929	3033494	2937	0.69	2	76.09	74
P. halophilus JCM 14014	3998285	3884	0.65	1	77.45	4
P. pantotrophus J46	4658858	4675	0.67	1	78.33	511
P. homiensis DSM 17862	3868263	3898	0.64	2	78.70	129
P. denitrificans DSM 415	5192961	5177	0.67	4	78.03	19
P. seriniphilus DSM 14827	4195486	4050	0.62	1	77.41	125
P. yeei ATCC BAA-599	4428895	4280	0.67	2	78.30	546
P. sediminis DSM 26170	3645751	3597	0.66	1	100.00	179
P. aminophilus JCM 7686	4917798	4642	0.63	4	75.31	987
P. zhejiangensis J6	4637247	4512	0.66	2	78.30	176

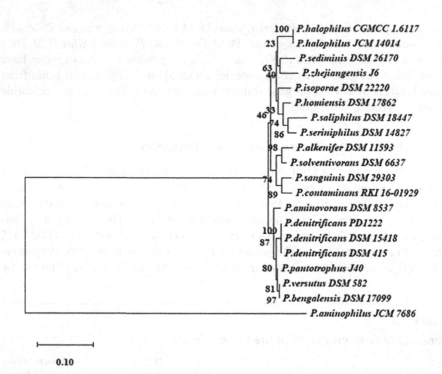

FIGURE 3.6 Phylogenetic tree depicting the relationships among *Paracoccus* species. The tree is based on Cpn60 amino acid sequences, which were aligned by ClustalW, and a neighbor-joining tree was generated using MEGA7. The tree is drawn to scale, with branch lengths in the same unit as those of the evolutionary distances used to infer the phylogenetic tree. The evolutionary distances were calculated using the JTT method. The bootstrap method was used as a test of the phylogeny with 500 replications. Bootstrap values are indicated at the nodes.

P. aminophilus JCM 7686 forms an out-group, while the rest of the *Paracoccus* species are present in a single clade. The 16 *Paracoccus* species further clustered into two subclades. The amino acid sequences of Cpn60s from three strains of *P. denitrificans* are highly conserved and form a subclade. Likewise, two strains, *P. halophilus* JCM 14040 and *P. halophilus* CGMCC 1.6117, are very closely related and clustered together with *P. yeei* ATCC-BBA 599. *P. denitrificans* is related to *P. pantotrophus*, *P. versutus*, and *P. bengalensis*, and forms another subclade. *P. isoporae*, *P. saliphilus*, *P. homiensis*, and *P. seriniphilus* form another group. *Paracoccus* strains isolated from different environments have close phylogenetic relationships.

3.9.2 CARBON METABOLISM FOR PHA PRODUCTION BY PARACOCCUS SPECIES

Some species of *Paracoccus* are methylotrophs, use C1 compounds, and have a role in carbon, nitrogen, and sulfur recycling. Genes for carbon metabolism have been identified on one of the four replicons in *P. aminophilus* JCM 7685. These genes are for L-arabinose use, C4-dicarboxylates transport, the glyoxylate shunt for acetate use, methylcitrate cycle genes for propionate use, methylotrophy (oxidation of methylated amines and the serine cycle for assimilation of C1 units), and stachydrine or D-amino acid use [109]. Strains of *Paracoccus* are known to produce PHB and PHBV copolymers. *P. pantotrophus*, *P. seriniphilus*, *P. denitrificans*, and *Paracoccus* sp. LL1 have been reported to produce PHB from glucose, glycerol, pentanol, corn stover hydrolysates, and CO_2 (10% in gas mixture). The maximum accumulation of up to 72% has been reported under batch culture conditions of glycerol and corn stover hydrolysate [19].

3.9.3 PHA Synthase Proteins in Paracoccus Species

Analysis of the *P. denitrificans* genome identified *phaA*, *B*, *C*, *P*, *R*, and *Z* genes related to PHA biosynthesis, which are present in two different clusters. The *phaA* (3-ketothiolase) and *phaB* (acetoacetyl-CoA reductase) are present in one cluster. The other cluster carries *phaC* (PHA synthase), *phaP* (PHA granule-associated phasin protein), and *phaR* (putative regulator of PHA synthesis) genes [112]. PHA production is higher under nitrogen-deficient conditions. *P. denitrificans* degrades PHA under carbon starvation conditions, and a PHA depolymerase gene (*phaZ*) was identified [113]. The analysis of 22 genomes of *Paracoccus* identified *phaC* genes encoding 49 PHA synthase proteins (Figure 3.7). PHA synthase proteins of *Paracoccus* belong to class I PHA synthases and produce *scl*-PHA. Genomic analyses revealed multiple copies of *phaC* genes in 15 *Paracoccus* strains, and 1–3 *phaC* genes are present in the genomes of different species. Ten *Paracoccus* strains have three *phaC* genes, five strains have two *phaC* genes, and the rest have a single *phaC* gene.

While *P. denitrificans* PD1222 has three *phaC* genes, one copy is present on chromosome I, chromosome II, and on the mega-plasmid. *P. denitrificans* DSM415 and *P. denitrificans* DSM 15148 carry three *phaC* genes, and the PhaC proteins encoded by them are not identical. The PHA synthase of *P. denitrificans* PD1222 present on chromosome I (Pden_0958) has a 33% amino acid sequence identity with the PHA synthase (Pden_4207) carried on chromosome II, and a 35% amino acid sequence identity with the PHA synthase carried on the mega-plasmid (Pden_5103). The PHA synthase (Pden_4207) present on chromosome II and the PHA synthase present on the mega-plasmid (Pden_5103) share 65% amino acid sequence identity. The PHA synthase on chromosome I (Pden_0958) has 173 additional amino acids at its N-terminus compared with the PHA synthase carried on chromosome II (Pden_4207), and 184 additional amino acids at its N-terminus compared with the PHA synthase on the mega-plasmid (Pden_5103). The genomes of *P. versutus* DSM 582, *P. alkenifer* DSM 11593, *P. halophilus* DSM 14040, *P. halophilus* CGMMC 1.6117, and *P. aminovorans* JCM 7686 encode two *phaC* genes. One of the *phaC* genes in *P. aminovorans* JCM 7686 is present on the chromosome, while the other is encoded on the mega-plasmid.

Phylogenetic analyses of PHA synthases identified three subclades. The first subclade has 25 PHA synthase proteins, the second clade has 23 PHA synthase proteins, and the third has one PHA synthase. The first clade is divided into four subclades. The first subclade is divided into two further subclades; *P. seriniphilus* DSM 14827 forms an out-group, while the rest of the 24 proteins are present in different groups.

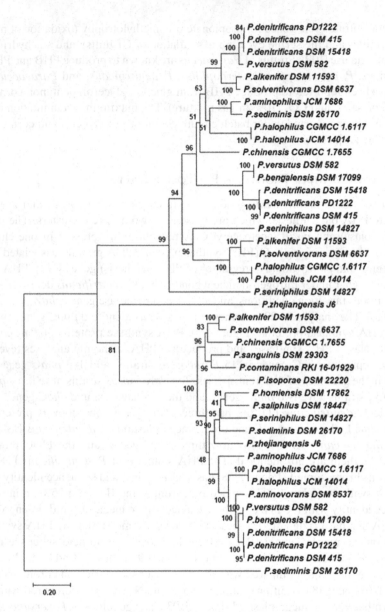

FIGURE 3.7 Phylogenetic tree depicting the relationship among PhaC of *Paracoccus* species. Amino acid sequences were aligned by ClustalW, and a neighbor-joining tree was generated using MEGA7 program. The tree is drawn to scale with branch lengths in the same unit as those of the evolutionary distances used to infer the phylogenetic tree. The evolutionary distances were calculated using JTT method. Bootstrap method was used as test of phylogeny with 500 replications. Bootstrap values are mentioned at the nodes.

Three *P. denitrificans* strains have nine PHA synthase proteins, which are identical and form three clusters in different groups. The first group is closely associated with PHA synthases from *P. versutus* DSM 582 and *P. pantotrophus* J46. The second and third PHA synthase proteins are also related to PHA synthases from *P. versutus* DSM 582 and *P. pantophila* J46. Likewise, the three PHA synthases from *P. halophilus* CGMCC 1.6117 and *P. halophilus* JCM 14014 are identical, but present in different groups. The first copy of PHA synthases from *P. halophilus* CGMCC 1.6117 and *P. halophilus* JCM 14014 are closely related to *P. yeei* ATCC-BBA 599. The second copy is related to *P. alkenifer* DSM 11593 and *P. solventivorans* DSM 6637. The third copy of the PHA synthase from *P. halophilus* JCM 14014 has high homology with the PHA synthases of *P. denitrificans*, *P. aminophilus*, *P. pantophilus*, *P. versutus*, and *P. bengalensis*.

3.9.4 PHA PRODUCTION IN PARACOCCUS SPECIES

Earlier studies reported PHA production in *Paracoccus* species in batch and shake-flask cultures using glucose, n-pentanol, glycerol, corn stover hydrolysate, and CO_2. The maximum production was 72.4% of CDM in *Paracoccus* sp. LL1, while *P. pantotrophus*, *P. denitrificans* NBRC13301, *P. denitrificans* PD01, and *P. seriniphilus* E71 are able to accumulate PHA only up to 50% of CDM [19].

3.10 THE PHA PRODUCTION MACHINERY IN *PSEUDOMONAS PUTIDA*, *CUPRIAVIDUS NECATOR*, *HALOMONAS* SPP., AND *PARACOCCUS* SPP.

The genes and proteins involved in PHA synthesis in *P. putida* and *C. necator* are well studied and have been discussed in previous review articles [43,59,60,114–116]. The genome of *Halomonas lutea* encodes only one PHA synthase gene (F568DRAFT_04081). Genes encoding two PHA synthases are encoded by the genomes of *P. putida* (PhaC1, PPUTLS46_005621, and PhaC2, PPUTLS46_005611); *P. denitrificans* has three PHA synthases (Pden_0957, Pden_4615, and Pden_5103). In contrast, *C. necator* encodes genes for two PHA synthases (H16_A1437 and H16_A2003) [117].

While the genes involved in PHA synthesis in *P. putida* and *C. necator* are clustered in operons, the organization of genes involved in PHA synthesis in *P. denitrificans* and *Halomonas lutea* is different; they are scattered all over the genomes. However, homologs of most of the PHA genes present in *C. necator* are present in *H. lutea* and *P. denitrificans* [112] (see Table 3.7). The *P. denitrificans* genome encodes three *pha*C genes (encoding PHA synthase), two *pha*Z genes (encoding PHA depolymerase), and one *pha*R gene (encoding a PHA repressor). Likewise, *H. lutea* encodes one *pha*C gene, one *pha*Z gene, and one *pha*R gene. The genomes of both *P. denitrificans* and *H. lutea* carry genes for multiple 3-ketothiolases (PhaA) and β-ketoacyl reductase (PhaB) enzymes.

Phylogenetic analyses of the *P. putida* LS46, *C. necator* H16, and *Halomonas lutea* DSM 23508 PHA synthase enzymes revealed some interesting relationships

TABLE 3.7

Amino Acid Sequence Identities Among Class I PHA Synthases of *Cupriavidus necator* H16 and *Halomonas Lutea* DSM 23508

Species	Locus tag	1	2	3	4	5	6
1) *Cupriavidus necator* H16	H16_A2003	100	43.4	46.4	28.1	31.8	31.7
2) *Paracoccus denitrificans* PD1222	Pden_4207		100	63.1	29.0	34.5	34.8
3) *Paracoccus denitrificans* PD1222	Pden_5103			100	29.3	35.0	34.4
4) *Paracoccus denitrificans* PD1222	Pden_0958				100	37.5	37.4
5) *Halomonas lutea* DSM 23508	F568DRAFT_01350					100	40.3
6) *Cupriavidus necator* H16	H16_A1437						100

(Figure 3.8). PHA synthases of *H. lutea* show 40.35% and 31.84% homology to *C. necator* PHA synthases (H16_A1437 and H16_A2003, respectively). The three PHA synthases of *P. denitrificans* (Pden_4207, Pden_5103, and Pden_0958) share 43.46%, 46.40%, and 28.17% amino acid sequence identity with the *C. necator* PHA synthase (H16_A2003), respectively. The amino acid sequence identities of the *P.*

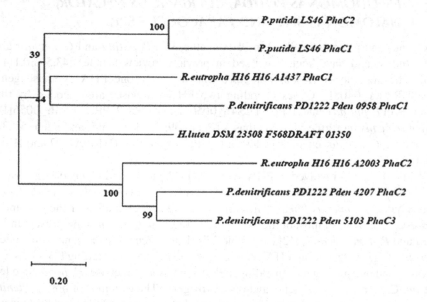

FIGURE 3.8 Phylogenetic tree depicting the relationships among PHA synthase proteins (PhaC) of *Pseudomonas putida* LS46, *C. necator* (*Ralstonia eutropha*) H16, and *Halomonas lutea* DSM 23508. Amino acid sequences were aligned by ClustalW, and a neighbor-joining tree was generated using MEGA7.

denitrificans PHA synthases (Pden_4207, Pden_5103, and Pden_0958) with the *C. necator* PHA synthase (H16_A1437) are low and range from 34.40% to 37.43%. The PHA synthase of *P. denitrificans* PD1222 encoded by Pden_0958 is closely related to the other PHA synthase (H16_A1437) of *C. necator*, indicating that this could be the main PHA synthase in *P. denitrificans*. Likewise, the PHA synthase protein of *H. lutea* (F568DRAFT_04081) shares 40.35% and 31.84% amino acid sequence identity with the two *C. necator* PHA synthases (H16_A1437 and H16_A2003), respectively.

3.11 DOMAIN ORGANIZATION AND STRUCTURAL COMPARISON OF PHAC FROM *CUPRIAVIDUS NECATOR*, *HALOMONAS LUTEA*, AND *PARACOCCUS DENITRIFICANS*

PHA synthases consist of two domains, the N-terminal domain, which plays an important role in stabilizing the PhaC dimer, and the C-terminal catalytic domain that possesses the conserved active site amino acid residues Cys, His, and Asp [118]. The catalytic His activates the nucleophilic Cys once the substrate acyl-CoA enters the catalytic pocket. The activated Cys then attacks the thiol group of the acyl-CoA, and the catalytic Asp is postulated to activate the 3-hydroxyl group of the substrate/acyl-CoA and attack the second incoming substrate for the elongation process [119,120]

The domain organization of PhaCs from *C. necator*, *H. lutea*, and *P. denitrificans* indicates a conserved α/β hydrolase-fold region, characterized as the α/β core subdomain (Figure 3.9), featuring a central mixed β-sheet flanked by α-helices on both

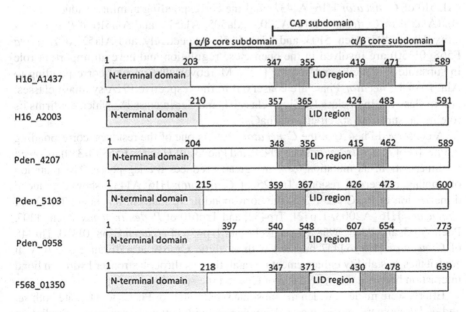

FIGURE 3.9 Schematic representation of PhaC domains from *Cupriavidus necator* (H16_A2003, H16_A1437), *Paracoccus denitrificans* (Pden_4207, Pden_5103, Pden_0958), and *Halomonas lutea* (F568_01350).

sides. This domain contains a catalytic pocket comprising a catalytic triad (Cys-Asp-His) at its core. A flexible catabolite activator protein (CAP) subdomain covers the α/β core subdomain from the top.

The *C. necator* PhaC enzyme was found to exist in equilibrium with the monomer and dimer in solution, with the dimer being the more catalytically active form [121–123]. The structural conformation of the CAP subdomain (closed/open) is the key indicator of the enzyme's active status. The closed form prevents the substrates from entering the catalytic pocket by covering the active site within the CAP subdomain, particularly, a short stretch of amino acid residues termed the LID region. The LID region in the partially opened PhaC of *C. necator* undergoes structural changes to allow substrate entry, as revealed by X-ray crystallography of its catalytic domain (residues 201–368 and 378–589; residues 369–377 are disordered) [124,125]. This contrasts with the closed configuration of the catalytic domain reported for the PhaC of *Chromobacterium* sp. [126]. The structure of *C. necator* PhaC represents the dimeric organization with the active site of each monomer separated by 33 Å across a dimeric interface, indicating that the polyhydroxybutyrate biosynthesis occurs at a single active site [124]. Thus, the CAP subdomain changes its conformation during catalysis, which involves rearrangement of the dimer to facilitate substrate entry as well as product formation [127].

Multiple amino acid sequence alignments of PHA synthases from *C. necator*, *H. lutea*, and *P. denitrificans* revealed the potential conserved amino acid residues involved in catalysis, substrate specificity, and possible dimerization and/or positioning of catalytic residues (Figure 3.10). The putative residues involved in altering the enzyme function are located near the catalytic triad, but not inside the catalytic pocket [127]. For instance, the alanine residue near the catalytic His of PhaC (Ala510 of *C. necator* H16_A1437) and the corresponding alanine residues in others (Ala514 of *C. necator* H16_A2003; Ala505, Ala516, and Ala516 of *P. denitrificans* Pden_4207, Pden_5103, and Pden_0958, respectively, and Ala521 of *H. lutea* F568_01350) are involved in the open-close regulation and have an important role in substrate specificity and activity [127]. Moreover, the residues corresponding to Ala510 of *C. necator* PhaC are conserved in the respective PHA synthase classes: Ala in class I, Gln in class II, Gly in class III, and Ser in class IV, which confirms its role in the substrate specificity of PhaC.

A conserved Phe420 of the *C. necator* PhaC is one of the residues (corresponding to Phe608 of *P. denitrificans* Pden_0958 and Phe431 of *H. lutea* F568_01350) involved in dimerization; its mutation to serine greatly reduces the lag phase. The mutation of another conserved residue, Trp425 of *C. necator* H16_A1437, showed reduced dimerization of PhaC [128,129]. The corresponding residues in others are Trp430 of *C. necator* H16_A2003, Trp421, Trp432, and Trp610 of *P. denitrificans*, Pden_4207, Pden_5103, and Pden_0958, respectively, and Trp436 of *H. lutea* F568_01350. Thr348 of *C. necator* H16_A1437 interacts with catalytic Cys319 at a distance of 3.6 Å. It contributes to stability enhancement of catalytic Cys through stronger hydrogen bond interaction between the main chain of Cys and the side chain of Thr.

Efforts were made to widen the substrate specificities of PhaCs to integrate both *scl* and *mcl* monomers into synthesized copolymers with better characteristics. Studies on the evolutionary engineering approach of class I PhaC from *C. necator* can be traced back to 2001 [130]. For instance, the fusion enzyme with the N-terminal portion (26%)

```
H16_A2003    WKNPDASARDFGLDTYLEAGLLTALNTVHARCDGAHVHAAG CLGGTLLATGAAMLARDA  345
Pden_4207    WRNPTADDRDLTLDDYRRLGVMAAVEAINAILPGRGIHAAGYC LGGTLLAIAAAAMAG-A   336
Pden_5103    WRNPTAEDRDLTLDDYRRRGVMAALEAINAILPQRKVHAVGYC LGGTLLSIAAADMAH-G   347
Pden_0958    WKNPDPSYGDTGMDDYVTA-YLEVMDRVLDLTDQKKLNVVGYC IAGTTLALTLSILKQ-R  528
F568_01350   WRNPGVEQSDITWADYMQMGPITAMEAIEQACGEKSVNLLSYC VGGTLTASTVAYLTSTR  351
H16_A1437    WRNPDASMAGSTWDDYIEHAAIRAIEVARDISGQDKINVLGF VGGTIVSTALAVLAA-R  335
             *;**    .    *   :.:::     ::  .:*:.**  :   : :
```

```
H16_A2003    AGGPLASMTLFASETDFHDPGELGLFIDKSSLATLDALMW-----SQGYLDGPQMKSAFQ  400
Pden_4207    RDDRLASVSLLAAQTDFTEPGDLALFVDHDQLHSLENAMR-----QRGYLTAEQMLGAFR  391
Pden_5103    DDDRLASLTLLAAQTDFSEPGELALFIDHSQVNLLESMMW-----NRGYLSADQMAGAFQ  402
Pden_0958    GDDRVNSATFFTALTDFADQGEFTAYLQEDFVSGIEEEAA-----RTGVLGAQLMTRTFS  583
F568_01350   RARKVRSVTYMATLLDFRDPGEIGVFLNETVLQGLERQME-----ADGYLDGRVMAFSFN  406
H16_A1437    GEHPAASVTLLTTLLDFADTGILDVFVDEGHVQLREATLGGGAGAPCALLRGLELANTFS  395
                   * : ::: ** : * :  :::. : :    . *. : :*
```

```
H16_A2003    MLNAQDLIWSRVMSEYLLGQRLRANDLVSWNRDTTRLPYRLHSECLHKLFLGNELAT-GK  459
Pden_4207    FLRSNDLVWSRMVHAYLMGEREPMTDILAWSSDSTRMPYRMQDEYLRRLYLENALAG-DR  450
Pden_5103    LLRSNDLIWSRMVHDYLMGERRPMIDLMAWNADSTRMPYRMHAEYLERLYLDNELAA-GR  461
Pden_0958    FLRANDLVWGPAIRSYMLGEMPPAFDLLFWNGDGTNLPGRMAVEYLRGLCQQNRFVKE-G  642
F568_01350   LLRENDLFWSFYVNNYLKGETPAAFDLLYWNTDGTNLPAATHGWYLRNMYLENRLVEPGG  466
H16_A1437    FLRPNDLVWNYVVDNYLKGNTPVPFDLLFWNGDATNLPGPWYCWYLRHTYLQNELKVPGK  455
             :*. :**.*.  . *: *:    *:: *. *  *.:*      *.     * :
```

```
H16_A2003    LCVGGQPVALSDLDLPLFVVGTEHD KVSPWRSVYKLHLLT-KAELTFLLTSGGH NAGIVS  518
Pden_4207    LRADGRPVALRNLRAPIFAVGTEHD VAPWRSVYQLHCLS-DAELTFVLADKI HLAGIIF  509
Pden_5103    FMVEGRPAAIQNIRVPMFVVGTEHD VAPWRSVFKIHYLS-NTELTFVLTSGGH NAGIVS  520
Pden_0958    FDLLGHRLHVGDVTVPLCAIACET D IAPWRDSWRGVAQMGSKDKTFILSESGH IAGIVN  702
F568_01350   IELDGVKIDLRKISTPSYFISTRD D IIAKWQSTYYGTQLP-KGPVTFVLGGSGH LAGIVN  525
H16_A1437    LTVCGVPVDLASIDVPTYIYGSRE D HIVPWTAAYASTALL-ANKLRFVLGASG HIAGVIN  514
             :  *    : .: *   . *:: *  :      *.*  .* **::
```

FIGURE 3.10 Sequence alignment of PhaC from *Cupriavidus necator* (H16_A2003, H16_A1437), *Paracoccus denitrificans* (Pden_4207, Pden_5103, Pden_0958), and *Halomonas lutea* (F568_01350). Catalytic amino acid residues (Cys-Asp-His) are highlighted in black. Phenylalanine and the conserved tryptophan residues involved in dimerization are marked in dark gray. The conserved alanine residues involved in regulating substrate specificity are shown in light gray.

of PhaC from *Aeromonas caviae* and the C-terminal portion (74%) of *C. necator* PhaC exhibit complete enzyme activity with broad substrate specificity [131]. The transfer of a *phaC1* gene from a metagenomic clone to *P. putida* LS46 broadens its substrate specificity and produces copolymers with different combinations of *scl-* and *mcl*-monomers, depending on the carbon source used to culture the recombinant bacterium [132].

Limited information on the three-dimensional structure of PHA synthases has limited our understanding of the molecular basis of the substrate specificity and the mechanisms involved in polymerization as well as chain length. The structural information of the intermediate complex of PHA synthases and its substrates will be extremely beneficial in identifying its substrate recognition and/or substrate specificity.

REFERENCES

1. Możejko-Ciesielska J, Kiewisz R. Bacterial polyhydroxyalkanoates: Still fabulous? *Microbiol Res* 2016; 192: 271–282.

2. Sudesh K, Abe H, Doi Y. Synthesis, structure and properties of polyhydroxyalkanoates: Biological polyesters. *Progr Polym Sci* 2000; 25: 1503–1555.
3. Albuquerque PBS, Malafaia CB. Perspectives on the production, structural characteristics and potential applications of bioplastics derived from polyhydroxyalkanoates. *Int J Biol Macromol* 2018; 107: 615–625.
4. Reinecke F, Steinbüchel A. *Ralstonia eutropha* Strain H16 as model organism for PHA metabolism and for biotechnological production of technically interesting biopolymers. *J Mol Microbiol Biotechnol* 2009; 16: 91–108.
5. Kahar P, Tsuge T, Taguchi K, et al. 2004 High yield production of polyhydroxyalkanoates from soybean oil by Ralstonia eutropha and its recombinant strain. *Polym Degrad Stab* 2004; 83: 79–86.
6. Chen GQ. A microbial polyhydroxyalkanoates (PHA) based bio and materials industry. *Chem Soc Rev* 2009; 38: 2434–2446.
7. Wang F, Lee SY. Poly (3-hydroxybutyrate) production with high productivity and high polymer content by a fed-batch culture of *Alcaligenes latus* under nitrogen limitation. *Appl Environ Microbiol* 1997; 63: 3703–3706.
8. Mezzina MP, Pettinari MJ. Phasins, multifaceted polyhydroxyalkanoate granule associated proteins. *Appl Environ Microbiol* 2016; 82: 5060–5067.
9. Van-Thuoc D, Quillaguamán J, Mamo G, et al. Utilization of agricultural residues for poly(3-hydroxybutyrate) production by *Halomonas boliviensis* LC1. *J Appl Microbiol* 2008; 104: 420–428.
10. Gao S, Zhang LH. The synthesis of poly-β-Hydroxybutyrate by moderately halophilic bacteria *Halomonas venusta*. *Adv Mat Res* 2014; 1033–1034: 306–310.
11. Mahansaria R, Choudhury JD, Mukherjee J. Polymerase chain reaction-based screening method applicable universally to environmental haloarchaea and halobacteria for identifying polyhydroxyalkanoate producers among them. *Extremophiles* 2015; 19: 1041–1054.
12. Obruca S, Sedlacek P, Mravec F, et al. The presence of PHB granules in cytoplasm protects non-halophilic bacterial cells against the harmful impact of hypertonic environments. *New Biotechnol* 2017; 39: 68–80.
13. Thomas T, Elain A, Bazire A, et al. Complete genome sequence of the halophilic PHA-producing bacterium *Halomonas* sp. SF2003: Insights into its biotechnological potential. *World J Microbiol Biotechnol* 2019; 35: 50.
14. Ücısık-Akkaya E, Ercan O, Yes SK, et al. Enhanced polyhydroxyalkanoate production by *Paracoccus pantotrophus* from glucose and mixed substrate. *Fresen Environ Bull* 2009; 18: 2013–2022.
15. López-Cortés A, Rodríguez-Fernández O, Latisnere-Barragán H, et al. Characterization of polyhydroxyalkanoate and the *phaC* gene of *Paracoccus seriniphilus* E71 strain isolated from a polluted marine microbial mat. *World J Microbiol Biotechnol* 2010; 26: 109–118.
16. Tanaka K, Mori S, Hirata M, et al. Autotrophic growth of *Paracoccus denitrificans* in aerobic condition and the accumulation of biodegradable plastics from CO_2. *Environ Ecol Res* 2016; 4: 231–236.
17. Yamane T, Chen X, Ueda S. Growth-associated production of poly(3-hydroxyvalerate) from n-pentanol by a methylotrophic bacterium *Paracoccus denitrificans*. *Appl Environ Microbiol* 1996; 62: 380–384.
18. Sawant SS, Salunke BK, Kim BS. Degradation of corn stover by fungal cellulase cocktail for production of polyhydroxyalkanoates by moderate halophile *Paracoccus* sp. LL1. *Bioresour Technol* 2015; 194: 247–255.
19. Kumar P, Jun HB, Kim BS. Co-production of polyhydroxyalkanoates and carotenoids through bioconversion of glycerol by *Paracoccus* sp. strain LL1. *Int J Biol Macromol* 2018; 107: 2552–2558.

20. Sharma PK, Fu J, Cicek N, et al. Kinetics of medium- chain-length polyhydroxyalkano-ate production by a novel isolate of *Pseudomonas putida* LS46. *Can J Microbiol* 2012; 58: 982–989.

21. Ouyang SP, Luo RC, Chen SS, et al. Production of polyhydroxyalkanoates with high 3-hydroxydodecanoate monomer content by *fadB* and *fadA* knockout mutant of *Pseudomonas putida* KT2442. *Biomacromolecules* 2007; 8: 2504–2511.

22. Poblete-Castro I, Binger D, Rodrigues A, et al. In-silico-driven metabolic engineering of *Pseudomonas putida* for enhanced production of poly-hydroxyalkanoates. *Metab Eng* 2013; 15: 113–23.

23. Chung AL, Jin HL, Huang LJ, et al. Biosynthesis and characterization of poly (3-hydroxydodecanoate) by β-oxidation inhibited mutant of *Pseudomonas entomophila* L48. *Biomacromolecules* 2011; 12: 3559–3566.

24. Tobin KM, Mcgrath JW, Mullan A, et al. Polyphosphate accumulation by *Pseudomonas putida* CA-3 and other medium chain-length polyhydroxyalkanoate-accumulating bacteria under aerobic growth conditions. *Appl Environ Microbiol* 2007; 73: 1383–1387.

25. Pham TH, Webb JS, Rehm BH. The role of polyhydroxyalkanoate biosynthesis by *Pseudomonas aeruginosa* in rhamnolipid and alginate production as well as stress tolerance and biofilm formation. *Microbiology* 2004; 150: 3405–3413.

26. Costa SG, Lépine F, Milot S, et al. Cassava wastewater as a substrate for the simultaneous production of rhamnolipids and polyhydroxyalkanoates by *Pseudomonas aeruginosa*. *J Ind Microbiol Biotechnol* 2009; 36: 1063–1072.

27. Marsudi S, Unno H, Hori K. Palm oil utilization for the simultaneous production of polyhydroxyalkanoates and rhamnolipids by *Pseudomonas aeruginosa*. *Appl Microbiol Biotechnol* 2008; 78: 955–961.

28. Choi MH, Xu J, Gutierrez M, et al. Metabolic relationship between polyhydroxyalkanoic acid and rhamnolipid synthesis in *Pseudomonas aeruginosa*: Comparative 13C NMR analysis of the products in wild-type and mutants. *J Biotechnol* 2011; 151: 30–42.

29. Tian W, Hong K, Chen GQ, et al. Production of polyesters consisting of medium chain length 3-hydroxyalkanoic acids by *Pseudomonas mendocina* 0806 from various carbon sources. *Anton Leeuw* 2000; 77: 31–36.

30. Guo W, Song C, Kong M, et al. Simultaneous production and characterization of medium-chain-length polyhydroxyalkanoates and alginate oligosaccharides by *Pseudomonas mendocina* NK-01. *Appl Microbiol Biotechnol* 2011; 92: 791–801.

31. Ni YY, Kim DY, Chung MG, et al. Biosynthesis of medium-chain-length poly(3-hydroxyalkanoates) by volatile aromatic hydrocarbons-degrading *Pseudomonas fulva* TY16. *Bioresour Technol* 2010; 101: 8485–8488.

32. Sharma PK, Munir RI, de Kievit T, et al. Synthesis of polyhydroxyalkanoates [PHAs] from vegetable oils and free fatty acids by wild-type and mutant strains of *Pseudomonas chlororaphis*. *Can J Microbiol* 2017; 63: 1009–1024.

33. Lee J, Choi MH, Kim TU, et al. Accumulation of polyhydroxyalkanoic acid containing large amounts of unsaturated monomers in *Pseudomonas fluorescens* BM07 utilizing saccharides and its inhibition by 2-bromooctanoic acid. *Appl Environ Microbiol* 2001; 67: 4963–4974.

34. He W, Tian W, Zhang W, et al. Production of novel polyhydroxyalkanoates by *Pseudomonas stutzeri* 1317 from glucose and soybean oil. *FEMS Microbiol Lett* 1998; 169: 45–49.

35. Lenz RW, Kim YB, Fuller RC. Production of unusual bacterial polyesters by *Pseudomonas oleovorans* through co-metabolism. *FEMS Microbiol Lett* 1992; 103: 207–214.

36. Kim DY, Kim YB, Rhee YH. Evaluation of various carbon substrates for the biosynthesis of polyhydroxyalkanoates bearing functional groups by *Pseudomonas putida*. *Int J Biol Macromol* 2000; 28: 23–29.

37. Imamura T, Kenmoku T, Honma T, et al. Direct biosynthesis of poly (3-hydroxyalkano-ates) bearing epoxide groups. *Int J Biol Macromol* 2001; 29: 295–301.
38. Scholz C, Fuller RC, Lenz RW. Growth and polymer incorporation of *Pseudomonas oleovorans* on alkyl esters of heptanoic acid. *Macromolecules* 1994; 27: 2886–2889.
39. Kim SN, Shim SC, Kim DY, et al. Photochemical crosslinking and enzymatic degradation of poly (3-hydroxyalkanoate)s for micropatterning in photolithography. *Macromol Rapid Commun* 2001; 22: 1066–1071.
40. Abraham GA, Gallardo A, San Roman J, et al. Microbial synthesis of poly (β-hydroxyalkanoates) bearing phenyl groups from *Pseudomonas putida*: Chemical structure and characterization. *Biomacromolecules* 2001; 2: 562–567.
41. Rai R, Yunos DM, Boccaccini AR, et al. Poly-3-hydroxyoctanoate P (3HO), a medium chain length polyhydroxyalkanoate homopolymer from *Pseudomonas mendocina*. *Biomacromolecules* 2011; 12: 2126–2136.
42. Tortajada M, da Silva LF, Prieto MA. Second-generation functionalized medium-chain-length polyhydroxyalkanoates: The gateway to high-value bioplastic applications. *Int Microbiol* 2013; 16: 1–15.
43. Sharma PK, Fu J, Zhang, X, et al. Genome features of *Pseudomonas putida* LS46, a novel polyhydroxyalkanoate producer and its comparison with other *P. putida* strains. *AMB Expr* 2014; 4: 1–18.
44. Udaondo Z, Molina L, Segura A, et al. Analysis of the core genome and pangenome of *Pseudomonas putida*. *Microbiology* 2016; 18: 3268–3283.
45. Prieto A, Escapa IF, Martínez V, et al. 2016. A holistic view of polyhydroxyalkanoate metabolism in *Pseudomonas putida*. *Environ Microbiol* 2016; 18: 341–357.
46. Noda I, Green PR, Satkowski MM, et al. Preparation and properties of a novel class of polyhydroxyalkanoate copolymers. *Biomacromolecules* 2005; 62: 580–586.
47. Nomura CT, Tanaka T, Gan Z, et al. 2004. Effective enhancement of short-chain-length-medium-chain-length polyhydroxyalkanoate copolymer production by coexpression of genetically engineered 3-ketoacyl-acyl-carrier-protein synthase III (*fab*H) and polyhydroxyalkanoate synthesis genes. *Biomacromolecules* 2004; 5: 1457–1464.
48. Matsusaki H, Abe H, Taguchi K, et al. 2000. Biosynthesis of poly (3-hydroxybutyrate-*co*-3-hydroxyalkanoates) by recombinant bacteria expressing the PHA synthase gene *pha*C1 from *Pseudomonas sp.* 61–3. *Appl Microbiol Biotechnol* 2000; 53: 401–409.
49. Liebergesell M, Rahalkar S, Steinbüchel A. Analysis of the *Thiocapsa pfennigii* polyhydroxyalkanoate synthase: Subcloning, molecular characterization and generation of hybrid synthases with the corresponding *Chromatium vinosum* enzyme. *Appl Microbiol Biotechnol* 2000; 54: 186–194.
50. Hill JE, Penny SL, Kenneth G, Crowell KG, et al. cpnDB: A chaperonin sequence database. *Genome Res* 2004; 14: 1669–1675.
51. Vancuren SJ, Hill JE. Update on cpnDB: A reference database of chaperonin sequences. *Database* 2019; 2019: 1–7. doi:10.1093/database/baz033.
52. Saitou N, Nei M. The neighbor-joining method: A new method for reconstructing phylogenetic trees. *Mol Biol Evol* 1987; 4: 406–425.
53. Kumar S, Stecher G, Tamura K. MEGA7: Molecular evolutionary genetics analysis version 7.0 for bigger datasets. *Mol Biol Evol* 2016; 33: 1870–1874.
54. De Eugenio LI, Galán B, Escapa IF, et al. The PhaD regulator controls the simultaneous expression of the *pha* genes involved in polyhydroxyalkanoate metabolism and turnover in *Pseudomonas putida* KT2442. *Environ Microbiol* 2010; 12: 1591–1603.
55. De Eugenio LI, Escapa IF, Morales V, et al. The turnover of medium chain length polyhydroxyalkanoates in *Pseudomonas putida* KT2442 and the fundamental role of PhaZ depolymerase for the metabolic balance. *Environ Microbiol* 2010; 12: 207–221.

56. Galán B, Dinjaski N, Maestro B, et al. Nucleoid-associated PhaF phasin drives intracellular location and segregation of polyhydroxyalkanoate granules in *Pseudomonas putida* KT2442. *Mol Microbiol* 2011; 79: 402–418.

57. Grage K, Jahns AC, Parlane N, et al. Bacterial polyhydroxyalkanoate granules: Biogenesis, structure, and potential use as nano-/micro-beads in biotechnological and biomedical applications. *Biomacromolecules* 2009; 10: 660–669.

58. Huisman GW, de Leeuw O, Eggink G, et al. Synthesis of poly-3-hydroxyalkanoates is a common feature of fluorescent pseudomonads. *Appl Environ Microbiol* 1989; 55: 1949–1954.

59. Rehm BH, Steinbüchel A. 1999. Biochemical and genetic analysis of PHA synthases and other proteins required for PHA synthesis. *Int Biol Macromol* 1999; 25: 3–19.

60. Rehm BH. Polyester synthases: Natural catalysts for plastics. *Biochem J* 2003; 376: 15–33.

61. Zou H, Shi M, Zhang T, et al. Natural and engineered polyhydroxyalkanoate (PHA) synthase: Key enzyme in biopolyester production. *Appl Microbiol Biotechnol* 2017; 101: 7417–7426.

62. Fu J, Sharma P, Spicer V, et al. Quantitative 'Omics analyses of medium chain length polyhydroxyalkanaote metabolism in *Pseudomonas putida* LS46 cultured with waste glycerol and waste fatty acids. *PLoS One* 2015; 10(11): e0142322.

63. Schweizer HP, Jump R, Po C. Structure and gene-polypeptide relationships of the region encoding glycerol diffusion facilitator (*glpF*) and glycerol kinase (*glp*K) of *Pseudomonas aeruginosa*. *Microbiology* 1997; 143: 1287–1297.

64. Castillo TD, Ramos JL, Rodrıguez-Herva JJ, et al. Convergent peripheral pathways catalyze initial glucose catabolism in *Pseudomonas putida*: Genomic and flux analysis. *J Bacteriol* 2007; 189: 5142–5152.

65. Stroud RM, Miercke LJ, O'Connell J, et al. Glycerol facilitator GlpF and the associated aquaporin family of channels. *Curr Opin Struct Biol* 2003; 13: 424–431.

66. Fu J, Sharma U, Sparling R, et al. Evaluation of medium-chain-length polyhydroxyalkanoate production by *Pseudomonas putida* LS46 using biodiesel by-product streams. *Can J Microbiol* 2014; 60: 461–468.

67. Escapa I, Del Cerro C, García J, et al. The role of GlpR repressor in *Pseudomonas putida* KT2440 growth and PHA production from glycerol. *Environ Microbiol* 2013; 15: 93–110.

68. Fu J, Sharma P, Spicer V, et al. Effects of impurities in biodiesel-derived glycerol on growth and expression of heavy metal ion homeostasis genes and gene products in *Pseudomonas putida* LS46. *Appl Microbiol Biotechnol* 2015; 99: 5583–5592.

69. Sánchez RJ, Schripsema J, Da Silva LF, et al. Medium-chain-length polyhydroxyalkanoic acids (PHA$_{mcl}$) produced by *Pseudomonas putida* IPT 046 from renewable sources. *Eur Polym J* 2003; 39: 1385–1394.

70. Kimber MS, Martin F, Lu Y, et al. The structure of (3R)-hydroxyacyl-acyl carrier protein dehydratase (FabZ) from *Pseudomonas aeruginosa*. *J Biol Chem* 2004; 279: 52593–52602.

71. Hoffmann N, Amara AA, Beermann BB, et al. Biochemical characterization of the *Pseudomonas putida* 3-hydroxyacyl ACP: CoA transacylase, which diverts intermediates of fatty acid de novo biosynthesis. *J Biol Chem* 2002; 277: 42926–42936.

72. Tsuge T, Taguchi K, Doi Y. Molecular characterization and properties of (R)-specific enoyl-CoA hydratases from *Pseudomonas aeruginosa*: Metabolic tools for synthesis of polyhydroxyalkanoates via fatty acid beta-oxidation. *Int J Biol Macromol* 2003; 31: 195–205.

73. Sato S, Kanazawa H, Tsuge T. Expression and characterization of (R)-specific enoyl coenzyme A hydratases making a channeling route to polyhydroxyalkanoate biosynthesis in *Pseudomonas putida*. *Appl Microbiol Biotechnol* 2011; 90: 951–959.

74. Ren Q, Sierro N, Witholt B, et al. FabG, an NADPH-dependent 3-ketoacyl reductase of *Pseudomonas aeruginosa*, provides precursors for medium-chain-length poly-3-hydroxyalkanoate biosynthesis in *Escherichia coli. J Bacteriol* 2000; 182: 2978–2981.
75. Sharma PK, Riffat MI, Plouffe J, et al. Polyhydroxyalkanotes (PHAs) polymer accumulation and *pha* gene expression in phenazine (phz⁻) and pyrrolnitrin (prn⁻) defective mutants of *P. chlororaphis* PA23. *Polymers* 2018; 10: 1203.
76. Solaiman DKY, Ashby RD, Foglia TA. Production of polyhydroxyalkanoates from intact triacylglycerols by genetically engineered *Pseudomonas. Appl Microbiol Biotechnol* 2001; 56: 664–669.
77. Legat A, Gruber C, Zangger K, et al. Identification of polyhydroxyalkanoates in *Halococcus* and other haloarchaeal species. *Appl Microbiol Biotechnol* 2010; 87: 1119–1127.
78. Cristea A, Baricz A, Leopold N, et al. Polyhydroxybutyrate production by an extremely halotolerant *Halomonas elongata* strain isolated from the hypersaline meromictic Fără Fund Lake (Transylvanian Basin, Romania). *J Appl Microbiol* 2018; 125(5): 1343–1357.
79. Raguénès G, Cozien J, Guezennec JG. *Halomonas profundus* sp. nov., a new PHA-producing bacterium isolated from a deep-sea hydrothermal vent shrimp. *J Appl Microbiol* 2008; 104: 1425–1432.
80. Yue HT, Ling C, Yang T, et al. A seawater-based open and continuous process for polyhydroxyalkanoates production by recombinant *Halomonas campaniensis* LS21 grown in mixed substrates. *Biotechnol Biofuels* 2014; 7: 108.
81. Diken E, Ozer T, Arikan M, et al. Genomic analysis reveals the biotechnological and industrial potential of levan producing halophilic extremophile, *Halomonas smyrnensis* AAD6T. *SpringerPlus* 2015; 4: 393.
82. Amjres H, Béjar V, Quesada E, et al. *Halomonas rifensis* sp. nov., an exopolysaccharide-producing, halophilic bacterium isolated from a solar saltern. *Int J Syst Evol Microbiol* 2011; 61: 2600–2605.
83. Llamas I, Amjres H, Mata JA, et al. The potential biotechnological applications of the exopolysaccharide produced by the halophilic bacterium *Halomonas almeriensis. Molecules* 2012; 17: 7103–7120.
84. Kulkarni SO, Kanekar PP, Jog JP, et al. Production of copolymer, poly (hydroxybutyrate-*co*-hydroxyvalerate) by *Halomonas campisalis* MCM B-1027 using agro-wastes. *Int J Biol Macromol* 2015; 72: 784–789.
85. Kulkarni SO, Kanekar PP, Nilegaonkar SS, et al. Production and characterization of a biodegradable poly (hydroxybutyrate-*co*-hydroxyvalerate) (PHB-*co*-PHV) copolymer by moderately halo-alkalitolerant *Halomonas campisalis* MCM B-1027 isolated from Lonar Lake, *India. Bioresour Technol* 2010; 101: 9765–9771.
86. Quillaguamán J, Guzmán H, Van-Thuoc D, et al. Synthesis and production of polyhydroxyalkanoates by halophiles: Current potential and future prospects. *Appl Microbiol Biotechnol* 2010; 85: 1687–1696.
87. Kucera D, Pernicová I, Kovalcik A, et al. Characterization of the promising poly(3-hydroxybutyrate) producing halophilic bacterium *Halomonas halophile. Bioresour Technol* 2018; 256: 552–556.
88. Chen Y, Chen XY, Du HT, et al. Chromosome engineering of the TCA cycle in *Halomonas bluephagenesis* for production of copolymers of 3-hydroxybutyrate and 3-hydroxyvalerate (PHBV). *Metab Eng* 2019; 54: 69–82.
89. Quillaguamán J, Doan-van T, Guzman H, et al. Poly(3-hydroxybutyrate) production by *Halomonas boliviensis* in fed-batch culture. *Appl Microbiol Biotechnol* 2008; 78: 227–232.
90. Kawata Y, Ando H, Matsushita I, et al. Efficient secretion of *(R)*-3-hydroxybutyric acid from *Halomonas* sp. KM-1 by nitrate fed-batch cultivation with glucose under microaerobic conditions. *Bioresour Technol* 2014; 156: 400–403.

91. Stanley A, Punil Kumar HN, Mutturi S, et al. Fed-batch strategies for production of PHA using a native isolate of *Halomonas venusta* KT832796 strain. *Appl Biochem Biotechnol* 2018; 184: 935–952.

92. Garcia-Torreiro M, Lu-Chau TA, Steinbuchel A, et al. Waste to bioplastic conversion by the moderate halophilic bacterium *Halomonas boliviensis*. *Chem Eng Trans* 2016; 49: 163–168.

93. Tan D, Wu Q, Chen JC, et al. Engineering halomonas TD01 for the low-cost production of polyhydroxyalkanoates. *Metab Eng* 2014; 26: 34–47.

94. Lemechko P, Le Fellic M, Bruzaud S. Production of poly(3-hydroxybutyrate-*co*-3-hydroxyvalerate) using agro-industrial effluents with tunable proportion of 3-hydroxyvalerate monomer units. *Int J Biol Macromol* 2019; 128: 429–434.

95. Tohme S, Hacıosmanoğlu GG, Eroğlu MS, et al. *Halomonas smyrnensis* as a cell factory for co-production of PHB and levan. *Int J Biol Macromol* 2018; 118(Pt A): 1238–1246.

96. Ouyang P, Wang H, Hajnal I, et al. Increasing oxygen availability for improving poly(3-hydroxybutyrate) production by *Halomonas*. *Metab Eng* 2018; 45: 20–31.

97. Cervantes-Uc JM, Catzin J, Vargas W, et al. Biosynthesis and characterization of polyhydroxyalkanoates produced by an extreme halophilic bacterium, *Halomonas nitroreducens*, isolated from hypersaline ponds. *J Appl Microbiol* 2014; 117: 1056–1065.

98. Kulkarni SO, Kanekar PP, Jog JP, et al. Characterization of copolymer, poly (hydroxybutyrate-*co*-hydroxyvalerate) (PHB-*co*-PHV) produced by *Halomonas campisalis* (MCM B-1027), its biodegradability and potential application. *Bioresour Technol* 2011; 102: 6625–6628.

99. Kulkarni SO, Kanekar PP, Nilegaonkar SS, et al. Production and characterization of a biodegradable poly (hydroxybutyrate-*co*-hydroxyvalerate) (PHB-*co*-PHV) copolymer by moderately haloalkalitolerant *Halomonas campisalis* MCM B-1027 isolated from Lonar Lake, India. *Bioresour Technol* 2010; 101(24): 9765–9771.

100. Simon-Colin G, Raguénès J, Cozien JG, et al. *Halomonas profundus* sp. nov., a new PHA producing bacterium isolated from a deepsea hydrothermal vent shrimp. *J Appl Microbiol* 2008; 104: 1425–1432.

101. Yin J, Chen JC, Wu Q, et al. Halophiles, coming stars for industrial biotechnology. *Biotechnol Adv* 2015; 33: 1433–1442.

102. Zheng Y, Chen JC, Ma YM, et al. Engineering biosynthesis of polyhydroxyalkanoates (PHA) for diversity and cost reduction. *Meta Eng* (in press). doi:10.1016/j.ymben.2019.07.004.

103. Shen R, Ning ZY, Lan YX, et al. Manipulation of polyhydroxyalkanoate granular sizes in *Halomonas bluephagenesis*. *Metabol Eng* 2019; 54: 117–126.

104. Li K, Wang S, Shi Y, et al. Genome sequence of *Paracoccus* sp. strain TRP, a chlorpyrifos biodegrader. *J Bacteriol* 2011; 193: 1786–1787.

105. Siddavattam D, Karegoudar TB, Mudde SK, et al. Genome of a novel isolate of *Paracoccus denitrificans* capable of degrading N,N-dimethylformamide. *J Bacteriol* 2011; 193: 5598–5599.

106. Lopez-Cortes A, Rodrıguez-Fernandez O, Latisnere-Barraga áH, et al. Characterization of polyhydroxyalkanoate and the *phaC* gene of *Paracoccus seriniphilus* E71 strain isolated from a polluted marine microbial mat. *World J Microbiol Biotechnol* 2010; 26: 109–118.

107. Copeland A, Lucas S, Lapidus A, et al. *Paracoccus denitrificans PD1222 Chromosome 1, CP00489 Complete Sequence*. US DOE Joint Genome Institute. /www.ncbi.nlm.nih .gov/ nuccore /NC_008686.1. Acc. No. CP00489, CP00490.1, CP000491.1.

108. Dziewit L, Czarneck J, Wibberg D, et al. Architecture and functions of a multipartite genome of the methylotrophic bacterium *Paracoccus aminophilus* JCM 7686, containing primary and secondary chromids. *BMC Genom* 2014; 15: 1–24.

109. Czarnecki J, Dziewit L, Puzyna M, et al. Lifestyle-determining extrachromosomal replicon pAMV1 and its contribution to the carbon metabolism of the methylotrophic bacterium *Paracoccus aminovorans* JCM 7685. *Environ Microbiol* 2017; 19: 4536–4550.
110. Lasek R, Szuplewska M, Mitura M, et al. Genome structure of the opportunistic pathogen *Paracoccus yeei* (alphaproteobacteria) and identification of putative virulence factors. *Front Microbiol* 2018; 9: 2553.
111. Baj J, Piechucka E, Bartosik D, et al. Plasmid occurrence and diversity in the genus *Paracoccus. Acta Microbiol Pol* 2000; 49: 265–270.
112. Kojima T, Nishiyama T, Maehara A, et al. Expression profiles of polyhydroxyalkanoate synthesis related genes in *Paracoccus denitrificans. J Biosci Bioeng* 2004; 97: 45–53.
113. Gao D, Maehara A, Yamane T, et al. Identification of the intracellular polyhydroxyalkanoate depolymerase gene of *Paracoccus denitriçcans* and some properties of the gene product. *FEMS Microbiol Lett* 2001; 196: 159–164.
114. Sharma PK, Fu J, Spicer V, et al. Global changes in the proteome of *Cupriavidus necator* H16 during poly-(3-hydroxybutyrate) synthesis from various biodiesel by-product substrates. *AMB Expr* 2016; 6: 36.
115. Kutralam-Muniasamy G, Pérez-Guevara F. Comparative genome analysis of completely sequenced *Cupriavidus* genomes provides insights into the biosynthetic potential and versatile applications of *Cupriavidus alkaliphilus* ASC-732. *Can J Microbiol* 2019; 65: 575–595.
116. Kutralam-Muniasamy G, Peréz-Guevara F. Genome characteristics dictate poly-R-(3)-hydroxyalkanoate production in *Cupriavidus necator* H16. *World J Microbiol Biotechnol* 2018; 34: 79.
117. Pohlmann A, Fricke WF, Reinecke F, et al. Genome sequence of the bioplastic-producing "Knallgas" bacterium *Ralstonia eutropha* H1. *Nat Biotechnol* 2006; 24: 1257–1262.
118. Jia Y, Yuan W, Wodzinska J, et al. Mechanistic studies on class I polyhydroxybutyrate (PHB) synthase from *Ralstonia eutropha*: Class I and III synthases share a similar catalytic mechanism. *Biochemistry* 2001; 40: 1011–1019.
119. Tian J, Sinskey AJ, Stubbe J. Detection of intermediates from the polymerization reaction catalyzed by a D302A mutant of class III polyhydroxyalkanoate (PHA) synthase. *Biochemistry* 2005; 44: 1495–1503.
120. Gerngross TU, Snell KD, Peoples OP, et al. Overexpression and purification of the soluble polyhydroxyalkanoate synthase from *Alcaligenes eutrophus*: Evidence for a required posttranslational modification for catalytic activity. *Biochemistry* 1994; 33: 9311–9320.
121. Wodzinska J, Snell KD, Rhomberg A. Polyhydroxybutyrate synthase: Evidence for covalent catalysis. *Am Chem Soc* 1996; 118: 6319–6320.
122. Buckley M, Lu-Chau TA, Lema JM. Effect of nitrogen and/or oxygen concentration on poly(3-hydroxybutyrate) accumulation by *Halomonas boliviensis. Bioproc Biosyst Eng* 2016; 39: 1365–1374.
123. Wittenborn EC, Jost M, Wei Y, et al. Structure of the catalytic domain of the class I polyhydroxybutyrate synthase from *Cupriavidus necator. J Biol Chem* 2016; 291: 25264–25277.
124. Kim J, Kim Y-J, Choi SY, et al. Crystal structure of *Ralstonia eutropha* polyhydroxyalkanoate synthase C-terminal domain and reaction mechanisms. *Biotechnol J* 2017; 12: 1600648.
125. Chek MF, Kim S-Y, Mori T, et al. Structure of polyhydroxyalkanoate (PHA) synthase PhaC from *Chromobacterium* sp. USM2, producing biodegradable plastics. *Sci Rep* 2017; 7: 5312.
126. Chek MF, Hiroe A, Hakoshima T, et al. PHA synthase (PhaC): Interpreting the functions of bioplastic-producing enzyme from a structural perspective. *Appl Microbiol Biotechnol* 2019; 103: 1131–1141.

127. Tsuge T, Saito Y, Narike M, et al. Mutation effects of a conserved alanine (Ala510) in type I polyhydroxyalkanoate synthase from *Ralstonia eutropha* on polyester biosynthesis. *Macromol Biosci* 2004; 4: 963–970.

128. Junker B, York G, Park C, et al. Genetic manipulation of polyhydroxybutyrate synthase activity in *Ralstonia eutropha*. In: *The 8th International Symposium of Biological Polyester, Abstract*, Cambridge, MA, 2000.

129. Taguchi S, Nakamura H, Hiraishi T, et al. *In vitro* evolution of a polyhydroxybutyrate synthase by intragenic suppression-type mutagenesis. *J Biochem* 2002; 131: 801–806.

130. Taguchi S, Maehara A, Takase K, et al. Analysis of mutational effects of polyhydroxybutyrate (PHB) polymerase on bacterial PHB accumulation using an *in vivo* assay system. *FEMS Microbiol Lett* 2001; 198: 65–71.

131. Matsumoto K, Takase K, Yamamoto Y, et al. Chimeric enzyme composed of polyhydroxyalkanoate (PHA) synthases from *Ralstonia eutropha* and *Aeromonas caviae* enhances production of PHAs in recombinant *Escherichia coli*. *Biomacromolecules* 2009; 10: 682–685.

132. Sharma PK, Munir RI, Blunt W, et al. Synthesis and physical properties of polyhydroxyalkanoate polymers with different monomer compositions by recombinant *Pseudomonas putida* LS46 expressing a novel PHA synthase (PhaC116) enzyme. *Appl Sci* 2017; 7: 242.

4 Molecular Basis of Medium-Chain Length- PHA Metabolism of *Pseudomonas putida*

Maria-T. Manoli, Natalia Tarazona, Aranzazu Mato, Beatriz Maestro, Jesús M. Sanz, Juan Nogales, and M. Auxiliadora Prieto

CONTENTS

4.1 *PSEUDOMONAS PUTIDA*, A MODEL BACTERIUM FOR THE PRODUCTION OF MEDIUM-CHAIN-LENGTH PHA

Pseudomonas spp. is a continuously growing genus (approximately 300 species) of Gram-negative, aerobic, bacillus, belonging to the γ-proteobacteria class. The first description of the genus came from Professor Migula of the Karlsruhe Institute in Germany 1894–1900 [1]. The etymology of the term "pseudomonas" is derived from the Greek terminology "pseudes" (ψευδές), which means false, and "monas" (μονάς), which means one single unit. They are rod-shaped, non-sporulating, and motile with polar flagella [1–3].

One of the early discoveries of the genus is its remarkable metabolic versatility and adaptability to endogenous and exogenous stresses. In 1926, L.E. Den Dooren first observed the ability of *Pseudomonas* to degrade a great range of substrates such as carbohydrates, organic acids, aromatics, amino acids, alcohols, and nitrogen compounds [2,4]. Furthermore, *Pseudomonas* spp. could tolerate harmful agents present in soils, such as disinfectants, detergents, organic solvents, and heavy metals [5,6], but not extreme changes in temperature and hydrogen ion concentration resulting in the absence of thermophilic and acidophilic strains in this genus [1]. Finally, some human and plant pathogens were distinguished between the varieties of *Pseudomonas* species due to the production of virulent factors such as *P. aeruginosa* and *P. syringae*, respectively [7,8].

P. putida is a "paradigm of a class of cosmopolitan bacteria" isolated from a variety of environments such as soils, plant rhizospheres, and water [6,9]. *P. putida* KT2440 (KT stands for Kenneth Timmis) strain [10–12] derives from a toluene-degrading organism initially isolated in Japan and designated as *P. arvilla* mt-2 (mt-2 stands for "meta-toluate degrader, isolate 2") and later renamed *P. putida* mt-2 [13]. This strain harbors the TOL plasmid pWW0, which encodes a pathway for toluene and *m/p*-xylene biodegradation [14]. Recently, genomic, phylogenetic, and catabolic re-assessment of the *P. putida* clade supports the delineation of four new species, *P. alloputida* sp. nov., *P. inefficax* sp. nov., *P. persica* sp. nov., and *P. shirazica* sp. nov. [15].

P. putida KT2440 is the best-characterized strain of its species. It was one of the first Gram-negative bacteria to be classified by the Food and Drug Regulation Administration as an HV1 (host-vector system safety level 1) certified strain, indicating that it is safe to be used in a P1 environment, but not as a food additive [16].

Since the publication and recent revision of the *P. putida* KT2440 genome [9,17], much effort has been invested in exploiting its capacity in biotechnological applications, such as the production of medium-chain-length (*mcl*) polyhydroxyalkanoates (PHA) (see Figure 4.1). Generally speaking, PHA can be classified according to the total number of carbon atoms within a PHA monomer (see Figure 4.2). For instance, short-chain-length (*scl*-PHA) such as poly(3-hydroxybutyrate) [PHB a.k.a. P(3HB)] contains 3–5 carbon atoms per monomer, while *mcl*-PHA monomers have 6–14 carbon atoms. The *mcl*-PHA are mainly produced and thoroughly studied by *Pseudomonas* spp. [18–20]. Microorganisms able to produce PHA with more than 14 carbon atoms are rarely found, such as that produced by a local isolate of *P. aeruginosa* MTCC [21]. Several *Pseudomonas* species can produce both *scl*-PHA and *mcl*-PHA [22,23]. Depending on the monomer composition and the length of the side chain, the mechanical and physicochemical properties of the resulting polymer, such as stiffness, brittleness, melting point, glass transition temperature, degree of crystallinity, and resistance to organic solvents, may vary. *Scl*-PHA tend to be stiff and brittle with a high degree of crystallinity (60%–80%), while *mcl*-PHA are elastic with low crystallinity (25%), low tensile strength, high elongation to break, low melting point, and glass transition temperature below room temperature [24,25]. The thermal and mechanical properties of two prototype PHA polymers are listed in Table 4.1.

P. putida is widely considered as a chassis for synthetic biology. There are tools available to solve synthetic biology challenges, such as the removal of non-desirable traits, enhance advantageous features, and the insertion of new activities, and understand the functionality of complex genetic and metabolic circuits [3,26].

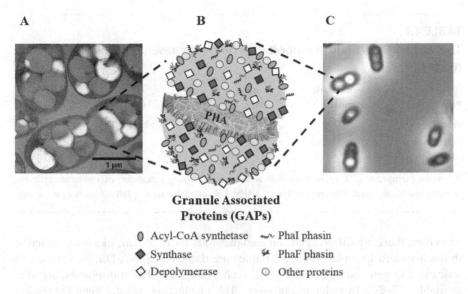

Granule Associated Proteins (GAPs)

◑ Acyl-CoA synthetase ⌇ PhaI phasin
◆ Synthase ⚘ PhaF phasin
◇ Depolymerase ○ Other proteins

FIGURE 4.1 PHA production and granule schematic structure. The presence of granules is observed by transmission electron microscopy (A) and phase contrast microscopy (C) pictures of *P. putida* KT2440 growing under PHA production conditions. B. Schematic representation of a PHA granule from *P. putida* KT2440. The hydrophobic polymer core (PHA) is surrounded by a layer of GAPs. The presence of phospholipids on the surface of the granule is under discussion.

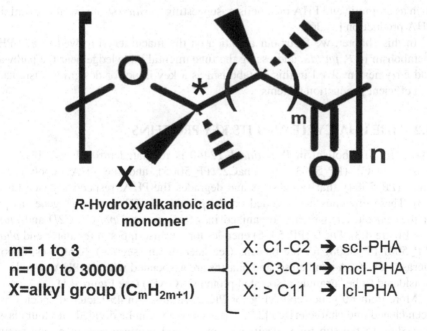

***R*-Hydroxyalkanoic acid monomer**

m= 1 to 3
n=100 to 30000
X=alkyl group (C_mH_{2m+1})

X: C1-C2 ➔ scl-PHA
X: C3-C11 ➔ mcl-PHA
X: > C11 ➔ lcl-PHA

FIGURE 4.2 PHA monomer chemical structure. Classification is based on the length of the aliphatic side chain X and has been divided into short-chain-length PHA, medium-chain-length PHA, and long-chain-length PHA depending on the number of carbons. All monomeric units have an asymmetric carbon center (*); *R*: enantiomer configuration.

TABLE 4.1

Thermal and Mechanical Properties of Some Representative Polymers from the PHA Family

PHA	Monomeric content	σ_t (MPa)	ε_b (%)	T_m (°C)	T_g (°C)	Ref.
P(3HB)	(100%)	40	3–8	173–180	5–9	[24]
P(3HHx-co-3HO)	(12%–88%)	9	380	61	−31	[25]

T_m, melting temperature; T_g, glass transition temperature; E, Young's modulus; σt, tensile strength; εb, elongation at break; 3HB: 3-hydroxy-butyrate; 3HHx: 3-hydroxy-hexanoate; 3HO: 3-hydroxy-octanoate.

Moreover, there are different customized plasmids for *P. putida*, like those included in the Standard European Vector Architecture database (SEVA-DB, http://seva.cnb .csic.es), and genome integration tools, such as Tn5 and Tn7 transposons, are also available [27–29]. In order to optimize PHA production, several gene expression and genome editing approaches have been used to reroute the carbon flux towards PHA production [30]. *P. putida*'s metabolism has been extensively characterized and seven genome-scale models (GEMs) are available including *i*JN746 [31], *i*JP850 [32], PpuMBEL1071 [33], *i*JP962 [34], *i*EB1050 [17], PpuQY1140 [35], and *i*JN1462 [36]. These GEMs have been widely used for studying key metabolic features of *P. putida*, such as aromatic and PHA metabolism, suggesting *in silico* strategies for optimizing PHA production [31,36].

In this chapter, we focus our attention on the machinery involved in *mcl*-PHA metabolism in *P. putida*, addressing the fundamental knowledge near the pathways and enzymes involved in this metabolism as a key issue for designing sustainable and efficient production systems.

4.2   THE PHA CYCLE AND ITS KEY PROTEINS

The PHA metabolism in *P. putida* KT2440 is mediated through two PHA syn-thases, PhaC1 (PP_5003) and PhaC2 (PP_5005), and the PHA depolymerase, PhaZ (PP_5004), that synthesizes and degrades the PHA, respectively (see Figure 4.3). These enzymes are encoded on the *phaC1*, *phaC2*, and *phaZ* genes as part of the *pha* cluster, which is organized into two operons, *phaC1ZC2D* and *phaFI* (see Figure 4.3). PhaD (PP_5006) encodes for a transcriptional regulator and *phaFI* (PP_5007-PP_5008) for the phasins (see later in this section). Synthases, depoly-merase, acyl-CoA synthetase, and phasins are associated with the PHA granule and considered to be "granule-associated proteins" (GAPs) (see Figure 4.1).

More than 60 genes codifying for PHA synthases in 45 bacterial species have been cloned and characterized [37]. These enzymes can be divided into four classes according to the substrate specificity, amino acid primary sequence, and subunit composition [38]. In general, the synthases of class I (e.g., *Cupriavidus necator*; formally known as *Ralstonia eutropha*), class III (e.g., *Allochromatium vinosum*),

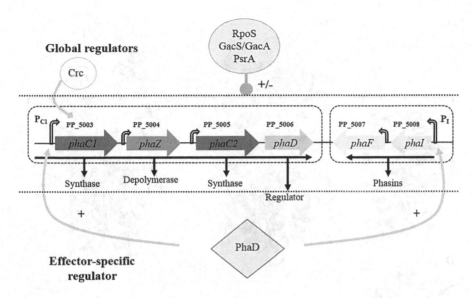

FIGURE 4.3 Regulation of *pha* gene expression in pseudomonads. Diagram of the *pha* gene cluster in *P. putida* KT2440 and its regulatory components. Both operons, *phaC1ZC2D* and *phaIF*, are under the control of the regulator, PhaD (effector-specific regulator). Despite increasing knowledge of PHA regulation, the specific roles of the global transcriptional regulators' (Crc, RpoS, and GacS/GacA) system on the synthesis and degradation of the polymer are still to be elucidated.

and class IV (e.g., *Bacillus megaterium*) preferentially use (R)-3-hydroxyacyl-CoA thioesters [(R)-HA-CoA] comprising 3–5 carbon atoms. The class II PHA synthases (PhaC1 and PhaC2) are mainly observed in pseudomonads and preferentially use CoA thioesters of various (R)-3-hydroxy acids comprising 6–14 carbon atoms. In *P. putida*, PHA synthases have been shown to differ in their substrate specificities [39].

A partial tertiary structure prediction of PhaC1 and PhaC2 synthases has been accomplished based on their sequence homology with several α/β hydrolases [39,40]. Residues 201–530 of both synthases were modeled with >90% accuracy as an α/β hydrolase fold core domain covered by a "lid" region (residues 330–425) that buries the predicted catalytic triad (Cys296, Asp451, and His479 for PhaC1, Cys296, Asp452, and His480 for PhaC2) (Figure 4.4A). This suggests an interfacial activation mechanism already described for several lipases, according to which the lid would open in the presence of the substrate, making the active site accessible for binding and catalytic processing. Whether lid opening is accompanied by the full dissociation of the dimer or, as suggested, for $PhaC_{Cs}$-CAT from *Chromobacterium* sp. USM2, the dimeric form is retained but in a more open conformation and remains to be investigated [41]. The Cys296 residue would act as a nucleophile, attacking the activated carboxyl group of the 3-OH-acyl-CoA, thereby displacing the CoA moiety and creating an intermediate acyl-enzyme that would subsequently react with the free hydroxyl group of another 3-OH-acyl-CoA molecule. Besides, Cys296 lies at the bottom of an elongated crevice that would accommodate the growing polymer. However, secondary structure predictions show some differences between the two

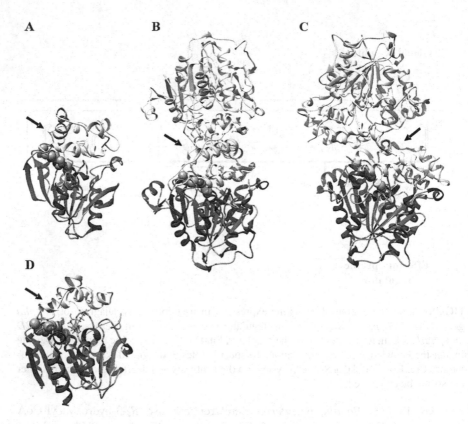

FIGURE 4.4 Three-dimensional structural features of synthases and depolymerases involved in PHA metabolism. A. Homology model of the PhaC1 synthase monomeric form from *P. putida* KT2440 [39]. Predicted catalytic residues are shown in van der Waals representation, and the arrow indicates the lid domain. B. Structure of dimeric PhaC from *C. necator* (PDB code 5HZ2) [42]; same indications as in A. C. Structure of dimeric PhaC$_{Cs}$-CAT from *Chromobacterium* sp. USM2 (PDB code 5XAV) [41]. Same indications as in A. D. Homology model of the PhaZ depolymerase from *P. putida* KT2440 [48]. Predicted catalytic residues are shown in van der Waals representation. The arrow indicates the lid domain, and a dimer of the substrate (3-hydroxyoctanoate) is depicted in stick representation.

synthases around residues 115–123, 316–324, the latter stretch located near the lid region [39]. These or other sparse changes might be of importance for lid movements and/or substrate accommodation, accounting for their differences in substrate specificity.

Numerous spontaneous mutants that block PHA accumulation have been isolated, all of them accounting for changes solely affecting the *phaC1* gene within the *pha* cluster [40]. Most single-residue PhaC1-inactivating mutations were mapped within or in the proximity of the predicted lid region, similar to those isolated in *C. necator*, reinforcing the importance of this region in catalytic activation [42]. Protein-protein docking procedures *in silico* predict both homodimeric and heterodimeric complexes interacting through an extensive surface involving the solvent-exposed regions of the lids of the two synthases. In this context, the inactivating mutations in *phaC1* near the lid might either affect the formation of a productive heterodimeric complex or subsequent interfacial activation steps. Besides, the existence of PhaC1-PhaC1 and PhaC2-PhaC2 homodimers is also possible through similar lid-lid interactions [40].

The three-dimensional structures of the catalytic domain of two highly similar class I PHA synthases, namely PhaC from *C. necator* [42,43] (PhaC$_{Re}$ hereafter), and from *Chromobacterium* sp. USM2 [41] (PhaC$_{Cs}$-CAT hereafter), have been recently solved (see Figure 4.4B, C), providing valuable information that supports the predicted *P. putida* PhaC models. Remarkably, in both cases, the protein experienced proteolysis cleavage during crystallization, and only the C-terminal moiety could be studied, comprising the catalytic domain, thus accounting for the *P. putida* PhaC1 and PhaC2 regions that were modeled (Figure 4.4A). The overall structure in the two cases shares the same α/β hydrolase fold predicted for the *P. putida* synthases, with a central core displaying a left-handed super-helical twist flanked by α-helices on both sides, and also covered by a lid domain. Furthermore, the major deviations with the *P. putida* PhaC models are found in the lid region.

Both PhaC$_{Re}$ and PhaC$_{Cs}$-CAT exist in solution in a monomer/dimer equilibrium, the dimer being the full catalytic active form [41–43]. The dimer interface is favored by lid-lid molecular interactions, similar to those predicted for the *P. putida* proteins, so that conformational changes involving rearrangement of the monomers would facilitate substrate entry and product formation and leaving. The catalytic triad comprises Cys291, Asp447, and His477 for *Chromobacterium* sp. and Cys319, His508, and Asp480 for *C. necator* and are located at the bottom of a crevice. The relative position of the three catalytic residues corresponds well with that of lipases, suggesting a similar mechanism by which these three residues catalyze the esterification reaction. The catalytic Cys would carry out a nucleophilic attack on the thioester of acyl-CoA, generating the acyl-Cys intermediate, whereas His would accelerate the reaction by deprotonating the thiol group of Cys, assisted by Asp. In any case, the catalytic model involves an open-closed equilibrium and comprises the use of a single active site.

Two types of enzymes responsible for the PHA depolymerization have been identified: the extracellular depolymerases and the intracellular ones. Several non-PHA producing microorganisms secrete extracellular depolymerase that allows them to obtain carbon and energy sources from the degradation of the exogenous PHA. The intracellular degradation of the *scl*-PHA was studied extensively in *C. necator* H16 and *Rhodospirillum rubrum*. Functional genomic analyses confirm the complexity of the PHB hydrolysis process [44,45].

The first intracellular *mcl*-PHA depolymerase was described in *P. putida* GPo1, and its enzymatic activity was confirmed *in vivo* on the defective *phaZ* mutant (GPo500), where PHA accumulated, and the mutant was not able to degrade the polymer in the stationary phase [46]. Similar experiments and conclusions were obtained from *P. putida* U and *P. putida* KT2442 [47–50]. The first biochemical characterization of an intracellular *mcl*-PHA depolymerase was realized for the PhaZ from *P. putida* KT2442 [48]. The PhaZ preferentially hydrolyzes *mcl*-PHA containing aliphatic and aromatic monomers. The enzyme behaves as a serine hydrolase that is inhibited by phenylmethylsulfonyl fluoride. The three-dimensional structure of PhaZ complex with a 3-hydroxyoctanoate dimer has been modeled. The enzyme presents an α/β hydrolase-type fold, a central core of an eight-stranded β-sheet and α-helices packed on both sides and capped with a lid structure (residues 133–192) built up from five α-helices connected by loops and covering a possible catalytic triad (Ser102, Asp221, and His248) near the "hinge" of the lid [48] (see Figure 4.4D). The binding surface of the core domain is highly hydrophobic, and the only polar residues surrounding the substrate are the catalytic triad and Asn35, which may take part in substrate recognition as described for the analogous Ser34 in the CumD hydrolase template [51]. Finally, the presence of a shallow surface across the core domain, followed by an elongated deep cavity facilitates the accommodation of *mcl*-hydrocarbon chains of the substrate [48].

The third type of key proteins in *mcl*-PHA machinery are the phasins; in *mcl*-PHA granules of pseudomonads, two phasins, PhaF and PhaI, have been identified as their major GAPs [52]. PhaF plays a crucial role in the spreading of PHA granules among daughter cells upon cell division by linking this event with chromosomal distribution [53]. A PhaF-deficient *P. putida* KT2440 strain showed the defective segregation of PHA granules and reduced PHA accumulation compared with the wild-type strain (33% vs. 63% PHA in cell dry mass). The organization of the granules into a needle array along the long axis of the cell, as observed in the wild-type strain, is also lost in the mutant [53]. Moreover, PhaF was demonstrated to function as a biosurfactant by separating PHA into granules in hydrophilic environments as it occurs in the cytoplasm of bacterial cells [54]. In this sense, the interfacial activity of PhaF and its PHA-binding domain confirmed the potential of these proteins to interact and stabilize the surface of very different hydrophobic materials [55]. For more detailed information about these proteins, see Chapter 2 of this volume, written by Mezzina et al.

PhaF is the major phasin in *Pseudomonas* species, and it is structurally organized in two domains. The N-terminal domain (termed as BioF in the case of the closely related *P. putida* GPo1 strain) shares sequence similarity with PhaI, and it is involved in PHA recognition and binding, while the C-terminal moiety binds DNA in a nonspecific fashion [52,53]. A structural model of PhaF has been elaborated based on sequence homology and *de novo* estimations, with the support of biophysical studies [56] (see Figure 4.5A). The model predicts an elongated shape of the phasin, with the N-terminal, PHA-binding domain acquiring a long amphipathic α-helix that interacts with the granule surface by hydrophobic interactions. However, the C-terminal domain, which contains eight AAKP-like tandem repeats, displays the characteristics of a highly positively charged, histone-like polypeptide. Both domains are linked by a leucine zipper sequence involved in protein tetramerization [56,57] (see Figure 4.5A and Figure 4.6), and are partly intrinsically disordered in the absence of their cognate ligands (PHA and DNA). Both coiled-coil-mediated oligomerization and intrinsic disorder may be common traits shared by a significant number of other phasins [56].

4.3 METABOLIC PATHWAYS INVOLVED IN *MCL*-PHA PRODUCTION IN *P. PUTIDA*

4.3.1 *Mcl*-PHA Synthesis from Structurally Related Substrates

In pseudomonads, structurally, PHA-related substrates like fatty acids are catabolized through the β-oxidation pathway. The fatty acids can be incorporated into PHA

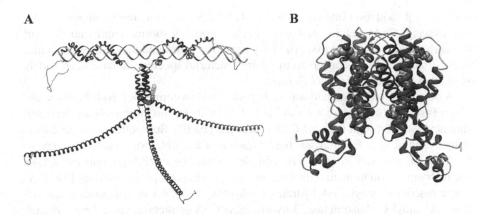

FIGURE 4.5 Three-dimensional structural features of the phasin PhaF and the transcriptional regulator PhaD. A. A theoretical model of the tetrameric form of the PhaF phasin from *P. putida* KT2440 bound to DNA [56]. B. Homology model of the PhaD regulator from *P. putida* KT2440 showing two docked molecules of 3-hydroxyoctanoyl-CoA in van der Waals representation [100].

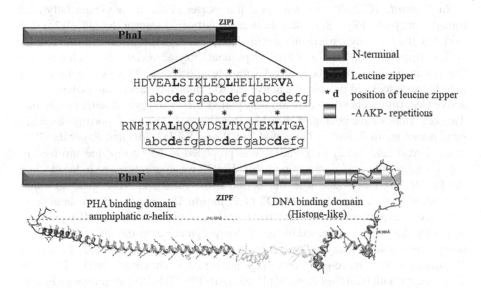

FIGURE 4.6 Phasins from *P. putida* KT2440: scheme of the PhaI and PhaF phasins structure showing three separate domains. The N-terminal PHA binding domain acquires an α-helix conformation (gray color). The leucine zipper motif of PhaI (ZIPI) and PhaF (ZIPF) consists of three heptad repeats (squared) with leucines occupying most of the *d* positions (black color) [57]. The structurally disordered C-terminal domain (DNA binding domain) comprises eight AAKP-like tandem repeats (shaded gray boxes) similar to those found in the DNA binding nucleoid-associated proteins of histone family H1. The predicted structure of the PhaF monomer is shown (bottom) [56].

directly as β-oxidation intermediates [(R)-HA-CoA], avoiding their complete oxidation to acetyl-CoA, yielding monomer chains of equal or shorter length than those of the carbon source applied [58]. The fatty acid catabolism and the involved enzymes in this route (Fad) are conserved in distinct bacterial species and have been widely studied in the bacterial model *E. coli*.

A detailed β-oxidation pathway in *P. putida* and its connection with PHA metabolism is shown (see Figure 4.7 and Table 4.2). In short, the fatty acids are activated into acyl-CoA through the acyl-CoA synthetase (FadD). Subsequently, an acyl-CoA dehydrogenase (FadE) catalyzes the formation of a double bond yielding an enoyl-CoA. In the next step, a tetrameric complex formed by FadBA proteins carries out the hydration, oxidation, and thiolysis processes. This complex comprises five enzymatic reactions (enoyl-CoA hydratase, 3-hydroxyacyl-CoA dehydrogenase, cis-Δ^3-trans-Δ^2-enoyl-CoA isomerase, 3-hydroxyacyl-CoA epimerase, and 3-ketoacyl-CoA thiolase) and it is responsible for the removal of two carbon units from the acyl-chain [59]. The intermediates of the β-oxidation enoyl-CoA, (S)-3-hydroxyacyl-CoA, and 3-ketoacyl-CoA can be converted into (R)-3-HA-CoA, which are the PHA synthase substrates through a stereospecific trans-enoyl-CoA hydratase (PhaJ), an epimerase (FadB), or a specific ketoacyl-CoA reductase (FabG), respectively [20] (see Figure 4.7 and Table 4.2).

In *P. putida* KT2440, two fatty acid transporter genes, a long-chain fatty acid transporter (*fadL*; PP_1689) and a short-chain fatty acid transporter (PP_3124) that initialize the pathway have been identified [60]. FadD (EC 6.2.1.3) is an ATP (adenosine triphosphate), CoA, and Mg^{2+} dependent enzyme widely distributed among the organisms and exhibits a broad substrate specificity. This is a cytoplasmic membrane-associated protein that activates exogenous fatty acids into metabolically active CoA thioesters when they are transported across the cytoplasmic membrane. Two key studies have been performed on *Pseudomonas* species reporting two *fadD* gene homologs in *P. putida* U (named FadD1 and FadD2) [61] and *P. putida* GPo1 (named Acs1 and Acs2) [62]. FadD1 was proposed to be an enzyme involved in physiological fatty acid degradation, while FadD2 was a cryptic gene induced when the FadD1 was inactivated [61]. Furthermore, the Acs1 and Acs2 share 94% and 92% identity with FadD1 and FadD2 of *P. putida* U, respectively. In fluorescent microscopy experiments, the Acs1 was mainly associated with the PHA granules activating the (R)-HAs released by the PHA depolymerase during the PHA degradation process, whereas the Acs2 was located in the cellular membrane [62]. Finally, in *P. putida* KT2440, three genes encoding putative FadD enzymes (FadD-I: PP_4549, 94% identity with FadD1 of *P. putida* U. FadD-II: PP_4550, 91% identity with FadD2 from *P. putida* U. FadDx: PP_2213) were identified, where the FadD-I was proposed to be the principal enzyme [60]. Finally, homologous genes have also been characterized in the *P. putida* CA-3 strain, where FadD is capable of activating long-chain phenylalkanoic and alkanoic acids [63].

The second step of fatty acid degradation in *P. putida* KT2440 is realized by acyl-CoA dehydrogenases, FadE. They belong to a large family of flavoproteins and show activity towards a broad range of substrates. Depending on the substrate specificity of FadE, the enzyme can be classified into short, medium, long, or very long-chain acyl-CoA dehydrogenases. However, the boundary of classification is not tight as substrate

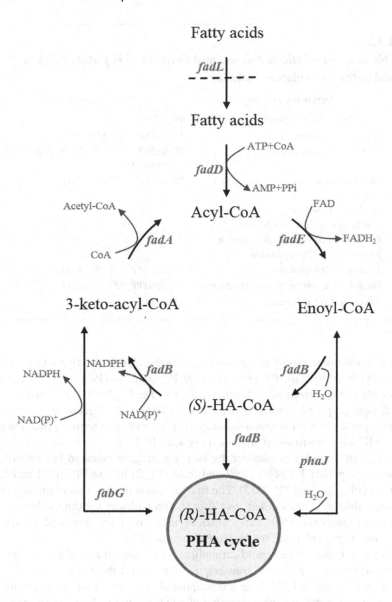

FIGURE 4.7 Fatty acid β-oxidation pathway and its connection with PHA metabolism. Diagram with the main metabolic steps involved in the β-oxidation pathway (black lines), the proposed connection with PHA metabolism, the responsible genes involved in these metabolic pathways (bold gray letters), and the reaction exchanges (gray color) are shown. Figure abbreviations: *(S)*-HA-CoA: *(S)*-3-hydroxyacyl-CoA, *(R)*-HA-CoA: *(R)*-3-hydroxyacyl-CoA.

specificities of particular enzymes may overlap [64]. The reaction of the dehydrogenation presents the rate-limiting step in the β-oxidation since FadE has the lowest activity among all β-oxidation enzymes in *E. coli* [65]. Initial work on *P. putida* KT2440 identified PP_2216 as a specific acyl-CoA dehydrogenase for short-chain aliphatic fatty acyl-CoAs [66] and the PP_0368 as an inducible phenylacyl-CoA dehydrogenase [67].

TABLE 4.2

Gene Names, Enzymatic Activities, and Gene ID of *P. putida* KT2440 Involved in the β-Oxidation Pathway

Gene	Enzymatic activity	Annotation
fadL	Long-chain fatty acid transporter (LCFAs)	PP_1689
fadD	Acyl-CoA synthetase	PP_2213/PP_4549/PP_4550
fadE	Acyl-CoA dehydrogenase	PP_0368/PP_1893/PP_2039/PP_2048/PP_2216/PP_2437
fadB	Multienzymatic activities of the FadBA complex: enoyl-CoA hydratase 3-hydroxyacyl-CoA dehydrogenase Cis-Δ³-trans-Δ²-enoyl-CoA isomerase 3-hydroxyacyl-CoA epimerase	PP_2047/PP_2136/PP_2214
fadA	3-ketoacyl-CoA thiolase	PP_2137/PP_2215/PP_4636
phaJ	Enoyl-CoA hydratase, *R* stereoselectivity	PP_0580/PP_4552/PP_4817
fabG	3-ketoacyl-CoA reductase	PP_1914

Recently, *in silico* analysis of its genome sequence revealed 21 putative acyl-CoA dehydrogenases (ACADs), four (PP_1893, PP_2039, PP_2048, and PP_2437) of which were functionally characterized by mutagenesis studies. The PP_1893 (FadE) and PP_2437 (FadE2) were proposed to directly participate in fatty acid degradation, while the 19 remaining putative ACADs have a redundant role or overlap in terms of function when *P. putida* KT2440 is grown on aliphatic fatty acids [68].

The FadBA complex is encoded by two sets of genes located in two different operons in *P. putida* KT2440, *fadB* and *fadA* (PP_2136 and PP_2137) and *fadBx* and *fadAx* (PP_2214 and PP_2215). The first set seems to play a more important role since their absence causes a defective β-oxidation pathway yielding polymers with longer chain monomers [69]. Various isoenzymes have been proposed to carry out some of the steps performed by the FadBA complex [70].

As far as it concerns fatty acid catabolism regulation, in *Pseudomonas* spp., the PsrA regulator was reported to transcriptionally control the *fad* genes [71,72]. For instance, in *P. putida* KT2440, the transcriptional repressor of the β-oxidation route, PsrA, negatively regulates the expression of genes related to fatty acid metabolism. It has been suggested that PsrA binds to the promoter region of the *fadBA* operon inhibiting its expression in *P. putida*. This explains why a *psrA* mutant strain showed increased β-oxidation and a reduction in its PHA content when using fatty acids as a growth and polymer precursor [72].

4.3.2 THE CONNECTION OF THE β-OXIDATION PATHWAY TO THE PHA CYCLE

In the β-oxidation pathway, acyl-CoA is sequentially oxidized into enoyl-CoA, *(S)*-3-hydroxyacyl-CoA, and 3-ketoacyl-CoA. All of these intermediates are proposed to be converted into a PhaC1 substrate, (R)-HA-CoA by a stereospecific trans-enoyl-CoA hydratase (PhaJ), an epimerase (FadB), and/or a specific ketoacyl-CoA reductase (FabG) (see Figure 4.7) [20,73].

PhaJ is one of the proposed enzymes linking β-oxidation and PHA synthesis and catalyzes a stereospecific hydration reaction from trans-2-enoyl-CoA to (R)-3-HA-CoA. Recent studies in pseudomonads have suggested that the PhaJ enzyme can contribute to the monomer supply for PHA biosynthesis from fatty acid precursors [74]. Four *phaJ* genes were identified in *P. aeruginosa*, where only PhaJ1 was found to be specific for *scl*-enoyl-CoAs (C4–C6) and the other three PhaJs were specific for *mcl*-enoyl-CoAs (C6–C12) [75]. From a database search, three homologs of *phaJ1*$_{pa}$, *phaJ3*$_{pa}$, and *phaJ4*$_{pa}$ were found in *P. putida* and named *phaJ1* (PP_4552), *phaJ3* (PP_0580), and *phaJ4* (PP_4817), respectively. Only *phaJ4* showed higher expression levels after growth on fatty acids, and it was shown to produce more PHA. Additionally, the PhaJ4 enzyme substrate specificity was shown to be *mcl*-enoyl-CoAs like PhaJ4 from *P. aeruginosa* [60,74–77].

Another link between β-oxidation and PHA synthesis was previously proposed as an NADPH-dependent 3-ketoacyl-ACP reductase, FabG. This is a reversible enzyme known to catalyze the conversion of 3-ketoacyl-CoA into (R)-HA-CoA and vice versa [78]. Initial studies on *E. coli* have shown that FabG$_{Ec}$ and FabG$_{pa}$ (from *P. aeruginosa*) were capable of generating (R)-HA-CoA precursors from 3-ketoacyl-CoA in the presence of a PHA synthase [79,80]. In pseudomonads, its contribution to PHA biosynthesis via the β-oxidation pathway is yet unclear. In fact, overexpression of *fabG* and *phaJ* in the *P. putida* strain resulted in controversial results, with negative and positive effects on PHA biosynthesis, respectively. The *fabG* negative effect could be due to the reversible conversion of (R)-HA-CoA into 3-ketoacyl-CoA [74].

Finally, FadB, a putative epimerase, was proposed to connect β-oxidation and PHA synthesis by converting (S)-HA-CoA to (R)-HA-CoA. Despite the assigned epimerase activity in *E. coli* [59], its role is not physiologically relevant in pseudomonads. Thus, FadBA does not seem to possess the putative epimerase function to provide the (R)-enantiomer of HA-CoA efficiently, and other linking enzymes are required to efficiently channel intermediates of β-oxidation towards *mcl*-PHA biosynthesis [76]. More studies need to be done to verify this.

4.3.3 DE NOVO FATTY ACID SYNTHESIS

Non-PHA related substrates, such as carbohydrates, can act as indirect precursors of PHA through their complete oxidation to acetyl-CoA and can be channeled towards the *de novo* synthesis pathway (see Figure 4.8 and Table 4.3). The genetic and biochemical characterization of the *de novo* fatty acid biosynthesis in pseudomonads has been principally analyzed in *P. aeruginosa*.

The *de novo* fatty acid intermediates are activated by the acyl carrier protein (ACP). The fatty acid synthesis starts with the acetyl-CoA carboxylation into malonyl-CoA by the acetyl-CoA carboxylase, the AccABCD complex [81]. Then, the malonyl-CoA is transesterified into malonyl-ACP by the malonyl-CoA: ACP transacylase, FabD [82]. The generated malonyl-ACP is further condensed into 3-ketoacyl-ACP by different 3-ketoacyl-ACP synthases. First, it is condensed by an acetyl-CoA molecule

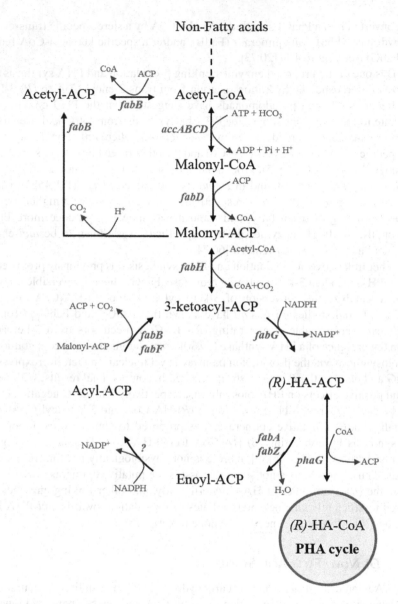

FIGURE 4.8 *De novo* fatty acid biosynthesis and its connection with PHA metabolism. Diagram with the main metabolic steps involved in *de novo* fatty acid biosynthesis (black lines), the proposed connection with the PHA metabolism, the responsible genes involved in these metabolic pathways (bold gray letters), and the reaction exchanges (gray color) are shown. Figure abbreviations: *(R)*-HA-ACP: *(R)*-3-hydroxyacyl-ACP, *(R)*-HA-CoA: *(R)*-3-hydroxyacyl-CoA.

TABLE 4.3
**Genes Names, Enzymatic Activities and Gene ID of *P. putida* KT2440
Involved in the *de novo* Fatty Acid Synthesis**

Gene	Enzymatic activity	Annotation
accABCD	*accABCD* complex: Acetyl-CoA carboxylase	PP_1607/PP_0559/PP_0558/ PP_1996
fabD	Malonyl-CoA transacylase	PP_1913
fabB	3-oxoacyl-ACP-synthase	PP_4175
fabF	beta-ketoacyl-ACP-synthase II	PP_1916/ PP_3303
fabH	3-ketoacyl-ACP-synthase III	PP_4379/PP_4545
fabG	3-oxoacyl-ACP-reductase subunit	PP_0581/PP_1914/PP_2540/ PP_2783
fabA	3R-3-hydroxydecanoyl-ACP dehydratase	PP_4174
fabZ	3-hydroxyacyl-ACP dehydratase	PP_1602
?	Enoyl-ACP reductase	PP_1852
phaG	3-hydroxyacyl-CoA-ACP transferase	PP_1408
?	Medium-chain-fatty acid 3-hydroxyacyl-CoA ligase	PP_0763

by FabH, then, in successive rounds of elongation, a new molecule of malonyl-ACP is condensed with the 3-acyl-ACP formed by FabB or FabF [83,84]. FabB has also been proposed to catalyze the decarboxylation of malonyl-ACP into acetyl-ACP [85]. The following step includes the reduction of 3-ketoacyl-ACP into (R)-3-hydroxyacyl-ACP by a 3-ketoacyl-ACP reductase, FabG [86], and the formation of a double bond into enoyl-ACP by 3-hydroxyacyl-ACP dehydratase, FabA, or FabZ [87]. Finally, one enoyl-ACP reductase, FabI, FabK, or FabL transforms the enoyl-ACP into 3-acyl-ACP [88]. The gene that codifies for enoyl-ACP reductase activity has not been identified in *P. putida* KT2440 [84].

4.3.4 DE NOVO FATTY ACID SYNTHESIS CONNECTION TO THE PHA CYCLE

Hydroxyacyl-ACP intermediates can be transformed into (R)-HA-CoAs by the specific transacylase PhaG, present in most pseudomonads [89]. PhaG is reported to be a 3-hydroxyacyl-ACP-CoA transferase able to transfer the 3-hydroxyacyl group from ACP to CoA moiety in *P. putida* KT2440 and *P. aeruginosa* [90–94]. However, PhaG is proposed to function as a 3-HA-ACP thioesterase. The existence of at least one 3-hydroxyacyl-CoA ligase is needed to convert 3-hydroxyalkanoate substrates into 3-HA-CoA thioester PHA precursors [95]. Based on homology studies, the PP_0763 gene was predicted to encode a medium-chain fatty acid CoA ligase; its transcription levels were highly upregulated when cells were grown under PHA-producing conditions [95]. In order to verify this, *phaG* (PP_1408) and PP_0763 genes from *P. putida* were cloned and co-expressed with the engineered *phaC1* (STQK) synthase from *Pseudomonas* sp. 61-3 in recombinant *E. coli*. The latter strain resulted in about four-fold higher PHA production from non-fatty acid carbon sources than that in *P. putida* KT2440 [95].

4.4 PHA METABOLISM REGULATION

The *pha* gene cluster is well conserved among the *mcl*-PHA producer strains [49] and contains two operons, *phaC1ZC2D* and *phaFI*, that are transcribed in opposite directions (see Figure 4.3). However, the intergenic regions vary depending on the species (as observed in *P. corrugata*, *P. mediterranea*, and *P. aeruginosa*), suggesting that the transcription control differs between the strains [49].

PHA are widely investigated concerning their structure, versatility, physico-chemical properties, biodegradability, and sustainable and optimizable production possibilities [96–99]. However, the molecular mechanisms behind the PHA metabolism regulation are not yet clear due to the complexity of this regulation system at different levels (genetic, enzymatic, and global regulators). Furthermore, PHA regulation could depend on the carbon source(s), the specificity of the PHA synthase, and the metabolic pathways involved. The PhaC synthase substrates derive from the β-oxidation and *de novo* fatty acid synthesis. Thus, we can assume that regulating factors affecting the enzymes involved in these pathways could be involved in the PHA synthesis regulation too.

The co-existence of polycistronic transcription units was proposed by the identification of five functional promoters upstream of the *phaC1*, *phaZ*, *phaC2*, *phaF*, and *phaI* genes by using *lacZ* translational fusions. However, in RT-qPCR experiments, the P_{C1} and P_I were demonstrated to be the most active promoters [100].

PhaD was proposed to transcriptionally activate the *pha* genes by binding to 25 and 29 bp target regions of the P_{C1} and P_I promoters, respectively [100]. Both sequences contain a single binding site formed by inverted half-sites of 6 bp separated by 8 bp. Unlike most characterized members of this family, PhaD functions as an activator and not as a repressor of transcription (see Figure 4.3). A 3D model of the PhaD structure predicts a dimeric structure with each monomer conformed by two domains built up from nine α-helices [100] (see Figure 4.5B). In turn, the N-terminal domain is composed of an HTH motif comprised of α2 and α3 helices, and in the C-terminal, the α8 and α9 helices might be involved in the dimerization of the protein through a four-helix bundle arrangement. Overall, a long, deep crevice enclosed by helices α5–α8 in the C-terminal domain could accommodate inside an elongated intermediate molecule of the fatty acid β-oxidation, hydroxyoctanoyl-CoA, that would act as the true inducer [100]. This model proposes the effector of PhaD to be a CoA intermediate of fatty acid β-oxidation or TCA (tricarboxylic acid cycle) metabolite. However, the true effector remains uncertain [100]. In different *Pseudomonas* strains, variable transcription levels of *phaF* and *phaFI* were observed in relation with the PhaD role [101]. A clear carbon source dependence was also demonstrated [50].

PhaF has been shown to regulate the expression of the *pha* genes, and to some extent, the entire transcriptomic profile. This effect is also believed to be derived from its DNA binding abilities and putative histone-like functionality [52,53]. The disruption of the *phaF* gene in *P. putida* KT2442 has been shown to affect PHA production (1.5-fold reduction) and the transcription of the *phaC1* and *phaI* genes. Experiments have shown that the transcription level of *phaC1* decreases approximately 3.5-fold with respect to the wild-type and that of *phaI* to reach its highest level much later in the stationary phase [53]. Even though there is no evidence for the specific binding of PhaF to the promoter regions of *pha* genes, preliminary transcriptomic analyses of a *phaF* mutant strain has suggested an indirect regulatory effect on the expression of these genes with the mild downregulation of the *pha* cluster. A wider effect on the *P. putida phaF* mutant transcriptome has also been observed, including upregulation of acetyl-CoA acetyltransferase (Q88E32), a key enzyme in fatty acid degradation, and other proteins involved in metabolic and regulatory processes [53].

Recent evidence suggests that the expression of the genes in the *pha* cluster is also subject to the influence of global regulatory systems, although most mechanisms are not fully understood (see Figure 4.3). De la Rosa et al. [102] reported PHA synthesis to be controlled by a catabolite repression control (Crc) protein, a global regulator that optimizes carbon metabolism by inhibiting the expression of genes involved in the use of non-preferred carbon sources. In balanced C/N conditions (which are not optimal for PHA accumulation), Crc was shown to inhibit the translation of *phaCl* mRNA via its interaction with a specific sequence, A*n*AA*n*AA, located immediately downstream of the AUG start codon (where *n* represents any nucleotide). This results in a reduction in the PHA content of the cell [102]. The RpoS global regulator has also been suggested to be involved in the regulation of the P_{C1} promoter. Experiments on *P. putida*, using the P_{C1}::*lacZ* translational fusion, showed that *rpoS* mutation increased the expression driven by the P_{C1} promoter. PHA synthesis is also affected by the two-component GacS/GacA global regulatory system (global antibiotic and cyanide control), as described in *P. putida* CA-3. GacS is a sensor kinase and GacA, its cognate regulator that activates the expression of small RNAs (sRNAs). These sRNAs and the sRNA-binding (RsmA) protein cooperatively regulate protein synthesis and together constitute the Gac/Rsm system. A *gacS* mutation in *P. putida* CA-3 prevents the translation of *phaCl* mRNA by a non-conventional mechanism that is yet to be fully understood [103]. As previously mentioned, a link between PHA accumulation and the transcriptional regulators of the central metabolic routes that provide substrates for PHA synthesis via the β-oxidation pathway has also been reported [72,104]. In *P. putida* KT2440, the transcriptional repressor of the β-oxidation route, PsrA, negatively regulates the expression of genes related to fatty acid metabolism by binding to the promoter region of the *fadBA* operon inhibiting its expression in *P. putida*. All these regulatory systems drive the PHA metabolism as a key cycle for maintaining the carbon and energy control in the cells. Very recently, the impact of this cycle on the robustness of *P. putida* has been highlighted.

4.4.1 The PHA Cycle as a Robustness Cycle

Cells facing carbon excess and nutrient limitation direct the PHA metabolism to store the extra carbon in the form of polyesters by PHA synthases. Under a balanced C/N ratio, PHA metabolism is directed to the hydrolysis pathway releasing of (*R*)-HAs monomers to the extracellular medium by the action of the PhaZ depolymerase. The released monomers are then activated to (*R*)-HA-CoAs by a granule-associated acyl-CoA synthetase (FadD) via an ATP-dependent reaction (see Figure 4.9). This metabolite is a substrate for PhaC as well as the β-oxidation cycle. Depending on the metabolic state of the cell, this (*R*)-HA-CoA is either

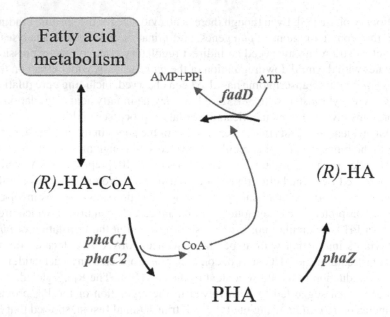

FIGURE 4.9 The PHA cycle. The key players are the PHA synthases (PhaC1, PhaC2), PHA depolymerase (PhaZ), and Acyl-CoA synthetase (FadD). Depending on the metabolic state of the cell, the (R)-HA-CoA can either be incorporated into a nascent PHA granule or be used for fatty acid metabolism. Figure abbreviations: (R)-HA-CoA: (R)-3-hydroxyacyl-CoA, (R)-HA: (R)-3-hydroxyacyl-carboxylic acid (free monomers).

incorporated again into nascent PHA polymer chains or used by the fatty acid metabolism to make the stored carbon available to the central metabolism. Thus, PHA metabolism is far from being a unidirectional metabolic process, where the PHA are either polymerized or depolymerized, but an "ongoing" cycle, where there is simultaneous polymerization and degradation of PHA polymer to facilitate the fast turnover of the polymer [50,105]. In fact, PhaZ is active even under conditions where it would appear not to be required [105].

Due to this cyclic operational mode, the PHA cycle has been traditionally considered as a futile cycle [50,105]. Futile cycles occur in microorganisms inducing a considerable energy burden for the cell, promoting the dissipation of energy. Such cycles fall into two main categories: those involving simultaneous phosphorylation and dephosphorylation and those involving energy-driven transport reactions in the opposite direction. However, based on this canonical definition of futile cycles, PHA cannot be considered as such. It was not clear what was the metabolic advantage of this cyclic operational mode of PHA metabolism, as a great amount of energy was wasted [50,105,106]. Interestingly, microbial systems operate far from optimally in terms of efficient use of ATP. Instead, they optimize multiple objectives in which growth, but also extra production of ATP, is prioritized [107]. It has been suggested that these extra ATP levels are required for fueling proper biological responses under (un)expected environmental perturbations, thus providing metabolic robustness. The inherent property ubiquitously observed in nature, allowing systems to maintain their functions despite external and internal perturbations, is understood as robustness [108,109]. Recent work from our group has identified a series of metabolic cycles providing robustness in *P. putida* KT2440 using the *in silico* genome-scale metabolic model *i*JN1411 [110]. Such cycles are more than conventional futile cycles as in addition to acting as dissipating energy, they likely act as metabolic capacitors connecting catabolism and anabolism with central metabolism. These metabolic capacitors provide a pre-processed source of energy and anabolic building blocks while balancing and optimizing the redox state. Thus, such buffering cycles provide metabolic robustness and fitness under changing environmental conditions. The PHA cycle was identified among these metabolic cycles, thus confirming the role of cyclic PHA metabolism as a mechanism providing robustness to *P. putida*. In addition, genes encoding PHA metabolism were found to be highly expressed in an exponential growth phase irrespective of the nutritional scenario [110]. This means that during unexpected and sudden environmental perturbations, such as carbon source depletion, the cells could adapt rapidly to the new situation providing a new source of carbon and energy such as (*R*)-HA-CoA without the need to induce complex cascades of gene expression [105]. Additionally, the PHA cycle was proposed to regulate fatty acid flux towards the fatty acid oxidation pathway [105]. This scenario effectively increases *P. putida* fitness under changing environmental conditions. Supporting this novel function of PHA metabolism to a greater extent, it has been demonstrated that an active PHA cycle has a great impact on a variety of physiological processes such as cell size and number of individuals as a function of carbon/nitrogen availability and biomass formation [50,106,111]. Therefore, PHA metabolism, far from being a single carbon source storage mechanism, has an essential role in maintaining optimal growth performance and homeostasis in *P. putida*.

4.5 CONCLUSIONS AND OUTLOOK

Fueled by recent studies, PHA metabolism is no longer considered as a residual and secondary metabolism; it is starting to be considered as a central metabolic and physiological hub in *P. putida*. The more is known about the complexity of PHA metabolism, the greater the need is to go further in its study by using cutting edge integrative and systems analyses. The complete understanding of the key involved proteins, regulatory systems, and the PHA metabolism as a metabolic robustness cycle in *P. putida*, driving its known stress endurance and robustness, is of crucial importance to increase the biotechnological applications of this unique bacterial workhorse. The possibility to construct optimized strains with higher metabolic robustness based on PHA cycle engineering will position *P. putida* as an ideal biocatalyst towards the revalorization of recalcitrant residues by significantly contributing to the largely expected circular economy.

ACKNOWLEDGMENTS

This work was supported by funding from the European Union's Horizon 2020 research and innovation program under grant agreement numbers 633962 (P4SB project) and 814418 (SinFonia project). We also acknowledge support from the

Community of Madrid (P2018/NMT4389), Instituto de Salud Carlos III (CIBER de enfermedades respiratorias) and the Spanish Ministry of Science, Innovation, and Universities (BIO2016-79323-R, BIO2017-83448-R).

REFERENCES

1. Palleroni, NJ. The *Pseudomonas* story: Editorial. *Environ Microbiol* 2010; 12(6): 1377–1383.

2. Stanier, RY., Palleroni, NJ., Doudoroff, M. The aerobic pseudomonads a taxonomic study. *J Gen Microbiol* 1966; 43(2): 159–271.

3. Nikel, PI., Martínez-García, E., De Lorenzo, V. Biotechnological domestication of pseudomonads using synthetic biology. *Nat Rev Microbiol* 2014; 12(5): 368–379.

4. Clarke, PH. The metabolic versatility of pseudomonads. *Anton Leeuw* 1982; 48(2): 105–130.

5. Jimenez, JI., Minambres, B., Garcia, JL., et al. Genomic analysis of the aromatic catabolic pathways from *Pseudomonas putida* KT2440. *Environ Microbiol* 2002; 4(12): 824–841.

6. dos Santos, VAPM., Heim, S., Moore, ERB., et al. Insights into the genomic basis of niche specificity of *Pseudomonas putida* KT2440. *Environ Microbiol* 2004; 6(12): 1264–1286.

7. Gellatly, SL., Hancock, REW. *Pseudomonas aeruginosa*: New insights into pathogenesis and host defenses. *Pathog Dis* 2013; 67(3): 159–173.

8. Morris, CE., Sands, DC., Vinatzer, BA., et al. The life history of the plant pathogen *Pseudomonas syringae* is linked to the water cycle. *ISME J* 2008; 2(3): 321–334.

9. Nelson, KE., Weinel, C., Paulsen, IT., et al. Complete genome sequence and comparative analysis of the metabolically versatile *Pseudomonas putida* KT2440. *Environ Microbiol* 2002; 4(12): 799–808.

10. Bagdasarian, M., Lurz, R., Rückert, B., et al. Specific-purpose plasmid cloning vectors. II. Broad host range, high copy number, RSF1010-derived vectors, and a host-vector system for gene cloning in *Pseudomonas*. *Gene* 1981; 16(1–3): 237–247.

11. Regenhardt, D., Heuer, H., Heim, S., et al. Pedigree and taxonomic credentials of *Pseudomonas putida* strain KT2440. *Environ Microbiol* 2002; 4(12): 912–915.

12. Timmis, KN. *Pseudomonas putida*: A cosmopolitan opportunist par excellence. *Environ Microbiol* 2002; 4(12): 779–781.

13. Nakazawa, T. Travels of a *Pseudomonas*, from Japan around the world. *Environ Microbiol* 2002; 4(12): 782–786.

14. Williams, PA., Murray, K. Metabolism of benzoate and the methylbenzoates by *Pseudomonas putida* (*arvilla*) mt-2: Evidence for the existence of a TOL plasmid. *J Bacteriol* 1974; 120(1): 416–423.

15. Keshavarz-Tohid, V., Vacheron, J., Dubost, A., et al. Genomic, phylogenetic and catabolic re-assessment of the *Pseudomonas putida* clade supports the delineation of *Pseudomonas alloputida* sp. nov., *Pseudomonas inefficax* sp. nov., *Pseudomonas persica* sp. nov., and *Pseudomonas shirazica* sp. nov. *Syst Appl Microbiol* 2019; 42(4): 468–480.

16. Kampers, LFC., Volkers, RJM., Martins dos Santos, VAP. *Pseudomonas putida* KT2440 is HV 1 certified, not GRAS. *Microb Biotechnol* 2019; 12(5): 845–848.

17. Belda, E., van Heck, RGA., José Lopez-Sanchez, M., et al. The revisited genome of *Pseudomonas putida* KT2440 enlightens its value as a robust metabolic chassis. *Environ Microbiol* 2016; 18(10): 3403–3424.

18. Lu, J., Tappel, RC., Nomura, CT. Mini-review: Biosynthesis of poly(hydroxyalkanoates). *Polym Rev* 2009; 49(3): 226–248.

19. Rai, R., Keshavarz, T., Roether, JA., et al. Medium chain length polyhydroxyalkanoates, promising new biomedical materials for the future. *Mat Sci Eng R* 2011; 72(3): 29–47.

20. Kniewel, R., Revelles Lopez, O., Prieto, MA. Biogenesis of medium-chain-length polyhydroxyalkanoates. In: Geiger O. (ed.), *Biogenesis of Fatty Acids, Lipids and Membranes*. Cham: Springer International Publishing, 2019, pp. 1–25.

21. Singh, AK., Mallick, N. SCL-LCL-PHA copolymer production by a local isolate, *Pseudomonas aeruginosa* MTCC 7925. *Biotechnol J* 2009; 4(5): 703–711.

22. Manso Cobos, I., Ibáñez García, MI., de la Peña Moreno, F., et al. *Pseudomonas pseudoalcaligenes* CECT5344, a cyanide-degrading bacterium with by-product (polyhydroxyalkanoates) formation capacity. *Microb Cell Fact* 2015; 14(1): 77.

23. Ayaka, H., Yuko, Y., Saki, G., et al. Biosynthesis of polyhydroxyalkanoate from steamed soybean wastewater by a recombinant strain of *Pseudomonas* sp. 61-3. *Bioengineering* 2017; 4(3): 68.

24. Anjum, A., Zuber, M., Zia, KM., et al. Microbial production of polyhydroxyalkanoates (PHAs) and its copolymers: A review of recent advancements. *Int J Biol Macromol* 2016; 89: 161–174.

25. Valappil, SP., Misra, SK., Boccaccini, AR., et al. Biomedical applications of polyhydroxyalkanoates, an overview of animal testing and *in vivo* responses. *Expert Rev Med Devices* 2006; 3(6): 853–868.

26. Nikel, PI., Chavarría, M., Danchin, A., et al. From dirt to industrial applications: *Pseudomonas putida* as a Synthetic Biology chassis for hosting harsh biochemical reactions. *Curr Opin Chem Biol* 2016; 34: 20–29

27. Silva-Rocha, R., Martínez-García, E., Calles, B., et al. The Standard European Vector Architecture (SEVA): A coherent platform for the analysis and deployment of complex prokaryotic phenotypes. *Nucleic Acids Res* 2013; 41(D1): D666–D675.

28. Martínez-García, E., Aparicio, T., Goñi-Moreno, A., et al. SEVA 2.0: An update of the Standard European Vector Architecture for de-/re-construction of bacterial functionalities. *Nucleic Acids Res* 2015; 43(D1): D1183–D1189.

29. Zobel, S., Benedetti, I., Eisenbach, L., et al. Tn7-based device for calibrated heterologous gene expression in *Pseudomonas putida*. *ACS Synth Biol* 2015; 4(12): 1341–1351.

30. Salvachúa, D., Rydzak, T., Auwae, R., et al. Metabolic engineering of *Pseudomonas putida* for increased polyhydroxyalkanoate production from lignin. *Microb Biotechnol* 2020; 13(1): 290–298.

31. Nogales, J., Palsson, BØ., Thiele, I. A genome-scale metabolic reconstruction of *Pseudomonas putida* KT2440: iJN746 as a cell factory. *BMC Syst Biol* 2008; 2(1): 79.

32. Puchałka, J., Oberhardt, MA., Godinho, M., et al. Genome-scale reconstruction and analysis of the *Pseudomonas putida* KT2440 metabolic network facilitates applications in biotechnology. *PLoS Comput Biol* 2008; 4(10): e1000210.

33. Sohn, SB., Kim, TY., Park, JM., et al. *In silico* genome-scale metabolic analysis of *Pseudomonas putida* KT2440 for polyhydroxyalkanoate synthesis, degradation of aromatics and anaerobic survival. *Biotechnol J* 2010; 5(7): 739–750.

34. Oberhardt, MA., Puchałka, J., Martins dos Santos, VAP., et al. Reconciliation of genome-scale metabolic reconstructions for comparative systems analysis. *PLoS Comput Biol* 2011; 7(3): e1001116.

35. Yuan, Q., Huang, T., Li, P., et al. Pathway-consensus approach to metabolic network reconstruction for *Pseudomonas putida* KT2440 by systematic comparison of published models. *PLoS One* 2017; 12(1): e0169437.

36. Nogales, J., Mueller, J., Gudmundsson, S., et al. High-quality genome-scale metabolic modelling of *Pseudomonas putida* highlights its broad metabolic capabilities. *Environ Microbiol* 2020; 22(1): 255–269.

37. Rehm, BHA. Polyester synthases: Natural catalysts for plastics. *Biochem J* 2003; 376(1): 15–33.
38. Zou, H., Shi, M., Zhang, T., et al. Natural and engineered polyhydroxyalkanoate (PHA) synthase: Key enzyme in biopolyester production. *Appl Microbiol Biotechnol* 2017; 101(20): 7417–7426.
39. Arias, S., Sandoval, A., Arcos, M., et al. Poly-3-hydroxyalkanoate synthases from *Pseudomonas putida* U: Substrate specificity and ultrastructural studies. *Microbial Biotechnol* 2008; 1(2): 170–176.
40. Obeso, JI., Maestro, B., Sanz, JM., et al. The loss of function of PhaC1 is a survival mechanism that counteracts the stress caused by the overproduction of poly-3-hydroxy-alkanoates in *Pseudomonas putida* Δ *fadBA*: PhaC1 is an essential protein in the biosynthesis of PHAs. *Environ Microbiol* 2015; 17(9): 3182–3194.
41. Chek, MF., Kim, S-Y., Mori, T., et al. Structure of polyhydroxyalkanoate (PHA) synthase PhaC from *Chromobacterium* sp. USM2, producing biodegradable plastics. *Sci Rep* 2017; 7(1): 5312.
42. Kim, J., Kim, Y-J., Choi, SY., et al. Crystal structure of *Ralstonia eutropha* polyhydroxyalkanoate synthase C-terminal domain and reaction mechanisms. *Biotechnol J* 2017; 12(1): 1600648.
43. Wittenborn, EC., Jost, M., Wei, Y., et al. Structure of the catalytic domain of the class I polyhydroxybutyrate synthase from *Cupriavidus necator*. *J Biol Chem* 2016; 291(48): 25264–25277.
44. Handrick, R., Reinhardt, S., Kimmig, P., et al. The 'Intracellular' poly(3-hydroxybutyrate) (PHB) depolymerase of *Rhodospirillum rubrum* is a periplasm-located protein with specificity for native PHB and with structural similarity to extracellular PHB depolymerases. *J Bacteriol* 2004; 186(21): 7243–7253.
45. Pohlmann, A., Fricke, WF., Reinecke, F., et al. Genome sequence of the bioplastic-producing "Knallgas" bacterium *Ralstonia eutropha* H16. *Nat Biotechnol* 2006; 24(10): 1257–1262.
46. Huisman, GW., Wonink, E., Meima, R., et al. Metabolism of poly(3-hydroxyalkanoates) (PHAs) by *Pseudomonas oleovorans*. Identification and sequences of genes and function of the encoded proteins in the synthesis and degradation of PHA. *J Biol Chem* 1991; 266(4): 2191–2198.
47. García, B., Olivera, ER., Miñambres, B., et al. Novel biodegradable aromatic plastics from a bacterial source: Genetic and biochemical studies on a route of the phenylacetyl-CoA catabolon. *J Biol Chem* 1999; 274(41): 29228–29241.
48. De Eugenio, LI., García, P., Luengo, JM., et al. Biochemical evidence that *phaZ* gene encodes a specific intracellular medium chain length polyhydroxyalkanoate depolymerase in *Pseudomonas putida* KT2442: Characterization of a paradigmatic enzyme. *J Biol Chem* 2007; 282(7): 4951–4962.
49. Prieto, MA., De Eugenio, LI., Galán, B., et al. Synthesis and degradation of poly-hydroxyalkanoates. In: *Pseudomonas*. Dordrecht: Springer Netherlands, 2007, pp. 397–428.
50. De Eugenio, LI., Escapa, IF., Morales, V., et al. The turnover of medium-chain-length polyhydroxyalkanoates in *Pseudomonas putida* KT2442 and the fundamental role of PhaZ depolymerase for the metabolic balance. *Environ Microbiol* 2010; 12(1): 207–221.
51. de Eugenio, LI., García, JL., García, P., et al. Comparative analysis of the physiological and structural properties of a medium chain length polyhydroxyalkanoate depolymerase from *Pseudomonas putida* KT2442. *Eng Life Sci* 2008; 8(3): 260–267.
52. Prieto, MA., Bühler, B., Jung, K., et al. PhaF, a polyhydroxyalkanoate-granule-associated protein of *Pseudomonas oleovorans* GPo1 involved in the regulatory expression system for *pha* genes. *J Bacteriol* 1999; 181(3): 858–868.

53. Galán, B., Dinjaski, N., Maestro, B., et al. Nucleoid-associated PhaF phasin drives intracellular location and segregation of polyhydroxyalkanoate granules in *Pseudomonas putida* KT2442. *Mol Microbiol* 2011; 79(2): 402–418.

54. Tarazona, NA., Machatschek, R., Schulz, B., et al. Molecular insights into the physical adsorption of amphiphilic protein PhaF onto copolyester surfaces. *Biomacromolecules* 2019; 20(9): 3242–3252.

55. Mato, A., Tarazona, NA., Hidalgo, A., et al. Interfacial activity of phasin PhaF from *Pseudomonas putida* KT2440 at hydrophobic–hydrophilic biointerfaces. *Langmuir* 2019; 35(3): 678–686.

56. Maestro, B., Galán, B., Alfonso, C., et al. A new family of intrinsically disordered proteins: Structural characterization of the major phasin PhaF from *Pseudomonas putida* KT2440. *PLoS One* 2013; 8(2): e56904.

57. Tarazona, NA., Maestro, B., Revelles, O., et al. Role of leucine zipper-like motifs in the oligomerization of *Pseudomonas putida* phasins. *Biochim Biophys Acta Gen Subj* 2019; 1863(2): 362–370.

58. Durner, R., Witholt, B., Egli, T. Accumulation of poly[(R)-3-hydroxyalkanoates] in *Pseudomonas oleovorans* during growth with octanoate in continuous culture at different dilution rates. *Appl Environ Microbiol* 2000; 66(8): 3408–3414.

59. Pramanik, A., Pawar, S., Antonian, E., et al. Five different enzymatic activities are associated with the multienzyme complex of fatty acid oxidation from *Escherichia coli*. *J Bacteriol* 1979; 137(1): 469–473.

60. Wang, Q., Nomura, CT. Monitoring differences in gene expression levels and polyhydroxyalkanoate (PHA) production in *Pseudomonas putida* KT2440 grown on different carbon sources. *J Biosci Bioeng* 2010; 110(6): 653–659.

61. Olivera, ER., Carnicero, D., García, B., et al. Two different pathways are involved in the beta-oxidation of n-alkanoic and n-phenylalkanoic acids in *Pseudomonas putida* U: Genetic studies and biotechnological applications. *Mol Microbiol* 2001; 39(4): 863–874.

62. Ruth, K., de Roo, G., Egli, T., et al. Identification of two acyl-CoA synthetases from *Pseudomonas putida* GPo1: One is located at the surface of polyhydroxyalkanoates granules. *Biomacromolecules* 2008; 9(6): 1652–1659.

63. Hume, AR., Nikodinovic-Runic, J., O'Connor, KE. FadD from *Pseudomonas putida* CA-3 is a true long-chain fatty acyl coenzyme A synthetase that activates phenylalkanoic and alkanoic acids. *J Bacteriol* 2009; 191(24): 7554–7565.

64. Ghisla, S., Thorpe, C. Acyl-CoA dehydrogenases. A mechanistic overview. *Eur J Biochem* 2004; 271(3): 494–508.

65. Lu, X., Zhang, J., Wu, Q., et al. Enhanced production of poly(3-hydroxybutyrate-*co*-3-hydroxyhexanoate) via manipulating the fatty acid beta-oxidation pathway in *E. coli*. *FEMS Microbiol Lett* 2003; 221(1): 97–101.

66. McMahon, B., Gallagher, ME., Mayhew, SG. The protein coded by the PP2216 gene of *Pseudomonas putida* KT2440 is an acyl-CoA dehydrogenase that oxidises only short-chain aliphatic substrates. *FEMS Microbiol Lett* 2005; 250(1): 121–127.

67. McMahon, B., Mayhew, SG. Identification and properties of an inducible phenylacyl-CoA dehydrogenase in *Pseudomonas putida* KT2440. *FEMS Microbiol Lett* 2007; 273(1): 50–57.

68. Guzik, MW., Narancic, T., Ilic-Tomic, T., et al. Identification and characterization of an acyl-CoA dehydrogenase from *Pseudomonas putida* KT2440 that shows preference towards medium to long chain length fatty acids. *Microbiology* 2014; 160(Pt 8): 1760–1771.

69. Ouyang, S-P., Luo, RC., Chen, S-S., et al. Production of polyhydroxyalkanoates with high 3-hydroxydodecanoate monomer content by *fadB* and *fadA* knockout mutant of *Pseudomonas putida* KT2442. *Biomacromolecules* 2007; 8(8): 2504–2511.

70. Olivera, ER., Carnicero, D., Jodra, R., et al. Genetically engineered *Pseudomonas*: A factory of new bioplastics with broad applications. *Environ Microbiol* 2001; 3(10): 612–618.

71. Kang, Y., Nguyen, DT., Son, MS., et al. The *Pseudomonas aeruginosa* PsrA responds to long-chain fatty acid signals to regulate the *fadBA*5-oxidation operon. *Microbiology* 2008; 154(6): 1584–1598.

72. Fonseca, P., de la Peña, F., Prieto, MA. A role for the regulator PsrA in the polyhydroxyalkanoate metabolism of *Pseudomonas putida* KT2440. *Int J Biol Macromol* 2014; 71: 14–20.

73. Ren, Q., Kessler, B., van der Leij, F., et al. Mutants of *Pseudomonas putida* affected in poly-3-hydroxyalkanoate synthesis. *Appl Microbiol Biotechnol* 1998; 49(6): 743–750.

74. Vo, MT., Lee, K-W., Jung, Y-M., et al. Comparative effect of overexpressed *phaJ* and *fabG* genes supplementing *(R)*-3-hydroxyalkanoate monomer units on biosynthesis of *mcl*-polyhydroxyalkanoate in *Pseudomonas putida* KCTC1639. *J Biosci Bioeng* 2008; 106(1): 95–98.

75. Tsuge, T., Taguchi, K., Seiichi., et al. Molecular characterization and properties of *(R)*-specific enoyl-CoA hydratases from *Pseudomonas aeruginosa*: Metabolic tools for synthesis of polyhydroxyalkanoates via fatty acid ß-oxidation. *Int J Biol Macromol* 2003; 31(4–5): 195–205.

76. Fiedler, S., Steinbüchel, A., Rehm, B. The role of the fatty acid β-oxidation multi-enzyme complex from *Pseudomonas oleovorans* in polyhydroxyalkanoate biosynthesis: Molecular characterization of the *fadBA* operon from *P. oleovorans* and of the enoyl-CoA hydratase genes *phaJ* from *P. oleovorans* and *Pseudomonas putida*. *Arch Microbiol* 2002; 178(2): 149–160.

77. Sato, S., Kanazawa, H., Tsuge, T. Expression and characterization of (R)-specific enoyl coenzyme A hydratases making a channeling route to polyhydroxyalkanoate biosynthesis in *Pseudomonas putida*. *Appl Microbiol Biotechnol* 2011; 90(3): 951–959.

78. Nomura, CT., Taguchi, K., Gan, Z., et al. Expression of 3-ketoacyl-acyl carrier protein reductase (*fabG*) genes enhances production of polyhydroxyalkanoate copolymer from glucose in recombinant *Escherichia coli* JM109. *Appl Environ Microbiol* 2005; 71(8): 4297–4306.

79. Taguchi, K., Aoyagi, Y., Matsusaki, H., et al. Co-expression of 3-ketoacyl-ACP reductase and polyhydroxyalkanoate synthase genes induces PHA production in *Escherichia coli* HB101 strain. *FEMS Microbiol Lett* 1999; 176(1): 183–190.

80. Ren, Q., Sierro, N., Witholt, B., et al. FabG, an NADPH-dependent 3-ketoacyl reductase of *Pseudomonas aeruginosa*, provides precursors for medium-chain-length poly-3-hydroxyalkanoate biosynthesis in *Escherichia coli*. *J Bacteriol* 2000; 182(10): 2978–2981.

81. Best, EA., Knauf, VC. Organization and nucleotide sequences of the genes encoding the biotin carboxyl carrier protein and biotin carboxylase protein of *Pseudomonas aeruginosa* acetyl coenzyme A carboxylase. *J Bacteriol* 1993; 175(21): 6881–6889.

82. Kutchma, AJ., Hoang, TT., Schweizer, HP. Characterization of a *Pseudomonas aeruginosa* fatty acid biosynthetic gene cluster: Purification of acyl carrier protein (ACP) and malonyl-coenzyme A:ACP transacylase (FabD). *J Bacteriol* 1999; 181(17): 5498–5504.

83. Campbell, JW., Cronan, JE. Bacterial fatty acid biosynthesis: Targets for antibacterial drug discovery. *Annu Rev Microbiol* 2001; 55(1): 305–332.

84. Schweizer, HP. Fatty acid biosynthesis and biologically significant acyl transfer reactions in *Pseudomonads*. In: Ramos J-L (ed.), *Pseudomonas*. Boston, MA: Springer US, 2012, pp. 83–109.

85. Cronan, Jr., JE., Rock, CO. Biosynthesis of membrane lipids. *EcoSal Plus* 2008; 3(1).

86. Sullivan, SA., Schweizer, HP., Hoang, TT., et al. β-Ketoacyl acyl carrier protein reductase (FabG) activity of the fatty acid biosynthetic pathway is a determining factor of 3-oxo-homoserine lactone acyl chain lengths. *Microbiology* 2002; 148(12): 3849–3856.

87. Heath, RJ., Rock, CO. Roles of the FabA and FabZ β-hydroxyacyl-acyl carrier protein dehydratases in *Escherichia coli* fatty acid biosynthesis. *J Biol Chem* 1996; 271(44): 27795–27801.

88. Hoang, TT., Schweizer, HP. Characterization of *Pseudomonas aeruginosa* enoyl-acyl carrier protein reductase (FabI): A target for the antimicrobial triclosan and its role in acylated homoserine lactone synthesis. *J Bacteriol* 1999; 181(17): 5489–5497.

89. Rehm, BH., Krüger, N., Steinbüchel, A. A new metabolic link between fatty acid de novo synthesis and polyhydroxyalkanoic acid synthesis. The *phaG* gene from *Pseudomonas putida* KT2440 encodes a 3-hydroxyacyl-acyl carrier protein-coenzyme a transferase. *J Biol Chem* 1998; 273(37): 24044–24051.

90. Rehm, BHA., Krüger, N., Steinbüchel, A. A new metabolic link between fatty acid *de Novo* synthesis and polyhydroxyalkanoic acid synthesis: The PhaG gene from *Pseudomonas putida* KT2440 encodes a 3-hydroxyacyl-acyl carrier protein-protein-coenzyme A transferase. *J Biol Chem* 1998; 273(37): 24044–24051.

91. Fiedler, S., Steinbuchel, A., Rehm, BHA. PhaG-mediated synthesis of poly(3-hydroxy-alkanoates) consisting of medium-chain-length constituents from nonrelated carbon sources in recombinant *Pseudomonas fragi*. *Appl Environ Microbiol* 2000; 66(5): 2117–2124.

92. Hoffmann, N., Steinbüchel, A., Rehm, BH. Homologous functional expression of cryptic *phaG* from *Pseudomonas oleovorans* establishes the transacylase-mediated polyhydroxyalkanoate biosynthetic pathway. *Appl Microbiol Biotechnol* 2000; 54(5): 665–670.

93. Hoffmann, N., Rehm, BHA. Regulation of polyhydroxyalkanoate biosynthesis in *Pseudomonas putida* and *Pseudomonas aeruginosa*. *FEMS Microbiol Lett* 2004; 237(1): 1–7.

94. Tobin, KM., O'Leary, ND., Dobson, ADW., et al. Effect of heterologous expression of *phaG* [(R)-3-hydroxyacyl-ACP-CoA transferase] on polyhydroxyalkanoate accumulation from the aromatic hydrocarbon phenylacetic acid in *Pseudomonas* species. *FEMS Microbiol Lett* 2007; 268(1): 9–15.

95. Wang, Q., Tappel, RC., Zhu, C., et al. Development of a new strategy for production of medium-chain-length polyhydroxyalkanoates by recombinant *Escherichia coli* via inexpensive non-fatty acid feedstocks. *Appl Environ Microbiol* 2012; 78(2): 519–527.

96. Chen, G-Q. A microbial polyhydroxyalkanoates (PHA) based bio- and materials industry. *Chem Soc Rev* 2009; 38(8): 2434–2446.

97. Rehm, BHA. Bacterial polymers: Biosynthesis, modifications and applications. *Nat Rev Microbiol* 2010; 8(8): 578–592.

98. Gao, X., Chen, J-C., Wu, Q., et al. Polyhydroxyalkanoates as a source of chemicals, polymers, and biofuels. *Curr Opin Biotechnol* 2011; 22(6): 768–774.

99. Li, Z., Yang, J., Loh, XJ. Polyhydroxyalkanoates: Opening doors for a sustainable future. *NPG Asia Mater* 2016; 8(4): e265–e265.

100. De Eugenio, LI., Galán, B., Escapa, IF., et al. The PhaD regulator controls the simultaneous expression of the *pha* genes involved in polyhydroxyalkanoate metabolism and turnover in *Pseudomonas putida* KT2442. *Environ Microbiol* 2010; 12(6): 1591–1603.

101. Sandoval, Á., Arias-Barrau, E., Arcos, M., et al. Genetic and ultrastructural analysis of different mutants of *Pseudomonas putida* affected in the poly-3-hydroxy-n-alkanoate gene cluster. *Environ Microbiol* 2007; 9(3): 737–751.

102. La Rosa, R., de la Peña, F., Prieto, MA., et al. The Crc protein inhibits the production of polyhydroxyalkanoates in *Pseudomonas putida* under balanced carbon/nitrogen growth conditions: Crc control of PHA production in *P. putida*. *Environ Microbiol* 2014; 16(1): 278–290.

103. Ryan, WJ., O'Leary, ND., O'Mahony, M., et al. GacS-dependent regulation of polyhydroxyalkanoate synthesis in *Pseudomonas putida* CA-3. *Appl Environ Microbiol* 2013; 79(6): 1795–1802.

104. Escapa, IF., del Cerro, C., García, JL., et al. The role of GlpR repressor in *Pseudomonas putida* KT2440 growth and PHA production from glycerol: *Pseudomonas putida* growth and PHA production from glycerol. *Environ Microbiol* 2013; 15(1): 93–110.

105. Ren, Q., De Roo, G., Ruth, K., et al. Simultaneous accumulation and degradation of polyhydroxyalkanoates: Futile cycle or clever regulation? *Biomacromolecules* 2009; 10(4): 916–922.

106. Escapa, IF., García, JL., Bühler, B., et al. The polyhydroxyalkanoate metabolism controls carbon and energy spillage in *Pseudomonas putida*. *Environ Microbiol* 2012; 14(4): 1049–1063.

107. Schuetz, R., Zamboni, N., Zampieri, M., et al. Multidimensional optimality of microbial metabolism. *Science* 2012; 336(6081): 601–604.

108. Kitano, H. Biological robustness. *Nat Rev Genet* 2004; 5(11): 826–837.

109. Kitano, H. Towards a theory of biological robustness. *Mol Syst Biol* 2007; 3:137.

110. Nogales, J., Gudmundsson, S., Duque, E., et al. Expanding the computable reactome in *Pseudomonas putida* reveals metabolic cycles providing robustness. *bioRxiv* 2017. doi:10.1101/139121.

111. Prieto, A., Escapa, IF., Martínez, V., et al. A holistic view of polyhydroxyalkanoate metabolism in *Pseudomonas putida*. *Environ Microbiol* 2016; 18(2): 341–357.

5 Production of Polyhydroxyalkanoates by *Paraburkholderia* and *Burkholderia* Species

A Journey from the Genes through Metabolic Routes to Their Biotechnological Applications

Natalia Alvarez-Santullano, Pamela Villegas,
Mario Sepúlveda, Ariel Vilchez, Raúl Donoso,
Danilo Pérez-Pantoja, Rodrigo Navia,
Francisca Acevedo, and Michael Seeger

CONTENTS

5.1 INTRODUCTION

Species of *Paraburkholderia* and *Burkholderia* genera are versatile bacteria that live in a wide range of ecological niches, including soils, plants, and the human respiratory tract [1,2]. *Burkholderia* and *Paraburkholderia* genera contain more than 120 and 75 species, respectively. The interactions of *Burkholderia* and *Paraburkholderia* strains with numerous hosts, such as plants, animals, invertebrates, and fungi, have been reported [3–5].

The *Burkholderia* genus was proposed in 1992 to describe the former rRNA group II Pseudomonads, based on the 16 rRNA sequences, DNA-DNA homology values, cellular lipid and fatty acid composition, and other phenotypic characteristics [6]. The *Burkholderia* genus was named in recognition of the research of Walter Burkholder. He first described the Gram-negative *Burkholderia cepacia* (formerly *Pseudomonas cepacia*) as the phytopathogen responsible for the bacterial rot of onions [7]. Since then, the taxonomy of the *Burkholderia* genus has been subject to modification. The species type *B. cepacia* has changed from a single species to a complex of 17 related species, currently called *B. cepacia* complex (BCC) [8]. In 2014, the division of the *Burkholderia* genus into two main groups was proposed [2,9]. Then, the *Burkholderia* genus includes only clinically relevant and phytopathogenic species, while the environmental species belonged to the newly proposed *Paraburkholderia* genus [2,9]. However, some *Paraburkholderia* strains have been isolated from patients [10]. *Paraburkholderia* species are straight or slightly curved rods with one or more polar flagella. The G + C content for *Paraburkholderia* species ranges from 58.9 to 65.0%, whereas for *Burkholderia*, it ranges from 65.7 to 68.5% [10]. From 16 new species of the *Burkholderia* genus reported after 2014, only 2 species belong to the *Burkholderia* genus, whereas 11 species belong to the *Paraburkholderia* genus [10]. Dobritsa and Samadpour [10] proposed reclassifying the remaining 3 new species into the new *Caballeronia* genus, including 12 species belonging to the *Burkholderia* and *Paraburkholderia* genera. The *Caballeronia* genus was named after Jesus Caballero-Mellado, who pioneered studies on plant-associated bacteria.

The ecological diversity of the *Paraburkholderia* and *Burkholderia* genera is associated with their large multi-replicon genomes (6–10 Mb), providing them with a metabolic versatility that allows, for example, the degradation of a wide range of aromatic compounds and xenobiotics [11–14]. Their genomes contain an array of insertion sequences that promote genomic plasticity and general adaptability [13,15]. Members of *Paraburkholderia* and *Burkholderia* genera have been studied for various biotechnological applications, including bioremediation of pollutants, plant-growth promotion, biocontrol of plant diseases, and synthesis of diverse compounds such as polyhydroxyalkanoates (PHA) and rhamnolipids [16–18]. Xenobiotic and aromatic-degrading strains have been applied in bioremediation processes [19]. *Paraburkholderia xenovorans* LB400 is capable of degrading a wide range of polychlorobiphenyls (PCBs) and diverse aromatic compounds [20–24]. *Paraburkholderia phenoliruptrix* AC1109 degrades the herbicide 2,4,5-trichlorophenoxyacetic acid. *Burkholderia vietnamiensis* G4, which belongs to the BCC, is capable of degrading trichloroethylene, benzene, phenol, toluene, naphthalene, and chloroform [12].

Paraburkholderia and *Burkholderia* strains that promote plant growth and protect plants from pests have been reported. *Burkholderia ambifaria* and *Paraburkholderia caribensis* are diazotrophic species that promote the growth of amaranth grain. *Burkholderia rinojensis* exhibits biocontrol activity against arthropod pests [25]. *Paraburkholderia tropica* converts insoluble mineral phosphorus into an available form for plant uptake. *Paraburkholderia phytofirmans* strain PsJN degrades auxin and protects *Arabidopsis thaliana* against the phytopathogen *Pseudomonas syringae* [17,26]. Some species of *Paraburkholderia* and *Burkholderia* may be beneficial for their hosts due to their capabilities to fix nitrogen, produce plant hormones or siderophores, and decrease ethylene levels. These species could be used in agriculture to promote plant growth and biocontrol of plant diseases [1,27]. However, specific *Burkholderia* strains are pathogens for both plants and animals [28]. The BCC includes human, animal, and plant pathogens, isolated from a variety of natural habitats (i.e., plant rhizosphere, soil, river water) and urban environments (i.e., playground). *Burkholderia pseudomallei* and *Burkholderia mallei,* which belong to BCC, are the agents responsible for melioidosis disease, which is a potentially lethal septic infection, and glanders disease, respectively [29]. Conversely, several species of BCC are beneficial to the natural environment. For example, *Burkholderia cepacia* AMMDR1 protects crop plants against fungal diseases such as root rot caused by *Aphanomyces euteiches* [30].

The synthesis of PHA by *Paraburkholderia* and *Burkholderia* species has high biotechnological potential. Due to the capabilities of *Paraburkholderia* and *Burkholderia* to synthesize diverse PHA and use low-cost substrates as carbon sources, specific strains have been applied for PHA production. Most species of *Paraburkholderia* and *Burkholderia* genera have class I PHA synthases. Strains belonging to these genera generally synthesize short-chain-length-PHA (*scl*-PHA), including homopolymers and copolymers. *Paraburkholderia sacchari* strain LMG 19450, *P. xenovorans* LB400, *B. cepacia* ATCC 17759, and *B. thailandensis* E264 are relevant strains for PHA production [16,31,32]. The aims of this study were to analyze the genes and metabolic pathways of *Paraburkholderia* and *Burkholderia* species for the synthesis of PHA from diverse substrates and discuss the biomedical and biotechnological applications of these biopolymers.

5.2 PHA SYNTHASES

The PHA synthase, which is encoded by the *phaC* gene, catalyzes the last step of the PHA anabolic pathway and is the key enzyme that determines the PHA structure. PHA synthases are classified into four groups based on their substrate specificity, primary structure, and the length of the PHA synthesized [33,34]. Class I and class II are the most studied PHA synthases and the more commonly present in bacterial species. Class I PHA synthase is widespread in *β-proteobacteria* and consists of a single type of subunit [16,33]. Class I PHA synthases can polymerize PHA monomers of 3–5 carbon chain lengths forming *scl*-PHA and has been extensively studied in the model PHA-producing strain *Cupriavidus necator* H16 [16,33,35]. Class II PHA synthases possess two types of subunit (PhaC1 or PhaC2) and synthesize medium-chain-length PHA (*mcl*-PHA) [33,36–38]. Class II PHA synthases have been widely described in *Pseudomonas* species. Class III PHA synthase is a heterodimer that requires the two subunits, PhaC and PhaE, and synthesizes *scl*-PHA (~40 kDa) [32]. Class III PHA synthases have been reported in the proteobacterium *Allochromatium vinosum* and *Haloarchaea* (e.g., *Haloarcula marismortui*) (Figure 5.1). Class IV PHA synthases are heterodimers of PhaC and PhaR (~20 kDa) that generally synthesize *scl*-PHA [33,39]. Class IV PHA synthases have been reported in *Bacillus* (e.g., *Bacillus megaterium*, *Bacillus cereus*). In *Paraburkholderia* and *Burkholderia* genera, a class I PHA synthase is generally present, whereas only a few strains possess class II PHA synthases (see Figure 5.1 and Figure 5.2).

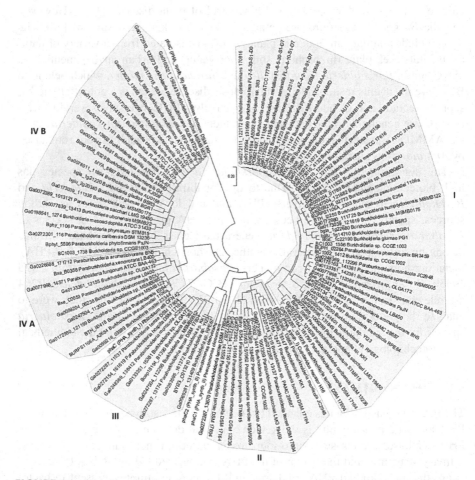

FIGURE 5.1 Neighbor-joining phylogenetic tree for PHA synthases from *Paraburkholderia* and *Burkholderia* genera. The phylogenetic tree for PHA synthases homologous to PhaC from *Cupriavidus necator* H16 was obtained by the neighbor-joining method using MEGA-X software based on sequence alignments calculated by CLUSTALW. PhaC synthases were retrieved from the Integrated Microbial Genomes and Microbiomes database considering similarity by the chance expectation values and scores of BLAST probing. Confidence in phylogenetic inference was assessed using non-parametric bootstrap resampling.

```
Cupriavidus necator H16            ---------MATGKGA-AASTQEGKSQPFKVTPGPFDPATWLEWSRQWQGTEGNG------   45
Paraburkholderia xenovorans LB400  MQQLFDAMMGAWRSLGTPPGSNGMPF-----PVFQMFQTGLFQMFSFPGMPDFARL----   51
Paraburkholderia sacchari LMG 19450 MQQMLDAWTQAWRSAMDAAQQSGMQH-----WGQAAGQPG----AFWPGMANGAASAAPN  51
                                        : :.       ..*                          :  *  .

Cupriavidus necator H16            ---------HAAASGIPGLDALAGV-----KIAPAQLGDIQQRYMKDFSALWQAMAEGKA   91
Paraburkholderia xenovorans LB400  ----------AAGSMPAMPSFAGLNIPSAAIPSERLQKLQADYSREAMELIQQAAT-SA   99
Paraburkholderia sacchari LMG 19450 PVPQAAFNAAMHPTAMRSALGALEAFKLPSASIEPTRLQHLQAEYSRDALELLREASA-AS 110
                                         1.   .: :1 ..     *   :*.:*  *  :: *:  : :

Cupriavidus necator H16            EATGFLHDRRFAGDAWRTNLPYRFAAAFYLLNARALTELADAVEADAKTRQRIRFAISQW  151
Paraburkholderia xenovorans LB400  TKAPELKDRRFSSDAWSSAPAYAFTAANYLLNARYLQEHVDALDIEPKVRERIRFAVQQW  159
Paraburkholderia sacchari LMG 19450 PKAPELKDRRFNADAWKATPAYAFTAANYLLNARYLHEHVDAIETDAKTRERIRFAVQQW  170
                                    : *;**** .*** :   *  *;*:;****** *  *;.**;:  :  *.*;****:*;.**

Cupriavidus necator H16            VDAMSPANFLATNPEAQRLLIESGGESLRAGVRNMMEDLTRGKISQTDESAFEVGRNVAV  211
Paraburkholderia xenovorans LB400  TAAASPSNFFALNPEAQKTLLDSNGESLRQGVMNLLGDLQRGKISQTDESRFVVGENLAH  219
Paraburkholderia sacchari LMG 19450 TAAASPSNFLALNPEAQHALLESHGESLRQGVMNLLGDHSRGKISQTDESRFVVGRNLAH  230
                                    .  *.:***;*  ***:;  ;:;: ***  ** :: *; *********  * **.*;*

Cupriavidus necator H16            TEGAVVFENEYFQLLQYKPLTDKVHARFLLNVPPCINKYYILDLQPESSLVRHVVEQGHT  271
Paraburkholderia xenovorans LB400  TEGSVVFENELMQLIQYKPKRTATVREKPLLIVPPCINKFYILDLQPENSLVAHGLDSGHQ 279
Paraburkholderia sacchari LMG 19450 TEGSVVFENELFQLIQYKPKTATVHEKPLLLVPPCINKYYILDLQPENSLVAYALEGQGQ  290
                                    ***;****** :*;:****  .*;  *;.  ****;**** .*** ; ::.*;

                                                                        Cys319
Cupriavidus necator H16            VFLVSWRNPDASMAGSTWDDYIEHAAIRAIEVARDISGQDKINVLGFCVGGTIVSTALAV  331
Paraburkholderia xenovorans LB400  VFLISWRNADQSIAHKTWDDYIGEGVLTAIETVSKISGREQINTLGFCVGGTMLATALAV  339
Paraburkholderia sacchari LMG 19450 VFLVSWHNGDASIAHKGWDDYIEEGVLTAIETTRSISGREQINTLGFCVGGTLATALAV  350
                                    ***;**;* *  *;:; *****  ...;  **... :**;**** ******:;; ****

Cupriavidus necator H16            LAARGEHPAASVTLLTTLLDFADTGILDVFVDEGHVQLREATLGGGAGAPCALLRGLELA  391
Paraburkholderia xenovorans LB400  AAARGQHPAASMTLLTAMLDFSDTGVLDVFVDDAHVKMKEQTIGGKNGTPPGLMRGIEFA  399
Paraburkholderia sacchari LMG 19450 AAARGEHPAASMTLLTTMLDFSDTGVLDVFIDEAHVQMKEQTIGGKNGGTPGLMRGVEFA  410
                                    ****;*****;****:.;:***;***;****;* ***; ;:  ** ::* :

Cupriavidus necator H16            NTFSFLRPNDLVWNYVVVDNYLKGNTPVPFDLLFWNGDATNLPGPWYCWYLRHTYLQNELK 451
Paraburkholderia xenovorans LB400  NTFSFLRPNDLVWNYVVVDNYLKGRTPVPFDLLYWNSDSTSLPGPMYVWYLRNTYLENRLK 459
Paraburkholderia sacchari LMG 19450 NTFSFLRPNDLVWNYVVVDNYLKGRTPAAFDLLYWNSDSTSLPGPMYAWYLRNTYLEHKLK 470
                                    *********************.**  ***:*;**.:*.;;* ;**** ;*****;*** *:;

                                         Asp480                              His508
Cupriavidus necator H16            VPGKLTVCGVPVDLASIDVPTYIYGSREDHIVPWTAAYASTALLANKLRFVLGASGHIAG  511
Paraburkholderia xenovorans LB400  EPGAVTTCGEPVDLSKIDVPTFIYGSREDHIVPWQTAYASVPLLSGPLKFVLGASGHIAG  519
Paraburkholderia sacchari LMG 19450 EPGALTVCGEEVDLSRIDVPTFIYGSREDHIVPWRSAYASVPLLSGPLKFVLGASGHIAG  530
                                    ** ;*.**  ***;.** *****;******;****;****  ;****. ;;******** *

Cupriavidus necator H16            VINPPAKNKRSHWTN----DALPESPQQWLAGAIEHHGSWWPDWTAWLAGQAGAKRAAPA  567
Paraburkholderia xenovorans LB400  VINPPAKKKRNFWMLEGDVKTLPENPEEWLDQAAEVPGSWWPEWTTWLDQYGGRKVKPRA  579
Paraburkholderia sacchari LMG 19450 VINPASKNKRSYNSVETDAKHLPGVADENFADAEEKPGSWWPTWIEWLDQFGGKKVKPRA  590
                                    **** ;*;**..*   .**   ;;*; * * ***** * **  .* * *

Cupriavidus necator H16            NYGNARYRAIEPAPGRYVKAKA  589
Paraburkholderia xenovorans LB400  AAGSAEFFVIEPAPGRYVRQRE  601
Paraburkholderia sacchari LMG 19450 QPCSAEFFEIEPAPGRYVQQRD  612
                                    *.*.:  *********;; ;
```

FIGURE 5.2 PHA synthases amino acid sequence of *Cupriavidus necator*, *Paraburkholderia xenovorans*, and *Paraburkholderia sacchari*. The 14 conserved amino acid residues identified in the PHA synthases' sequences are highlighted in boxes. The lipase box-like sequences are highlighted by dotted lines. The alignment was made using MAFFT.

Class I PHA synthases have an average molecular mass of 63,000 Da, while the average molecular mass of PHA synthases of *Burkholderia* and *Paraburkholderia* species is 68,226 Da (Protein Molecular Weight Calculator, Science Gateway). Madison and Huisman [40] identified 15 fully conserved residues within 26 PHA polymerases analyzed, which are involved in PHA biosynthesis. The PHA synthases of *Burkholderia* and *Paraburkholderia* species contain 14 of these conserved residues, which represent ~2.3% of the total number of amino acids. Generally, PHA synthases contain a lipase box-like sequence, Gly-X-Cys-X-Gly, in catalytic sites, which is similar to the lipase box sequence, Gly-X-Ser-X-Gly, of some lipases [36]. The PHA synthases of *Paraburkholderia* and *Burkholderia* species (i.e., *P. sacchari*, *P. xenovorans*, *Burkholderia glumae*, *Burkholderia thailandensis*, *Burkholderia contaminans*, and *B. cepacia*) contain a lipase box-like sequence in the same position as the *C. necator* PHA synthase. The Cys contained in the lipase box-like sequence is the catalytic amino acid, bounding covalently and forming the intermediate Cys-S-3HB. The catalytic triad, C-H-D, is crucial for the activity of PHA synthases. In the PHA synthase sequence of *C. necator*, this triad is Cys_{319}, Asp_{480}, and His_{508} [34]. These amino acids are present in the PHA synthases of species of *Burkholderia* and *Paraburkholderia* species (Figure 5.2).

5.3 GENOMIC ANALYSIS OF *PHA* GENES ON *PARABURKHOLDERIA* AND *BURKHOLDERIA* SPECIES

Currently, 180 genomes belong to the *Burkholderia* and *Paraburkholderia* strains listed in the JGI database (https://img.jgi.doe.gov/index.html). Since the *phaC* gene is the key gene in PHA synthesis, we surveyed the *Burkholderia* and *Paraburkholderia* genera using the BLASTp tool from the Integrated Microbial Genomes and Microbiomes database (https://img.jgi.doe.gov/cgi-bin/m/main.cgi?section=Find Genes Blast&page=geneSearchBlast) to look for homologs. The PHA synthase from *C. necator* H16 (accession number: P23608) was used as a query to carry out the search. Only proteins displaying ≥30% amino acid identity with H16 PhaC were selected for further analysis. We obtained 346 putative homologs of PhaC and all *Burkholderia* and *Paraburkholderia* strains possessed at least one copy of the *phaC* gene in their genome. Furthermore, the most common gene organization is the *phaCABR* gene cluster that is present in 178 out of the 180 *Burkholderia* and *Paraburkholderia* genomes reviewed, with PHA synthases encoded in these clusters possessing amino acid identity around 59 to 65% with a class I PHA synthase from *C. necator* H16. The most deeply studied *Paraburkholderia* and *Burkholderia* species for PHA synthesis, such as *P. sacchari*, *P. xenovorans*, *B. thailandensis*, and *B. cenocepacia*, contain this gene organization, which is similar to the *pha* gene organization in *C. necator* strain H16 (see Figure 5.3). Only two strains do not possess this gene organization, indicating that the *phaCABR* cluster is highly conserved in these bacterial genera.

Remarkably, a phylogenetic tree constructed for PHA synthases illustrates four different groups of PhaC proteins from *Paraburkholderia* and *Burkholderia* genomes reviewed, which are more similar to PHA synthases of class I and II (Figure 5.1). The more distantly related class III PhaC from *Allochromatium vinosum* was used as a reference outgroup. The PHA synthases of strains possessing the *phaCABR* cluster were grouped in the same clade in the phylogenetic tree (Figures 5.1 and 5.3). The only strains without the archetypic *phaCABR* gene organization include the opportunistic human pathogen, *B. pseudomallei* MSHR435 [41], and *Burkholderia* sp. strain MSMB1588 isolated from soil in Australia (https://www.ncbi.nlm.nih.gov/biosample/SAMN03449293), both without functional evidence for PHA synthesis. An additional phylogenetic tree of PhaC amino acid sequences of the *Paraburkholderia* and *Burkholderia* genomes reviewed was generated, including only representative species. The 346 PhaC sequences were sorted into four groups. For every group, we selected representative strains of *Burkholderia* and *Paraburkholderia* genomes to build a PhaC sequence phylogenetic tree matching their genomic context arrangement (Figure 5.3). It should be noted that group I corresponds to the *phaCABR* cluster organization, and the encoded PhaC homologs are closely related to *C. necator* H16 PhaC (Figure 5.3).

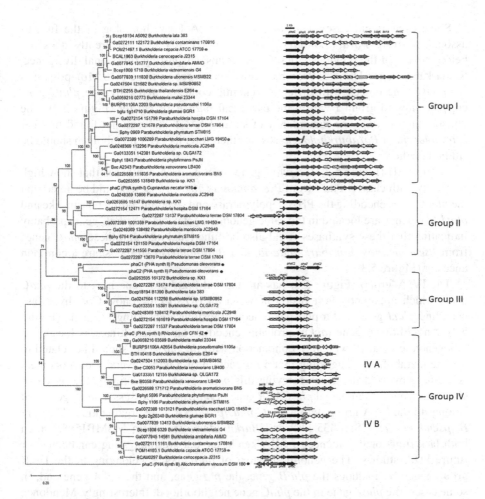

FIGURE 5.3 The *pha* gene organization in representative *Burkholderia* and *Paraburkholderia* species. Neighbor-joining phylogenetic tree for selected PhaC *Burkholderia/Paraburkholderia* homologs of *Cupriavidus necator* H16. Branches tagged with dots indicate sequences with confirmed functionality not belonging to *Burkholderia/Paraburkholderia* species that have been incorporated for additional phylogenetic comparisons. The *phaC* genes are in black, while *phaA*, *phaB*, *phaR*, and *bktB* genes are in gray. The functions of the different types of genes included in the clusters are indicated above the genes. The sizes of the genes are to scale.

Remarkably, the *phaC* belonging to group II has only a phasin-encoding *phaP* gene and/or *phaJ* gene in their genomic context. All these strains possess a second *phaC* gene copy located in the archetypic *phaCABR* cluster of group I (Figure 5.3). It has been reported that PhaJ converts the β-oxidation pathway intermediates to (*R*)-3-hydroxyacyl-CoA for *mcl*-PHA synthesis, while *phaP* encodes a phasin surface protein that covers PHA storage granules [38,42]. These data suggest that PhaC from this group may also participate in PHA synthesis, increasing the monomer diversity of these strains.

Some exceptional cases were also observed. A PhaC homolog of the fungal-associated strain *Paraburkholderia terrae* DSM17804 [43] apparently does not belong to any of the previously described groups, suggesting functional divergence. Remarkably, strain DSM17804, which contains a large genome (~10 Mb), possesses six *phaC* gene copies with different genomic contexts, and only one *phaC* gene copy belongs to group I within the canonical *phaCABR* cluster (Figure 5.3). The unusual high *phaC* gene redundancy of strain DSM 17804 was not found in other *Burkholderia* or *Paraburkholderia* strains; therefore, its PHA metabolism should be studied further.

In group III (Figure 5.3), the *phaC* genes encode PHA synthases that show high similarity with class II PhaC from *Pseudomonas oleovorans*. It should be noted that the *phaZ* genes encoding the PHA depolymerase and genes encoding phasin-like and *phaJ*-like genes are located in the same genomic context as the *phaC* gene, probably indicating that these synthases could also be functional and that these *phaC* genes from *Pseudomonas*, *Paraburkholderia*, and *Burkholderia* strains share a common ancestor (Figure 5.3).

The IV A group (Figure 5.3) has an unusual *phaJ-pta* gene next to the *phaC* gene, which apparently is a fusion between an (*R*)-specific enoyl-CoA hydratase encoding *phaJ* gene and a phosphate acetyltransferase encoding *pta* gene (Figure 5.3). An additional gene (*ackA*) probably encoding an acetate kinase is located in this genomic context, which, in conjunction with the *pta* gene, should be related to acetate metabolism, suggesting that a production of a wide diversity of monomers could have evolved in strains carrying this *phaC* gene. Interestingly, the only two strains lacking the canonical *phaCABR* gene organization, possess a *phaC* gene that belongs to the IV A group with a *phaC-phaJ/pta-ackA* gene organization. They are *B. pseudomallei* MSHR435 [40] and *Burkholderia* sp. strain MSMB1588, which both lack functional evidence of PHA synthesis, but are interesting candidates for future PHA studies. The genomic context of *phaC* genes belonging to the IV B group generally includes the *phaB* gene, the *pta* gene, and the *ackA* gene and, in some cases, the *phaZ* gene in the *phaC* gene neighborhood. Interestingly, Mendonça et al. [32] reported that *P. sacchari* LMG 19450, which has a *phaC* gene copy of the second group, synthesizes short- and medium-length chain copolymers [P(3-hydroxybutyrate-*co*-3-hydroxyhexanoate)] [P(3HB-*co*-3HHx)] from glucose and hexanoic acid, suggesting the possible participation of any of these *phaC* gene copies in its biopolymer production profile. *B. contaminans* 170816 has a *phaC* gene that belongs to this group. However, no functional evidence of PHA synthesis by strain 170816 has been reported, although other *B. contaminans* strains, such as IPT 553 and Kad1 synthesize *scl*-PHA and *mcl*-PHA [44,45]. However, the genomes of the IPT 553 and Kad1 strains have not been sequenced yet. Therefore, the genomic search for the *phaC* gene copies related to these strains is still not possible.

The *bktB* gene responsible for 3HV monomer synthesis is located in 128/178 (72%) of *Burkholderia* and *Paraburkholderia* strains that possess the archetypic cluster for PHA synthesis (*phaCABR*). Interestingly, *Paraburkholderia aromaticivorans* BN5 [45,46] is the only strain that has two copies of the *bktB* gene located close to the *phaC* gene; however, PHA synthesis by this strain has not been reported yet. One of these copies is located near the *phaCABR* gene cluster, and the other one is positioned close to a second *phaC* copy, which has a non-canonical neighborhood. Remarkably, the *bktB* gene copy is the one associated with PHA synthesis that is located close to the *phaC* gene copy but is not included in the *phaCABR* cluster in the genomes analyzed. The genes in the neighborhood are arranged as *bktBhbd-phaCphaJ-ptaackAfabIphaB* (Figure 5.3).

Finally, it should be noted that only 17% of the *Paraburkholderia* and *Burkholderia* strains studied possess only a single *phaC* gene. A unique *phaC* gene was observed in *Burkholderia anthina*, *B. cenocepacia* (AU 1054, DDS 22E-1, H11, HI2424, MC0-3), *B. cepacia* (GG4, JBK9), *Burkholderia gladioli* (ATCC 10248, pv. gladioli KACC 11889), *Burkholderia lata*, *B. mallei* (2002734299, 2002734306, NCTC 10229), *Burkholderia multivorans* CEPA002, *Burkholderia plantarii* (ATCC 43733, PG1), *B. pseudomallei* MSHR435, *Burkholderia pyrrocinia* DSM 10685, *Burkholderia* sp. (CCGE1001, HB1, MSMB1588, NRF60-BP8, PAMC 26561), *Burkholderia stabilis* FERMP-21014, *Burkholderia stagnalis* MSMB735WGS, *Burkholderia ubonensis* (MSMB1157, MSMB2035), and *P. phenoliruptrix* BR3459 strains.

5.4 METABOLIC ROUTES OF PHA SYNTHESIS

PHA anabolic pathways are associated with the central metabolic pathways such as glycolysis, the pentose phosphate pathway, the Entner–Doudoroff (ED) pathway, fatty acid β-oxidation, fatty acid synthesis, and the Krebs cycle [46,47]. The monomer composition and, therefore, the properties of PHA depend on the carbon source supplied, the metabolic pathway, and specificity of the enzymes for the substrates (see Table 5.1).

5.4.1 P(3HB) HOMOPOLYMER SYNTHESIS

The most common PHA monomer synthesized by *Paraburkholderia* and *Burkholderia* species is poly(3-hydroxybutyrate) [P(3HB), a.k.a. PHB]. P(3HB) can be synthesized from sole or mixed carbon sources and complex substrates. P(3HB) is synthesized from two acetyl-CoA that are condensed by 3-ketothiolase (acetyl-CoA acetyltransferase, PhaA) into acetoacetyl-CoA. This product is reduced by aceto-acetyl-CoA reductase (PhaB) into 3-hydroxybutyryl-CoA, which is polymerized by PHA synthase (PhaC) into P(3HB) (see Figure 5.4).

The production of P(3HB) has been reported in *P. sacchari*, *P. xenovorans*, *B. cepacia*, *B. thailandensis*, and *B. contaminans* strains (Table 5.1). *P. sacchari* type strain LMG 19450 (IPT 101) (formerly *Burkholderia sacchari*) was isolated from sugarcane plantation soil and can grow on sucrose, synthesizing PHA and reaching high cell densities [55,79,72]. *P. sacchari* LMG 19450 may use a mix of glucose and glycerol as carbon sources for P(3HB) production (4.8 g/L) [52]. *P. sacchari* LMG 19450 is capable of synthesizing P(3HB) from glucose, xylose, sucrose, arabinose, glycerol, mannose, galactose, and other renewable carbon sources (Table 5.1). *Burkholderia* sp. AB4 transforms glycerol into P(3HB), whereas *Burkholderia* sp. F24 converts xylose into P(3HB) [77,78]. *P. xenovorans* LB400 produces up to 40% P(3HB) of cell dry mass (CDM) from glucose under nitrogen limitation [16]. In addition, members of genus *Burkholderia* have been reported to synthesize P(3HB). *B.*

TABLE 5.1

PHA Production by *Paraburkholderia* and *Burkholderia* Strains

Strain	Substrate	Carbon source	Products	PHA concentration [g/L]	Scale	Ref.
Paraburkholderia sacchari LMG 19450(IPT 101)	bagasse hydrolysate	Xyl + Glu + Ara	P(3HB)	2.73	SF	[48]
	MM	Xyl + Glu	P(3HB)	34.8	Br	[49]
	WSH	Glu + gamma-butyrolactone	P(3HB), P(3HB-co-4HB)	4.3–101	SF-Br[a]	[49]
	WSH	Xyl + Glu + Ara	P(3HB)	45–95.2	Br[a]	[50]
	MM	Glu + gamma-butyrolactone	P(3HB-co-4HB)	2.1–70	SF-Br	[51]
		Glu + gly	P(3HB)	4.5	Br[a]	[52]
		Xyl + Glu + levulinic acid	P(3HB-co-3HV-co-4HV)	0.9–2.3	SF	[53]
	MM	Suc + PA	P(3HB-co-3HV)	6.8	Br	[54]
	Synthetic HH	Glu + hexanoic acid or Glu + fatty acids	P(3HB-co-3HV)	0.4–0.9	SF	[55]
		Glu + Xyl	P(3HB), xylitol, xylonic acid	2.6–88	SF-Br[a]	[56]
	Softwood HH	mannose, Xyl, Glu, galactose or Ara	P(3HB)	4.7	SF-Br[a]	[57]
	Waste paper hydrolysate	Cellulose + hemicellulose + lignin	P(3HB)	1.6	SF	[58]
	Cheese whey and/ or glucose, lactose	Glu or lactose	P(3HB)	0.1–1.8	SF	[59]

(Continued)

TABLE 5.1 (CONTINUED)
PHA Production by *Paraburkholderia* and *Burkholderia* Strains

Strain	Substrate	Carbon source	Products	PHA concentration [g/L]	Scale	Ref.
Paraburkholderia sacchari IPT 189 (DSM 17165 mutant strain)[b]	MM	Xyl	P(3HB)	–	SF	[60]
	MM	Suc	P(3HB)	63	Br[a]	[61]
Paraburkholderia sacchari LFM 828 (DSM 17165 mutant strain)[c]	MM	Glu + Ara + Xyl	P(3HB)	3.6	SF	[62]
Paraburkholderia xenovorans LB400	Glucose	Glu	P(3HB)	40% w/w	SF	[16]
Burkholderia cepacia ATCC 17759	Waste glycerol	Gly	P(3HB)	7.4	Br	[63]
	Sugar maple or HH	–	P(3HB)	–	SF	[64]
Burkholderia cepacia	MM	Gly + levulinic acid	P(3HB), P(3HB-co-3HV)	–	SF	[65]
Burkholderia cepacia JCM15050	MM	Palm oil sub-products	P(3HB)	1	SF	[66]
	RM	Glu	P(3HB-co-3H4MV)[e]	–	SF	[67]
	Crude palm kernel oil	–	P(3HB), P(3HB-co-3HHx)	–	SF	[68]
	Rice husk	Glu	–	2.0–7.8	SF-Br	[69]
Burkholderia cepacia	–	Gluconate	PHB	49–63% w/w	SF	[70]
Burkholderia cepacia IPT119, IPT400, and IPT64	Soybean hydrolysate	–	P(3HB)	0.4–0.5	SF	[71]
Burkholderia cepacia D1	MM	Glu + VA	P(3HB-co-3HV)	0.4–1.6	SF	[72]
Burkholderia cepacia IPT438	MM	waste gly	3HB, HTD, HPD, HHD	1.3–1.5	SF	[73]

(Continued)

TABLE 5.1 (CONTINUED)
PHA Production by *Paraburkholderia* and *Burkholderia* Strains

Strain	Substrate	Carbon source	Products	PHA concentration [g/L]	Scale	Ref.
Burkholderia thailandensis E264 (ATCC 700388)	Used cooking oil	Fatty acids	P(3HB), rhamnolipids	7.5	Br	[18]
Burkholderia thailandensis (E264 mutant strain)[d]	RM	Glu	P(3HB)	–	SF	[74]
Burkholderia sp. AIU M5M02	MM	Man, Fru, Glu, Gly, Maltose. Suc, Xyl, or Starch	P(3HB)	0.05–1.5	SF	[75]
Burkholderia sp. EMB5B	RM	Glu	P(3HB) macrocyclic pentolides	–	SF	[76]
Burkholderia sp. F24	HH	Xylose + levulinic acid	P(3HB), P(3HB-*co*-3HV)	3.3	SF	[77]
Burkholderia sp. AB4	RM	Gly	P(3HB) granules	–	SF	[78]
Burkholderia contaminans IPT553	MM	Glu. Glu + casein, Suc or Suc + casein	P(3HB), P(3HDd)	0.8–3.2	SF	[44]

WSH: wheat straw hydrolysates; MM: mineral medium; RM: rich medium; HH: hemicellulosic hydrolysates; SF: shake flash, Br: Bioreactor, [a]: fed-batch; Glu: glucose; Gly: glycerol; Xyl: xylose; PA: propionic acid; VA: valeric acid; Suc: Sucrose; Man: Mannitol; Fru: Fructose; 3HTD: 3-hydroxytetradecanoate; 15HPD: 15-hydroxypentadecanoate; 11HHD: 11-hydroxyhexadecanoate. Data was obtained through a Web of Science search (November 2019), crossing the terms "polyhydroxyalkanoate or polyhydroxybutyrate" with "*Burkholderia* or *Paraburkholderia*" considering the respective synonyms.

[a] fed-batch culture; [b] mutant strain with overexpression of xylose transport; [c] mutant strain deficient in catabolite repression mechanisms; [d] strain deficient in rhamnolipid biosynthesis.

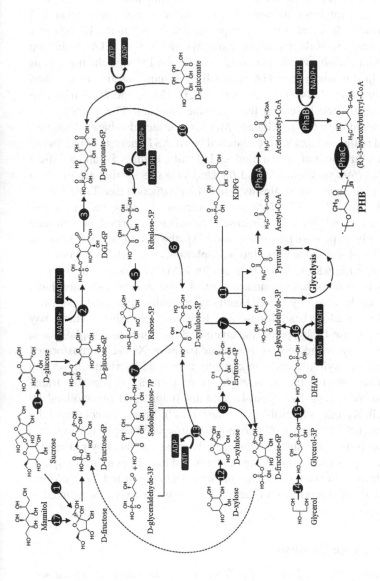

FIGURE 5.4 Metabolic pathways for polyhydroxybutyrate synthesis from sugars and glycerol in model *Paraburkholderia* and *Burkholderia* strains. This analysis is based on the genomes of *Paraburkholderia sacchari* LMG 19450, *Paraburkholderia xenovorans* LB400, *Burkholderia cepacia* ATCC 25416, and *Burkholderia thailandensis* E264. Arrows represent reactions; the numbers indicate the specific enzymes. Dashed arrows represent missing reactions in the genome of at least one *Paraburkholderia* or *Burkholderia* strain. The reactions were retrieved from curated genomic data in the KEGG and MetaCyc databases. 1. α-glucosidase; 2. glucose-6P dehydrogenase; 3. 6-phosphogluconolactonase; 4. phosphogluconate dehydrogenase; 5. ribulose-5P isomerase; 6. ribulose-5P epimerase; 7. transketolase; 8. transaldolase; 9. gluconokinase; 10. phosphogluconate dehydratase; 11. 2-dehydro-3-deoxy-phosphogluconate aldolase; 12. xylose isomerase; 13. xylulokinase; 14. glycerol kinase; 15. glycerol-3P dehydrogenase; 16. triosephosphate dehydrogenase; 17. mannitol dehydrogenase. PhaA, 3-ketothiolase. PhaB, acetoacetyl-CoA reductase. PhaC, PHA synthase. KDPG, 2-dehydro-3-deoxy-D-gluconate-6P; DGL-6P, D-glucono-1,5-lactone-6P.

cepacia ATCC17759 is able to use glycerol as a carbon source for PHA production [78]. *B. contaminans* IPT 553 (129B) synthesizes 37% PHA of CDM (2.3 g/L) from glucose, containing the monomers 3-hydroxybutyrate (3HB) and 3-hydroxydodecanoate (3HDd) (94.85 and 5.09 mol %, respectively) [44].

Previous studies on PHA production in *Paraburkholderia* and *Burkholderia* indicate that carbohydrates (monosaccharides, oligosaccharides, and polysaccharides) are metabolized through different pathways to produce these polymers. This chapter shows an *in silico* analysis of the metabolic pathways involved in PHA production in the most important *Paraburkholderia* and *Burkholderia* PHA-producing strains (Figures 5.4, 5.5). In this study, the PHA metabolism of model strains *P. sacchari* LMG19450, *P. xenovorans* LB400, *B. cepacia* ATCC 25416, and *B. thailandensis* E264 were reconstructed based on their genome analyses. The genomes and enzymes were retrieved from the KEGG and MetaCyc curated databases. Enzymes that were not found in the databases were searched by BLAST analysis of the Swiss-Prot database with a threshold of 30% amino acid identity. Figures 5.4 and 5.5 illustrate the enzymes of *Paraburkholderia* and *Burkholderia* species that are involved in the metabolism of the main substrates used in PHA production (see Table 5.1).

In *Paraburkholderia* and *Burkholderia* strains, sucrose is hydrolyzed by sucrose hydrolase into glucose and fructose. Glucose catabolism is carried out through enzymes of the glycolysis, pentose phosphate, and ED pathways (Figure 5.4). Glucose is phosphorylated by glucokinase into glucose-6-phosphate, which undergoes oxidation by glucose-6-phosphate dehydrogenase with NADPH production and subsequent hydrolysis by 6-phosphogluconolactonase to yield gluconate-6-phosphate, which can be further degraded by the ED pathway or pentose phosphate pathway. Fructose is phosphorylated by fructokinase into fructose-6-phosphate, which may be isomerized into glucose-6-phosphate and channeled into gluconate-6-phosphate, entering the ED pathway or the pentose phosphate pathway. Xylose is transformed by xylose isomerase into xylulose and consequently converted by xylulokinase into xylulose 5-phosphate, entering the pentose phosphate pathway. Glycerol is transported through the membrane by glycerol facilitator (GlpF) and metabolized by glycerol kinase (GlpK) into glycerol-3-phosphate (G3P), which is converted by glycerol-3-phosphate dehydrogenase (GlpD) into dihydroxyacetone phosphate that is channeled into glycolysis. The substrate gluconate is phosphorylated by gluconokinase into gluconate-6-phosphate, which can be channeled into the pentose phosphate pathway or ED pathway. Most of the genomes of *Paraburkholderia* and *Burkholderia* strains reported for P(3HB) production (Table 5.1) possess the enzymes depicted in Figure 5.4.

5.4.2 PHA COPOLYMER SYNTHESIS

PHA copolymers are synthesized through different metabolic pathways. Precursors of PHA copolymer synthesis may be generated from β-oxidation of fatty acids or fatty acid biosynthesis (Figure 5.5). Trans-2-enoyl-CoA (β-oxidation intermediate) and (*R*)-3-hydroxyacyl-ACP (fatty acid biosynthesis intermediate) can be converted into (*R*)-3-hydroxyacyl-CoA by enoyl-CoA hydratase (PhaJ) and acyl-ACP-CoA transacylase (PhaG), respectively. (*R*)-3-hydroxyacyl-CoA is polymerized into a

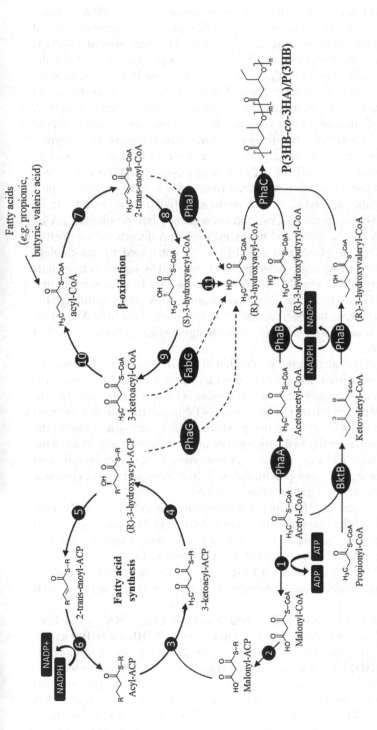

FIGURE 5.5 Metabolic pathways for polyhydroxyalkanoate synthesis from fatty acids in model *Paraburkholderia* and *Burkholderia* strains. This analysis is based on the genomes of *Paraburkholderia sacchari* LMG 19450, *Paraburkholderia xenovorans* LB400, *Burkholderia cepacia* ATCC 25416, and *Burkholderia thailandensis* E264. Arrows represent reactions; the numbers indicate the specific enzymes. Dashed arrows represent missing reactions in the genome of at least one *Paraburkholderia* or *Burkholderia* strain. The reactions were retrieved from curated genomic data in the KEGG and MetaCyc databases. 1. acetyl-CoA carboxylase; 2. malonyl-CoA-ACP transacylase (FabD); 3. β-ketoacyl-ACP-synthase (FabB), 3-ketoacyl-ACP-synthase (FabF); 4. 3-ketoacyl-ACP-reductase (FabG); 5. β-hydroxyacyl-ACP-dehydratase (FabA, FabZ); 6. enoyl-ACP reductase (FabI; FabK; FabV); 7. acyl-CoA dehydrogenase; 8. enoyl-CoA hydratase; 9. NAD-dependent (S)-3-hydroxyacyl-CoA dehydrogenase; 10. β-ketoacyl thiolase; 11. 3-hydroxy-butyryl-CoA epimerase. PhaJ, (R)-specific enoyl-CoA hydratase. FabG, NADPH-dependent 3-ketoacyl-CoA reductase. PhaG, (R)-3-hydroxyacyl-ACP-CoA transacylase. PhaC, 3-ketothiolase. PhaB, acetoacetyl-CoA reductase. PhaC, PHA synthase.

PHA copolymer by PHA synthase. The use of co-substrates to obtain PHA with different monomer compositions has been investigated extensively. Structurally related substrates, such as propionate, valerate, and hexanoate, are precursors of scl-PHA. Enoyl-CoA hydratase PhaJ provides (R)-3-hydroxyacyl-CoA monomers from the β-oxidation of fatty acids. However, the monomer supplied by PhaJ depends on its chain-length substrate specificity [39]. The genomic analyses of Paraburkholderia and Burkholderia strains showed that PhaJ is present in almost all strains, except B. cepacia ATCC 25416. In contrast, the enzyme PhaG is present in B cepacia ATCC 25416 and B. thailandensis E264, but not in Paraburkholderia strains. The enzymes that were not present in all analyzed genomes are depicted as dashed arrows in Figure 5.5. The synthesis of P(3HB-co-3HV) from glucose and propionate has been studied [54,77]. Propionyl-CoA is synthesized from a few substrates, such as propionic acid, fatty acids, succinyl-CoA, and methionine [80]. Succinyl-CoA can be converted into (R)-3-methylmalonyl-CoA and consequently into propionyl-CoA by methylmalonyl-CoA mutase (Sbm) and methylmalonyl-CoA decarboxylase (YgfG), respectively [46]. Propionyl-CoA and acetyl-CoA are condensed by ketothiolase Bktb into 3-ketovaleryl-CoA, which subsequently is reduced by ketoreductase PhaB into 3-hydroxyvaleryl-CoA. Finally, 3-hydroxyvaleryl-CoA is polymerized into P(3HB-co-3HV). 3HV can be synthesized from succinyl-CoA obtained through the Krebs cycle. The bktB gene has been identified in C. necator H16 [81]. Almost all the Paraburkholderia and Burkholderia strains belonging to group I (see Figure 5.3) possess the bktB gene.

Mendonça et al. [55] evaluated the production of P(3HB-co-3HV) copolymer by P. sacchari LMG19450 from different co-substrates. Strain LMG19450 uses propionic, valeric, heptanoic, nonanoic, and undecanoic acids as co-substrates to synthesize P(3HB-co-3HV) copolymer. The highest 3HV molar fraction (65.3% mol) synthesized by strain LMG 19450 was observed when valeric acid was added to the culture medium. P. sacchari IPT189, which is a mutant that has partially blocked the propionic acid oxidation pathway, was able, at low propionic acid concentration, to increase the 3HV yield, under nitrogen limitation [82]. B. cepacia is able to produce P(3HB-co-3HV) from glycerol and levulinic acid [65].

Copolymers composed of 3HB and 4-hydroxybutyrate P(3HB-co-4HB) possess better physical and thermal properties than P(3HB) [83]. This precursor is also synthesized through the reduction of 1,4-butanediol by alcohol dehydrogenase into 4-hydroxybutanoate, which is transformed by a hydroxyacyl-CoA synthase into 4-hydroxybutyryl-CoA [35]. P. sacchari LMG19450 can produce P(3HB-co-4HB) from γ-butyrolactone, reaching copolymer concentrations of 37 g/L with 5% 4HB monomer [49].

The synthesis of copolymers of 3HB and 3-hydroxyacyl$_{mcl}$ (HA$_{mcl}$) has been studied. Poly(3-hydroxybutyrate-co-3-hydroxyhexanoate) [P(3HB-co-HHx)] exhibits similar mechanical properties to low-density polyethylene. P. sacchari can synthesize P(3HB-co-HHx) from glucose and hexanoic acid [32]. The biosynthesis of P(3HB-co-HHx) starts with the conversion of two molecules of acetyl-CoA into acetoacetyl-CoA by ketothiolase (PhaA), with a subsequent reduction by acetyl-CoA reductase (PhaB) into 3-hydroxybutyryl-CoA (3HB-CoA). Three-hydroxyhexanoyl-CoA (3HH-CoA) units are synthesized through fatty acid biosynthesis or β-oxidation

(Figure 5.5). Finally, the PHA copolymer is produced by copolymerization of 3HB-CoA and 3HA-CoA by PHA synthase.

The simultaneous synthesis of *scl*-PHA and *mcl*-PHA by bacteria has been studied to improve biopolymer properties [44]. By adding hydrolyzed casein to the culture medium, *B. contaminans* strain IPT553 was able to synthesize the copolymer P(3HB-*co*-3HDd) (91.55 and 8.29 mol-%), yielding biomass of 4.3 g/L. The highest cell densities (4.9 g/L) of strain IPT553 were observed when the medium was supplied with sucrose, accumulating P(3HB) and P(3HDd) (98.83 and 0.97 mol-%, respectively). The highest amount of PHA (69% of CDM) was observed when sucrose was supplied with hydrolyzed casein, reaching biomass of 4.6 g/L and accumulating P(3HB) and P(3HDd) [44].

5.5 PHA PRODUCTION FROM LOW-COST SUBSTRATES

Up to 50% of the high production cost of commercial PHA is based on carbon sources. A wide range of low-cost substrates has been analyzed as feedstock for PHA production, including waste glycerol, agro-industrial residues, and food-based derivatives [84]. Waste glycerol is the main by-product (~10%) of industrial biodiesel production. Glycerol has been studied as a low-cost carbon source for PHA production by several strains of *Burkholderia*, *Paraburkholderia*, *Cupriavidus*, *Pseudomonas*, and *Halomonas* genera [85]. *B. cepacia* ATCC 17759 reached 5.8 g/L of CDM and accumulated up to 81.9% P(3HB) using 3% (v/v) biodiesel waste glycerol [63]. Using waste glycerol, *B. cepacia* IPT 438 was able to accumulate up to 67% PHA of CDM and reached 2.41 g/L CDM [73].

Lignocellulosic materials (e.g., agricultural and forestry by-products and residues) are low-cost renewable carbon sources that are not used by the food industry. Lignocellulose, which is composed of cellulose, hemicellulose, and lignin, is the most abundant renewable resource in the world. The cellulose and hemicellulose contained in lignocellulose are natural polysaccharide polymers that can be hydrolyzed into glucose, xylose, and other sugars. The conversion of xylose into added-value products has been reported [86]. Agro-industrial waste, such as agro-industrial by-products and forest and wood industry residues with a high xylose content, are potential carbon sources for PHA production. *P. sacchari* LMG 19450 synthesizes up to 62% P(3HB) using bagasse hydrolysate, which contains mainly xylose, glucose, and arabinose [48]. Strain LMG 19450 converted wheat straw hydrolysates (32.4 g/L glucose, 12.9 g/L xylose, and 4.5 g/L arabinose) into 45 to 60% P(3HB) of CDM and reaching 100–140 g/L CDM [50]. A synthetic medium composed of the main hemicellulose sugars (i.e., glucose, mannose, galactose, xylose, arabinose) and potential inhibitors (i.e., acetic acid, 5-hydroxymethylfurfural, furfural, and vanillin) was used for PHB production (5.72 g/L) by *P. sacchari* LMG 19450. However, the inhibitors stopped growth after 24 h [87]. *B. cepacia* ATCC 17759 can produce PHA from lignocellulosic materials [31].

Cheese whey (4.5% lactose, 0.8% protein, and 1% salts) is the main waste from the dairy industry and has been used as a low-cost carbon source for PHA production [88]. The production of PHA from glucose and cheese whey as a co-substrate by *P. sacchari* LMG 19450 has been reported [59]. *B. thailandensis* E264 can use cooking sunflower oil to simultaneously synthesize P(3HB) (7.5 g/L) and rhamnolipids [18].

5.6 PROPERTIES OF PHA SYNTHESIZED BY *PARABURKHOLDERIA* AND *BURKHOLDERIA* SPECIES

Biodegradability is one of the major advantages of PHA polymers. The variation of the monomer composition and molecular structure changes the physicochemical and thermal properties of PHA. Zhu et al. [65] studied the physicochemical and thermal properties of P(3HB) synthesized by *B. cepacia* ATCC 17759 from biodiesel glycerol. The P(3HB) molecular mass varies depending on the microorganism and the culture conditions, ranging from 50 to 3,000 kDa. The molecular mass (M_w) of P(3HB) synthesized by *B. cepacia* ATCC 17759 from xylose as a carbon source decreased when glycerol was added as a co-substrate. The P(3HB) produced by strain ATCC 17759 from xylose and glycerol presented a higher decomposition temperature (281.5°C), providing a wider separation between the temperature required for injection molding (melting temperature) and thermal degradation of the biopolymer. In contrast, P(3HB) homopolymer produced from xylose has a remarkably high melting temperature (T_m) (about 177°C), which is close to its thermal decomposition temperature (T_{decomp}).

The copolymer P(3HB-*co*-3HV) is less stiff and brittle than P(3HB). The PHA produced by strain ATCC 17759 from wood hydrolysate has a T_m of 174.4°C, a glass transition temperature (T_g) of 7.3°C, and a T_{decomp} lower than using xylose and glycerol as sole carbon sources [64]. The synthesis of copolymers of 3HB and HA$_{mcl}$ [P(3HB-*co*-*mcl*-3HA)] has been studied extensively due to their improved properties. Films made of P(3HB-*co*-*mcl*-3HA) (6 mol-% *mcl*-3HA) showed an elongation at break of 680% and tensile strength of 17 MPa, indicating an improvement in brittleness. P(3HB-*co*-*mcl*-3HA) (6 mol-% *mcl*-3HA) has similar mechanical properties to low-density polyethylene.

5.7 BIOMEDICAL AND BIOTECHNOLOGICAL APPLICATIONS

PHA are increasingly attracting interest in commercial applications due to their versatility, thermoplasticity, biodegradability, and biocompatibility. These bioplastics are used for a variety of pharmaceutical and therapeutic applications, such as drug release systems or as sutures, implants, repair patches, and electrospun scaffolds for tissue or bone regeneration. PHA have partially replaced petrochemical polymers for packaging and coating applications in industrial and agricultural areas. The use of bacterial PHA for biomedical and biotechnological applications will be briefly presented.

5.7.1 BIOMEDICAL APPLICATIONS

5.7.1.1 Tissue Engineering

In the last two decades, the use of PHA in tissue engineering (TE) has been studied due to their biodegradability and biocompatibility properties [89–92]. The advances in PHA-based materials for TE are mainly focused on the development of biodegradable materials for scaffolds. The use of scaffolds is important in treatment for the regeneration of tissue such as bones [93], tendons, ligaments, nerves [94], skin [95], or some organ tissue, such as cardiac organs [96,97]. Scaffolds provide mechanical support and a proper environment for cell propagation and restoring damaged tissues; fiber mats and films are the most common application forms. Electrospun microfibers were produced by Acevedo et al. [98] using biobased P(3HB) synthesized by *P. xenovorans* LB400 from the substrates xylose and mannitol [98]. PHA electrospun scaffolds possess different properties to films. The mechanical and thermal properties of PHA electrospun fibers have been reviewed [99].

Table 5.2 indicates different types of PHA used in scaffold design, highlighting the variety of applications of PHA-based scaffolds.

TABLE 5.2
The Application of PHA for Scaffold Development

PHA[a]	Other structural compounds	PHA source	Target tissue or application	Ref.
P(3HB-*co*-3HHx)/ P(3HB)	–	Commercial	Cartilage	[100]
P(3HB-*co*-3HHx)/ P(3HB)	–	Commercial	Cartilage	[101,102]
P(3HB-*co*-4HB)	–	Commercial	Blood vessels	[103]
P(3HB-*co*-3HHx), P(3HB-*co*-4HB)	–	*Delftia acidovorans*	Subcutaneous implantations	[104]
P(3HB) and P(3HB-*co*-3HHx)	Lipolase and hyaluronan	Commercial	Connective tissue	[105]
P(3HB), P(3HB-*co*-3HV), P(3HB-*co*-3HHx), P(3HB-*co*-4HB)	–	Commercial	Central nervous system	[106]
P(3HB), P(3HB-*co*-3HV)B, PHUA, PHUE, PHOUE	Chemical modification of PHUE and PHOUE with ozone and POSS, respectively	Commercial, *Ralstonia eutropha*, or *Pseudomonas putida*	Ligament and tendon	[107]
P(3HB-*co*-3HV)	–	Commercial	Ulcers, burns, epithelial tissue regeneration	[108]
P(3HB), P(3HB-*co*-3HV)	Collagen and peptides (YIGSR, GRGDS, and p20)	Commercial	Nerve	[109]
P(3HB), P(3HB-*co*-3HV)	–	Commercial	Vascular	[110]
P(3HB), PHO	Single wall carbon nanotubes	Commercial or *Pseudomonas oleovorans*	Neural cell growth	[111]
PHO	Polycaprolactone (PCL)	*Pseudomonas mendocina* CH50	Cardiac patches	[112]
P(3HB) and PHO	–	Not described	Neural cells growth	[113]

[a] PHUA: poly(3-hydroxyundecanoate), PHUE: poly(3-hydroxyundecenoate), PHOUE: poly(3-hydroxyoctanoate-*co*-3-hydroxyundecenoate), POSS: polyhedral oligomeric silsesquioxane, PHO: poly(3-hydroxyoctanoate).

Additional TE applications have been reported. Chen et al. [114] studied P(3HB) (synthesized by recombinant *E. coli* XL1) and commercial P(3HB-*co*-3HV) for surgical film development applied in hernia repair. Pramanik et al. [115] synthesized an electro-conductive, thermally stable, and magnetically active composite of Fe_2O_3 embedded P(3HB-*co*-3HV)-grafted reduced graphene oxide in order to use it as an image sensor for TE.

5.7.1.2 Drug Delivery

PHA have been applied for the development of drug delivery systems that enhance the stability and bioavailability of drugs, providing a controlled release. PHA can be used in different forms, either nanoparticles, micro/nanospheres or fibers, tablets, micelles, or block copolymers, encapsulating hormones, anticancer molecules, antibiotics, anti-inflammatory compounds, antioxidants, and other drugs [116–118]. Murueva et al. [119] prepared microparticles from different PHA loaded with doxorubicin (Dox) that are used in cancer treatment. The drug release reached 30% after 550 h, showing a similar cytostatic effect against tumor cells than free Dox. Microparticles containing antitumor drugs (paclitaxel and 5-fluorouracil) released antitumor drugs over 300 h, inhibiting tumor cell proliferation [120]. Jiang et al. [121] developed a PHA-based thermogel using P(3HB-*co*-3HHx), polyethylene glycol, polypropylene glycol, and docetaxel (DTX). A release of 10% DTX per day was observed, enhancing the *in vivo* DTX effects against melanomas compared with free DTX. PHA combined with polyvinyl alcohol (PVA) microspheres containing tetracycline (TC) for potential use in periodontal disease treatment has been described [122]. Low molecular weight PVA combined with P(3HB) enables a slow TC release for prolonged action. The addition of sulperazone and duocid to different P(3HB) and P(3HB-*co*-3HV) implantable rods for the treatment of osteomyelitis allowed a slow release for two weeks [123].

5.7.2 BIOTECHNOLOGICAL APPLICATIONS

Additional biotechnological applications of PHA have been described. The most common field of application is the packaging sector. A packaging material requires proper mechanical and thermal properties and gas and fragrance barriers; it must also possess an environmentally sustainable life cycle. P(3HB) and P(3HB-*co*-3HV) are the most used PHA in packaging materials (Table 5.3). PHA are commonly used as the main matrix blended with another compound. Alternatively, PHA have been used as a coating layer in packaging materials, favoring a proper hydrophobic surface to keep water away. The properties of these biocomposites range from moisture and gas barriers to antiviral activities [126, 129, 132].

Other applications for PHA have also been proposed. On the one hand, the addition of PHA may improve the physical properties of materials. P(3HB-*co*-4HB) fibers obtained by polymer solution casting were included in calcium phosphate cement (CPC) to improve its mechanical properties and reinforce its structure [134]. Qiang et al. [135] incorporated P(3HB-*co*-4HB) into a wood plastic composite (WPC). Pine fibers, polylactic acid (PLA), and PHA were mixed and extruded; this process bolstered the WPC and modified its mechanical properties. This biocomposite could be useful for the car industry and civil engineering. On the other hand, specific natural

TABLE 5.3

Studies Reporting the Use of PHA in Novel Packaging Materials

PHA[a]	Other structural components	Produced material	Ref.
P(3HB), P(3HB-co-3HV) (Biopol)	Thermoplastic starch (TPS)	Extruded biodegradable cast films	[124]
P(3HB), P(3HB-co-3HV) (3% V)	Zein, polyethylene glycol (PEG)	Multi-layer of compression molded P(3HB) and PHBV-PEG, with electrospun zein	[125]
P(3HB-co-3HV) (12% 3HV)	Keratin	Film	[126]
P(3HB-co-3HV) (3% 3HV)	Beer spent grain fibers (BSGF)	Film	[127]
P(3HB-co-3HV) (12% 3HV)	Nanokeratin	Films coated by electrospun fibers	[128]
P(3HB-co-3HV) (18% 3HV)	Silver nanoparticles	Films of PHBV coated with PHBV/AgNP electrospun fibers	[129]
P(3HB), P(3HB-co-3HV) (3% 3HV)	Cellulose nanofibrils (CNFs), and lignocellulose nanofibrils (LCNFs)	Monolayer of CNFs or LCNFs coated by electrospun P(3HB) or PHBV	[130]
P(3HB-co-4HB)	Cellulosic fiber matrix	Hydrophobic natural fiber composite	[131]
P(3HB-co-3HV) (5% 3HV)	Thermoplastic starch (TPS)	Multi-layer material films	[132]
P(3HB-co-3HV) (5% 3HV)	Natural fibers from wood industry and pea wastes	Extruded composite	[133]

[a] Percentage of 3-hydroxyvalerate (3HV) in the copolymer is expressed in parenthesis.

materials have been added to PHA as fillers in order to improve their biodegradability and decrease their cost. Crop containers are examples of the proposed application for PHA owing to its biodegradability and mechanical properties. Distillers drain grains with solubles (a by-product of the distillation process), which are added to a blend of P(3HB) to develop a crop container with a reduced cost, enhanced dynamic viscosity, and elongation modulus compared with a PHA container [136]. The addition of fibers of the seagrass *Posidonia oceanica* to P(3HB-co-4HB) and P(3HB-co-3HV), resulted in a biocomposite with enhanced biodegradability in marine environments [137].

Electrochemical materials have been developed using PHA. A carbon microstructure with electrochemical properties was developed using a PHA/ferrocene/chloroform precursor [138]. A biodegradable polymer gel electrolyte (PGE) made of P(3HB-co-3HV) for lithium batteries showed thermal stability and promising electrochemical performance [139]. The use of PHA in feedstock material for 3D printers has been described [140,141]. Wu et al. reported the fabrication of biodegradable low-cost 3D printing filaments using PHA, maleic acid, and fibers from palm waste in an extruder. Valentini et al. [141] prepared a biocomposite from P(3HB-*co*-3HHx) and cellulose nanofibers (CNFs); this extruded composite was characterized and used in fused deposition modeling. The hydrophobicity of PHA has also been useful for different purposes. Sudesh et al. [142] reported the ability of PHA films to absorb oil, depending on its porosity and smoothness. Therefore, PHA may be used in facial oil-blotting film, fast food wrappers, wastewater treatment, and oil spill control [142].

Paraburkholderia and *Burkholderia* strains are attractive bacteria for PHA production due to their metabolic versatility. Although P(3HB) is the most studied polymer in these genera, several *scl*-PHA and *mcl*-PHA copolymers have been studied with a wide range of applications in high-value products for health care, biomedicine, and packing.

ACKNOWLEDGMENTS

The authors acknowledge financial support by PhD Conicyt (MS) and Anillo GAMBIO (AV) fellowships, CONICYT PIA Ring Genomics and Applied Microbiology for Bioremediation and Bioproducts (GAMBIO), ACT172128 Chile (MS, FA, RN, DPP, RD, NA, PV, AV), Fondecyt 1200756 (MS), and DIUFRO DI18-0087 (FA) grants.

REFERENCES

1. Coenye T, Vandamme P. Diversity and significance of *Burkholderia* species occupying diverse ecological niches. *Environ Microbiol* 2003; 5: 719–729.
2. Sawana A, Adeolu M, Gupta RS. Molecular signatures and phylogenomic analysis of the genus *Burkholderia*: Proposal for division of this genus into the emended genus *Burkholderia* containing pathogenic organisms and a new genus *Paraburkholderia* gen. nov. harboring environmental species. *Front Gen* 2014; 5: 1–22.
3. Verstraete B, Janssens S, Smets E, et al. Symbiotic ß-proteobacteria beyond legumes: *Burkholderia* in *Rubiaceae*. *PLoS One* 2013; 8: e55260.
4. Stopnisek N, Zuhlke D, Carlier A. Molecular mechanisms underlying the close association between soil *Burkholderia* and fungi. *ISME J* 2016; 10: 253–264.
5. Xu Y, Buss EA, Boucias DG, et al. Culturing and characterization of gut symbiont *Burkholderia* spp. from the Southern chinch bug, *Blissus insularis* (Hemiptera: Blissidae). *Appl Environ Microbiol* 2016; 82: 3319–3330.
6. Yabuuchi E, Kosako Y, Oyaizu H, et al. Proposal of *Burkholderia* gen. nov. and transfer of seven species of the genus *Pseudomonas* homology Group II to the new genus, with the type species *Burkholderia cepacia* (Palleroni and Holmes 1981). *Microbiol Immunol* 1992; 36: 1251–1275.
7. Chiarini L, Bevivino A, Dalmastri C, et al. *Burkholderia cepacia* complex species: Health hazards and biotechnological potential. *Trends Microbiol* 2006; 14: 277–286.
8. Vial L, Chapalain A, Groleau MC, et al. The various lifestyles of the *Burkholderia cepacia* complex species: A tribute to adaptation. *Environ Microbiol* 2011; 13: 1–12.
9. Estrada-de los Santos P, Rojas-Rojas FU, Tapia-García EY, et al. To split or not to split: An opinion on dividing the genus *Burkholderia*. *Ann Microbiol* 2016; 66: 1303–1314.
10. Dobritsa AP, Samadpour M. Transfer of eleven species of the genus *Burkholderia* to the genus *Paraburkholderia* and proposal of *Caballeronia* gen. nov. to accommodate twelve species of the genera *Burkholderia* and *Paraburkholderia*. *Int J Syst Evol Microbiol* 2016; 66: 2836–2846.

11. Cámara B, Herrera C, Gonzalez M, et al. From PCBs to highly toxic metabolites by the biphenyl pathway. *Environ Microbiol* 2004; 6: 842–850.

12. O'Sullivan L, Mahenthiralingam E. Under the microscope: Biotechnological potential within the genus *Burkholderia*. *Lett Appl Microbiol* 2005; 41: 8–11.

13. Chain P, Denef V, Konstantinos T, et al. *Burkholderia xenovorans* LB400 harbors a multi-replicon, 9,73-Mbp genome shaped for versatility. *PNAS* 2006; 103: 15280–15287.

14. Peeters C, Meier-Kolthoff J, Verheyde B, et al. Phylogenomic study of *Burkholderia glathei*-like organisms, proposal of 13 novel *Burkholderia* species and emended descriptions of *Burkholderia sordidicola*, *Burkholderia zhejiangensis*, and *Burkholderia grimmiae*. *Front Microbiol* 2016; 7: 887.

15. Lessie TG, Hendrickson W, Manning BD, et al. Genomic complexity and plasticity of *Burkholderia cepacia*. *FEMS Microbiol Lett* 1996; 144: 117–128.

16. Urtuvia V, Villegas P, Fuentes S, et al. *Burkholderia xenovorans* LB400 possesses a functional polyhydroxyalkanoate anabolic pathway encoded by the *pha* genes and synthesizes poly (3-hydroxybutyrate) under nitrogen-limiting conditions. *Int J Microbiol* 2018; 21: 47–57.

17. Donoso R, Leiva-Novoa P, Zúñiga A, et al. Biochemical and genetic bases of indole-3-acetic acid (auxin phytohormone) degradation by the plant-growth-promoting rhizobacterium *Paraburkholderia phytofirmans* PsJN. *Appl Environ Microbiol* 2017; 83: e01991–e02061.

18. Kourmentza C, Costa J, Azevedo Z, et al. *Burkholderia thailandensis* as a microbial cell factory for the bioconversion of used cooking oil to polyhydroxyalkanoates and rhamnolipids. *Bioresour Technol* 2018; 247: 8229–837.

19. Fuentes S, Méndez V, Aguila P, et al. Bioremediation of petroleum hydrocarbons: Catabolic genes, microbial communities, and applications. *Appl Biotechnol Microbiol* 2014; 98: 4781–4794.

20. Seeger M, Timmis K, Hofer B. Degradation of chlorobiphenyls catalyzed by the bph-encoded biphenyl-2,3-dioxygenase and biphenyl-2,3-dihydrodiol-2,3-dehydrogenase of *Pseudomonas* sp. LB400. *FEMS Microbiol Lett* 1995; 133: 259–264.

21. Seeger M, Zielinski M, Kenneth T, et al. Regiospecificity of deoxygenation of di- to pentachlorobiphenyls and their degradation to chlorobenzoates by the bph-encoded catabolic pathway of *Burkholderia* sp. strain LB400. *Appl Environ Microbiol* 1999; 65: 3614–3621.

22. Méndez V, Agulló L, González M, et al. The homogentisate and homoprotocatechuate central pathways are involved in 3- and 4-hydroxyphenylacetate degradation by *Burkholderia xenovorans* LB400. *PLoS One* 2011; 10: e17583.

23. Chirino B, Strahsburger E, Agulló L, et al. Genomic and functional analyses of the 2-aminophenol catabolic pathway and partial conversion of its substrate into picolinic acid in *Burkholderia xenovorans* LB400. *PLoS One* 2013; 8: e75746.

24. Agulló L, Romero-Silva MJ, Domenech M, et al. p-Cymene promotes its catabolism through the p-Cymene and the p-cumate pathways, activates a stress response and reduces the biofilm formation in *Burkholderia xenovorans* LB400. *PLoS One* 2017; 12(1): e0169544.

25. Cordova-Kreylos AL, Fernandez LE, Koivunen M, et al. Isolation and characterization of *Burkholderia rinojensis* sp. nov., a non-*Burkholderia cepacia* complex soil bacterium with insecticidal and miticidal activities. *Appl Environ Microbiol* 2013; 79: 7669–7678.

26. Timmermann T, Armijo G, Donoso R, et al. *Paraburkholderia phytofirmans* PsJN protects *Arabidopsis thaliana* against a virulent strain of *Pseudomonas syringae* through the activation of induced resistance. *Mol Plant Microbe Interact* 2017; 30: 215–230.

27. Compant S, Nowak J, Coenye T, et al. Diversity and occurrence of *Burkholderia* spp. in the natural environment. *FEMS Microbiol Rev* 2008; 32: 607–626.

28. Mahenthiralingam E, Baldwin A, Dowson CG. *Burkholderia cepacia* complex bacteria: Opportunistic pathogens with important natural biology. *J Appl Microbiol* 2008; 104(6): 1539–1551.
29. Paganin P, Tabacchioni S, Chiarini L. Pathogenicity and biotechnological applications of the genus *Burkholderia. CEJB* 2011; 6: 997–1005.
30. Heungens K, Parke JL. Postinfection biological control of oomycete pathogens of pea by *Burkholderia cepacia* AMMDR1. *Phytopathology* 2001; 91: 383–391.
31. Pan W, Nomura C, Nakas J. Estimation of inhibitory effects of hemicellulosic wood hydrolysate inhibitors on PHA production by *Burkholderia cepacia* ATCC 17759 using response surface methodology. *Bioresour Technol* 2012; 125: 275–282.
32. Mendonça TT, Tavares RR, Cespedes LG, et al. Combining molecular and bioprocess techniques to produce poly(3-hydroxybutyrate-*co*-3-hydroxyhexanoate) with controlled monomer composition by *Burkholderia sacchari. Int J Biol Macromol* 2017; 98: 654–663.
33. Zou H, Shi M, Zhang T, et al. Natural and engineered polyhydroxyalkanoate (PHA) synthase key enzyme in biopolyester production. *Appl Microbiol Biotechnol* 2017; 101: 7417–7426.
34. Mezzolla V, D'Urso O, Poltronieri P. Role of PhaC type I and type II enzymes during PHA biosynthesis. *Polymers* 2018; 10: 1–12.
35. Meng D, Shen R, Yao H, et al. Engineering the diversity of polyesters. *Curr Opin Biotechnol* 2014; 29: 24–33.
36. Rehm B. Polyester synthases: Natural catalysts for plastics. *Biochem J* 2003; 33: 15–33.
37. Jiang X, Luo X, Zhou X. Two polyhydroxyalkanoate synthases from distinct classes from the aromatic degrader *Cupriavidus pinatubonensis* JMP134 exhibit the same substrate preference. *PLoS One* 2015; 11: e0142332.
38. Kniewel R, Lopez OR, Prieto MA. Biogenesis of medium-chain-length polyhydroxyalkanoates. In: Geiger O, Ed., *Biogenesis of Fatty Acids, Lipids and Membranes.* Switzerland: Springer Nature, 2018; pp. 457–481.
39. Tsuge T, Fukui T, Matsusaki H, et al. Molecular cloning of two (R)-specific enoyl-CoA hydratase genes from *Pseudomonas aeruginosa* and their use for polyhydroxyalkanoate synthesis. *FEMS Microbiol* 1999; 184: 193–198.
40. Madison L, Huisman G. Metabolic engineering of poly(3-hydroxyalkanoates): From DNA to plastic. *Microbiol Mol Biol Rev* 1999; 63: 21–53.
41. Morris JL, Fane A, Rush CM, et al. Neurotropic threat characterization of *Burkholderia pseudomallei* strains. *Emerg Infect Dis* 2015; 21: 58–63.
42. Mezzina MP, Pettinari MJ. Phasins, multifaceted polyhydroxyalkanoate granule-associated proteins. *Appl Environ Microbiol* 2016; 82: 5060–5067.
43. Haq I, Zwahlen R, Yang P, et al. The response of *Paraburkholderia terrae* strains to two soil fungi and the potential role of oxalate. *Front Microbiol* 2018; 9: 989.
44. Matias F, Brandt CA, da Silva ES, et al. Polyhydroxybutyrate and polyhydroxydodecanoate produced by *Burkholderia contaminans* IPT553. *J Appl Microbiol* 2017; 123: 124–133.
45. Al-Kaddo KB, Mohamad F, Murugan P, et al. Production of P (3HB-*co*-4HB) copolymer with high 4HB molar fraction by *Burkholderia contaminans* Kad1 PHA synthase. *Biochem Eng J* 2020; 153: 107394.
46. Lu J, Tappel R, Nomura C. Mini-review: Biosynthesis of poly(hydroxyalkanoates). *J Macromol Sci Pt C Polym Rev* 2009; 49: 226–248.
47. Li M, Eskridge K, Wilkins M. Optimization of polyhydroxybutyrate production by experimental design of combined ternary mixture (glucose, xylose and arabinose) and process variables (sugar concentration, molar C:N ratio). *Bioproc Biosyst Eng* 2019; 42: 1495–1506.
48. Silva L, Taciro M, Michelin R, et al. Poly-3-hydroxybutyrate (P3HB) production by bacteria from xylose, glucose and sugarcane bagasse hydrolysate. *J Ind Microbiol Biotechnol* 2004; 31: 245–254.

49. Cesário M, Raposo R, de Almeida M, et al. Production of poly(3-hydroxybutyrate-co-4-hydroxybutyrate) by *Burkholderia sacchari* using wheat straw hydrolysates and gamma-butyrolactone. *Int J Biol Macromol* 2014; 71: 59–67.

50. Cesário M, Raposo R, de Almeida M, et al. Enhanced bioproduction of poly-3-hydroxybutyrate from wheat straw lignocellulosic hydrolysates. *New Biotechnol* 2014; 31: 104–113.

51. Raposo R, de Almeida M, de Fonseca M, et al. Feeding strategies for tuning poly (3-hydroxybutyrate-co-4-hydroxybutyrate) monomeric composition and productivity using *Burkholderia sacchari*. *Int J Biol Macromol* 2017; 105: 825–833.

52. Rodriguez-Contreras A, Koller M, Dias M, et al. Influence of glycerol on poly(3-hydroxybutyrate) production by *Cupriavidus necator* and *Burkholderia sacchari*. *Biochem Eng J* 2014; 94: 50–57.

53. Ashby R, Solaiman D, Nuñez A, et al. *Burkholderia sacchari* DSM 17165: A source of compositionally-tunable block-copolymeric short-chain poly(hydroxyalkanoates) from xylose and levulinic acid. *Bioresour Technol* 2018; 253: 333–342.

54. Rocha R, da Silva L, Taciro M, et al. Production of poly(3-hydroxybutyrate-co-3-hydroxyvalerate) P(3HB-co-3HV) with a broad range of 3HV content at high yields by *Burkholderia sacchari* IPT 189. *World J Microbiol Biotechnol* 2008; 24: 427–431.

55. Mendonça TT, Gomez JGC, Schripsema J, et al. Exploring the potential of *Burkholderia sacchari* to produce polyhydroxyalkanoates. *J Appl Microbiol* 2013; 116: 815–829.

56. Raposo R, de Almeida M, de Oliveira M, et al. A *Burkholderia sacchari* cell factory: Production of poly-3-hydroxybutyrate, xylitol and xylonic acid from xylose-rich sugar mixtures. *New Biotechnol* 2017; 34: 12–22.

57. Dietrich K, Dumont M, Orsat V, et al. Consumption of sugars and inhibitors of softwood hemicellulose hydrolysates as carbon sources for polyhydroxybutyrate (PHB) production with *Paraburkholderia sacchari* IPT 101. *Cellulose* 2019; 26: 7939–7952.

58. Al-Battashi H, Annamalai N, Al-Kindi S, et al. Production of bioplastic (poly-3-hydroxybutyrate) using waste paper as a feedstock: Optimization of enzymatic hydrolysis and fermentation employing *Burkholderia sacchari*. *J Clean Prod* 2019; 214: 236–247.

59. de Andrade C, Nascimento V, Cartez-Vega WR, et al. Exploiting cheese whey as co-substrate for polyhydroxyalcanoates synthesis from *Burkholderia sacchari* and as raw material for the development of biofilms. *Waste Biomass Valori* 2019; 10: 1609–1616.

60. Guaman L, Barba-Ostria C, Zhang F, et al. Engineering xylose metabolism for production of polyhydroxybutyrate in the non-model bacterium *Burkholderia sacchari*. *Microb Cell Fact* 2018; 17: 74.

61. Pradella J, Taciro M, Mateus A. High-cell-density poly (3-hydroxybutyrate) production from sucrose using *Burkholderia sacchari* culture in airlift bioreactor. *Bioresour Technol* 2010; 101: 8355–8360.

62. Lopes M, Gosset G, Rocha R, et al. PHB Biosynthesis in catabolite repression mutant of *Burkholderia sacchari*. *Curr Microbiol* 2011; 63: 319–326.

63. Zhu C, Nomura C, Perrotta J, et al. Production and characterization of poly-3-hydroxybutyrate from biodiesel-glycerol by *Burkholderia cepacia* ATCC 17759. *Biotechnol Prog* 2010; 64: 124–560.

64. Pan W, Perrotta JA, Stipanovic A, et al. Production of polyhydroxyalkanoates by *Burkholderia cepacia* ATCC 17759 using a detoxified sugar maple hemicellulosic hydrolysate. *J Ind Microbiol Biotechnol* 2012; 39: 459–469.

65. Zhu CJ, Nomura CT, Perrotta JA, et al. The effect of nucleating agents on physical properties of poly-3-hydroxybutyrate (PHB) and poly-3-hydroxybutyrate-co-3-hydroxyvalerate (PHB-co-HV) produced by *Burkholderia cepacia* ATCC 17759. *Polym Test* 2012; 31: 579–585.

66. Chee J, Tan Y, Samian M, et al. Isolation and characterization of a *Burkholderia* sp USM (JCM15050) capable of producing polyhydroxyalkanoate (PHA) from triglycerides, fatty acids and glycerols. *J Polym Environ* 2010; 18: 584–592.
67. Lau N, Tsuge T, Sudesh K. Formation of new polyhydroxyalkanoate containing 3-hydroxy-4-methylvalerate monomer in *Burkholderia* sp. *Appl Microbiol Biotechnol* 2011; 89: 1599–1609.
68. Chee J, Lau N, Samian M, et al. Expression of *Aeromonas caviae* polyhydroxyalkanoate synthase gene in *Burkholderia* sp USM (JCM15050) enables the biosynthesis of SCL-MCL PHA from palm oil products. *J Appl Microbiol* 2012; 112: 45–54.
69. Heng K, Hatti-Kaul R, Adam F, et al. Conversion of rice husks to polyhydroxyalkanoates (PHA) via a three-step process: Optimized alkaline pre-treatment, enzymatic hydrolysis, and biosynthesis by *Burkholderia cepacia* USM (JCM 15050). *JCTB* 2017; 92: 100–108.
70. Rodrigues M, Vicente E, Steinbuchel A. Studies on polyhydroxyalkanoate (PHA) accumulation in a PHA synthase I-negative mutant of *Burkholderia cepacia* generated by homogenotization. *FEMS Microbiol Lett* 2000; 193: 179–185.
71. Rodrigues R, Nunes J, Lordelo L, et al. Assessment of polyhydroxyalkanoate synthesis in submerged cultivation of *Cupriavidus necator* and *Burkholderia cepacian* strains using soybean as substrate. *Braz J Chem Eng* 2019; 36: 73–83.
72. Hsieh W, Chen W, Yang S, et al. 3D tissue culture and fermentation of poly(3-hydroxybutyrate-*co*-3-hydroxyvalerate) from *Burkholderia cepacia* D1. *J Taiwan Inst Chem Eng* 2011; 42: 883–888.
73. Ribeiro P, da Silva A, Menezes J, et al. Impact of different by-products from the biodiesel industry and bacterial strains on the production, composition, and properties of novel polyhydroxyalkanoates containing achiral building blocks. *Ind Crop Prod* 2015; 69: 212–223.
74. Funston S, Tsaousi K, Smyth T. Enhanced rhamnolipid production in *Burkholderia thailandensis* transposon knockout strains deficient in polyhydroxyalkanoate (PHA) synthesis. *Appl Microbiol Biotechnol* 2017; 101: 8443–8454.
75. Yamada M, Yukita A, Hanazumi Y, et al. Poly(3-hydroxybutyrate) production using mannitol as a sole carbon source by *Burkholderia* sp AIU M5M02 isolated from a marine environment. *Fisher Sci* 2018; 84: 405–412.
76. Petta T, Raichardt L, Melo I, et al. Bioassay-guided isolation of a low molecular weight PHB from *Burkholderia* sp with phytotoxic activity. *Appl Biochem Biotechnol* 2013; 170: 1689–1701.
77. Lopes M, Gomez J, Taciro M. Polyhydroxyalkanoate biosynthesis and simultaneous remotion of organic inhibitors from sugarcane bagasse hydrolysate by *Burkholderia* sp. *J Ind Microbiol Biotefchnol* 2014; 41: 1353–1363.
78. Sacco L, Castellane T, Lopes E, et al. Properties of polyhydroxyalkanoate granules and bioemulsifiers from *Pseudomonas* sp and *Burkholderia* sp isolates growing on glucose. *Appl Biochem Biotechnol* 2016; 178: 990–1001.
79. Brämer CO, Vandamme P, da Silva LF, et al. *Burkholderia sacchari* sp. nov., a polyhydroxyalkanoate-accumulating bacterium isolated from soil of a sugarcane plantation in Brazil. *Int J Syst Evol Microbiol* 2001; 51: 1709–1713.
80. Catalán A, Malan AK, Ferreira F, et al. Propionic acid metabolism and poly-3-hydroxybutyrate-*co*-3-hydroxyvalerate production by a prpC mutant of *Herbaspirillum seropedicae* Z69. *J Biotechnol* 2018; 286: 36–44.
81. Reinecke F, Steinbüchel A. *Ralstonia eutropha* strain H16 as model organism for PHA metabolism and for biotechnological production of technically interesting biopolymers. *J Mol Microbiol Biotechnol* 2008; 16: 91–108.
82. Silva LF, Gomez J, Oliveira MS. Propionic acid metabolism and poly-3-hydroxybutyrate-*co*-3-hydroxyvalerate (P3HB-*co*-3HV) production by *Burkholderia* sp. *J Biotechnol* 2000; 76: 165–174.

83. Martin D, Williams S. Medical applications of poly-4-hydroxybutyrate: A strong flexible absorbable biomaterial. *Biochem Eng J* 2003; 16: 97–105.

84. Anjum A, Zuber M, Zia K, et al. Microbial production of polyhydroxyalkanoates (PHAs) and its copolymers: A review of recent advancements. *Int J Biol Macromol* 2016; 89: 161–174.

85. Kumar P, Ray S, Patel S. Bioconversion of crude glycerol to polyhydroxyalkanoate by *Bacillus thuringiensis* under non-limiting nitrogen conditions. *Int J Biol Macromol* 2015; 78: 9–16.

86. Chen H. Biotechnology of lignocellulose: Theory and practice. In: Chen H, Ed., *Biotechnology of Lignocellulose: Theory and Practice*. Beijing, China: Springer, 2014; pp. 1–511.

87. Dietrich K, Dumont MJ, Schwinghamer T, et al. A model study to assess softwood hemicellulose hydrolysates as the carbon source for PHB production in *Paraburkholderia sacchari* IPT 101. *Biomacromolecules* 2018; 19: 188–200.

88. Castilho L, Mitchell D, Freire D. Production of polyhydroxyalcanoates (PHAs) from waste materials and by-products by submerged and solid-state fermentation. *Bioresour Technol* 2009; 100: 5996–6009.

89. Chen GQ, Wu Q. The application of polyhydroxyalkanoates as tissue engineering materials. *Biomaterials* 2005; 26: 6565–6578.

90. Misra K, Valappil SP, Roy I, Boccaccini AR. Polyhydroxyalkanoate (PHA)/inorganic phase composites for tissue engineering applications. *Biomacromolecules* 2006; 7: 2249–2258.

91. Zubairi SI, Mantalaris A, Bismarck A, et al. Polyhydroxyalkanoates (PHAs) for tissue engineering applications: Biotransformation of palm oil mill effluent (POME) to value-added polymers. *J Teknol* 2016; 78: 13–29.

92. Chen GQ, Zhang J. Microbial polyhydroxyalkanoates as medical implant biomaterials. *Art Cells Nanomed Biotechnol* 2018; 46: 1–18.

93. Chocholata P, Kulda V, Babuska V. Fabrication of scaffolds for bone-tissue regeneration. *Materials* 2019; 12: 568.

94. Biazar E. Polyhydroxyalkanoates as potential biomaterials for neural tissue regeneration. *Int J Polym Mater Polym Biomater* 2014; 63: 898–908.

95. Li XT, Zhang Y, Chen GQ. Nanofibrous polyhydroxyalkanoate matrices as cell growth supporting materials. *Biomaterials* 2008; 29: 3720–3728.

96. Dubey P, Boccaccini AR, Roy I. Novel cardiac patch development using natural biopolymers. In: Yang Y, Xu H, Yu X, Eds., *Lightweight Materials from Biopolymers and Biofibers*. ACS Symposium Series, 2014; pp. 159–175.

97. Bagdadi AV, Safari M, Dubey P, et al. Poly(3-hydroxyoctanoate), a promising new material for cardiac tissue engineering. *J Tissue Eng Regen Med* 2018; 12: e495–e512.

98. Acevedo F, Villegas P, Urtuvia V, et al. Bacterial polyhydroxybutyrate for electrospun fiber production. *Int J Biol Macromol* 2018; 106: 692–697.

99. Sanhueza C, Acevedo F, Rocha S, et al. Polyhydroxyalkanoates as biomaterial for electrospun scaffolds. *Int J Biol Macromol* 2019; 124: 102–110.

100. Deng Y, Zhao K, Zhang XF, et al. Study on the three-dimensional proliferation of rabbit articular cartilage-derived chondrocytes on polyhydroxyalkanoate scaffolds. *Biomaterials* 2002; 23: 4049–4056.

101. Zhao K, Deng Y, Chen JC. Polyhydroxyalkanoate (PHA) scaffolds with good mechanical properties and biocompatibility. *Biomaterials* 2003; 24: 1041–1045.

102. Zhao K, Deng Y, Chen G-Q. Effects of surface morphology on the biocompatibility of polyhydroxyalkanoates. *Biochem Eng J* 2003; 16: 115–123.

103. Cheng ST, Chen ZF, Chen GQ. The expression of cross-linked elastin by rabbit blood vessel smooth muscle cells cultured in polyhydroxyalkanoate scaffolds. *Biomaterials* 2008; 29: 4187–4194.

104. Ying TH, Ishii D, Mahara A, et al. Scaffolds from electrospun polyhydroxyalkano-ate copolymers: Fabrication, characterization, bioabsorption and tissue response. *Biomaterials* 2008; 29: 1307–1317.

105. Wang YW, Wu Q, Chen GQ. Reduced mouse fibroblast cell growth by increased hydro-philicity of microbial polyhydroxyalkanoates via hyaluronan coating. *Biomaterials* 2003; 24: 4621–4629.

106. Xu XY, Li XT, Peng SW, et al. The behaviour of neural stem cells on polyhydroxyal-kanoate nanofiber scaffolds. *Biomaterials* 2010; 31: 3967–3975.

107. Rathbone S, Furrer P, Lübben J. Biocompatibility of polyhydroxyalkanoate as a poten-tial material for ligament and tendon scaffold material. *J Biomed Mater Res Pt A* 2010; 93: 1391–1403.

108. Ellis G, Cano P, Jadraque M, et al. Laser microperforated biodegradable microbial polyhydroxyalkanoate substrates for tissue repair strategies: An infrared microspec-troscopy study. *Anal Bioanal Chem* 2011; 399: 2379–2388.

109. Masaeli E, Wieringa PA, Morshed M, et al. Peptide functionalized polyhydroxyalkano-ate nanofibrous scaffolds enhance Schwann cells activity. *Nanomedicine* 2014; 10: 1559–1569.

110. Yao CL, Chen JH, Lee CH. Effects of various monomers and micro-structure of poly-hydroxyalkanoates on the behavior of endothelial progenitor cells and endothelial cells for vascular tissue engineering. *J Polym Res* 2018; 25: 187.

111. Russell RA, Foster LJR, Holden PJ. Carbon nanotube mediated miscibility of poly-hydroxyalkanoate blends and chemical imaging using deuterium-labelled poly(3-hydroxyoctanoate). *Eur Polym J* 2018; 105: 150–157.

112. Constantinides C, Basnett P, Lukasiewicz B, et al. *In Vivo* tracking and 1H/19F mag-netic resonance imaging of biodegradable polyhydroxyalkanoate/polycaprolactone blend scaffolds seeded with labeled cardiac stem cells. *ACS Appl Mater Interfaces* 2018; 10: 25056–25068.

113. Lizarraga-Valderrama LR, Taylor CS, Claeyssens F, et al. Unidirectional neuronal cell growth and differentiation on aligned polyhydroxyalkanoate blend microfibres with varying diameters. *J Tissue Eng Regen Med* 2019; 13: 1–14.

114. Chen Y, Tsai YH, Chou IN, et al. Application of biodegradable polyhydroxyalkano-ates as surgical films for ventral hernia repair in mice. *Int J Polym Sci* 2014; 2014: 789681.

115. Pramanik N, De J, Basu RK, et al. Fabrication of magnetite nanoparticle doped reduced graphene oxide grafted polyhydroxyalkanoate nanocomposites for tissue engineering application. *RSC Adv* 2016; 6: 46116–46133.

116. Michalak M, Kurcok P, Hakkarainen M. Polyhydroxyalkanoate-based drug delivery systems. *Polym Int* 2017; 66: 617–622.

117. Li Z, Loh XJ. Recent advances of using polyhydroxyalkanoate-based nanovehicles as therapeutic delivery carriers. *Wiley Interdiscip Rev Nanomed Nanobiotechnol* 2017; 9: 19–22.

118. Nigmatullin R, Thomas P, Lukasiewicz B, et al. Polyhydroxyalkanoates, a family of natural polymers, and their applications in drug delivery. *J Chem Technol Biotechnol* 2015; 90: 1209–1221.

119. Murueva AV, Shishatskaya EI, Kuzmina AM, et al. Microparticles prepared from bio-degradable polyhydroxyalkanoates as matrix for encapsulation of cytostatic drug. *J Mater Sci Mater Med* 2013; 24: 1905–1915.

120. Shershneva A, Murueva A, Nikolaeva E, et al. Novel spray-dried PHA microparticles for antitumor drug release. *Dry Technol* 2018; 36: 1387–1398.

121. Jiang L, Luo Z, Loh XJ, et al. PHA-based thermogel as a controlled zero-order che-motherapeutic delivery system for the effective treatment of melanoma. *ACS Appl Biol Mater* 2019; 2: 3591–3600.

122. Panith N, Assavanig A, Lertsiri S, et al. Development of tunable biodegradable polyhydroxyalkanoates microspheres for controlled delivery of tetracycline for treating periodontal disease. *J Appl Polym Sci* 2016; 133: 1–12.

123. Türesin F, Gürsel I, Hasirci V. Biodegradable polyhydroxyalkanoate implants for osteomyelitis therapy: *In vitro* antibiotic release. *J Biomater Sci Polym Ed* 2001; 12: 195–207.

124. Parulekar Y, Mohanty AK. Extruded biodegradable cast films from polyhydroxyalkanoate and thermoplastic starch blends: Fabrication and characterization. *Macromol Mater Eng* 2007; 292: 1218–1228.

125. Fabra MJ, Sánchez G, López-Rubio A, et al. Microbiological and ageing performance of polyhydroxyalkanoate-based multilayer structures of interest in food packaging. *LWT Food Sci Technol* 2014; 59: 760–767.

126. Pardo-Ibañez P, Lopez-Rubio A, Martínez-Sanz M, et al. Keratin-polyhydroxyalkanoate melt-compounded composites with improved barrier properties of interest in food packaging applications. *J Appl Polym Sci* 2014; 131: 1–10.

127. Cunha M, Berthet MA, Pereira R, et al. Development of polyhydroxyalkanoate/beer spent grain fibers composites for film blowing applications. *Polym Compos* 2015; 36: 1859–1865.

128. Fabra MJ, Pardo P, Martinez-Sanz M, et al. Combining polyhydroxyalkanoates with nanokeratin to develop novel biopackaging structures. *J Appl Polym Sci* 2016; 133: 42695.

129. Castro-Mayorga JL, Randazzo W, Fabra MJ, et al. Antiviral properties of silver nanoparticles against norovirus surrogates and their efficacy in coated polyhydroxyalkanoates systems. *LWT Food Sci Technol* 2017; 79: 503–510.

130. Cherpinski A, Torres-Giner S, Cabedo L, et al. Post-processing optimization of electrospun submicron poly(3-hydroxybutyrate) fibers to obtain continuous films of interest in food packaging applications. *Food Addit Contam Part A Chem Anal Control Expo Risk Assess* 2017; 34: 1817–1830.

131. Zhao C, Li J, He B, et al. Fabrication of hydrophobic biocomposite by combining cellulosic fibers with polyhydroxyalkanoate. *Cellulose* 2017; 24: 2265–2274.

132. Dilkes-Hoffman LS, Pratt S, Lant PA, et al. Polyhydroxyalkanoate coatings restrict moisture uptake and associated loss of barrier properties of thermoplastic starch films. *J Appl Polym Sci* 2018; 135: 1–9.

133. Cinelli P, Mallegni N, Gigante V, et al. Biocomposites based on polyhydroxyalkanoates and natural fibres from renewable byproducts. *Appl Food Biotechnol* 2019; 6: 35–43.

134. Ogasawara T, Sawamura T, Maeda H, et al. Enhancing the mechanical properties of calcium phosphate cements using short-length polyhydroxyalkanoate fibers. *J Ceram Soc Japan* 2016; 124: 180–183.

135. Qiang T, Yu D, Gao H. Wood flour/polylactide biocomposites toughened with polyhydroxyalkanoates. *J Appl Polym Sci* 2012; 124: 1831–1839.

136. Lu H, Madbouly SA, Schrader JA, et al. Novel bio-based composites of polyhydroxyalkanoate (PHA)/distillers dried grains with solubles (DDGS). *RSC Adv* 2014; 4: 39802–39808.

137. Seggiani M, Cinelli P, Mallegni N, et al. New bio-composites based on polyhydroxyalkanoates and *Posidonia oceanica* fibres for applications in a marine environment. *Materials (Basel)* 2017; 10: 326.

138. Chen Q, Xiao S, Zhang R, et al. Spindle-like hierarchical carbon structure grown from polyhydroxyalkanoate/ferrocene/chloroform precursor. *Carbon (NY)* 2016; 103: 346–351.

139. Dall'Asta V, Berbenni V, Mustarelli P, et al. biomass-derived polyhydroxyalkanoate biopolymer as safe and environmental-friendly skeleton in highly efficient gel electrolytes for lithium batteries. *Electrochim Acta* 2017; 247: 63–70.

140. Wu CS, Liao HT, Cai YX. Characterisation, biodegradability and application of palm fibre-reinforced polyhydroxyalkanoate composites. *Polym Degrad Stab* 2017; 140: 55–63.
141. Valentini F, Dorigato A, Rigotti D, et al. Polyhydroxyalkanoates/fibrillated nanocellulose composites for additive manufacturing. *J Polym Environ* 2019; 27: 1333–1341.
142. Sudesh K, Loo CY, Goh LK, et al. The oil-absorbing property of polyhydroxyalkanoate films and its practical application: A refreshing new outlook for an old degrading material. *Macromol Biosci* 2007; 7: 1199–1205.

6 Genetic Engineering as a Tool for Enhanced PHA Biosynthesis from Inexpensive Substrates

Lorenzo Favaro, Tiziano Cazzorla,
Marina Basaglia, and Sergio Casella

CONTENTS

6.1 INTRODUCTION

In order to achieve cost-effective large-scale production of PHA, in-depth research is still required. Besides efficient natural strains converting substrates into PHA, it has become clear during the last few years that the engineering of microorganisms is needed to increase the efficiency of PHA producers and, above all, obtain new proficient bacteria able to use inexpensive, although complex, carbon sources. Indeed, the carbon source required for a hypothetical industrial process could even account for 50% of the total cost [1].

The agricultural production chain, as well as industrial by-products, can be considered interesting sources of a variety of waste and by-product streams potentially suitable as feedstock for microbial production of polyhydroxyalkanoates (PHA), providing, sometimes, the double advantage of saving on disposal costs while producing value-added goods [2,3]. Unfortunately, wild-type strains for the direct and efficient conversion of low-cost waste streams into PHA are not available in nature, and therefore the engineering of strains has become important. Therefore, the economic feasibility of efficient PHA biotechnological processes demands inexpensive carbon sources and, possibly, the integration of producing plants where the by-product is generated.

Of course, a similar industrial-scale production is not only dependent on the cost of the substrate, but a number of other critical factors may affect the economic feasibility of the whole process, such as the growth ability of the selected strains, the microbial content of the accumulated PHA and its yield, as well as the downstream process, for example [4].

6.2   ENGINEERING TECHNIQUES APPLIED TO OBTAIN RECOMBINANT STRAINS FOR PHA PRODUCTION

Although many reviews and chapters have dealt with PHA production from organic waste streams by using engineered bacteria, no insight into the engineering techniques applied to obtain such recombinant strains has been proposed. This section is the first survey of the main engineering methods employed so far to develop engineered bacteria strains for PHA production from low-cost substrates. The vectors, their ancestors, as well as the promoter sequences and origin of replication, have been investigated (see Table 6.1 and Table 6.2).

In the case of the PHA-producers platform (see Table 6.1), the most commonly used species is *Cupriavidus necator*, which has been engineered by means of both episomal plasmids and chromosomal integration. Many research projects made use of the plasmid pBBRI MCS or a few derivatives (pBBR1MCS2, pBBR1MCS3, and pBBR1MCS5). Such plasmids have a CmR vector of about 4.7 kb, containing 16 unique cloning sites within the *lacZα* gene. It is relatively stable both *in vitro* (>10 days) and *in vivo* (>4 weeks) without antibiotic selection [49]. The high plasmid stability is strictly linked to the *oriV*, the origin of replication, with low copy numbers (up to 10) per cell [50]. So far, researchers have used strong promoters, mostly of phagic origin, such as T3 or from *C. necator* itself (*phaABC*) or *Escherichia coli* (*tac*), to allow a constitutive expression of the heterologous gene(s). Nevertheless, the limited number of characterized promoters poses a significant challenge during the engineering of *C. necator* for biotechnological applications, and two recent pioneering papers paved the way for the development of a novel constitutive promoters toolbox that will serve the biotechnology community working on *C. necator* [51,52].

In other cases, chromosomal integration of the gene(s) has been pursued to guarantee the stability of the recombinants under the transcriptional control of strong promoters. However, the copy number of the targeted gene(s) is generally lower than those reported for the episomal vectors.

As reported in Table 6.2, the engineering of non-PHA producers with the operon, *phaABC*, has been achieved by applying a number of different vectors. In contrast, no chromosomal integration has been reported so far. This is mostly because *E. coli* was by far the most commonly used host strain. As such, many stable plasmids have been developed in the last century for the genetic engineering of this microorganism. In most cases, the average copy number was found to be 20, and the *phaABC* operon was cloned under the regulation of the strong phagic promoters, T3 and T7.

TABLE 6.1

Most-Used Plasmids and Their Major Traits Applied to the Engineering of PHA Producers to Convert Organic Waste Streams into PHA

PHA From	Recipient	GMM[a]	Plasmid	Copy Number[b]	Promoter	Ancestor
Whey permeate [5]	C. necator DSM545	mREPT	pSUP102 [6]	chromosomal	phaABC	pACYC184
Sucrose [7]	C. necator NCIMB11599;437-540	NCIMB11599 (pKM212-SacC) 437-540 (pKM212-SacC)	pKM212-SacC [8]	10 (oriV)	tac	pBBR1MCS2
Sucrose [9]	C. necator 142SR	142SR (pCUV5-cscAB)	pCUV5-cscAB [9]	10 (oriV)	PlacUV5RBS	pCUP3
Sucrose [9]	C. necator 005dZG	005dZG(pCTRC-cscAB)	pCTRC-cscAB [9]	10 (oriV)	PtrcRBS	pCUP3
Soybean oil [10–13]	A. eutrophus PHB-4	PHB-4 (pJRDEE32d13)	pJRDEE32d13 [13]	10 (oriV)	PphaCAc	pJRD215
Waste animal fat [14]	C. necator Re2058, Re2160	Re2058(pCB113) Re2160(pCB113)	pCB113 [15]	10 (oriV)	T3	pBBR1MCS2
Soybean oil [16]	C. necator NSDGΔA	MF03	pK18mobsacB [17]	chromosomal	phaABC	pK18
Palm kernel oil [18]	C. necator PHB-4	PHB-4 harboring phaCcs	pBBR1MCS-C2 [18]	10 (oriV)	T3	pBBR1MCS2
Spent palm oil [19]	C. necator PHB-4	PHB-4 harboring phaC_{USMAA24}	pMBHC2.5 [20]	10 (oriV)	T3	pBBR1MCS3
Udder, lard, tallow [21]	D. acidovorans DSM39	DSM39 (pBBR1MCS-5-lipH-lipC)	pBBR1MCS-5-lipH-lipC [21]	10 (oriV)	T3	pBBR1MCS5
Crude glycerol [22]	C. necator H16	H16 (pBBR-glpFK)	pBBR1-glpFK [22]	10 (oriV)	Pglp	pBBR1
Crude glycerol [22]	C. necator H16	H16_glpFK	pK18ms-glpFK-A2858 [22]	chromosomal	PA2858	pK18
Xylose [23]	C. necator NCIMB11599	NCIMB11599 (pKM212-XylAB)	pKM212-XylAB [8]	10 (oriV)	tac	pBBR1MCS2

a Genetically modified microorganism, b Origin of replication is reported within brackets

TABLE 6.2

Most-Used Plasmids and Their Major Traits Applied in the Engineering of Non-PHA Producers to Convert Organic Waste Streams into PHA

PHA From	Recipient	GMM	Plasmid	Copy Number[a]	Promoter[b]	Ancestor
Whey [24]	*E. coli* SP314	SP314 (pJP24)	pJP24[24]	150:200 (pUC)	T5	pQE32
Bovine whey powder solution [25,26]	*E. coli GCSC 4401, GCSC 6576*	*GCSC 4401* (pSYL107) *6576* (pSYL107)	pSYL107[27]	15:20 (ColE1)	T7/T3	pSYL105
Crude glycerol [28]	*E. coli (JM109)*	*HBP01*	*pBAD18-phaAB* *pWQ02* *pWQ04*[28]	15:20 (pbr322) 15:20 (pbr322) 10 (p15A)	P_{BAD}-AraC T7 T7	pBAD18-kan pHP 301 pACYCDuet-1
Processed bovine whey powder solution [29–31]	*E. coli* CGSC4401, CGSC3121, CGSC 2507, DSM499, KCTC2223	CGSC4401 (pJC4), CGSC3121 (pJC4), CGSC 2507 (pJC4), DSM499 (pJC4), KCTC2223 (pJC4)	pJC4[32]	15:20 (pbr322)	T7/sp6	pGEM-7ZF
Cheese whey [33]	*E. coli* MG1655	CML3-1 and P8-X8	pMAB26 [34]	15:20 (r6k)	tac	pCNB5
Sucrose [35]	*K. aerogenes* 2688, *E.coli* JMU213	2688 (pJM9131) JMU213 (pJM9131)	pJM9131 [35]	15:20 (PBR322)	T7	p4A
Molasses, sucrose [36]	*E. coli*	*phaC1*	pDRIVE [37]	15:20 (pbr322)	T7	pDRIVE

(Continued)

TABLE 6.2 (CONTINUED)
Most-Used Plasmids and Their Major Traits Applied in the Engineering of Non-PHA Producers to Convert Organic Waste Streams into PHA

PHA From	Recipient	GMM	Plasmid	Copy Number[a]	Promoter[b]	Ancestor
Crude glycerol [38]	E. coli HMS174(DE3)	HMS174(DE3) pCOLADuet-1::dhaB1B2::pduP::phaC1	pCOLADuet-1::dhaB1B2::pduP::phaC1 [38]	20-40 (ColA)	T7	pCOLAduet-1
Xylose [39]	E. coli TG1	TG1(pSYL107)	pSYL107[40]	15:20 (ColE1)	T7/T3	pBluescript SK (−)
Cellulose hydrolysate [41]	E. coli LS5218	LS5218 (pGEM-phaABC)	pGEM-phaABC [41]	15:20 (pbr322)	T7/SP6	pGEM
Xylan, Eucalyptus hydrosylate [42,43]	E. coli JM109, BW25113	JM109 [pTV118NpctphaC1(ST′QK)] BW2511 [pTV118NpctphaC1(ST′QK)]	pTV118NpctphaC1(ST′QK) [42]	15:20 (pbr322)	lac	pTV118N
Xylan from beechwood [44]	E. coli LS5218	LS5218 (pTV118NpctphaC1/ pBBRXBB2)	pBBRXBB2 [44]	10 (oriV)	T7/T3	pBBR1-MCS2
Hydrolyzed corn starch [45]	E. coli JM101, DH10B	JM101 (pBHR68), DH10B (pBHR68)	pBHR68 [46]	15/20 (colE1)	T7/T3	pBluescript SK (−)
Starch [47]	E. coli BL21(DE3)	SKB99 (pLW487 / pTAmyl)	pLW487 [47]	15-20 (pbr322)	trc	pEP2
Starch [47]	E.coli BL21(DE3)	SKB99 (pLW487 / pTAmyl)	pTAmyl [47]	10 (p15A)	T7	pET24ma
Sucrose [48]	E. coli WΔcscR	WΔcscR (pAet41)	pAeT41 [49]	150–200 (pUC)	lac	pUC18

[a] Origin of replication is reported within brackets, [b] Promoter column with "/" reports the availability of two regulatory sequences

Most of the techniques reported above have been used to verify, at least at a laboratory scale, the possibility of making appropriate microbes to produce PHA from inexpensive carbon sources.

In this section, the most encouraging results from whey, molasses, lipids, and starchy and lignocellulosic materials are reported and discussed. Tables 6.3, 6.4, and 6.5 report the most efficient engineered microbes for PHA production from organic waste streams.

6.3   THE USE OF WHEY AS A CARBON SOURCE

Both the EU (>65%) and North America (24%) strongly contribute to the approximate 120 million tons of whey produced globally per year [5,49] as the principal by-product of the dairy industry (about 90% of the volume of handled milk). Of this, 94% is water, 4.5% lactose, and less than 1% is protein, ash, and fat [53,54]. High BOD (about 60,000 ppm) and COD (up to 80,000 ppm) make whey as a waste to be adequately disposed, giving rise to relevant management problems, even if some part of this by-product could be used for the production of human goods and animal feed [55].

Unfortunately, only a few wild-type microorganisms can directly convert lactose into PHA without its preliminary hydrolysis into glucose and galactose, and at low-efficiency levels [56].

Attempts to increase PHA yields from whey were already made in the early nineties by engineering strains of *C. necator* and *Pseudomonas saccharophila*, since these bacteria, although unable to cleave lactose into glucose and galactose, have some strains recognized as excellent PHA producers [57]. Highly proficient strains of *C. necator* were later genetically modified by introducing the *lacZ*, *lacI*, and *lacO* genes of *E. coli*, encoding for β-galactosidase, *lac* repressor, and *lac* operator, respectively. As reported in Table 6.3, the *lacZ* gene was inserted within a depolymerase (*phaZ1*) sequence, thus obtaining a recombinant strain producing polymer from lactose and whey permeate, while reducing cell depolymerization of the polymer by 30–40% [5].

The opposite strategy would be the translocation of the PHA biosynthesis pathway into suitable hosts efficient in hydrolyzing complex substrates to obtain simple carbon sources but unable to produce and accumulate PHA under natural conditions. This will be possible only if the correct structural genes result in enzymatically active synthases that should be adequately supported by the correct pathways and substrates. Since several PHA synthase pathways have been described [58], a key factor that controls the composition of PHA monomers is known to be the "specificity" of the synthase (PhaC). That is why the earliest and successful PHA genes used have been from *A. eutrophus* (*C. necator*) [59,60]. However, non-producer *E. coli* is the most studied host for PHA gene cloning in view of polymer production (Table 6.4). Indeed, knowledge of related metabolic engineering tools makes heterologous gene expression reasonably simple in this host [61,62], which also lacks an intracellular depolymerization system, and whose culturing, polymer extraction, and purification procedures are well recognized [62]. Most importantly, it is well known that *E. coli* is able to convert lactose directly into galactose and glucose. Therefore, the goal

TABLE 6.3

PHA Production from Organic Waste Streams by Engineered *C. necator* Strains

Strain	Type of PHA	Operation Mode	Substrate	PHA Concentration [g/L]	PHA Content [wt.-% in CDM]
C. necator mREPT [6]	3HB	Flask	Whey permeate	1.4	22
*C. necator NCIMB11599 [7]	3HB	Batch reactor	Sucrose	2.0	73
*C. necator 437-540 [7]	3HB-*co*-LA	Batch reactor	Sucrose	0.1	19
*C. necator 142SR [9]	3HB-*co*-3HHx	Fed-batch reactor	Sucrose	113	81
*C. necator EO1 [52]	3HB	Flask	Waste rapeseed oil	7.6	88
*C. necator EO1 [52]	3HB-*co*-3HV	Batch reactor	Waste rapeseed oil[a]	16.0	86
*C. necator (p*JRDEE32d13*) [10]	3HB-*co*-3HHx	Fed-batch reactor	Soybean oil	102	74
*C. necator (p*JRDEE32d13*) [11]	3HB-*co*-3HHx	Flask	Palm kernel oil	3.7	87
*C. necator (p*JRDEE32d13*) [12]	3HB-*co*-3HV-*co*-3HHx	Flask	Palm kernel oil[b]	5.7	80
*C. necator Re2058 [15]	3HB-*co*-3HHx	Batch reactor	Palm oil	20.7	71
*C. necator Re2160 [15]	3HB-*co*-3HHx	Batch reactor	Palm oil	20.5	66
*C. necator Re2058 (p*CB113*) [14]	3HB-*co*-3HHx	Batch reactor	Waste frying oil, waste animal fat	27.0	60
C. necator MF03 [16]	3HB-*co*-3HHx	Flask	Soybean oil	3.8	79
C. necator PHB-4 harboring *phaC*$_{Cs}$ [18]	3HB-*co*-3HV-*co*-3HHx	Flask	Palm kernel oil[b]	9.6	83
C. necator PHB-4 harboring *phaC*$_{Cs}$ [19]	3HB-*co*-3HHx	Flask	Spent palm oil	6.1	72
*C. necator H16 *glpK*$_{Ec}$ [22]	3HB	Flask	Crude glycerol	1.4	64
*C. necator pKM212-XylAB [23]	3HB	Batch reactor	Sunflower stalk hydrolysate	7.9	73

* Reported as *R. eutropha*; [a] Propionic acid; [b] Sodium valerate was used as 3HV Precursor

TABLE 6.4

PHA Production from Organic Waste Streams by Engineered *E. coli* Strains

Strain	Composition of PHA	Operation Mode	Substrate	PHA Concentration [g/L]	PHA Content [wt.-% in CDM]
E. coli ΔarcA [24]	P(3HB)	Batch reactor	Whey	51.1	73
E. coli GCSC4401 (pSYL107) [25]	P(3HB)	Flask	Bovine whey powder solution	4.5	79
E. coli GCSC6576 (pSYL107) [25]	P(3HB)	Flask	Bovine whey powder solution	5.2	81
E. coli GCSC6576 (pSYL107) [26]	P(3HB)	Fed-batch reactor	Bovine whey powder	69.0	87
E. coli CGSC 4401 (pJC4) [29]	P(3HB)	Fed-batch reactor	Processed bovine whey powder solution	59.6	58
E. coli CGSC 4401 (pJC4) [29]	P(3HB)	Fed-batch reactor with controlled DOC	Processed bovine whey powder solution	96	80
E. coli CGSC 4401 (pJC4) [30]	P(3HB)	Fed-batch reactor with cell recycle membrane	Processed bovine whey powder solution	168	87
E. coli CGSC 4401 (pJC4) [31]	P(3HB)	Fed-batch reactor	Processed bovine whey powder solution	35.7	51
E. coli P8-X8[33]	P(3HB)	Fed-batch reactor	Cheese whey	19	39
E. coli (phaC1) [36]	P(3HB)	Fed-batch reactor	Sugar cane molasses	3.0	75
E. coli (phaC1) [36]	P(3HB)	Fed-batch reactor	Sucrose	2.5	65
E. coli WΔcscR [48]	P(3HB)	Fed-batch reactor	Sucrose	47.7	46
E. coli HMS174(DE3)/ pCOLADuet-1::dhaB1B2::pduP::phaC1 [38]	P(3HP-*co*-3HB)	Fed-batch reactor	Crude glycerol	16.2	89
E. coli HBP01 [28]	P(3HP-*co*-3HB)	Fed-batch reactor	Crude glycerol	10.1	46
E. coli TG1 [39]	P(3HB)	Flask	Soybean hydrolysate	4.4	74

(Continued)

TABLE 6.4 (CONTINUED)
PHA Production from Organic Waste Streams by Engineered *E. coli* Strains

Strain	Composition of PHA	Operation Mode	Substrate	PHA Concentration [g/L]	PHA Content [wt.-% in CDM]
E. coli LS5218 [41]	P(3HB)	Flask	Cellulose hydrolysate	3.3	59
E. coli LS5218 (p*TVSTQKAB/ pBBRXBB2*) [44]	P(3HB)	Flask	Xylan from beechwood	1.1	33
E. coli [p*TV118NpctC1AB(STQK)*] [42]	P(3HB-co-LA)	Flask	Xylan	0.8	29
E. coli [p*TV118NpctC1AB(STQK)*] [43]	P(3HB-co-LA)	Flask	*Eucalyptus* hydrolysate	5.4	62
E. coli JM101 and DH10B [45]	P(3HB)	Flask	Hydrolyzed corn starch	1.0	50
E. coli SKB99 [47]	P(3HB)	Flask	Starch	1.2	57

was firstly reached by cloning PHB-biosynthesis genes [25,26]. Later, a concentrated whey substrate containing high levels of lactose (280 g/L) was used for a fed-batch culture of *E. coli* containing *A. latus* PHB biosynthetic genes [29]. *E. coli* was also used as a host strain for PHA genes coming from *Azotobacter* sp., thus producing the polymer for more than 70% of CDM directly from milk whey, even taking advantage of the absence of the lactose repressor [24]. *C. necator* PHB-synthesis genes fused to a lactose-inducible promoter were also cloned in *E. coli*, thus resulting in remarkable production of the polymer [39].

6.4   THE USE OF MOLASSES AS A CARBON SOURCE

Molasses is a low-value by-product of the sugar industry. Its composition depends on the original source (cane or beet) and refining (sulfured or unsulfured), but 50% of its weight is due to sugars, mainly sucrose and also glucose and fructose [63]. Although pure sucrose is difficult to obtain, store, and transport, molasses is used for a number of industrial fermentations by yeast [64–66], and the problem of the cost of waste disposal persists.

Original attempts to obtain PHA from molasses was made by using *Azotobacter vinelandii* [67], followed by many other efforts with *Alcaligenes latus* (*Azohydromonas lata*), *Pseudomonas corrugata*, and species of *Bacillus*, until recent years [68–74].

However, high PHA production was obtained by genetically engineered sucrose-using *Escherichia coli*, *Klebsiella oxytoca*, and *K. aerogenes*, if properly provided with PHA genes [35]. For instance, in *E. coli*, the *phaC1* from *Pseudomonas* sp. was successfully expressed, thus obtaining 75% polymer concentration [36], as well as PHB biosynthesis genes from *C. necator* giving 80% of CDM and high biomass in a fed-batch culture mode [75]. However, in fed-batch cultivation, the sugar concentration is usually maintained at low levels between feeding pulses, and when sucrose concentration drops below 2 g/L, a *cscR* factor represses the *csc* regulon transcription thus hampering bacterial growth [76]. The obtainment of a *cscR* knock-out *E. coli* mutant [77] overcame this problem, and the *phbABC* operon from *C. necator* allowed more than 1.5-fold PHB production [77].

In contrast, few PHA producers were engineered for sucrose utilization (Table 6.3 and 6.5). For instance, the heterologous *sacC* gene enabled a sucrose utilization pathway in *C. necator* [7]. At the same time, a poly(3-hydroxybutyrate-*co*-lactate) was obtained from sucrose by the same bacterial species expressing genes from *Pseudomonas* sp., *Clostridium propionicum*, *Mannheimia succiniciproducens*, and *E. coli* [8]. Very recently, a recombinant strain of *C. necator* able to produce poly(3-hydroxybutyrate-*co*-3-hydroxyexanoate) [11] has been further provided with *csc* genes from *E. coli* to grow on sucrose; the resulting strain produced vast amounts of the copolymer at high concentrations on CDM [9].

6.5   THE USE OF LIPIDS AS A CARBON SOURCE

A number of industries produce great volumes of lipid-rich waste. Besides slaughterhouses and food processing activities, waste cooking oil and animal fats are

TABLE 6.5

PHA Production From Organic Waste Streams by Other Engineered Microbes

Strain	Type of PHA	Operation Mode	Substrate	PHA Concentration [g/L]	PHA Content [wt.-% in CDM]
K. aerogenes 2688 (pJM9131) [35]	P(3HB)	Flask	Sucrose	3.0	50
K. aerogenes 2688 (pJM9131) [35]	P(3HB)	Fed-batch reactor	Sucrose	24.0	70
B. licheniformis M2-12 [56]	P(3HB-co-3HV-co-3HHx)	Batch reactor	Palm oil mill effluent	16.2	89
D. acidovorans (pBBR1MCS-5-lipH-lipC) [21]	P(3HB-co-3HV-co-4HB)	Flask	Udder	N.A.	27
D. acidovorans (pBBR1MCS-5-lipH-lipC) [21]	P(3HB-co-3HV-co-4HB)	Flask	Lard	N.A.	39
D. acidovorans (pBBR1MCS-5-lipH-lipC) [21]	P(3HB-co-3HV-co-4HB)	Flask	tallow	N.A.	15

"N.A." stands for not available.

produced in large volumes, often causing water and land pollution together with disposal problems [78,79]. The possibility of using this material as a substrate for PHA production is supported by the above considerations, together with the PHA yield from oil and fatty acid, which is ≥ 0.65 g PHA/g, while only 0.32–0.48 g PHA/g for glucose [82]. Wild-type *P. aeruginosa* and *C. necator* can accumulate *mcl*-PHA from residual cooking oil and other waste oils [53,80–84], while the same *mcl*-PHA can be produced by *Erwinia* sp. from crude palm oil [85]. Other residual fats from rendering industries can be converted into PHA by *P. resinovorans* and *R. eutropha* (*C. necator*) [83,86].

Once again, the development of more efficient lipid utilizing strains required some genetic modification (Table 6.3 and 6.5). Obruca and colleagues obtained a mutant strain of *C. necator* with an increased NADPH/NADP$^+$ ratio, thus resulting in the accumulation from waste frying rapeseed oil of almost 90% PHB on CDM. Mutants of *Bacillus licheniformis* produced 3-hydroxyvalerate (3HV) [87] and 3-hydroxyhexanoate (3HHx) copolymers from palm oil [53]. Even more interesting, due to thermal and mechanical properties, copolymers with a high content of 4HB were obtained from oily substrates and by-products [88]. A PHA heteropolymer with 4–5 mol-% 3HHx was obtained from olive oil, corn oil, palm oil [89], and later from soybean oil [10], palm kernel oil, and palm acid oil [11,12] in a PHA-negative *C. necator* mutant containing a heterologous PHA-synthase gene. In a series of studies [38], high percentages on CDM of several copolymers were obtained by recombinant *C. necator* strains expressing PHA synthases with broad substrate specificity. Both *scl-* and *mcl-*monomers were incorporated into the polymer, and Mifune et al. [16] obtained higher levels of 3HHx by the insertion of *phaJ* from *A. caviae*. The same *phaJ* gene was also isolated from *P. aeruginosa* and cloned into a *C. necator* strain containing a PHA synthase from *Rhodococcus aetherivorans*. The recombinant strains obtained accumulated a copolymer with a 17–30% content of 3HHx from palm oil [15]. Similar 3HHx contents were obtained by a recombinant *C. necator* strain from waste animal fats within a protein hydrolysis plant and residual frying oil [14]. PHA synthase gene (*phaC*) was moved from *Chromobacterium* sp. to *C. necator*, thus allowing the production of the interesting terpolymer poly(3-hydroxybutyrate-*co*-3-hydroxyvalerate-*co*-3-hydroxyhexanoate) with a wide range of 3HV monomer compositions, starting from crude kernel oil and 3HV precursors [18]. Moreover, *mcl*-PHA were obtained from lard by *P. putida* and *P. oleovorans* genetically modified to metabolize triacylglycerols [90], and *Delftia acidovorans* engineered for the co-expression of *lipC* and *lipH* lipase genes from *P. stutzeri* resulted in its ability to grow on fatty substrates from slaughterhouses, thus producing PHA with high molar fractions of 4HB [21].

A key component of lipids is glycerol, and among several bacterial genera, both *Cupriavidus* sp. and *Pseudomonas* sp. can metabolize glycerol and accumulate PHA under aerobic conditions. Currently, glycerol is generated in considerable amounts as a major residue from biodiesel manufacturing, and the recent drop in its price makes this an appealing source for microbial production of PHA. However, the presence of impurities seems to affect both growth and polymer properties [91–93]. For this purpose, many different bacterial strains were found to accumulate the polymer from glycerol [94–98], and 3HV-containing copolymers were also obtained [99–101]. Even in the case of this by-product, some genetic modification was performed on *C. necator* by the introduction of genes involved in aerobic metabolism of *E. coli*. In this case, the overexpression of the glycerol kinase (*glp*) genes enhanced polymer accumulation [22]. More genetic modification was carried out on *P. putida* by the deletion of the transcription repressor gene, *glpR*, to reduce the lag phase and increase *mcl*-PHA accumulation [102].

Homopolymers such as P(4HB), P(3HV), and P(3HP), as well as random copolymers and block copolymers, were obtained by engineering *E. coli* strains from glycerol (see Table 6.4) [73,33,103–106]. In addition, *E. coli* was genetically modified by introducing glycerol dehydratase genes from *C. butyricum*, propionaldehyde dehydrogenase from *Salmonella enterica*, and PHA synthase from *C. necator* [38] for the synthesis of an interesting homopolymer such as poly(3-hydroxypropionate) [P(3HP)]. As these monomers are incorporated into PHA, the resulting copolymer will present lower crystallinity and fragility [35]. The addition of glucose to waste glycerol allowed a higher titer of P(3HP) by another recombinant *E. coli* strain containing the heterologous genes propionaldehyde dehydrogenase (*pduP*), glycerol dehydratase and its factors *dhaB123* and *gdrAB* from *K. pneumoniae*, and *phaC1* from *C. necator* [28].

6.6 THE USE OF STARCHY MATERIALS AS A CARBON SOURCE

Besides the high amounts of starch and starch derivatives consumed in the human diet as the main carbohydrate [36,37], almost one-third of the rest is used for other applications such as paper and other derived products, pharmaceuticals, textiles, bio-fuels, and bioplastics [107–112]. The use of hydrolyzed corn starch in place of glucose could reduce the cost of PHB by around 25% [1]. In some cases, some selected bacteria can directly convert raw starch into a polymer and/or a copolymer [111–113].

Once again, strains of *E. coli* have been used as heterologous gene recipients. Poor yields were obtained from hydrolyzed corn starch by cloning *C. necator* PHA synthase genes [44], while in a strain expressing amylase genes from *Paenibacillus* sp. a higher yield was reached [46]. In the efficient amylolytic *Aeromonas* sp., the whole *phaCAB* operon was also inserted to achieve the one-step conversion of starch to PHA, but the results were not particularly encouraging [114]. Since the gram-positive *Corynebacterium glutamicum* is endotoxin-free, it was used in place of *E. coli*, by cloning the operon *phbCAB* of *C. necator* first, and by also inserting alfa amylase from *S. bovis* to use soluble starch [115].

6.7 THE USE OF LIGNOCELLULOSIC MATERIALS AS A CARBON SOURCE

Although lignocellulosic materials are known to be rather recalcitrant to pre-treatments needed to obtain fermentable sugars, lignin, cellulose, and hemicellulose are certainly the most available and cheapest renewable resources in the world [116]. Unfortunately, together with its recalcitrance, this material releases some toxic by-products during hydrolysis, such as 5-hydroxymethylfurfural (5-HMF), thus threatening bacterial growth [117]. However, other than *C. necator* [118,119], a number of bacteria have been isolated as PHA producing from sugarcane bagasse hydrolysate and xylose [120]. While *C. necator* can grow on hydrolysates of different biomass origin, it lacks the enzymes to metabolize pentoses [121, 122], and many efforts are in progress by several research groups to obtain strains able to use xylose and arabinose (Table 6.3 and 6.4). These sugars were metabolized by recombinant strains of *C. necator* expressing *E. coli* genes for arabinose uptake and metabolism [122] and xylose transporters [123], but with a modest polymer production. More recently, high P(3HB) content was obtained from sunflower stalk hydrolysate by the same bacterial species expressing xylose isomerase and xylulokinase genes from *E. coli* [23]. Moreover, the mixture of different sugars, such as glucose and xylose, may induce diauxic growth, resulting in a general slowing down of the process. In this case, although not completely stable yet, a phosphotransferase system (PTS) mutant of *E. coli* was found to use both the above sugars simultaneously [124]. Other attempts at hydrolysate/xylose utilization have been made directly in recombinant *E. coli* strains, one holding *C. necator* PHA biosynthesis genes [39], and another, strain LS5218-STQKABGK [125], showing superior resistance to 5-HMF, one of the major inhibitors released in the pretreated cellulosic material [40]. In order to proceed toward the one-step conversion of cellulosic biomass into PHA, *E. coli* has been further engineered by inserting endoxylanase *xylB* from *S. coelicolor* and *xynB* from *B. subtilis* [43], PHA synthase from *Pseudomonas* sp., propionyl-CoA transferase (PCT) from *Megasphaera elsdenii*, and *phaA* and *phaB* from *C. necator* [41]. Efficient production of P(3HB-*co*-LA) was also achieved by the same recombinant strain from *Eucalyptus* hydrolysate [42]. Thus, consolidated bioprocessing of biomasses, typical of biofuel production [126,127], could be expected in the near future for biopolymer production too.

6.8 CONCLUSIONS AND OUTLOOK

The results reported above indicate that many different approaches are in progress globally to obtain new materials at affordable costs. This concept has to necessarily consider the use of cheap feedstocks for chemical, but especially microbiological, conversion. However, residual, low-cost biomasses from a large array of anthropic activities and their compositions are extremely variable. The types of biomass discussed above are probably the most suitable to be processed into PHA because of their availability, carbon content, and low cost.

Moreover, there are other reasonably interesting new feedstocks to be investigated, such as volatile fatty acids (VFA), from the fermentation of agriculture residues or syngas originating from gasification processes. In the latter case, *E. coli* was used again as a gene donor, and a recombinant strain of *Rhodospirillum rubrum* was constructed and found able to produce, besides PHA, the industrially interesting coproduct, H_2 [128–130]. However, there are numerous metabolic pathways developed mainly in recombinant *E. coli* for PHA biosynthesis from unrelated carbon sources [131,132]. In all the contexts, old and new feedstocks, the use of recombinant DNA techniques are becoming more and more important to increase the possibility of obtaining added-value products from waste materials.

Recently, the development of the proficient gene-editing CRISPR/Cas9 technology paved the way for novel and extremely interesting research tools. The genome-editing methods based on CRISPR/Cas9, CRISPR-Cas12a, and/or CRISPRi have been recently reviewed as becoming increasingly crucial for the regulation of metabolic flux to PHA, the development of strong PHA synthetic pathways, and further host strain optimization [133]. The simultaneous integration of genes into multiple loci of the host genome via CRISPR/Cas9 will support researchers in the near future to edit microbial genomes more quickly, targeting both several PHA syntheses and multiple substrate-utilization pathways to be promptly improved.

ACKNOWLEDGMENTS

This work was partially financed by Padova University with the following research projects CPDA137517/13, BIRD187814/18, DOR1715524/17, DOR1728499/17, DOR1827441/18, DOR1824847/18, and DOR1931153/19.

REFERENCES

1. Choi J, Lee SY. Factors affecting the economics of polyhydroxyalkanoate production by bacterial fermentation. *Appl Microbiol Biotechnol* 1999; 51: 13–21.

2. Du C, Sabirova J, Soetaert W, et al. Polyhydroxyalkanoates production from low-cost sustainable raw materials. *Curr Chem Biol* 2012; 6: 14–25.

3. Koller M, Atlić A, Dias M, et al. Microbial PHA production from waste raw materials. In: Chen GGQ (Ed.), *Plastics from Bacteria: Natural Functions and Applications*. Berlin Heidelberg: Springer, 2010; pp. 85–119.

4. Kaur G, Roy I. Strategies for large-scale production of polyhydroxyalkanoates. *Chem Biochem Eng Q* 2015; 29: 157–172.

5. Povolo S, Toffano P, Basaglia M, et al. Polyhydroxyalkanoates production by engineered *Cupriavidus necator* from waste material containing lactose. *Bioresour Technol* 2010; 101: 7902–7907.

6. Simon RUPAP, Priefer U, Pühler A. A broad host range mobilization system for *in vivo* genetic engineering: Transposon mutagenesis in gram negative bacteria. *Biotechnology* 1983; 1: 784–791.

7. Park SJ, Jang YA, Noh W, et al. Metabolic engineering of *Ralstonia eutropha* for the production of polyhydroxyalkanoates from sucrose. *Biotechnol Bioeng* 2015; 112: 638–643.

8. Park SJ, Jang YA, Lee H, et al. Metabolic engineering of *Ralstonia eutropha* for the biosynthesis of 2-hydroxyacid-containing polyhydroxyalkanoates. *Metab Eng* 2013; 20: 20–28.

9. Arikawa H, Matsumoto K, Fujiki T. Polyhydroxyalkanoate production from sucrose by *Cupriavidus necator* strains harboring *csc* genes from *Escherichia coli* W. *Appl Microbiol Biotechnol* 2017; 101: 7497–7507.

10. Kahar P, Tsuge T, Taguchi K, et al. High yield production of polyhydroxyalkanoates from soybean oil by *Ralstonia eutropha* and its recombinant strain. *Polym Degrad Stab* 2004; 83: 79–86.

11. Loo C-Y, Lee W-H, Tsuge T, et al. Biosynthesis and characterization of poly (3-hydroxy-butyrate-*co*-3-hydroxyhexanoate) from palm oil products in a *Wautersia eutropha* mutant. *Biotechnol Lett* 2005; 27: 1405–1410.

12. Bhubalan K, Lee W-H, Loo C-Y, et al. Controlled biosynthesis and characterization of poly (3-hydroxybutyrate-*co*-3-hydroxyvalerate-*co*-3-hydroxyhexanoate) from mixtures of palm kernel oil and 3HV-precursors. *Polym Degrad Stab* 2008; 93: 17–23.

13. Fukui T, Doi Y. Cloning and analysis of the poly (3-hydroxybutyrate-*co*-3-hydroxyhex-anoate) biosynthesis genes of *Aeromonas caviae*. *J Bacteriol* 1997; 179(15): 4821–4830.

14. Riedel SL, Jahns S, Koenig S, et al. Polyhydroxyalkanoates production with *Ralstonia eutropha* from low quality waste animal fats. *J Biotechnol* 2015; 214: 119–127.

15. Budde CF, Riedel SL, Willis LB, et al. Production of poly (3-hydroxybutyrate-*co*-3-hy-droxyhexanoate) from plant oil by engineered *Ralstonia eutropha* strains. *Appl Environ Microbiol* 2011; 77(9): 2847–2854.

16. Mifune J, Nakamura S, Fukui T. Engineering of *pha* operon on *Cupriavidus necator* chromosome for efficient biosynthesis of poly (3-hydroxybutyrate-*co*-3-hydroxyhex-anoate) from vegetable oil. *Polym Degrad Stab* 2010; 95: 1305–1312.

17. Schäfer A, Tauch A, Jäger W, et al. Small mobilizable multi-purpose cloning vectors derived from the *Escherichia coli* plasmids pK18 and pK19: Selection of defined deletions in the chromosome of *Corynebacterium glutamicum*. *Gene* 1994; 145(1): 69–73.

18. Bhubalan K, Rathi DN, Abe H, et al. Improved synthesis of P (3HB-*co*-3HV-*co*-3HHx) terpolymers by mutant *Cupriavidus necator* using the PHA synthase gene of *Chromobacterium* sp. USM2 with high affinity towards 3HV. *Polym Degrad Stab* 2010; 95(8): 1436–1442.

19. Sudesh K, Bhubalan K, Chuah J-A, et al. Synthesis of polyhydroxyalkanoate from palm oil and some new applications. *Appl Microbiol Biotechnol* 2011; 89: 1373–1378.

20. Kek YK, Chang CW, Amirul AA, et al. Heterologous expression of *Cupriavidus* sp. USMAA2-4 PHA synthase gene in PHB⁻ 4 mutant for the production of poly (3-hydroxy-butyrate) and its copolymers. *World J Microbiol Biotechnol* 2010; 26(9): 1595–1603.

21. Romanelli MG, Povolo S, Favaro LS, et al. Engineering *Delftia acidovorans* DSM39 to produce polyhydroxyalkanoates from slaughterhouse waste. *Int J Biol Macromol* 2014; 71: 21–27.
22. Fukui T, Mukoyama M, Orita I, et al. Enhancement of glycerol utilization ability of *Ralstonia eutropha* H16 for production of polyhydroxyalkanoates. *Appl Microbiol Biotechnol* 2014; 98(17): 7559–7568.
23. Kim HS, Oh YH, Jang Y-A, et al. Recombinant *Ralstonia eutropha* engineered to utilize xylose and its use for the production of poly (3-hydroxybutyrate) from sunflower stalk hydrolysate solution. *Microb Cell Fact* 2016; 15. doi:10.1186/s12934-016-0495-6.
24. Nikel PI, Pettinari MJ, Galvagno MA. et al. Poly (3-hydroxybutyrate) synthesis by recombinant *Escherichia coli arcA* mutants in microaerobiosis. *Appl Environ Microbiol* 2006; 72: 2614–2620.
25. Lee S, Middelberg A, Lee Y. Poly (3-hydroxybutyrate) production from whey using recombinant *Escherichia coli*. *Biotechnol Lett* 1997; 19: 1033–1035.
26. Wong H, Lee SY. Poly-(3-hydroxybutyrate) production from whey by high-density cultivation of recombinant *Escherichia coli*. *Appl Microbiol Biotechnol* 1998; 50: 30–33.
27. Lee SY. Suppression of filamentation in recombinant *Escherichia coli* by amplified FtsZ activity. *Biotechnol Lett* 1994; 16: 1247–1252.
28. Wang Q, Yang P, Xian M, et al. Production of block copolymer poly (3-hydroxybutyrate)-block-poly (3-hydroxypropionate) with adjustable structure from an inexpensive carbon source. *ACS Macro Lett* 2013; 2: 996–1000.
29. Ahn WS, Park SJ, Lee SY. Production of poly (3-hydroxybutyrate) by fed-batch culture of recombinant *Escherichia coli* with a highly concentrated whey solution. *Appl Environ Microbiol* 2000; 66: 3624–3627.
30. Ahn WS, Park SJ, Lee SY. Production of poly (3-hydroxybutyrate) from whey by cell recycle fed-batch culture of recombinant *Escherichia coli*. *Biotechnol Lett* 2001; 23: 235–240.
31. Park SJ, Park JP, Lee SY. Production of poly (3-hydroxybutyrate) from whey by fed-batch culture of recombinant *Escherichia coli* in a pilot-scale fermenter. *Biotechnol Lett* 2002; 24: 185–189.
32. Choi J-I, Lee SY, Han K. Cloning of the *Alcaligenes latus* polyhydroxyalkanoate biosynthesis genes and use of these genes for enhanced production of poly (3-hydroxybutyrate) in *Escherichia coli*. *Appl Environ Microbiol* 1998; 64: 4897–4903.
33. Pais J, Farinha I, Freitas F, Serafim LS, et al. Improvement on the yield of polyhydroxyalkanotes production from cheese whey by a recombinant *Escherichia coli* strain using the proton suicide methodology. *Enzyme Microb Technol* 2014; 55: 151–158.
34. de Lorenzo V, Eltis L, Kessler B, et al. Analysis of *Pseudomonas* gene products using lacIq/Ptrp-lac plasmids and transposons that confer conditional phenotypes. *Gene* 1993; 123: 17–24.
35. Zhang H, Obias V, Gonyer K, et al. Production of polyhydroxyalkanoates in sucrose-utilizing recombinant *Escherichia coli* and *Klebsiella* strains. *Appl Environ Microbiol* 1994; 60: 1198–1205.
36. Saranya V, Shenbagarathai R. Production and characterization of PHA from recombinant *E. coli* harbouring *phaC1* gene of indigenous *Pseudomonas* sp. LDC-5 using molasses. *Braz J Microbiol* 2011; 42: 1109–1118.
37. Sujatha K, Shenbagarathai R. A study on medium chain length-polyhydroxyalkanoate accumulation in *Escherichia coli* harbouring *phaC1* gene of indigenous *Pseudomonas* sp. LDC-5. *Lett Appl Microbiol* 2006; 43: 607–614.
38. Andreeßen B, Lange AB, Robenek H, et al. Conversion of glycerol to poly (3-hydroxypropionate) in recombinant *Escherichia coli*. *Appl Environ Microbiol* 2010; 76: 622–626.

39. Lee SY. Poly (3-hydroxybutyrate) production from xylose by recombinant *Escherichia coli*. *Bioprocess Eng* 1998; 18: 397–399.

40. Nduko JM, Suzuki W, Matsumoto KI, et al. Polyhydroxyalkanoates production from cellulose hydrolysate in *Escherichia coli* LS5218 with superior resistance to 5-hydroxymethylfurfural. *J Biosci Bioeng* 2010; 113: 70–72.

41. Taguchi S, Yamada M, Matsumoto KI, et al. A microbial factory for lactate-based polyesters using a lactate-polymerizing enzyme. *PNAS* 2008; 105: 17323–17327.

42. Takisawa K, Ooi T, Matsumoto KI, et al. Xylose-based hydrolysate from eucalyptus extract as feedstock for poly (lactate-*co*-3-hydroxybutyrate) production in engineered *Escherichia coli*. *Process Biochem* 2017; 54: 102–105.

43. Salamanca-Cardona L, Ashe CS, Stipanovic AJ, et al. Enhanced production of polyhydroxyalkanoates (PHAs) from beechwood xylan by recombinant *Escherichia coli*. *Appl Microbiol Biotechnol* 2014; 98: 831–842.

44. Fonseca GG, de Arruda-Caulkins JC, Antonio RV. Production and characterization of poly-(3-hydroxybutyrate) from recombinant *Escherichia coli* grown on cheap renewable carbon substrates. *Waste Manage Res* 2008; 26: 546–552.

45. Atwood JA, Rehm BH. Protein engineering towards biotechnological production of bifunctional polyester beads. *Biotechnol Lett* 2009; 31: 131–137.

46. Bhatia SK, Shim YH, Jeon JM, et al. Starch based polyhydroxybutyrate production in engineered *Escherichia coli*. *Bioproc Biosyst Eng* 2015; 38: 1479–1484.

47. Arifin Y, Sabri S, Sugiarto H, et al. Deletion of *cscR* in *Escherichia coli* W improves growth and poly-3-hydroxybutyrate (PHB) production from sucrose in fed batch culture. *J Biotechnol* 2011; 156: 275–278.

48. Peoples O, Sinskey AJ. Poly-beta-hydroxybutyrate (PHB) biosynthesis in *Alcaligenes eutrophus* H16. Identification and characterization of the PHB polymerase gene (*phbC*). *J Biol Chem* 1989; 264: 15298–15303.

49. Elzer PH, Kovach ME, Phillips RW, et al. *In vivo* and *in vitro* stability of the broad-host-range cloning vector pBBR1MCS in six *Brucella* species. *Plasmid* 1995; 33(1): 51–57.

50. Smillie C, Garcillán-Barcia MP, Francia MV, et al. Mobility of plasmids. *Microbiol Mol Biol Rev* 2010; 74: 434–452.

51. Alagesan S, Hanko EK, Malys N, et al. Functional genetic elements for controlling gene expression in *Cupriavidus necator* H16. *Appl Environ Microbiol* 2018; 84: e00878–e00918.

52. Johnson AO, Gonzalez-Villanueva M, Tee KL, et al. An engineered constitutive promoter set with broad activity range for *Cupriavidus necator* H16. *ACS Synth Biol* 2018; 7: 1918–1928.

53. Obruca S, Snajdar O, Svoboda Z, et al. Application of random mutagenesis to enhance the production of polyhydroxyalkanoates by *Cupriavidus necator* H16 on waste frying oil. *World J Microbiol Biotechnol* 2013; 29: 2417–2428.

54. Sangkharak K, Prasertsan P. The production of polyhydroxyalkanoate by *Bacillus licheniformis* using sequential mutagenesis and optimization. *Biotechnol Bioprocess Eng* 2013; 18: 272–279.

55. Prazeres AR, Carvalho F, Rivas J. Cheese whey management: A review. *J Environ Manage* 2012; 110: 48–68.

56. Favaro L, Basaglia M, Casella S. Improving polyhydroxy-alkanoates production from inexpensive carbon sources by genetic approaches: A review. *Biofuel Bioprod Biorefin* 2019; 13: 208–227.

57. Pries A, Steinbüchel A, Schlegel HG. Lactose-and galactose-utilizing strains of poly (hydroxyalkanoic acid)-accumulating *Alcaligenes eutrophus* and *Pseudomonas saccharophila* obtained by recombinant DNA technology. *Appl Microbiol Biotechnol* 1990; 33: 410–417.

58. Meng D-C, Shen R, Yao H, et al. Engineering the diversity of polyesters. *Curr Opin Biotechnol* 2014; 29: 24–33.
59. Schubert P, Steinbüchel A, Schlegel HG. Cloning of the *Alcaligenes eutrophus* genes for synthesis of poly-beta-hydroxybutyric acid (PHB) and synthesis of PHB in *Escherichia coli. J Bacteriol* 1988; 170: 5837–5847.
60. Slater SC, Voige W, Dennis D. Cloning and expression in *Escherichia coli* of the *Alcaligenes eutrophus* H16 poly-beta-hydroxybutyrate biosynthetic pathway. *J Bacteriol* 1988; 170: 4431–4436.
61. Aldor IS, Keasling JD. Process design for microbial plastic factories: Metabolic engineering of polyhydroxyalkanoates. *Curr Opin Biotechnol* 2003; 14: 475–483.
62. Shiloach J, Fass R. Growing *E. coli* to high cell density – A historical perspective on method development. *Biotechnol Adv* 2005; 23: 345–357.
63. Yadav R, Solomon S. Potential of developing sugarcane by-product based industries in India. *Sugar Tech* 2006; 8: 104–111.
64. Gopal AR, Kammen DM. Molasses for ethanol: The economic and environmental impacts of a new pathway for the lifecycle greenhouse gas analysis of sugarcane ethanol. *Environ Res Lett* 2009; 4: 044005.
65. Saxena R, Adhikari D, Goyal H. Biomass-based energy fuel through biochemical routes: A review. *Renew Sust Energ Rev* 2009; 13: 167–178.
66. Solomon S. Sugarcane by-products based industries in India. *Sugar Tech* 2011; 13: 408–416.
67. Page WJ. Suitability of commercial beet molasses fractions as substrates for polyhydroxyalkanoate production by *Azotobacter vinelandii* UWD. *Biotechnol Lett* 1992; 14: 385–390.
68. Braunegg G, Genser K, Bona R, Eds., et al. Production of PHAs from agricultural waste material. In: *Macromolecular Symposia 1999* (Vol. 144, No. 1). Weinheim, Germany: Wiley-VCH Verlag GmbH & Co. KGaA, 1999; pp. 375–383.
69. Xie C-H, Yokota A. Reclassification of *Alcaligenes latus* strains IAM 12599T and IAM 12664 and *Pseudomonas saccharophila* as *Azohydromonas lata* gen. nov., comb. nov., *Azohydromonas australica* sp. nov. and *Pelomonas saccharophila* gen. nov., comb. nov., respectively. *Int J Syst Evol Microbiol* 2005; 55: 2419–2425.
70. Solaiman DK, Ashby RD, Hotchkiss Jr AT, et al. Biosynthesis of medium-chain-length poly (hydroxyalkanoates) from soy molasses. *Biotechnol Lett* 2006; 28: 157–162.
71. Santimano M, Prabhu NN, Garg S. PHA production using low-cost agro-industrial wastes by *Bacillus* sp. strain COL1/Afi. *Res J Microbiol* 2009; 4: 89–96.
72. Anjali M, Sukumar C, Kanakalakshmi A, et al. Enhancement of growth and production of polyhydroxyalkanoates by *Bacillus subtilis* from agro-industrial waste as carbon substrates. *Compos Interfaces* 2014; 21: 111–119.
73. Wu Q, Huang H, Hu G, et al. Production of poly-3-hydroxybutyrate by *Bacillus* sp. JMa5 cultivated in molasses media. *Anton Leeuw* 2001; 80: 111–118.
74. Kanjanachumpol P, Kulpreecha S, Tolieng V, et al. Enhancing polyhydroxybutyrate production from high cell density fed-batch fermentation of *Bacillus megaterium* BA-019. *Bioprocess Biosyst Eng* 2013; 36: 1463–1474.
75. Liu F, Li W, Ridgway D, et al. Production of poly-β-hydroxybutyrate on molasses by recombinant *Escherichia coli. Biotechnol Lett* 1998; 20: 345–348.
76. Jahreis K, Bentler L, Bockmann J, et al. Adaptation of sucrose metabolism in the *Escherichia coli* wild-type strain EC3132†. *J Bacteriol* 2002; 184: 5307–5316.
77. Sabri S, Nielsen LK, Vickers CE. Molecular control of sucrose utilization in *Escherichia coli* W, an efficient sucrose-utilizing strain. *Appl Environ Microbiol* 2013; 79: 478–487.
78. Balat M, Balat H. Progress in biodiesel processing. *Appl Energ* 2010; 87: 1815–1835.
79. Talebian-Kiakalaieh A, Amin NAS, Mazaheri H. A review on novel processes of biodiesel production from waste cooking oil. *Appl Energ* 2013; 104: 683–710.

80. Verlinden RA, Hill DJ, Kenward MA, et al. Production of polyhydroxyalkanoates from waste frying oil by *Cupriavidus necator*. *AMB Expr* 2011; 1. doi:10.1186/2191-0855-1-11.

81. Fernández D, Rodríguez E, Bassas M, et al. Agro-industrial oily wastes as substrates for PHA production by the new strain *Pseudomonas aeruginosa* NCIB 40045: Effect of culture conditions. *Biochem Eng J* 2005; 26: 159–167.

82. López-Cuellar M, Alba-Flores J, Rodríguez JG, et al. Production of polyhydroxyalkanoates (PHAs) with canola oil as carbon source. *Int J Biol Macromol* 2011; 48: 74–80.

83. Taniguchi I, Kagotani K, Kimura Y. Microbial production of poly (hydroxyalkanoate) s from waste edible oils. *Green Chem* 2003; 5: 545–548.

84. Obruca S, Marova I, Snajdar O, et al. Production of poly (3-hydroxybutyrate-*co*-3-hydroxyvalerate) by *Cupriavidus necator* from waste rapeseed oil using propanol as a precursor of 3-hydroxyvalerate. *Biotechnol Lett* 2010; 32: 1925–1932.

85. Majid M, Akmal D, Few L, et al. Production of poly (3-hydroxybutyrate) and its copolymer poly (3-hydroxybutyrate-*co*-3-hydroxyvalerate) by *Erwinia* sp. USMI-20. *Int J Biol Macromol* 1999; 25: 95–104.

86. Cromwick A-M, Foglia T, Lenz R, The microbial production of poly (hydroxyalkanoates) from tallow. *Appl Microbiol Biotechnol* 1996; 46: 464–469.

87. Yu L, Dean K, Li L. Polymer blends and composites from renewable resources. *Progr Polym Sci* 2006; 31: 576–602.

88. Fukui T, Doi Y. Efficient production of polyhydroxyalkanoates from plant oils by *Alcaligenes eutrophus* and its recombinant strain. *Appl Microbiol Biotechnol* 1998; 49: 333–336.

89. Solaiman D, Ashby R, Foglia T. Production of polyhydroxyalkanoates from intact triacylglycerols by genetically engineered *Pseudomonas*. *Appl Microbiol Biotechnol* 2001; 56: 664–669.

90. Chatzifragkou A, Papanikolaou S. Effect of impurities in biodiesel-derived waste glycerol on the performance and feasibility of biotechnological processes. *Appl Microbiol Biotechnol* 2012; 95: 13–27.

91. Tanadchangsaeng N, Yu J. Microbial synthesis of polyhydroxybutyrate from glycerol: Gluconeogenesis, molecular weight and material properties of biopolyester. *Biotechnol Bioeng* 2012; 109: 2808–2818.

92. Yang F, Hanna MA, Sun R. Value-added uses for crude glycerol-a byproduct of biodiesel production. *Biotechnol Biofuels* 2012; 5: 1–10.

93. Ibrahim M, Steinbüchel A. *Zobellella denitrificans* strain MW1, a newly isolated bacterium suitable for poly (3-hydroxybutyrate) production from glycerol. *J Appl Microbiol* 2010; 108: 214–225.

94. Ashby RD, Solaiman DK, Foglia TA. Bacterial poly (hydroxyalkanoate) polymer production from the biodiesel co-product stream. *J Polym Environ* 2004; 12: 105–112.

95. Kawata Y, Aiba S-I. Poly (3-hydroxybutyrate) production by isolated *Halomonas* sp. KM-1 using waste glycerol. *Biosci Biotechnol Biochem* 2010; 74: 175–177.

96. Mothes G, Schnorpfeil C, Ackermann JU. Production of PHB from crude glycerol. *Eng Life Sci* 2007; 7: 475–479.

97. Cavalheiro JM, de Almeida MCM, Grandfils C, et al. Poly (3-hydroxybutyrate) production by *Cupriavidus necator* using waste glycerol. *Process Biochem* 2009; 44: 509–515.

98. Cavalheiro JM, Raposo RS, de Almeida MCM, et al. Effect of cultivation parameters on the production of poly (3-hydroxybutyrate-*co*-4-hydroxybutyrate) and poly (3-hydroxybutyrate-4-hydroxybutyrate-3-hydroxyvalerate) by *Cupriavidus necator using* waste glycerol. *Bioresour Technol* 2012; 111: 391–397.

99. Volodina E, Raberg M, Steinbüchel A. Engineering the heterotrophic carbon sources utilization range of *Ralstonia eutropha* H16 for applications in biotechnology. *Crit Rev Biotechnol* 2015; 36(6): 1–14.

100. Hermann-Krauss C, Koller M, Muhr A, et al. Archaeal production of polyhydroxy-alkanoate (PHA) co-and terpolyesters from biodiesel industry-derived by-products. *Archaea* 2013. doi:10.1155/2013/129268.

101. Escapa I, Del Cerro C, García J, et al. The role of GlpR repressor in *Pseudomonas putida* KT2440 growth and PHA production from glycerol. *Environ Microbiol* 2013; 15: 93–110.

102. Li Z-J, Shi Z-Y, Jian J, et al. Production of poly (3-hydroxybutyrate-*co*-4-hydroxybu-tyrate) from unrelated carbon sources by metabolically engineered *Escherichia coli*. *Metab Eng* 2010; 12: 352–359.

103. Yang JE, Choi YJ, Lee SJ, et al. Metabolic engineering of *Escherichia coli* for biosynthesis of poly (3-hydroxybutyrate-*co*-3-hydroxyvalerate) from glucose. *Appl Microbiol Biotechnol* 2014; 98: 95–104.

104. Meng D-C, Shi Z-Y, Wu L-P, et al. Production and characterization of poly (3-hydroxy-propionate-*co*-4-hydroxybutyrate) with fully controllable structures by recombinant *Escherichia coli* containing an engineered pathway. *Metab Eng* 2012; 14: 317–324.

105. Tripathi L, Wu L-P, Meng D, et al. Biosynthesis and characterization of diblock copo-lymer of P (3-hydroxypropionate)-block-P (4-hydroxybutyrate) from recombinant *Escherichia coli*. *Biomacromolecules* 2013; 14: 862–870.

106. Bergthaller W. Starch world markets and isolation of starch. In: *Chemical and Functional Properties of Food Saccharides*. CRC Press, 2003; pp. 117–136.

107. Eliasson A-C, Ed. *Starch in Food: Structure, Function and Applications*. CRC Press, 2006.

108. Favaro L, Jooste T, Basaglia M, et al. Designing industrial yeasts for the consolidated bioprocessing of starchy biomass to ethanol. *Bioengineered* 2013; 4: 97–102.

109. Favaro L, Cagnin L, Basaglia M, et al. Production of bioethanol from multiple waste streams of rice milling. *Bioresour Technol* 2017; 244: 151–159.

110. Kim BS, Chang HN. Production of poly (3-hydroxybutyrate) from starch by *Azotobacter chroococcum*. *Biotechnol Lett* 1998; 20: 109–112.

111. Halami PM. Production of polyhydroxyalkanoate from starch by the native isolate *Bacillus cereus* CFR06. *World J Microbiol Biotechnol* 2008; 24: 805–812.

112. González-García Y, Rosales MA, González-Reynoso O, et al. Polyhydroxybutyrate production by *Saccharophagus degradans* using raw starch as carbon source. *Eng Life Sci* 2011; 11: 59–64.

113. Huang TY, Duan K J, Huang SY, Chen C W. Production of polyhydroxyalkanoates from inexpensive extruded rice bran and starch by *Haloferax mediterranei*. *J Ind Microbiol Biotechnol* 2006; 33: 701–706.

114. Chien CC, Ho LY. Polyhydroxyalkanoates production from carbohydrates by a genetic recombinant *eromonas* sp. *Lett Appl Microbiol* 2008; 47: 587–593.

115. Song Y, Matsumoto Ki, Tanaka T, et al. Single-step production of polyhydroxybutyrate from starch by using α-amylase cell-surface displaying system of *Corynebacterium glutamicum*. *J Biosci Bioeng* 2013; 115: 12–14.

116. Clark JH, Deswarte F, Eds. *Introduction to Chemicals from Biomass*. John Wiley & Sons, 2015.

117. Jönsson LJ, Alriksson B, Nilvebrant N-O. Bioconversion of lignocellulose: Inhibitors and detoxification. *Biotechnol Biofuels* 2013; 6. doi:10.1186/1754-6834-6-16.

118. Annamalai N, Sivakumar N. Production of polyhydroxybutyrate from wheat bran hydrolysate using *Ralstonia eutropha* through microbial fermentation. *J Biotechnol* 2016; 237: 13–17.

119. Obruca S, Benesova P, Marsalek L, et al. Use of lignocellulosic materials for PHA production. *Chem Biochem Eng Q* 2015; 29: 135–144.

120. Cesário MTF, de Almeida MCMD. Lignocellulosic hydrolysates for the production of polyhydroxyalkanoates. In: *Microorganisms in Biorefineries*. Springer, 2015; pp. 79–104.

121. Pohlmann A, Fricke WF, Reinecke F, et al. Genome sequence of the bioplastic-producing "Knallgas" bacterium *Ralstonia eutropha* H16. *Nat Biotechnol* 2006; 24: 1257–1262.

122. Lu X, Liu G, Wang Y, et al. Engineering of an L-arabinose metabolic pathway in *Ralstonia eutropha* W50. *Acta Microbiol Sin* 2013; 53: 1267–1275.

123. Liu K, Liu G, Zhang Y, et al. Engineering of a D-xylose metabolic pathway in *Ralstonia eutropha* W50. *Acta Microbiol Sin* 2014; 54: 42–52.

124. Li R, Chen Q, Wang PG, et al. A novel-designed *Escherichia coli* for the production of various polyhydroxyalkanoates from inexpensive substrate mixture. *Appl Microbiol Biotechnol* 2007; 75: 1103–1109.

125. Tappel RC, Pan W, Bergey NS, et al. Engineering *Escherichia coli* for improved production of short-chain-length-*co*-medium-chain-length poly[(R)-3-hydroxyalkanoate] (SCL-*co*-MCL PHA) copolymers from renewable nonfatty acid feedstocks. *ACS Sust Chem Eng* 2014; 2: 1879–1887.

126. Favaro L, Viktor M, Rose S, et al. Consolidated bioprocessing of starchy substrates into ethanol by industrial *Saccharomyces cerevisiae* strains secreting fungal amylases. *Biotechnol Bioeng* 2015; 112: 1751–1760.

127. Lynd LR, Van Zyl WH, McBride JE, et al. Consolidated bioprocessing of cellulosic biomass: An update. *Curr Opin Biotechnol* 2005; 16: 577–583.

128. Do YS, Smeenk J, Broer KM, et al. Growth of *Rhodospirillum rubrum* on synthesis gas: Conversion of CO to H_2 and poly-β-hydroxyalkanoate. *Biotechnol Bioeng* 2007; 97: 279–286.

129. Choi D, Chipman DC, Bents SC, et al. A techno-economic analysis of polyhydroxyalkanoate and hydrogen production from syngas fermentation of gasified biomass. *Appl Biochem Biotechnol* 2010; 160: 1032–1046.

130. Klask C, Raberg M, Heinrich D, et al. Heterologous expression of various PHA synthase genes in *Rhodospirillum rubrum*. *Chem Biochem Eng Q* 2015; 29: 75–85.

131. Wang Q, Zhuang Q, Liang Q, et al. Polyhydroxyalkanoic acids from structurally-unrelated carbon sources in *Escherichia coli*. *Appl Microbiol Biotechnol* 2013; 97: 3301–3307.

132. Kang Z, Gao C, Wang Q, et al. A novel strategy for succinate and polyhydroxybutyrate co-production in *Escherichia coli*. *Bioresour Technol* 2010; 101: 7675–7678.

133. Zhang X, Lin Y, Wu Q, et al. Synthetic biology and genome-editing tools for improving PHA metabolic engineering. *Trends Biotechnol* 2019. doi:10.1016/j.t.

7 Biosynthesis and Sequence Control of *scl*-PHA and *mcl*-PHA

Camila Utsunomia, Nils Hanik, and Manfred Zinn

CONTENTS

7.1 INTRODUCTION

A unique feature of the class of polyhydroxyalkanoates (PHA) is the fact that the number of available monomers that can constitute the polymer backbone exceeds 140 [1], thus forming a large family of biodegradable plastics with a huge variety of material properties based on the composition and sequence of copolymers made from different available monomers. The low substrate specificity of the enzymes involved in the polymer formation is the reason for this variety. It is praised as a good potential for finding and tuning material properties according to the desired processing techniques and the final application. However, it might as well depict a burden when it comes to the control of the final properties of these materials produced from different carbon substrates. The minute adjustment of a broad spectrum of cultivation parameters might be necessary to gain control of the exact polymer structure and distribution when it comes to constant material properties.

In the context of economically interesting and ecologically sustainable resources, one must expect that the composition of carbon sources for monomer production is somewhat stable and underlies batch-to-batch variabilities or continuously changes over time (e.g., seasonal influences). It is, therefore, of importance to gain an in-depth understanding of the intracellular as well as the extracellular influences on the polymer composition as well as the effect of the composition and monomer sequence on the resulting chemical, thermal, and mechanical properties. A lot of recent research has been dedicated to this topic, and the following chapter is intended to give an overview of the most relevant findings.

Starting with an introduction and an overview of the key features of PHA biosynthesis (i.e., the polymer producing organisms and their characteristics), this section covers substrate related and non-related monomers of PHA, work reported on the elucidation of the spatial structure, and genetic modifications on the key enzymes involved in the polymerization as well as the mechanistic model and its kinetics.

In the second part of this chapter, the focus is on the current concepts for creating sequence-controlled PHA. Special attention is paid to the syntheses of block copolymers with promising resulting material properties, as illustrated by many synthetic approaches (*in vitro* syntheses). The chapter ends with an introduction to different approaches to gaining control of the monomer sequence in living microorganisms (*in vivo* syntheses). The reader is advised that this chapter is not to be understood as an exhaustive review of the existing literature but supposed to give an idea about the major concepts in this field of research.

7.2   THE KEY FACTORS OF PHA BIOSYNTHESIS

PHA are designated as short-chain-length (*scl*), medium-chain-length (*mcl*), and long-chain-length (*lcl*) [2] PHA depending on the number of carbons in the monomer units (i.e., C_3 to C_5, C_6 to C_{14}, and $> C_{14}$, respectively) [3]. The type of PHA produced is dictated by the substrate provided, the bacterial strain, and the existing PHA metabolic pathway, with special importance of the PHA synthase enzyme catalyzing the polymerization reaction.

7.2.1   SUBSTRATES FOR PHA PRODUCTION

Substrates for PHA production are usually restricted to small molecules taken up by simple and facilitated diffusion or actively assisted by membrane transporters embedded in the cell envelope. Typical substrates are simple sugars (i.e., hexoses, pentoses, and disaccharides), triacylglycerols, and hydrocarbons. Monosaccharides (e.g., glucose, fructose) and disaccharides (e.g., sucrose, lactose) can be fermented directly to produce PHA, while polysaccharides (e.g., starch) are not fermentable unless hydrolyzed first [4]. Triacylglycerols, molecules composed of three fatty acids attached to a glycerol backbone, are the main components of oils from plants and fats from animals. Direct use of triacylglycerols requires triacylglycerol-utilizing bacteria secreting lipases in the fermentation media for the release of fatty acids [5]. Alternatively, fatty acids (C_6–C_{14}) can be directly fed to the cells. The fatty acids are then transported into the cell through the cell membrane, where they are catabolized via enzymes of the β-oxidation pathway. Hydrocarbons, in addition, comprising gaseous n-alkanes, gaseous 1-alkenes (C_1–C_6), long-chain paraffinic or olefinic hydrocarbons up to C_{44}, and aromatic hydrocarbons can be used for cell growth and production of *mcl*-PHA [4,6,7]. In addition, hydrogen-oxidizing bacteria, such as *Cupriavidus necator* (formerly *Ralstonia eutropha*), *Ideonella* sp. O-1, and *Rhodospirillum rubrum*, autotrophically synthesizes PHA using CO_2/CO as a carbon source and H_2 as a source of energy metabolism [8,9]. Therefore, the gas phase of the thermodynamic decomposition of organic material (e.g., municipal solid waste), known as syngas, which is mainly composed of CO, H_2, CO_2, and N_2, is a suitable substrate for PHA production [10] (see also Chapter 8 of this volume).

PHA are also synthesized by feeding structurally related carbon sources to the bacteria, which are considered as unusual substrates. Four-hydroxybutyrate (4HB) is a linear monomer unit polymerized into poly(4-hydroxybutyrate) (P4HB) by diverse bacteria from structurally related 4HB, γ-butyrolactone, or 1,4-butanediol [11]. P4HB is an FDA-approved material available on the market for medical applications due to its proven biocompatibility and biodegradability [12]. PHA that are not found in nature containing monomers from 2-hydroxycarboxylate, such as poly(2-hydroxybutyrate) (P2HB), poly(glycolate) (PGL) [13,14], or poly(3-hydroxypropionate) (PHP) from 3-hydroxypropionic acid (3HP) [15], can also be synthesized from structurally related carbon sources. It is only by genetic engineering, that bacteria turned out to be capable of converting complex substrates (e.g., sugars) into such unusual monomers and polymerizing them. Functionalized PHA containing aromatic constituents, such as 3-hydroxy-5-phenylvalerate [16], 3-hydroxy-5-(*p*-methylphenyl)valerate [17], and 3-hydroxy-*p*-cyanophenoxyhexanoate [18], are produced, especially in *Pseudomonas* species, when the respective acids are fed to the bacteria.

7.2.2 THE MICROBIAL PRODUCERS OF PHA

Natural PHA producers can be easily found in various ecological niches, such as soil and water [19]. Due to the high concentration of organic compounds, activated sludge from wastewater treatment plants [20] is an excellent source for the isolation of PHA producers. Generally, with the ample availability of a carbon substrate, stress conditions, and limitation of an essential element for cell growth (e.g., nitrogen, phosphorous, oxygen), bacteria possessing PHA synthases accumulate PHA intracellularly as energy storage material in the form of PHA granules, also referred to as carbonosomes [21,22]. When the starvation of nutrients is relieved, the polymer is degraded by enzymes, known as depolymerases and hydrolases, and the degradation products are used as carbon and energy sources by the bacteria [23,24]. Genetic modification of producers and non-producers of PHA (e.g., *Escherichia coli*) has been extensively carried out, aiming at improving the polymer production [25], expanding the range of PHA monomer composition [26], and controlling the monomer sequence in the polymer [27].

Cupriavidus necator and *Pseudomonas putida* are two representative producers of *scl*-PHA and *mcl*-PHA, respectively. Poly(3-hydroxybutyrate) (PHB) is the most common PHA polymer in nature, and *C. necator* is capable of accumulating up to 90% PHB of its cell dry mass [28]. This bacterium has been found to possess genome features completely dedicated to PHA production from 3-hydroxyacyl-coenzyme A (3HA-CoA), thus not showing sub-products that would decrease the carbon flux toward PHA synthesis [24]. The PHB metabolic pathway starts with the reversible condensation reaction of two molecules of acetyl-CoA (AcCoA) to form acetoacetyl-CoA (AcacCoA) catalyzed by 3-ketothiolase, encoded by the *phaA* gene. The *phaB* gene expresses an AcacCoA reductase, which converts AcacCoA to 3-hydroxybutyryl-CoA (3HB-CoA) using NADPH. Finally, the enzyme PHA synthase (PhaC) polymerizes 3HBCoA monomers into PHB, releasing CoA [29] (Figure 7.1). The β-oxidation pathway can also lead to PHA production from oil substrates and fatty acids in *C. necator*. In the genome of *C. necator* H16, numerous genes encoding for enzymes involved in fatty acid oxidation have been identified, including

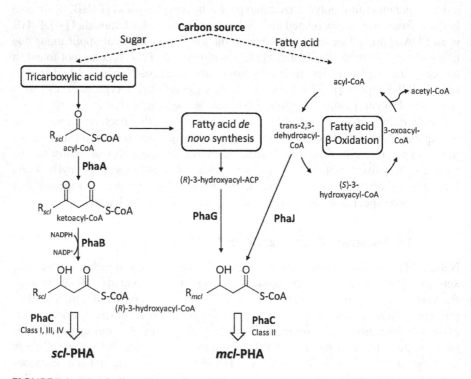

FIGURE 7.1 Metabolic pathways for the biosynthesis of *scl*- and *mcl*-PHA. PhaA: 3-ketothiolase, PhaB: acetoacetyl-CoA reductase, PhaJ: (*R*)-enoyl-CoA hydratase, PhaG: 3-hydroxy-acyl-CoA-acyl carrier protein transferase, and PhaC: PHA synthase. *Scl*: short-chain-length; *Mcl*: medium-chain-length.

acyl-CoA ligases and also enoyl-CoA hydratases catalyzing the generation of (S) and (R)-specific 3HA-CoA intermediates [24].

In 1983, De Smet et al. [7] discovered the very first bacterium of the genus *Pseudomonas*, *P. oleovorans*, was capable of producing *mcl*-PHA from hydrocarbons. These bacteria were well known as efficient degraders of recalcitrant and inhibiting xenobiotics. Currently, *Pseudomonas* species are extensively studied *mcl*-PHA producers [30–32]. It is noteworthy that the central carbon metabolic pathways of *P. putida* were found to differ in key aspects from the generally conserved pathways of bacteria [33]. Most bacteria use glucose as the preferred carbon source. In pseudomonads, glucose assimilation is suppressed in the presence of succinate and other intermediates of the tricarboxylic acid cycle (TCA) due to carbon catabolite repression [34]. Additionally, *P. putida* can also metabolize fatty acids, polyols (e.g., glycerol), amino acids, and aromatic compounds [33]. Biosynthetic pathways of *mcl*-PHA are mostly a direct branch of the fatty acid β-oxidation cycle in which the fatty acids are degraded by the removal of a C_2-unit each cycle, producing AcCoA. The remaining acyl-CoA is further oxidized to 3-ketoacyl-CoA via (S)-3HA-CoA intermediates by well-known pathways [35]. Enoyl-CoA is converted by (R)-enoyl-CoA hydratase (*phaJ* gene) generating (R)-3HA-CoA, which is then polymerized by PhaC into PHA. Alternatively, PHA monomers are obtained via fatty acid *de novo* biosynthesis, where substrates, such as gluconate, glucose, acetate, and ethanol, are metabolized to AcCoA. Here, a 3HA-CoA-acyl carrier protein transferase (PhaG) catalyzes the transfer of the acyl moiety from 3-hydroxyacyl-ACP to produce (R)-3HA-CoA, which is available for PHA polymerization by PhaC [36] (see Figure 7.1).

By feeding structurally related carbon sources to a variety of bacteria for the production of unnatural PHA, other metabolic pathways are involved in the PHA biosynthesis [37,38]. Moreover, as a patchwork involving heterologous expression of genes, new PHA metabolic pathways could be constructed. One example is the synthesis of lactate-based PHA copolymers from simple sugars (e.g., glucose and xylose) in recombinant *E. coli*. Lactyl-CoA (LACoA) is converted from sugars via the expression of genes encoding for propionyl-CoA transferase. The polymerization of LACoA was only possible by employing a PhaC mutant (STQK) originally from *Pseudomonas* sp. 61-3 [39]. Also, in *E. coli*, the expression of genes encoding succinate degradation from *Clostridium kluyveri* (*sucD*, *4hbD*, *orfZ*) and with the *phaC* from *C. necator*, P4HB was successfully synthesized from glucose [40].

Another important aspect of PHA producers is their condition of having non-growth-associated or growth-associated PHA production. In most cases, under nutrient-rich conditions, exponential bacterial growth is observed until the depletion of a growth-essential nutrient (e.g., nitrogen and phosphate). With an almost constant concentration of catalytically active biomass and excess carbon, PHA is synthesized and linearly accumulated. PHA production ends once the bacterium runs out of substrate or intracellular space to store the polymer. This profile characterizes non-growth-associated PHA production [41,42]. However, some bacteria, such as *Azohydromonas lata* (formerly known as *Alcaligenes latus)* and *Paracoccus denitrificans*, efficiently accumulate PHA under nutrient-rich conditions [43], thus possessing growth-associated PHA production.

7.2.3 CLASSIFICATION OF PHA SYNTHASES

PhaCs are the core of the PHA metabolic pathway since they directly influence the PHA monomer composition, the molecular weight, polydispersity, and productivity [44]. Based on primary amino acid sequence, substrate specificity, and subunit composition, PHA synthases can be classified into four classes (Figure 7.2) [45]: (i) class I and class II are enzymes consisting of only one type of subunit (PhaC) with molecular masses between 61 and 73 kDa. Class I PhaCs (e.g., in *C. necator*, *A. lata*, and *Delftia acidovorans*) preferentially use CoA thioesters of various *scl-(R)*-3-hydroxy fatty acids, whereas class II PhaCs (e.g., in *Pseudomonas aeruginosa*, *P. putida*, and *P. oleovorans*) comprises two types of PhaCs, PhaC1 and PhaC2, and preferentially use a CoA thioester of various *mcl-(R)*-3-hydroxy fatty acids [45]. (ii) Class III PHA synthases (e.g., in *Allochromatium vinosum*) comprise enzymes consisting of two different types of subunit. The PhaC subunit (molecular mass of approximately 40 kDa) presents an amino acid sequence similarity of 21–28% to class I and II PhaCs, suggesting structural similarity, and the PhaE subunit (molecular mass of approximately 40 kDa) with no similarity to PHA synthases. Class III PHA synthases preferentially polymerize *scl-(R)*-3-hydroxy fatty acids. (iii) Class IV PHA synthases (e.g., in *Bacillus megaterium*) also have two subunits, where PhaE is replaced by PhaR (molecular mass of approximately 20 kDa), and commonly polymerize *scl-(R)*-3-hydroxy fatty acids (Figure 7.3). The catalytic domains of class II and class IV PhaCs have 40 and 30% identity to that of *C. necator*'s PhaC, indicating a homology based on sequence conservation between them. Besides their particularities, all PhaCs are thought to possess the same catalytic mechanism and a conserved active site architecture [46].

As mentioned above, PhaCs are capable of polymerizing more than 140 different monomers, demonstrating their broad substrate specificity [1]. Class IV PhaCs,

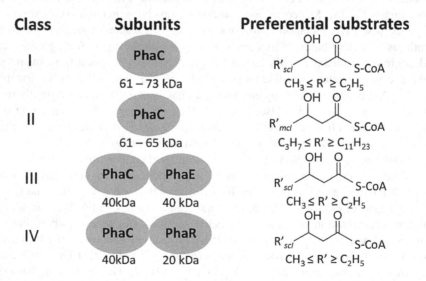

FIGURE 7.2 Classification of PHA synthases (PhaC) based on their subunit composition, preferential substrates, and primary amino acid sequence. R_{scl}: short-chain-length side chain. R_{mcl}: medium-chain-length side chain.

Processive polymerization mechanism   Ping-pong polymerization mechanism

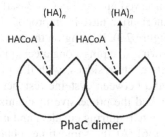 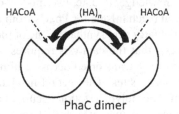

- PHA elongation at a single catalytic site
- PHA elongation with chain transfer between two sets of active sites

Adapted from Chek *et al.* (2017)

FIGURE 7.3   The two putative models of PHA polymerization catalyzed by PHA synthase (PhaC).

however, have a high substrate specificity toward 3HB [47]. Notably, some bacteria have been reported to synthesize PHA copolymers containing both *scl-* and *mcl-*monomers. Class I PhaCs from bacteria, such as *C. necator* [48], *Aeromonas caviae* [49], *A. hydrophila* [50], and *Chromobacterium* sp. USM2 [51], as well as class II PhaCs from *Pseudomonas* sp. 61-3 (PhaC1) [52] and *P. stutzeri* (PhaC2) [53], can polymerize *scl-co-mcl*-PHA copolymers. These polymers have much better material properties and industry-desired characteristics (e.g., flexibility and higher elongation at break) than *scl-* or *mcl*-PHA copolymers. However, the polymerization of the secondary monomer unit is less efficient and has a negative impact on PHA productivity. Therefore, much effort was directed to address the activity/substrate-specificity related positions in PhaCs to improve the production of tailor-made PHA, as reviewed by Chek et al. [54]. The generation of a beneficial mutant PhaC from *A. caviae* resulted in a recombinant strain used for the industrial production of poly(3-hydroxybutyrate-*co*-3-hydroxyhexanoate) (PHBHHx) by KANEKA Co. Ltd., Japan. However, without genetically modifying PhaC1 and PhaC2 from *Rhodococcus aetherivorans* I24, a polymer with 30 mol-% 3HHx was accumulated to up to 66% of its cell dry mass. This was realized by optimizing the upstream metabolic process in *C. necator* expressing the enoyl-CoA hydratase gene (*phaJ*) from *P. aeruginosa* and changing the activity of AcacCoA reductase (*phaB*) [55]. In its first report, however, *R. aetherivorans* I24 was found to accumulate only low amounts (18% of its cell dry mass) of poly(3-hydroxybutyrate-*co*-3-hydroxyvalerate) (PHBV) from toluene [56].

7.2.4   CRYSTAL STRUCTURES OF CLASS I PHA SYNTHASES
AND THE POLYMERIZATION MECHANISM

PHA synthases have been considered as part of the lipase superfamily, which, in turn, is a member of the α/β hydrolase superfamily of proteins. PhaCs possess the signature lipase box sequence, Gx_1Sx_2G (x_1 and x_2 are any amino acid residues, and the essential active site serine of the lipase is replaced with a cysteine in PhaC), which is found in the C-terminal region and is conserved among PhaCs. Based on this knowledge, to date, two models of PHA polymerization have been proposed as general mechanisms for PhaCs from all classes (Figure 7.3). Relying on the mechanism of fatty acid synthases, one of the models, known as the ping-pong mechanism, suggests that the catalysis requires two sets of active site residues, with the PHA chain being transferred across a dimer interface and between cysteine residues of their active sites. The second mechanism, often called the processive mechanism, claims that polymerization occurs in a single active site requiring covalent and noncovalent intermediates during the catalytic cycle [46]. At the same time, obtaining the crystal structure of PhaCs was attempted to gain a better understanding of the biochemistry, factors determining the polymer molecular weight, polydispersity, substrate specificity, and to enhance the precision of the proposed polymerization mechanisms without success until 2016. The disclosure of the X-ray crystal structure of the catalytic domain of class I PhaCs by three research groups brought new insights into the molecular mechanisms of the PHA synthases.

In 2016, Wittenborn et al. [46] reported the first crystal structure of the catalytic domain of *C. necator*'s PhaC (designated here as *Cn*PhaC-CAT) to 1.80 Å resolution. The active site cysteine (Cys^{319}) was replaced by alanine resulting in *Cn*PhaC(C319A) to improve protein stability without the need for detergents. During crystallization, the full-length protein went through proteolysis, resulting in a structure comprising the *Cn*PhaC-CAT containing the residues 201–368 and 378–589 (residues 369–377 are disordered). Resembling the architecture of lipases, but also an archaeal aminopeptidase and bacterial haloperoxidases, *Cn*PhaC-CAT has an α/β-hydrolase fold presenting a central mixed β-sheet flanked by α-helices on both sides. The monomeric forms of PhaC are known to exist in equilibrium with the dimer. However, the dimer is the catalytically active form [57]. The crystal structure also revealed that the residues composing the catalytic site of the enzyme, Cys^{319}-His^{508}-Asp^{480}, are arranged in a cavity with a distance of approximately 10 Å from the nearest surface of the protein. In contrast, for the dimer, the catalytic sites are separated by ~33 Å. A channel with ~18 Å in length was found to be a possible entrance route for 3HB-CoA to access the catalytic site of *Cn*PhaC-CAT from the protein surface. Two arginine residues (Arg^{398}), one residue from each PhaC forming the dimer, are found to be close to the opening of the channel at the protein surface. Arg^{398} is strictly conserved in class I PhaCs, and the two residues together could play a role in binding the CoA nucleotide 5′-pyrophosphate, which can correlate with the increased activity of the dimeric form. Another conserved residue, His^{481}, was found in the surroundings of the channel opening, and it is thought to be linked to substrate recognition by stabilizing the 3′-phosphate of CoA. A putative egress route for the product formed by a series of hydrophobic residues leading from the active site to the surface of the protein at a ~95° angle to the proposed entrance channel was also predicted. A narrow hydrophobic conduit of ~12.5 Å starting from the β-sheet core of the catalytic domain widening into a small solvent pocket near the surface of the protein in the vicinity of N-terminal residues was thought to be involved in PHB chain termination. The narrowest point of the egress channel was predicted to have ~0.7 Å diameter. Since the average radius of a PHB polymer is ~2 Å, an expansion of the channel is likely to be necessary to accommodate the growing polymer chain. This could occur by a simple rearrangement of the hydrophobic side chains forming the egress channel combined with a larger conformational change mediated by a series of conserved structural motifs identified in the channel vicinity.

Once the active sites of the dimer are found inside each monomer and are separated by 33 Å, there is direct structural evidence that the polymerization mechanism very likely involves catalysis at a single active site following the processive model. Modeling of 3HB-CoA into the structure of the CnPhaC-CAT using Arg[398] and His[481] as anchoring points for 5'- and 3'-CoA phosphates, respectively, suggests that the active site cavity is large enough to facilitate catalysis for both PHA initiation and elongation. However, the position of the Cys[319] side chain that is closest to His[508] is not ideal for the nucleophilic attack necessary for the initiation of polymer synthesis because it is positioned too far from the substrate channel. This suggests that at least two Cys[319] side chain conformations may be necessary to fulfill the catalytic cycle. A short distance of 2.2 Å to the 3HB-CoA thioester was found when Cys[319] was modeled to point toward the substrate channel. Interaction of the 3HB-CoA carbonyl oxygen with the Val[320] amide group could serve as a partial oxyanion hole to stabilize the Cys-3HB tetrahedral intermediate. An amide group, Cys[246] or Ile[247], located on a loop near the nucleophile elbow, could also be used in the formation of the oxyanion hole. Once the 3HB-Cys adduct is formed, protonated CoA would exit the active site, and a second 3HB-CoA would enter. The entrance of a new 3HB-CoA into the active site would likely push the 3HB-Cys adduct to the rear of the active site pocket, positioning it at the putative egress channel. For elongation, the second 3HB-CoA has to be oriented to allow for the nucleophilic attack of the 3HB hydroxyl group on the 3HB-Cys thioester. During polymer elongation, however, the functional groups playing a role in oxyanion hole formation are not apparent from the CnPhaC-CAT structure. A model dictated by the required geometry of the thioester bond resulted in 3HB-Cys placed at a distance of ~2.8 Å from the hydroxyl group of 3HB-CoA, such that nucleophilic attack on the 3HB-Cys thioester is possible. In such an orientation, the Asp[480] carboxylate is as far as ~6.7 Å from the hydroxyl group of the 3HB moiety, which hampers the proton transfer. Therefore, Asp[480] would not be the general base catalyst for the direct deprotonation and may play an indirect role. Asp[480] is positioned on the hydrogen bond with His[508] in such a way that His[508] could serve as the general base catalyst analogous to its role in the catalytic triads of other systems. In this model, His[508] is 2.8 Å away from the 3HB hydroxyl moiety. The proposed mechanism by Wittenborn et al. [46] is summarized as (see also Figure 7.3):

1) Nucleophilic attack of Cys[319], deprotonated by His[481], on the 3HB-CoA thioester, forms a 3HB-Cys tetrahedral intermediate.
2) A second 3HB-CoA enters the active site, and its hydroxyl group is deprotonated by His[508]. The newly formed 3HB alkoxide attacks the 3HB-Cys thioester, generating a noncovalent intermediate $(HB)_n$-SCoA and freeing up the catalytic Cys[319] residue. The noncovalent intermediate quickly acylates the active site Cys[319]; the growing polymer chain is transferred back to Cys[319], and CoA is released.

The crystal structure of *C. necator*'s PhaC catalytic domain (defined here as *Cn*PhaC-CAT) at 1.8 Å resolution was also reported by Kim et al. in 2017 [44]. Additional insights into the polymerization mechanism were obtained with the mutagenesis of several amino acid residues playing a role in the substrate binding and dimerization of PhaC. The polymerization mechanism proposed, however, diverged from that of Wittenborn et al. [46]. During the process of protein crystallization, Kim et al. observed proteolysis of PhaC, resulting in two fragments with molecular weights of ~45 and ~20 kDa. The fragments were assigned as the C- and N-terminal domains of *Cn*PhaC. The structure of *Cn*PhaC-CAT has a core subdomain and a dimerization subdomain. The core subdomain presents an α/β hydrolase fold and an extended C-terminal region. Such an extension is found only in class I and II PhaCs. In dimeric *Cn*PhaC-CAT, the helix-turn-helix (HTH) motif of one *Cn*PhaC-CAT subunit is inserted into the D-loop of the other subunit, mediating the hydrophobic interaction between the two monomers. The importance of the hydrophobic residues located at the dimerization interface in the dimer formation was verified by replacing several of them with bulky hydrophilic residues. In all mutants generated, the dimerization was dramatically decreased compared with the wild-type protein as well as the activity of the PHA synthase. As observed by Wittenborn et al. [46], here, the catalytic sites of each *Cn*PhaC-CAT monomer forming the dimer were also found to be 33.4 Å apart. However, two putative substrate-binding tunnels approximately 13 Å in length were identified in the vicinity of the two catalytic sites, suggesting that the polymerization reaction occurs independently at each site. Molecular docking simulations of 3HB-CoA and 3HB-trimer demonstrated that both substrates fit into the substrate-binding tunnel. This indicated that the monomer and growing polymer chain share the same substrate-binding tunnel. Molecular docking identified residues involved in stabilizing the substrates, and the function of some residues was confirmed by site-directed mutagenesis. The tunnel entrances of the two subunits of dimeric *Cn*PhaC-CAT are located at the dimerization interface, and when 3HB-CoA binds to one subunit, it is stabilized by the neighboring subunit. Thus, 3HB-CoA may function as a mediator for dimerization by helping the enzyme to form a dimer more tightly than its apo-form. In addition, it has been reported that the lag phase observed at the onset of PHA polymerization is caused by the low affinity of PhaC for the nucleotide moiety of CoA. When 3HB-trimer was added to an *in vitro* polymerization assay, the lag phase was reduced. This corroborates the presumption that monomers and growing polymer chains share the same substrate-binding tunnel and that both induce PhaC dimerization. Based on the simulation results and the determined structure, Kim et al. proposed the ping-pong mechanism as the most likely model in which two substrates share the same substrate-binding tunnel for PHB polymerization by *Cn*PhaC. The two main steps of this model are summarized as follows (see also Figure 7.3):

1) The 3HB-CoA substrate enters the active site comprising the catalytic triad Cys^{319}-His^{508}-Asp^{480}. The deprotonated thiol group of Cys^{319} serves as a nucleophile that attacks the carbonyl carbon of 3HB-CoA. The result is the covalent bonding of the 3HB moiety to Cys^{319} and the release of the protonated CoA molecule.

2) A growing PHB_n polymer substrate (not CoA activated) enters the active site. Deprotonation in between the terminal hydroxyl group of the growing polymer chain involving Asp[480] and His[508] generates a second nucleophile. This nucleophile attacks the carbonyl carbon of the 3HB-moiety that is already covalently bonded to Cys[319]. The formed PHB_{n+1} polymer leaves the tunnel, the catalytic triad restores its original electrostatic state, and the tunnel becomes available for 3HB-CoA to enter again.

In the proposed mechanism, different from that of Wittenborn et al. [46], two sets of active sites are necessary for PHA chain elongation with chain transfer between them across the dimer interface. During the initial phase of polymerization, the binding affinity of the second substrate is thought to be weaker since the substrate-binding tunnel is occupied by the covalent bond 3HB-Cys. A possible explanation is that the positioning of the terminal hydroxyl group might push the second substrate out of the tunnel. Consequently, this would prevent the proper binding of the nucleotide moiety of CoA and reduce the binding affinity of the following substrate to the binding site. Along with the polymerization, this phenomenon is thought to become more pronounced because, at some point, the nucleotide moiety of CoA would completely leave its binding pocket.

A few months later, in 2017, the crystal structure of the catalytic domain of *Chromobacterium* sp. USM2 PHA synthase (*Cs*PhaC-CAT, residues 175–567) was determined at 1.48 Å resolution [58]. This class I PhaC has the unusual capability of polymerizing PHA, including *scl-* and *mcl-*monomers, producing PHA such as PHBHHx copolymer and poly(3-hydroxybutyrate-*co*-3-hydroxyvalerate-*co*-3-hydroxyhexanoate) (PHBVHHx) terpolymer. *Cs*PhaC has a 46% amino acid sequence identity with the *Cn*PhaC-CAT. The obtained crystal is formed by two *Cs*PhaC-CAT molecules in the asymmetric unit, and these molecules form a dimer. *Cs*PhaC-CAT is folded into a globular structure that belongs to the α/β hydrolase superfamily and comprises the α/β core and CAP subdomains. The CAP subdomains mediate the intermolecular interaction at the dimer interface. Different from the canonical α/β hydrolase fold of lipases, the structure of *Cs*PhaC-CAT has five additional β-strands and two α-helices besides the presence of a major deviation in the helical packing of the CAP subdomain. Accordingly, the α/β core subdomains of *Cs*PhaC-CAT and *Cn*PhaC-CAT resemble each other, and the relative positions of the catalytic triad residues are well conserved. However, the active site of *Cs*PhaC-CAT englobing the catalytic triad (Cys[291]-His[477]-Asp[447]) is covered by the CAP subdomain. Thus, different from the partially open form found for *Cn*PhaC-CAT [44,46], no path was identified from the surface of *Cs*PhaC-CAT for substrate access to the active site located on the interior of the enzyme. Moreover, one disulfide bond is formed in between the non-conserved residues Cys[382] and Cys[438] in the *Cn*PhaC-CAT structure. It was hypothesized that the formation of a disulfide bond induces a conformational change of the CAP subdomain of *Cn*PhaC-CAT and stabilizes the partially open conformation, which is not observed in the closed conformation of the *Cs*PhaC-CAT structure. A possible dynamic structural change during the catalytic action of *Cs*PhaC-CAT may take place mediated by clusters of water molecules existing at the interface between the CAP and α/β core subdomains. With respect to the PHA polymerization on the catalytic triad of CsPhaC-CAT, His[477] is thought to be assisted by Asp[447] in the deprotonation of the thiol group of Cys[291] carrying out the nucleophilic attack of the acyl-CoA (e.g., 3HB-CoA). The same mechanism seems to take place in the formation of a general base catalyst by speeding up the deprotonation of the 3-hydroxyl group of acyl-CoA. The catalytic Cys[291] residues of CsPhaC-CAT dimer are 28.1 Å apart, and no path between them was found in the CsPhaC-CAT crystal structure. This indicates that the PHA polymerization by CsPhaC follows the processive model in a single active site as proposed by Wittenborn et al. [46] for CnPhaC. The ping-pong mechanism suggested by Kim et al. for CnPhaC requires two thiol groups located at a distance short enough to shuttle the growing PHB$_n$ chain back and forth between the two thiols. This mechanism could occur with the help of a Cys residue near Cys[291]. However, mutations of candidate Cys residues did not significantly affect CnPhaC activity. By comparing the structures of CnPhaC and CsPhaC, Chek et al. [58] suggested that the catalytic domain of PHA synthases, in general, exists in an open-closed equilibrium in solution by means of conformational changes in the CAP subdomain.

The PhaC of *C. necator* without the N-terminal domain (1–199 amino acid residues) presented negligible activity in *in vivo* and *in vitro* assays [44]. This fact motivated Kim et al. to assess the function of the N-terminus of CnPhaC (designated here as CnPhaC-ND) in a follow-up publication in 2017 [59]. They reported the first 3D reconstructed model of a full-length *C. necator* PhaC (CnPhaC-F) and the complex of CnPhaC with a granule-associated protein (CnPhaM) based on small-angle X-ray scattering (SAXS) analyses. The SAXS analyses showed that CnPhaC-ND is located on the opposite side of the dimerization subdomain to CnPhaC-CAT, suggesting that the N-terminus might be essential for the binding of CnPhaC-F to the PHA granule. Accordingly, when His-tagged PhaCs were coexpressed with CnPhaAB in *E. coli*, both CnPhaC-ND and CnPhaC-F and not CnPhaC-CAT were localized in the granule fraction. CnPhaM and a PhaM with a truncated C-terminal PAKKA motif that is needed for binding to nucleoids (CnPhaM-C) were used for studying the interaction of CnPhaM with CnPhaC-ND. In *in vitro* polymerization, both CnPhaM and CnPhaM-C increased the PHB polymerization activity of CnPhaC-F without affecting the activity of CnPhaC-CAT. Moreover, protein pull-down assays showed that CnPhaC-ND is required for binding to CnPhaM-C and CnPhaM. Based on these results, the authors proposed that CnPhaM not only localizes CnPhaC to the PHB granule, but it also enhances the enzyme activity. By extending the structure of the CnPhaC-ND by CnPhaM activation, the binding capacity of CnPhaC-ND to the growing PHA polymer is reinforced, and the accessibility of the growing polymer chain to the active site of PhaC is increased.

Modeling of a class I PhaC from *Aquitalea* sp. USM4 (defined here as AqPhaC) based on the crystal structure of CnPhaC-CAT contributed to deciphering the mechanism of PHA polymerization [60]. *Aquitalea* sp. USM4 can accumulate PHA-incorporating monomers such as 3-hydroxyvalerate (HV) and 3-hydroxy-4-methylvalerate. The AqPhaC model revealed a large cavity around the catalytic triad. A channel with three branches (I–III) was found at the cavity and, thus, associated with the catalytic center. The large cavity observed in CnPhaC-CAT [46] was formed by branches I and III, which were proposed as the channels for substrate entrance and product egress, respectively. The hydrophobic branch III forms a dead-end inside the protein core requiring large conformational changes to allow a growing PHA chain to pass through. Branch II, however, is 19.5 Å long and directly open to the solvent. Only small conformational changes would be required for branch II to serve as the tunnel for product exit. Similarly, a three-branched channel was also found when CsPhaC was modeled. The dead-end branch III was thought to be involved in binding the 3HA moiety of the 3HA-CoA substrate. Branch III of AqPhaC was found to be ~14 Å long and able to accommodate mcl-3HA-CoAs (e.g., 3-hydroxydodecanoyl-CoA). However, substantial conformational changes would be needed to allow the passage of polymers with such long 3HA-CoAs. By modeling the class II PhaC1 and PhaC2 from $P.\ aeruginosa$ (PaPhaC1, PaPhaC2), the three-branch channel was also observed, but with an elongated branch III with ~22 Å. In addition, the presence of a much smaller Ala^{295} residue in PaPhaC1 instead of the highly conserved Phe^{318} in Class I PhaCs, seems to enable branch III of PaPhaC1 to create the necessary space for an mcl-3HA-CoA. Indeed, C_{19}3HA-CoA was successfully docked in branch III of PaPhaC1 and PaPhaC2. The finding of the three-branched channel supports the processive model mechanism of PHA polymerization. In this study, because the substrate site is more than 6 Å away from Cys^{319}, it was proposed that the transfer of a PHA$_n$ chain from the catalytic Cys^{319} to a 3HA-CoA would require a ~180° rotation of the 3HA moiety to bring the carbonyl carbon into the catalytic site. In addition, the authors indicated a physical explanation for the lag phase of PHA polymerization. When the first monomer, 3HA-CoA, enters the active site, it can go through branch III, but also to the wider branch II. The lag phase is reduced by using 3HA oligomers as priming units because they covalently bind to the enzyme and increase the stability of the substrate at branch III more efficiently than a single monomer. In addition, by improving the binding of the upcoming 3HA-CoAs, the entrance of monomers into branch II is hampered. An overview of the key points of the studies on the structure of class I PhaCs [44,46,58,60] is presented in Table 7.1.

With the disclosure of the crystal structure of class I PhaCs, the rational design of the PHA synthase by modifying key amino acid residues for improving enzyme activity and expanding substrate specificity has a clearer path. It has been noted that beneficial mutations are found in residues near the catalytic triad that are not inside the catalytic pocket [54]. However, precise knowledge of the influence of PhaCs on copolymer compositions (e.g., $in\ vivo$ substrate specificity) remains to be completed.

7.2.5 Kinetics of PHA Synthesis

Back in 1992, Doi and Kawaguchi [61] proposed a scheme describing the basis of the PHB polymerization kinetic consisting of:

1) Initiation: On the active site of PhaC, a reaction of 3HB-CoA with PhaC takes place, an enzyme/substrate complex is formed, and CoA is released.
2) Propagation: Once initiated, the polymer chain is elongated by the incorporation of new 3HB monomers while freeing up the CoA.
3) Termination: The end of polymerization was thought to involve a chain transfer reaction between PhaC and water.

TABLE 7.1

Compilation of Findings from the Obtained X-Ray Crystal Structure of *C. Necator* and *Chromobacterium* Sp. USM2 PhaCs and the Modeled Phac of *Aquitalea* Sp. USM4

Bacterium	Polymerization Mechanism	Paper Highlights	Ref.
C. necator	Processive	Residues of the catalytic triad are located in a cavity ~10 Å far from the protein surface. Active sites of each PhaC monomer of the dimer are separated by ~33 Å. Putative substrate entrance route is ~18 Å long. Putative product egress route is at a ~95° angle to the proposed entrance channel. For elongation, His^{508} instead of Asp^{480} is the general base catalyst for deprotonation of HB-CoA's hydroxyl group.	[46]
C. necator	Non-processive, ping-pong	Active sites of each monomer of the PhaC dimer are separated by ~33.4 Å. Two putative substrate-binding tunnels of 13 Å length are located near the two catalytic sites of the PhaC dimer. HB-CoA and growing polymer chain seems to fit and share the same substrate-binding tunnel, HB-CoA; its oligomers are suggested to mediate PhaC dimerization.	[44]
Chromobacterium sp. USM2	Processive	The catalytic triad is covered by a subdomain (CAP) of PhaC. The catalytic domain of PhaC has a closed form different from *C. necator*'s PhaC. No path for substrate access to the active site was identified. Catalytic Cys residues from each PhaC monomer are 28.1 Å distance from each other.	[58]
Aquitalea sp. USM4	Processive	A three-branched (I–III) channel found in the large cavity around the catalytic triad. Channels: I. Substrate entrance; II. Product egress; III. Accommodation of the hydroxyacyl (HA) moiety of the HA-CoA monomer. In the elongation, a ~180° rotation of the HA moiety of HA-CoA is required. The binding of HB-CoA and its oligomer decreases the lag phase by hampering the mistaken entrance of HB-CoA to branch II.	[60]

Over the years, many studies were conducted to better understand the kinetics of PHA synthesis. However, it has been very challenging due to the difficulty in reproducing the process of PHA polymerization *in vitro*. Therefore, mathematical models, as well as experimental evidence, have been used to gain further knowledge. As an example, the dynamics of PHA molecular weight distribution was addressed in a population-based model [62]. In a steady-state moment, when the polymer molecular weight is not changing even though the amount of polymer produced increases over time, the active polymer chain distribution is thought to be time-invariant while the inactive chain distribution rises linearly. The modeling takes into account that the chain extension rate could be either constant or linear considering cells in this state. In the linear rate of extension, it was considered that the elongation rate could decrease due to a reduction in PhaC activity caused by the increase in polymer molecular weight. In both models, the same trend was noted, steady-state active chain distribution was found to be a monotonically decreasing function of the polymer molecular weight. Also, as a function of polymer molecular weight, the termination rate was observed to increase until it reached a maximum value succeeded by a slight decrease. It was thought that the binding of the chains with PhaC becomes tighter above a molecular weight threshold, thus resisting termination. The knowledge from this model was further used to design a rational carbon source switch strategy for the biosynthesis of PHA block copolymers (see also Section 7.3 on the sequence control of PHA copolymers). To date, the triggers for the termination reaction resulting in polymers with relatively uniform molecular weight are still poorly understood.

The elongation rate of polymerization is much faster than the initiation rate. For the PhaC of *C. necator*, the turnover number was found to be 188 s^{-1} [63]. Therefore, a mixture of PhaCs already producing large polymers with PhaCs not initiated with HA-CoA is commonly obtained during *in vitro* investigations. In addition, PhaCs from different classes have different kinetics of CoA release. The PhaC of *C. necator*, for example, has a lag phase succeeded by a linear phase of elongation. For class III PhaEC from *A. vinosum*, a fast linear phase followed by a slower linear phase is observed. In a series of publications by Stubbe's group [64–66] studying initiation and elongation of the PHA synthases of *C. necator*, *A. vinosum*, and *Caulobacter crescentus* (class III), brand new understandings on the enzyme kinetics were discovered. First, the results of *in vitro* assays support the processive mechanistic model, which involves both covalent and noncovalent intermediates at a single catalytic site and is the most likely mode of PHA synthesis catalyzed by PhaC. In addition, they could empirically observe that PhaC itself is sufficient for promoting termination where CoA would be the chain terminator. Moreover, the conformation of the complex formed by PhaC and the growing polymer chain seems to play an important role in determining the fate of the polymer chain, namely extension or termination. The knowledge gained over the years on the kinetics of PHA synthesis and the catalytic mechanism of PhaC was essential to fine-tuning the PHA monomer composition and further understanding the synthesis of sequence-controlled PHA, such as PHA block copolymers, as will be depicted in section 7.3 of this chapter.

7.3   SEQUENCE CONTROL OF *SCL*-PHA AND *MCL*-PHA

This section will summarize recent advances on the topic of block copolymeric PHA production. This is further distinguished as *in vitro* and *in vivo* syntheses depending on whether bacterial PHA is isolated and purified prior to subsequent chemical modification or whether the synthesis of sequence-controlled microstructures takes place in the living organisms.

7.3.1   *IN VITRO* SYNTHESES OF BLOCK COPOLYMERIC PHA

The importance of sequence control of copolymeric PHA has been demonstrated by a great deal of synthetic approaches that are based on the chemical modification of PHA isolated from producer strains and subsequently modified chemically or by means of enzymatic reactions. This section gives a general overview of the existing concepts of *in vitro* syntheses for the formation of block copolymeric PHA.

7.3.1.1   PHA-Diol Syntheses

The first reports on block copolymeric PHA go back to the work of Kumagai et al. in 1993 [67]. Macroinitiation by poly(ethylene glycol) (PEG) of tin catalyzed ring-opening polymerization (ROP) of racemic β-butyrolactone produced an A-B-A type triblock copolymer comprising atactic poly[(R, S)-3-hydroxybutyrate] (A: aPHB) and PEG (B) segments. These block copolymeric structures were employed in binary blends with microbial and isotactic PHB produced from butyric acid by *C. necator.* The miscibility, resulting mechanical properties, as well as the enzymatic degradation of produced blend films was studied. While miscibility in the amorphous phase of PHB was promoted, the blended films became more flexible and tough. Depending on the blend composition, enzymatic degradation with extracellular PHB depolymerase from *A. faecalis* T1 took place with up to 48% weight loss in 0.1 M aqueous phosphate buffer solution (pH 7.4) at 37°C within 5 h. The degradation rates were suggested to depend on the degree of crystallinity. With the same approach of macro initiation of ROP of racemic β-butyrolactone, Chen et al. prepared biodegradable nanoparticles from the synthesized triblock copolymers [68] and showed their application as drug carriers [69].

The description of a synthetic route to telechelic diols from PHB and PHBV by Hirt et al. [70] enabled easy access to well-defined lower molecular weight building blocks from microbial, high molecular weight PHA. This eventually leads to several developments of block copolymeric syntheses by either condensation of these building blocks (PHA-diol) with synthetic or biobased polymers or by using the PHA-diol as macroinitiator.

7.3.1.2   PHA-Polyurethanes

The diisocyanate mediated formation of block copolyesterurethanes was realized using carbamate chemistry. Here, diols from PHBV acted as "hard segments" due to their ability to crystalize, while different types of amorphous diol-polyester built "soft segments." The mechanical properties of the resulting block copolymers were characterized [71], and their biocompatibility [72] and biodegradation [73] were further investigated. These materials, also recognized under the registered trademark Degrapol®, were further suggested for use in absorbable biocompatible block copolymers [74,75], porous membranes [76], and scaffolds for artificial heart valves and vascular structures [77].

The synthesis of "all PHA" polyester urethanes by the combination of *scl*-PHA as crystallizing hard segments and *mcl*-PHA as amorphous soft segments was carried out to obtain P(3-hydroxyoctanoate-*b*-PHB) (PHO-*b*-HB) [78], PHB4HB-*b*-PHHxHO [79], and PHB4HB-*b*-PHBHHx [80]. These materials were suggested in the application of wound healing applications as hemostatic materials as they showed promising results when investigating platelet adhesion and activation, cell growth promotion capacity, and cell viability.

A similar route to telechelic diols from PHA to the one from Hirt et al. [70] was carried out using a transesterification between microbial PHA and 1,4-butanediol catalyzed by *p*-toluenesulfonic acid. The subsequent reaction with PEG and diisocyanates gave polyether-*b*-polyester multiblock copolymers that were investigated regarding their mechanical properties as well as their *in vitro* degradation [82]. The amphiphilic nature of these systems promoted by the hydrophobic polyester and the hydrophilic polyether blocks was further investigated by Li et al. [83] using triblock copolymers synthesized by coupling methoxy PEG monopropionic acid (mPEG-acid) to PHA-diol under Steglich conditions; the spherical micelles with a diameter of 20–200 nm were self-assembled in solution upon critical micelle concentrations (cmcs) in the range of 1.3×10^{-5} to 1.1×10^{-3} g/mL.

7.3.1.3 PHA Block Copolyester

The combination of different PHA block segments was realized by Dai et al. [84] using an enzyme-catalyzed ester interchange reaction between divinyl adipate, as the diester component, and the diols of PHB and PHO. The synthesized block copolymers exhibited a melting temperature in the range between 136 and 153°C and a low glass transition temperature between −37 and −39°C.

Vastano et al. [85] produced PHBHHx-*b*-PEG copolymers with and without unsaturated itaconic acid moieties for further chemical modification using lipase-catalyzed transesterification. The PHBHHx-*b*-PEG copolymer was able to form a stable suspension in a hydrophilic environment, which did not show morphological changes between −20 and 70°C.

Upon the polycondensation reaction of PHB-diol (M_n = 1740–2740 g mol⁻¹) with carboxylic-acid-functionalized PEG (PEG-diacid; M_n = 600–3500 g mol⁻¹) as macromonomers, Li et al. [86] were able to isolate and characterize alternating amphiphilic multiblock copolymers that formed separate crystalline phases. Atomic force microscopy (AFM) imaging revealed laterally regular lamellar structures of the investigated thin films made from PHB-*alt*-PEG block copolymers perpendicular to the substrate surface used for imaging.

7.3.1.4 Macroinitiation by PHA-Diols

PHA-diols were used as macroinitiators to prepare di- or multiblock copolymers. The enzymatic ROP of ε-caprolactone catalyzed by *Candida antarctica* lipase B (CALB) in the presence of PHB-diol as a macroinitiator showed selective initiation by the primary alcohol group of the PHB-diol. In contrast, the secondary alcohol group remained as an end group in the final polymers. This synthesis provided access to PHB-b-poly(ε-caprolactone) diblock copolymers with a glass transition temperature around −60°C [87].

Methanolysis of mcl-PHA and subsequent activation of the remaining secondary hydroxyl group of the mcl-PHA methyl ester oligomers with triethylaluminum produced an aluminum alkoxide macroinitiator that was used by Timbart et al. [88] to initiate the ROP of ε-caprolactone.

The same synthetic approach was used by Schreck and Hillmyer [89]. Macroinitiation of the ROP of L-lactide (LLA) by an aluminum alkoxide species of PHBHHx resulted in a microphase-separated PHBHHx-b-PLLA copolymer that was used as a compatibilizer in melt blending of PLLA and PHBHHx. While the addition of a block copolymer promoted the dispersion of PHBHHx in PLLA, the impact properties of the produced ternary blends investigated by means of a notched Izod impact resistance analysis (Izod I.S.) did not improve significantly. This result was explained by low interfacial adhesion at the particle-matrix interface due to small PHBHHx segment sizes in the diblock copolymer as well as the brittleness and crystallinity of the PHBHHx particles that limited their deformation and dissipation of impact loads.

Graft copolymers were produced by Macit et al. [90]. Poly(3-hydroxyalkanoate)-g-[poly(tetrahydrofuran)-b-poly(methyl methacrylate)] [PHA-g-(PTHF-b-PMMA)] was synthesized by a combination of a "grafting-from" approach and subsequent radical polymerization. First, PHA was chlorinated by bubbling the gaseous halogen through a solution of PHA in a chloroform/carbon tetrachloride mixture under direct sunlight. The obtained chlorinated PHA were employed as initiator systems in the presence of hexafluoroantimonate (AgSbF$_6$) in the cationic ROP of tetrahydrofuran monomers, producing poly(3-hydroxyalkanoate)-g-poly(tetrahydrofuran) block copolymers (PHA-g-PTHF). The residual hydroxyl groups of the PTHF side chains were further employed in combination with a Ce(IV)-salt redox system to radically polymerize methyl methacrylate monomers, eventually leading to the above-mentioned multiblock graft copolymer.

Carboxyfunctional PHA can be produced from partial saponification of bacterial polyesters in the presence of 5 M aqueous potassium hydroxide solution containing 18-crown-6 complexing agent. Beta-elimination takes place under basic conditions and leads to an unsaturated chain end instead of the secondary alcohol function of the PHA. Adamus et al. [91] employed this reaction to use the obtained low molecular weight PHA as macroinitiators for the polymerization of β-butyrolactone producing diblock copolymers of natural PHA with atactic PHB via anionic ROP. The PHO-b-poly[(R, S)-HB] (PHO-b-a-PHB) was investigated as a blend compatibilizer for blends of atactic PHB with PHA. The two glass transition temperatures for each of the homopolymers in the blend at −2.5 and −41.3°C disappeared after the addition of 10 wt.-% of the diblock copolymer and a new glass transition temperature at −11.7°C was observed indicating a compatibilization effect. In addition, PHO-b-a-PHB was coated onto vascular prostheses using electrospinning technology. While the elasticity of the prostheses was maintained, the permeability could be reduced. Results similar to those obtained by the coating with poly(D,L-lactide-co-glycolide)

(PLGA), lead to the desired effect of providing sufficient surgical tightness to the textile material.

7.3.1.5 Solvent-Free Transesterification of PHA

Catalyzed transesterification in the melt was used by Ravenelle and Marchessault [92] to produce amphiphilic diblock copolymers from bacterial PHB and monomethoxy poly(ethylene glycol) (mPEG). In this method, the PHB is undergoing simultaneous pyrolysis and transesterification in a one-step process that is catalyzed in the presence of bis(2-ethylhexanoate) tin, leading to amphiphilic diblock copolymers. The self-assembling properties of these structures were further investigated, showing nanoparticle formation with different morphologies. Micelles with a crystalline PHB core and amorphous mPEG shell were obtained when slowly evaporating the organic solvent in an aqueous emulsion. Nanometer-scale rods were obtained by fast evaporation and high shear [93]. Table 7.2 summarizes representative works on the synthetic synthesis of block copolymers containing PHA segments.

7.3.2 *In Vivo* Syntheses of Block Copolymeric PHA

In contrast with the *in vitro* syntheses of block copolymeric PHA, the *in vivo* synthesis requires an in-depth understanding of the bacterial polymerization mechanism and kinetics of the formation and degradation of the carbon storage polymers in living microorganisms. Furthermore, the identification of the produced microstructures plays an important role since intermediates are not always straightforward to isolate. Considering the complete microbial population, the different mechanistic steps of the polymers' formation often proceed in parallel instead of a purely sequentially fashion due to the lack of synchronization of the individual cells. While the reader is referred to Section 7.2, in which the polymerization mechanism, the involved enzymes, and regulating genes are introduced, in Section 7.3.2.1 a summary is given of the major aspects that need to be considered when analyzing *in vivo* produced copolymers followed by different approaches to achieve *in vivo* syntheses of PHA block copolymers.

7.3.2.1 Sequencing PHA Copolymers Produced in Bacteria (*In Vivo*)

While microstructure analyses of synthetic block copolymers are well established and the techniques commonly are well adaptable to the class of PHA, the microstructure analysis of PHA produced in bacteria often seems to remain difficult. Few reports deal with an in-depth analysis of the sequence of the produced polymers. One major issue is the inhomogeneous system of bacterial cultures in terms of kinetics and synchronization. As long as the reaction parameters during polymer synthesis are homogeneous, it is often possible to predict and control the polymer's microstructure. Control of a homogeneous temperature, concentrations, time of addition of monomers, or a catalytic active substance leads in many cases to polymers composed of a homogeneous microstructure. The subsequent characterization may be supported by the possibility of isolating intermediates during the synthesis. For example, in living polymerization syntheses, a first homopolymer block can be produced, and an aliquot of this reaction can be characterized in terms

TABLE 7.2

Summary of the Described *In Vitro* Syntheses of Block Copolymers Containing PHA Segments

	Block Copolymer	Molecular Architecture	Synthetic Approach	Application	Ref.
Macroinitiation	aPHB-*b*-PEG	PHB-PEG-PHB triblock copolymer	Macroinitiation of ROP of β-butyrolactone	Impact modifier, biodegradable nanoparticles as drug carrier	[67,69]
	PHB-*b*-PCL/ PHO-*b*-PCL	Diblock copolymer	Macroinitiation of ROP of ε-caprolactone		[84, 88]
	PHB-*b*-PLLA	Diblock copolymer	Macroinitiation of ROP of L-lactide	Blend compatibilizer	[89]
	PHA-b-aPHB	Diblock copolymer	Macroinitiation from PHA (PHB, PHBV, and PHO) of ROP of β-butyrolactone	blend compatibilizer and cardiovascular engineering	[91]
Polycondensation	PHB-*b*-PEG	Multiblock copolymer	Polycondensation of macrodiols to Polyesterurethane	Impact modifier	[82]
	PHB-*b*-PEG	Multiblock copolymer (alternating)	Polycondensation of macrodiols and macrodiacids	Laterally regular lamellar structures	[108]
	PHB-*b*-PCL	Multiblock copolymer	Polycondensation of macrodiols to polyesterurethane		[81]
	PHBV-*b*-PCL	Multiblock copolymer	Polycondensation of macrodiols to Polyesterurethane	impact modifier	[71]
	PHBV-*b*-Diorez® / Degrapol®	Multiblock copolymer	Polycondensation of macrodiols to polyesterurethane	Porous membranes, scaffolds for artificial heart valve and vascular structures	[75–77]
	PHO-*b*-PHB	Multiblock copolymer	Polycondensation of macrodiols with terephthaloyl chloride		[78]

(Continued)

TABLE 7.2 (CONTINUED)
Summary of the Described *In Vitro* Syntheses of Block Copolymers
Containing PHA Segments

	Block Copolymer	Molecular Architecture	Synthetic Approach	Application	Ref.
	P(3HB-*co*-4HB)-*b*-PHA	Multiblock copolymer	Polycondensation of macrodiols P(3HB)-*co*-4HB, P(3HHx-*co*-3HO), and P(3HB-*co*-3HHx) to polyesterurethane	Hemostatic materials in wound healing	[79, 80]
Esterification	PHB-*b*-mPEG	PEG-PHB-PEG triblock copolymer	Ester bond formation between PHB-diols and mPEG-acid	Spherical micelles for drug delivery	[83]
	PHB-*b*-mPEG	Diblock copolymer	Solvent-free transesterification	Nanoparticles with different morphologies (e.g., micelles and rods)	[93]
Enzymatically catalyzed	PHB-*b*-PHO	Multiblock copolymer (random and alternating)	Enzymatically catalyzed transesterification to block copolymers		[84]
	P3HBHHx-*b*-PEG	Diblock copolymer	Enzymatically catalyzed transesterification to block copolymers	Latent reactive polymers and stable emulsions	[85]
Grafting	PHA-*g*-(PTHF-*b*-PMMA)	Multiblock graft copolymer	Grafting from PHA (PHBV and PHBHHx) and subsequential macroinitiation		[90]

of molecular weight average, monomer sequence, and composition (i.e., purity in the case of a homopolymer). The same reaction can be continued after monomer consumption with the addition of a second monomer. By the definition of living polymerization, reactions that would lead to lower molecular weight homopolymers resulting from incomplete block transfer or immature termination (i.e., the polymerization stops before the second monomer is polymerized) or chain transfer reactions (i.e., the reactive site at the polymerizing chain end is transferred to a new chain polymerizing only the second monomer) are absent.

In this example, the living character of the polymerization leads to well-defined block copolymers with a very narrow molecular weight distribution (PDI) and a defined monomer distribution, thus homogeneous polymers. For the resulting microstructure, nuclear magnetic resonance (NMR) spectroscopy is a very useful tool, as the resonance signals directly depend on the molecular structure of the synthesized molecules. More than just interpreting comonomer composition, the sensitivity of NMR spectroscopy to the chemical environment of atoms allows the analysis of comonomer composition distribution (CCD) by identifying dyads (two adjacent monomers) or triads (two adjacent dyads) in the monomer sequence [94]. The dyad sequence distribution (D) is calculated, and its value becomes one for statistically random copolymers. At the same time, it is close to zero for alternating copolymers and becomes much larger than one for "blocky" copolymers. Here, it should be pointed out that in the literature, the term "blocky" might be used to indicate that the copolymer is a true block copolymer, a mixture of copolymers with different compositions, or a mixture of homopolymers since the dyad sequence distribution does not allow further differentiation [95].

NMR spectroscopy was performed by many researchers in the context of the characterization of PHA copolymers [96–101]. However, in the case of the analysis of a block copolymer, as the number of NMR signals corresponding to a dyad containing two different monomers decreases (i.e., D increases), the sensitivity of the spectroscopic technique does not resolve these signals anymore. Thus, it becomes impossible to differentiate between a blend of homopolymers and block copolymers. In the synthetic example above, this case can simply be excluded due to the polymerization mechanism and the possible analysis of structures produced along with the synthesis. This might not be the case for the bacterial PHA synthesis. Here, several parallel analytical methods need to be performed to exclude the possibility of the formation of more inhomogeneous mixtures of polymers. Molecular weight analyses [e.g., size exclusion chromatography (SEC)], melting, and crystallization behavior determined by thermal analyses [e.g., differential scanning calorimetry (DSC)] are helpful to support the analyses but do not give direct information of the structural architecture of the polymer.

By combining NMR spectroscopy, electrospray ionization mass spectrometry (ESI-MS), SEC, and DSC, Žagar et al. elucidated the effect of nonuniform CCD on the analysis of dyad and triad sequence distribution [95]. Even with the addition of thermal analysis (DSC), fractionation, according to the composition prior to the determination of CCD, was found to be necessary to obtain exact results.

Lütke-Eversloh et al. performed microstructure analysis on PHA copolyesters containing thioesters produced by cultivating cells of *C. necator* H16 in medium containing 3-mercaptobutyric acid (3MB) or 3-mercaptopropionic acid (3MP) and gluconate as carbon sources [102]. Based on NMR analysis, the structure was interpreted as slightly "blocky" rather than random. The interpretation of existing microblock structures in these thioester containing copolymers produced in *C. necator* H16 was questioned by Impallomeni et al. [103], who applied a more integral approach, similar to the one by Žagar et al. [95], to elucidate the polymeric microstructure.

Calculated theoretical dyad intensities for mixtures of random copolymer with homopolymers were in excellent agreement with the experimentally obtained results. This suggests that the earlier analyzed thioester containing copolymer did not contain blocky segments but consisted of a blend of homopolymer and a random copolymer.

7.3.2.2 Sequence Control through the Application of Different Feed Strategies

Kelley et al. reported on the production of layered PHA granules in *C. necator* H16 by sequentially providing fructose and valeric acid and fructose only [104]. The polyester that accumulated in the granula of the microorganisms showed a layered structure like an onion consisting of layers of PHBV and layers of PHB. The different polymer types were differentiated by staining with ruthenium tetroxide (RuO$_4$) as a contrasting agent. With a higher staining efficiency for the PHBV, dark and bright layers in the granules were formed during 24 hours of accumulation. The results were further underlined by DSC analysis of the produced polymers. The formation of block copolymers was hypothesized, taking the accumulation rates of the different substrates into account, for an experiment where the amount of valeric acid consumed per produced polymer does not exceed the kinetic chain length. Intermittent feeding was suggested to be timed to obtain alternating regions of PHB and PHBV. From this hypothesis, a mathematical model of a population balance framework describing the dynamics of the molecular weight distribution of polymer chains that change over time was developed [62,105]. Expressing the molecular weight distributions of active and inactive chains by taking the initiation rate, the molecular weight of a chain, and the monomer concentration into account, an optimum number of switches between fructose and fructose/valerate co-feed for maximum diblock copolymer synthesis was estimated and an experimental attempt was undertaken [106]. Successive fractionation led to 7–14% of polymer that did not show any further different solubility, and the polymer phase separated after annealing into co-continuous phases. Using the same mathematical model for the optimization of the sequential feed strategy, Pederson et al. [107] included online mass spectrometry feedback control. Monitoring the valerate consumption kinetics based on the respiratory quotient (RQ) of *C. necator* allowed the automated control of valerate feeding during polymer accumulation. A total amount of approximately 30% of the polymer sample exhibited melting characteristics and nearest neighbor statistics (i.e., dyad analysis by ^{13}C NMR spectroscopy) indicative of block copolymers. While the authors were able to perform further testing on the produced polymers (DSC, rheology, dynamic mechanical analysis, and tensile testing) with results in good agreement with expected results for block copolymeric materials, they stated that their approach for the block copolymer synthesis is complicated by the fact that the synthesis does not occur synchronously on all polymer chains because of random chain initiation and termination events. They further expressed their difficulties in calculating the block copolymer yield from the analytical results due to the polymer chain heterogeneity, broad molecular weight distribution, and marginal chemical difference between the comonomers.

A combination of *scl*- and *mcl*-PHA was produced in the genetically modified *P. putida* KTOY06ΔC strain. Using the sequential feeding approach and by monitoring the substrate consumption through the dissolved oxygen concentration, Li et al. [108] were able to produce and isolate a ternary copolymer consisting of 3HB, 3HV, and 3-hydroxyheptanoate (3HP) repeating units. The physical characterization of the obtained materials by means of DSC, rheology, and tensile testing showed results different from blends of PHB and poly(3-hydroxyvalerate-*co*-3-hydroxyheptanoate) (PHVHP), which, supported by NMR spectroscopy, led to the conclusion of the successful synthesis of a PHB-*b*-PHVHp block copolymer. The same strategy, including the mutant strain *P. putida* KTOY06ΔC, was shown to enable the synthesis of a combination of PHB and PHHx. Tripathi et al. [109] used fractionation methods, NMR-spectroscopy-based dyad analysis, and DSC to elucidate the microstructure of the produced copolymers.

Employing *P. putida* KTHH06 in combination with a sequential feeding strategy, Hu et al. [110] produced copolymers of 3HB and 4HB from sodium butyrate and γ-butyrolactone, respectively. The obtained D value for the dyad analysis was found to be around 356, and, together with 2D NMR ^{1}H-^{13}C HMBC experiments showing covalent bond formation between 3HB and 4HB monomers, the results were taken as solid evidence of the existence of the PHB-*b*-P4HB block copolymer. Further thermal analysis by DSC and mechanical testing (tensile testing) allowed them to differentiate between the produced polymers and random copolymer as well as blends of homopolymers.

7.3.2.3 Sequence Control through Genetic Modification

After the successful construction of a mutated PHA synthase based on the wild-type synthase (*Ps*PhaC1) from thermotolerant *Pseudomonas* sp. SG4502, Tajima et al. investigated the potential of the isolated two-point mutated enzyme *Ps*PhaC1 (STQK) to produce copolymers of poly(lactic acid) (PLA) and PHB with a block sequence [111]. Priming the enzyme *in vitro* with (*R*)-3-hydroxybutyryl coenzyme A (3HB-CoA) and the subsequent addition of (*R*)-lactyl coenzyme A (LACoA) after the consumption of 3HB-CoA copolymer with 30 mol-% of 3HB and 70 mol-% of lactic acid was isolated. The comparison with synthetically synthesized poly(3-hydroxybutyrate)-*b*-poly(lactic acid) (PHB-*b*-PLLA) by ^{1}H NMR spectroscopy suggested the existence of block sequences. Furthermore, a single mutation in the PHA synthase from *C. necator* (*Cn*PhaC) was conducted, and the polymerizing activity was evaluated *in vivo* using recombinant *E. coli* LS5218 [112]. NMR sequence analysis in combination with fractionation by SEC and solvent extraction indicated the production of a block copolymer with 7 mol-% of lactic acid and 93 mol-% of 3HB comonomers. However, the authors stated that they were not successful in distinguishing between a tapered block and a multiblock microstructure of the produced polymers.

By investigating the polyester production of a chimeric enzyme composed of the PHA synthases from *A. caviae* (N-terminal) and *C. necator* (C-terminal), designated here as PhaC$_{Ac-Cn}$ in *E. coli* harboring propionyl coenzyme A transferase (PCT) and PhaC$_{Ac-Cn}$, Matsumoto et al. not only identified higher productivity than one of the parent enzymes but also the formation of a block sequence generated under the simultaneous consumption of two monomer precursors [113]. Based on the systematic investigation of enzymatic activities *in vitro* and *in vivo* as well as the analysis of intracellular intermediates in combination with a structural analysis of the produced polymers by NMR spectroscopy (triad and dyad analyses) accompanied by fractionation studies and thermo-analysis, the authors proposed a model including four steps, hypothesizing that the reason for the sequence control is a rapid fluctuation of the intracellular 3HB-CoA concentration. In the first step, PCT synthesizes a larger pool of 2-hydroxybutyryl-CoA (2HB-CoA) monomer due to higher substrate specificity toward 2-hydroxybutyrate (2HB) than 3HB and no initiation of polymer accumulation triggered by 2HB-CoA. In the second step, 3HB-CoA is accumulated during a prolonged lag phase of the synthase enzyme until, in the next step, 3HB-CoA initiates polymerization with a much faster propagation rate than for 2HB-CoA until, in step 4, a P2HB segment is synthesized due to a low 3HB-CoA level. The composition of the block copolymers produced from the above-described method could be influenced by the ratio of sodium 3HB to sodium 2HB ranging from 25 to 90% of 2HB fraction.

7.3.2.4 Sequence Control through the Influence of Genetic Regulation

The control of the expression level of enzymes involved in the polymer accumulation in PHA accumulating organisms was developed by Aldor et al. [114], introducing an approach for the control of the monomer composition in environments, where the composition of the extracellular substrates is not adjustable. In a recombinant strain of *Salmonella enterica serovar Typhimurium*, a gene encoding for propionyl coenzyme A synthetase was placed under the control of the isopropyl-β-thiogalactopyranoside (IPTG)-inducible taclacUV5 promoter ($P_{taclacUV5}$) while the PHA synthesis operon (phaBCA) from *Acinetobacter sp.* RA3849 was coexpressed under the control of the arabinose-inducible araBAD promoter (P_{BAD}). This system, also referred to as the "dial-a-composition" system was grown in medium containing fixed substrate concentrations and the composition of the copolymer was varied between 2 and 25 mol-% 3-hydroxyvalerate (3HV) by controlling the IPTG level in the medium.

The limitations on the amount of 3HV content in the produced copolymers were considered to be based on the high activity of enzymes forming AcCoA strongly proliferating the comonomer synthesis of 3HB. Hence, Iadevaia and Mantzaris suggested tighter control by the simultaneous manipulation of gene expression levels for genes involved in the production of both comonomers [115]. For such a tightly regulated system, the authors suggested the construction of an artificial genetic toggle network to influence the expression levels of the synthetase enzymes responsible for the formation of AcCoA and propionyl coenzyme A (PrCoA) (i.e., Acs and PrpE enzymes) encoded by *acs* and *prpE* genes to modulate the intracellular monomer precursor concentrations. The proposed genetic toggle network originally constructed by Gardner et al. [116] is composed of two promoter-repressor pairs. The protein expressed by each gene downregulates the transcriptional activity of the other gene by binding to the promoter that controls the expression of the other gene. It was postulated that the inclusion of an IPTG-inducible promoter in the genetic toggle network would allow the control of the composition of PHA copolymers produced by *E. coli* cell populations through manipulation of extracellular inducer concentrations and allow a composition of the substrates. A mathematical model was developed to describe the coupling between the dynamics of molecular weight distribution of

PHA copolymers with those of monomer formation through metabolic processes, including the artificial regulation of copolymer composition, to evaluate the potential of such a regulated gene expression system. The results from the performed *in silico* study based on the described model indicated the feasibility of the approach with a varying mass fraction of 3HB between 95 and 5% depending on extracellular inducer concentrations, thus improving the range of theoretically available PHBV compositions.

Block copolymer synthesis was brought into this context by a feasibility study by the same authors evaluating the synthesis of block copolymers driven by an oscillatory genetic network [117]. Such an artificial genetic network was developed by Elowitz and Leibler [118] and termed as "repressilator." It consists of three promoter-repressor pairs mutually interacting in negative feedback loops, including IPTG induction of one of the three involved promoters to allow for external influence on the network behavior. Putting the regulation of the above-mentioned genes (i.e., *acs* and *prpE*) under control of the so-called repressilator was hypothesized to lead to oscillatory levels of Acs and PrpE enzymes, which would drive the formation of the PHB-*b*-PHV block copolymer structures (Figure 7.4). A mathematical model of this system and the simulation results predicted efficiently driving the periodic incorporation of block structures into actively elongating polymer chains. The results further suggested that the oscillatory patterns produced by the repressilator are significantly affected by the choice of extracellular inducer concentration as well as various molecular characteristics, such as promoter strength and repressor protein half-lives. Although Portle et al. found an implementation of this system in *E. coli* to be affected by initially non-predicted interactions between a repressor and the nonspecific promoters of the other two repressors, the *in silico* studies showed the

Composition Control of PHBV Copolymers

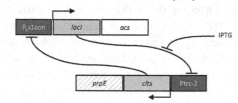

Artificial genetic toggle network to influence the expression levels of the genes *acs* and *prpE* responsible for the formation of the comonomer precursors Acetyl coenzyme A (AcCoA) and propionyl coenzyme A (PrCoA). The toggle is composed of two promoter–repressor pairs and an IPTG-inducible promoter that allows the control of the composition of PHA copolymers without changing the composition of the substrates.

Sequence Control of PHBV Copolymers

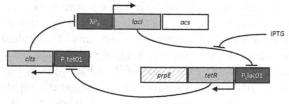

Putting the regulation of *acs* and *prpE* under control of the so called repressilator was hypothesized to lead to oscillatory levels of the comonomer forming enzymes, which would drive the formation of the P3HB-*b*-3HV block copolymer structures. The repressilator consists of three promoter–repressor pairs mutually interacting in negative feedback loops including IPTG induction to allow for external influence of the network behavior.

FIGURE 7.4 Schematic representation of the *in silico* studied artificial genetic networks. Both systems control the transcription of the key enzymes Acs and PrpE for acetyl-CoA and propionyl-CoA formation, respectively.

potential of artificial genetic networks in achieving sequence control during *in vivo* syntheses of PHA block copolymers [119].

REFERENCES

1. Kim DY, Kim HW, Chung MG, et al. Biosynthesis, modification, and biodegradation of bacterial medium-chain-length polyhydroxyalkanoates. *J Microbiol* 2007; 45(2): 87–97.
2. Singh AK, Mallick N. Enhanced production of SCL-LCL-PHA co-polymer by sludge-isolated *Pseudomonas aeruginosa* MTCC 7925. *Lett Appl Microbiol* 2008; 46(3): 350–357.
3. Lu J, Tappel RC, Nomura CT. Mini-review: Biosynthesis of poly(hydroxyalkanoates). *J Macromol Sci Pt C Polym Rev* 2009; 49(3): 226–248.
4. Jiang G, Hill DJ, Kowalczuk M, et al. Carbon sources for polyhydroxyalkanoates and an integrated biorefinery. *Int J Mol Sci* 2016; 17(7): 1157.
5. Solaiman DKY, Ashby RD, Foglia TA, et al. Conversion of agricultural feedstock and coproducts into poly(hydroxyalkanoates). *Appl Microbiol Biotechnol* 2006; 71(6): 783–789.
6. Ni YY, Kim DY, Chung MG, et al. Biosynthesis of medium-chain-length poly(3-hydroxyalkanoates) by volatile aromatic hydrocarbons-degrading *Pseudomonas fulva* TY16. *Bioresour Technol* 2010; 101(21): 8485–8488.
7. De Smet MJ, Eggink G, Witholt B, et al. Characterization of intracellular inclusions formed by *Pseudomonas oleovorans* during growth on octane. *J Bacteriol* 1983; 154(2): 870–878.
8. Volova TG, Kiselev EG, Shishatskaya EI, et al. Cell growth and accumulation of poly-hydroxyalkanoates from CO_2 and H_2 of a hydrogen-oxidizing bacterium, *Cupriavidus eutrophus* B-10646. *Bioresour Technol* 2013; 146: 215–222.
9. Tanaka K, Miyawaki K, Yamaguchi A, et al. Cell growth and (3HB) accumulation from CO_2 of a carbon monoxide-tolerant hydrogen-oxidizing bacterium, *Ideonella* sp. O-1. *Appl Microbiol Biotechnol* 2011; 92(6): 1161–1169.
10. Karmann S, Follonier S, Egger D, et al. Tailor-made PAT platform for safe syngas fermentations in batch, fed-batch and chemostat mode with *Rhodospirillum rubrum*. *Microb Biotechnol* 2017; 10(6): 1365–1375.
11. Saito Y, Nakamura S, Hiramitsu M, et al. Microbial synthesis and properties of poly(3-hydroxybutyrate-*co*-4-hydroxybutyrate). *Polym Int* 1996; 39(3): 169–174.
12. Williams SF, Rizk S, Martin DP. Poly-4-hydroxybutyrate (P4HB): A new generation of resorbable medical devices for tissue repair and regeneration. *Biomed Tech (Berl)* 2013; 58(5): 439–452.
13. Matsumoto K, Terai S, Ishiyama A, et al. One-pot microbial production, mechanical properties, and enzymatic degradation of isotactic P[(*R*)-2-hydroxybutyrate] and its copolymer with (*R*)-lactate. *Biomacromolecules* 2013; 14(6): 1913–1918.
14. Matsumoto K, Shiba T, Hiraide Y, et al. Incorporation of glycolate units promotes hydrolytic degradation in flexible poly(glycolate-*co*-3-hydroxybutyrate) synthesized by engineered *Escherichia coli*. *ACS Biomat Sci Eng* 2016; 3(12): 3058–3063.
15. Nakamura S, Kunioka M, Doi Y. Biosynthesis and characterization of bacterial poly(3-hydroxybutyrate-*co*-3-hydroxypropionate). *J Macromol Sci Pt A Chem* 1991; 28(Sup1): 15–24.
16. Curley JM, Hazer B, Lenz RW, et al. Production of poly(3-hydroxyalkanoates) containing aromatic substituents by *Pseudomonas oleovorans*. *Macromolecules* 1996; 29(5): 1762–1766.
17. Hany R, Brinkmann M, Ferri D, et al. Crystallization of an aromatic biopolyester. *Macromolecules* 2009; 42(16): 6322–6326.

18. Kim O, Gross RA, Rutherford DR. Bioengineering of poly(β-hydroxyalkanoates) for advanced material applications: Incorporation of cyano and nitrophenoxy side chain substituents. *Canad J Microbiol* 1995; 41(13): 32–43.

19. Amirul AA, Yahya ARM, Sudesh K, et al. Biosynthesis of poly(3-hydroxybutyrate-*co*-4-hydroxybutyrate) copolymer by *Cupriavidus* sp. USMAA1020 isolated from Lake Kulim, Malaysia. *Bioresour Technol* 2008; 99(11): 4903–4909.

20. Inoue D, Suzuki Y, Uchida T, et al. Polyhydroxyalkanoate production potential of heterotrophic bacteria in activated sludge. *J Biosci Bioeng* 2016; 121(1): 47–51.

21. Jendrossek D, Pfeiffer D. New insights in the formation of polyhydroxyalkanoate granules (carbonosomes) and novel functions of poly(3-hydroxybutyrate). *Environ Microbiol* 2014; 16(8): 2357–2373.

22. Sudesh K, Abe H. *Practical Guide to Microbial Polyhydroxyalkanoates*. Shrewsbury, UK: Smithers Rapra Technology, 2010, 160 pp.

23. Zinn M. Tailor-made synthesis of polyhydroxyalkanoate. *Eur Cell Mat* 2003; 5: 38–39.

24. Kutralam-Muniasamy G, Perez-Guevara F. Genome characteristics dictate poly-R-(3)-hydroxyalkanoate production in *Cupriavidus necator* H16. *World J Microbiol Biotechnol* 2018; 34(6): 79.

25. Obruca S, Snajdar O, Svoboda Z, et al. Application of random mutagenesis to enhance the production of polyhydroxyalkanoates by *Cupriavidus necator* H16 on waste frying oil. *World J Microbiol Biotechnol* 2013; 29(12): 2417–2428.

26. Sheu D-S, Lee C-Y. Altering the substrate specificity of polyhydroxyalkanoate synthase 1 derived from *Pseudomonas putida* GPo1 by localized semirandom mutagenesis. *J Bacteriol* 2004; 186(13): 4177–4184.

27. Wang Q, Yang P, Xian M, et al. Production of block copolymer poly(3-hydroxybutyrate)-block-poly(3-hydroxypropionate) with adjustable structure from an inexpensive carbon source. *ACS Macro Lett* 2013; 2(11): 996–1000.

28. Park DH, Kim BS. Production of poly(3-hydroxybutyrate) and poly(3-hydroxybutyrate-*co*-4-hydroxybutyrate) by *Ralstonia eutropha* from soybean oil. *New Biotechnol* 2011; 28(6): 719–724.

29. Pena C, Castillo T, Garcia A, et al. Biotechnological strategies to improve production of microbial poly-(3-hydroxybutyrate): A review of recent research work. *Microb Biotechnol* 2014; 7(4): 278–293.

30. Hartmann R, Hany R, Pletscher E, et al. Tailor-made olefinic medium-chain-length poly[(*R*)-3-hydroxyalkanoates] by *Pseudomonas putida* GPo1: Batch versus chemostat production. *Biotechnol Bioeng* 2006; 93(4): 737–746.

31. Hoffmann N, Rehm BHA. Regulation of polyhydroxyalkanoate biosynthesis in *Pseudomonas putida* and *Pseudomonas aeruginosa*. *FEMS Microbiol Lett* 2004; 237(1): 1–7.

32. Hanik N, Utsunomia C, Arai S, et al. Influence of unusual co-substrates on the biosynthesis of medium-chain-length polyhydroxyalkanoates produced in multistage chemostat. *Front Bioeng Biotechnol* 2019; 7: 301.

33. Poblete-Castro I, Becker J, Dohnt K, et al. Industrial biotechnology of *Pseudomonas putida* and related species. *Appl Microbiol Biotechnol* 2012; 93(6): 2279–2290.

34. Wolff JA, MacGregor CH, Eisenberg RC, et al. Isolation and characterization of catabolite repression control mutants of *Pseudomonas aeruginosa* PAO. *J Bacteriol* 1991; 173(15): 4700–4706.

35. Fiedler S, Steinbüchel A, Rehm BH. The role of the fatty acid β-oxidation multienzyme complex from *Pseudomonas oleovorans* in polyhydroxyalkanoate biosynthesis: Molecular characterization of the *fadBA* operon from *P. oleovorans* and of the enoyl-CoA hydratase genes *phaJ* from *P. oleovorans* and *Pseudomonas putida*. *Arch Microbiol* 2002; 178(2): 149–160.

36. Rehm BHA, Krüger N, Steinbüchel A. A new metabolic link between fatty acid *de novo* synthesis and polyhydroxyalkanoic acid synthesis. *J Biol Chem* 1998; 273(37): 24044–24051.
37. Sudesh K, Abe H, Doi Y. Synthesis, structure and properties of polyhydroxyalkanoates: Biological polyesters. *Progr Polymer Sci* 2000; 25(10): 1503–1555.
38. Masood F. Polyhydroxyalkanoates in the food packaging industry, in *Nanotechnology Applications in Food*. Elsevier, 2017, pp. 153–177.
39. Taguchi S, Yamada M, Matsumoto K, et al. A microbial factory for lactate-based polyesters using a lactate-polymerizing enzyme. *Proc Natl Acad Sci* 2008; 105(45): 17323–17327.
40. Zhou XY, Yuan XX, Shi ZY, et al. Hyperproduction of poly(4-hydroxybutyrate) from glucose by recombinant *Escherichia coli*. *Microb Cell Fact* 2012; 11(1): 54.
41. Zinn M, Witholt B, Egli T. Occurrence, synthesis and medical application of bacterial polyhydroxyalkanoate. *Adv Drug Del Rev* 2001; 53(1): 5–21.
42. Koller M. A review on established and emerging fermentation schemes for microbial production of polyhydroxyalkanoate (PHA) biopolyesters. *Fermentation* 2018; 4(2): 30.
43. Kourmentza C, Placido J, Venetsaneas N, et al. Recent advances and challenges towards sustainable (PHA) production. *Bioengineering (Basel)* 2017; 4(2): E55.
44. Kim J, Kim YJ, Choi SY, et al. Crystal structure of *Ralstonia eutropha* polyhydroxyalkanoate synthase C-terminal domain and reaction mechanisms. *Biotechnol J* 2017; 12(1).
45. Rehm BHA. Polyester synthases: Natural catalysts for plastics. *Biochem J* 2003; 376(Pt 1): 15.
46. Wittenborn EC, Jost M, Wei Y, et al. Structure of the catalytic domain of the blass I polyhydroxybutyrate synthase from *Cupriavidus necator*. *J Biol Chem* 2016; 291(48): 25264–25277.
47. Tsuge T, Hyakutake M, Mizuno K. Class IV polyhydroxyalkanoate (PHA) synthases and PHA-producing *Bacillus*. *Appl Microbiol Biotechnol* 2015; 99(15): 6231–6240.
48. Dennis D, McCoy M, Stangl A, et al. Formation of poly(3-hydroxybutyrate-*co*-3-hydroxyhexanoate) hy PHA synthase from *Ralstonia eutropha*. *J Biotechnol* 1998; 64(2–3): 177–186.
49. Doi Y, Kitamura S, Abe H. Microbial synthesis and characterization of poly(3-hydroxybutyrate-*co*-3-hydroxyhexanoate). *Macromolecules* 1995; 28(14): 4822–4828.
50. Lee SH, Oh DH, Ahn WS, et al. Production of poly(3-hydroxybutyrate-*co*-3-hydroxyhexanoate) by high-cell-density cultivation of *Aeromonas hydrophila*. *Biotechnol Bioeng* 2000; 67(2): 240–244.
51. Bhubalan K, Yong KH, Kam YC, et al. Cloning and expression of the PHA synthase gene from a locally isolated *Chromobacterium* sp. USM2. *Mal J Microbiol* 2010; 6: 81–90.
52. Kato M, Bao HJ, Kang CK, et al. Production of a novel copolyester of 3-hydroxybutyric acid and medium-chain-length 3-hydroxyalkanoic acids by *Pseudomonas* sp. 61–3 from sugars. *Appl Microbiol Biotechnol* 1996; 45(3): 363–370.
53. Chen JY, Song G, Chen GQ. A lower specificity PhaC2 synthase from *Pseudomonas stutzeri* catalyses the production of copolyesters consisting of short-chain-length and medium-chain-length 3-hydroxyalkanoates. *Anton Leeuw* 2006; 89(1): 157–167.
54. Chek MF, Hiroe A, Hakoshima T, et al. PHA synthase (PhaC): Interpreting the functions of bioplastic-producing enzyme from a structural perspective. *Appl Microbiol Biotechnol* 2019; 103(3): 1131–1141.
55. Budde CF, Riedel SL, Willis LB, et al. Production of poly(3-hydroxybutyrate-*co*-3-hydroxyhexanoate) from plant oil by engineered *Ralstonia eutropha* strains. *Appl Environ Microbiol* 2011; 77(9): 2847–2854.

56. Hori K, Kobayashi A, Ikeda H, et al. *Rhodococcus aetherivorans* IAR1, a new bacterial strain synthesizing poly(3-hydroxybutyrate-*co*-3-hydroxyvalerate) from toluene. *J Biosci Bioeng* 2009; 107(2): 145–150.
57. Stubbe J, Tian J. Polyhydroxyalkanoate (PHA) homeostasis: The role of the PHA synthase. *Nat Prod Rep* 2003; 20(5): 445–457.
58. Chek MF, Kim SY, Mori T, et al. Structure of polyhydroxyalkanoate (PHA) synthase PhaC from *Chromobacterium* sp. USM2, producing biodegradable plastics. *Sci Rep* 2017; 7(1): 5312.
59. Kim YJ, Choi SY, Kim J, et al. Structure and function of the N-terminal domain of *Ralstonia eutropha* polyhydroxyalkanoate synthase, and the proposed structure and mechanisms of the whole enzyme. *Biotechnol J* 2017; 12(1).
60. Teh AH, Chiam NC, Furusawa G, et al. Modelling of polyhydroxyalkanoate synthase from *Aquitalea* sp. USM4 suggests a novel mechanism for polymer elongation. *Int J Biol Macromol* 2018; 119: 438–445.
61. Kawaguchi Y, Doi Y. Kinetics and mechanism of synthesis and degradation of poly (3-hydroxybutyrate) in *Alcaligenes eutrophus*. *Macromolecules* 1992; 25(9): 2324–2329.
62. Mantzaris NV, Kelley AS, Daoutidis P, et al. A population balance model describing the dynamics of molecular weight distributions and the structure of PHA copolymer chains. *Chem Eng Sci* 2002; 57(21): 4643–4663.
63. Ushimaru K, Sangiambut S, Thomson N, et al. New insights into activation and substrate recognition of polyhydroxyalkanoate synthase from *Ralstonia eutropha*. *Appl Microbiol Biotechnol* 2013; 97(3): 1175–1182.
64. Wodzinska J, Snell KD, Rhomberg A, et al. Polyhydroxybutyrate synthase: Evidence for covalent catalysis. *J Am Chem Soc* 1996; 118(26): 6319–6320.
65. Buckley RM, Stubbe J. Chemistry with an artificial primer of polyhydroxybutyrate synthase suggests a mechanism for chain termination. *Biochemistry* 2015; 54(12): 2117–2125.
66. Jia K, Cao R, Hua DH, et al. Study of class I and class III polyhydroxyalkanoate (PHA) synthases with substrates containing a modified side chain. *Biomacromolecules* 2016; 17(4): 1477–1485.
67. Kumagai Y, Doi Y. Synthesis of a block copolymer of poly(3-hydroxybutyrate) and poly(ethylene glycol) and its application to biodegradable polymer blends. *J Environ Polym Degrad* 1993; 1(2): 81–87.
68. Chen C, Yu CH, Cheng YC, et al. Preparation and characterization of biodegradable nanoparticles based on amphiphilic poly(3-hydroxybutyrate)-poly(ethylene glycol)-poly(3-hydroxybutyrate) triblock copolymer. *Eur Polym J* 2006; 42(10): 2211–2220.
69. Chen C, Yu CH, Cheng YC, et al. Biodegradable nanoparticles of amphiphilic triblock copolymers based on poly(3-hydroxybutyrate) and poly(ethylene glycol) as drug carriers. *Biomaterials* 2006; 27(27): 4804–4814.
70. Hirt TD, Neuenschwander P, Suter UW. Telechelic diols from poly[(*R*)-3-hydroxybutyric acid] and poly{[(*R*)-3-hydroxybutyric acid]-*co*-[(*R*)-3-hydroxyvaleric acid]}. *Macromol Chem Phys* 1996; 197(5): 1609–1614.
71. Hirt TD, Neuenschwander P, Suter UW. Synthesis of degradable, biocompatible and though block-copolyesterurethanes. *Macromol Chem Phys* 1996; 197: 4253–4268.
72. Saad B, Hirt TD, Welti M, et al. Development of degradable polyesterurethanes for medical applications: *In vitro* and *in vivo* evaluations. *J Biomed Mater Res* 1997; 36(1): 65–74.
73. Saad B, Ciardelli G, Matter S, et al. Degradable and highly porous polyesterurethane foam as biomaterial: Effects and phagocytosis of degradation products in osteoblasts. *J Biomed Mat Res* 1998; 39(4): 594–602.

74. Lendlein A, Neuenschwander P, Suter UW. Tissue-compatible multiblock copolymers for medical applications, controllable in degradation rate and mechanical properties. *Macromol Chem Phys* 1998; 199(12): 2785–2796.
75. Neuenschwander P. Absorbable biocompatible block copolymer. *EP20030016148 20030716.*
76. Neuenschwander P. Porous membrane comprising a biocompatible block-copolymer. *EP20060011457 20060602.*
77. Neuenschwander P. Scaffold for artificial heart valve and vascular structure. *JP20130142581 20130708.*
78. Andrade AP, Witholt B, Chang DL, et al. Synthesis and characterization of novel thermoplastic polyester containing blocks of poly[(R)-3-hydroxyoctanoate] and poly [(R)-3-hydroxybutyrate]. *Macromolecules* 2003; 36(26): 9830–9835.
79. Chen Z, Cheng S, Xu K. Block poly(ester-urethane)s based on poly(3-hydroxybutyrate-*co*-4-hydroxybutyrate) and poly(3-hydroxyhexanoate-*co*-3-hydroxyoctanoate). *Biomaterials* 2009; 30(12): 2219–2230.
80. Chen Z, Cheng S, Li Z, et al. Synthesis, characterization and cell compatibility of novel poly(ester urethane)s based on poly(3-hydroxybutyrate-*co*-4-hydroxybutyrate) and poly(3-hydroxybutyrate-*co*-3-hydroxyhexanoate) prepared by melting polymerization. *J Biomater Sci Polym Ed* 2009; 20(10): 1451–1471.
81. Saad GR. Synthesis and characterization of biodegradable poly(esterurethanes) based on bacterial poly(R-3-hydroxybutyrate). *J Appl Pol Sci* 2002; 83: 703–718.
82. Zhao Q, Cheng GX, Li HM, et al. Synthesis and characterization of biodegradable poly(3-hydroxybutyrate) and poly(ethylene glycol) multiblock copolymers. *Polymer* 2005; 46(23): 10561–10567.
83. Li J, Ni X, Li X, et al. Micellization phenomena of biodegradable amphiphilic triblock copolymers consisting of poly(b-hydroxyalkanoic acid) and poly(ethylene oxide). *Langmuir* 2005; 21(19): 8681–8685.
84. Dai S, Xue L, Zinn M, et al. Enzyme-catalyzed polycondensation of polyester macrodiols with divinyl adipate: A green method for the preparation of thermoplastic block copolyesters. *Biomacromolecules* 2009; 10(12): 3176–3181.
85. Vastano M, Pellis A, Immirzi B, et al. Enzymatic production of clickable and PEGylated recombinant polyhydroxyalkanoates. *Green Chem* 2017; 19(22): 5494–5504.
86. Li X, Liu KL, Li J, et al. Synthesis, characterization, and morphology studies of biodegradable amphiphilic poly[(R)-3-hydroxybutyrate]-*alt*-poly(ethylene glycol) multiblock copolymers. *Biomacromolecules* 2006; 7(11): 3112–3119.
87. Dai S, Li Z. Enzymatic preparation of novel thermoplastic di-block copolyesters containing poly[(R)-3-hydroxybutyrate] and poly(epsilon-caprolactone) blocks via ring-opening polymerization. *Biomacromolecules* 2008; 9(7): 1883–1893.
88. Timbart L, Renard E, Langlois V, et al. Novel biodegradable copolyesters containing blocks of poly(3-hydroxyoctanoate) and poly(epsilon-caprolactone): Synthesis and characterization. *Macromol Biosci* 2004; 4(11): 1014–1020.
89. Schreck KM, Hillmyer MA. Block copolymers and melt blends of polylactide with Nodax™ microbial polyesters: Preparation and mechanical properties. *J Biotechnol* 2007; 132(3): 287–295.
90. Macit H, Hazer B, Arslan H, et al. The synthesis of PHA-*g*-(PTHF-*b*-PMMA) multiblock/graft copolymers by combination of cationic and radical polymerization. *J Appl Pol Sci* 2009; 111(5): 2308–2317.
91. Adamus G, Sikorska W, Janeczek H, et al. Novel block copolymers of atactic PHB with natural PHA for cardiovascular engineering: Synthesis and characterization. *Eur Polym J* 2012; 48(3): 621–631.

92. Ravenelle F, Marchessault RH. One-step synthesis of amphiphilic diblock copolymers from bacterial poly([R]-3-hydroxybutyric acid). *Biomacromolecules* 2002; 3(5): 1057–1064.
93. Ravenelle F, Marchessault RH. Self-assembly of poly([R]-3-hydroxybutyric acid)-*block*-poly(ethylene glycol) diblock copolymers. *Biomacromolecules* 2003; 4(3): 856–858.
94. Rendall JC. *Polymer Sequence Determination*. Academic Press, 1977.
95. Žagar E, Krzan A, Adamus G, et al. Sequence distribution in microbial poly(3-hydroxybutyrate-*co*-3-hydroxyvalerate) *co*-polyesters determined by NMR and MS. *Biomacromolecules* 2006; 7(7): 2210–2216.
96. Bloembergen S, Holden DA, Hamer GK, et al. Studies of composition and crystallinity of bacterial poly(b-hydroxybutyrate-*co*-b-hydroxyvalerate). *Macromolecules* 1986; 19(11): 2865–2871.
97. Mitomo H, Barham PJ, Keller A. Crystallization and morphology of poly(β-hydroxybutyrate) and its copolymer. *Polym J* 1987; 19(11): 1241–1253.
98. Doi Y, Kunioka M, Tamaki A, et al. Nuclear magnetic-resonance studies on bacterial copolyesters of 3-hydroxybutyric acid and 3-hydroxyvaleric acid. *Makromol Chem* 1988; 189(5): 1077–1086.
99. Organ SJ, Barham PJ. Phase-separation in a blend of poly(hydroxybutyrate) with poly (hydroxybutyrate-*co*-hydroxyvalerate). *Polymer* 1993; 34(3): 459–467.
100. Ishihara Y, Shimizu H, Shioya S. Mole fraction control of poly(3-hydroxybutyric-*co*-3-hydroxhvaleric) acid in fed-batch culture of *Alcaligenes eutrophus*. *J Ferment Bioeng* 1996; 81(5): 422–428.
101. Wang Y, Yamada S, Asakawa N, et al. Comonomer compositional distribution and thermal and morphological characteristics of bacterial poly(3-hydroxybutyrate-*co*-3-hydroxyvalerate)s with high 3-hydroxyvalerate content. *Biomacromolecules* 2001; 2(4): 1315–1323.
102. Lütke-Eversloh T, Kawada J, Marchessault RH, et al. Characterization of microbial polythioesters: Physical properties of novel copolymers synthesized by *Ralstonia eutropha*. *Biomacromolecules* 2002; 3(1): 159–166.
103. Impallomeni G, Steinbüchel A, Lütke-Eversloh T, et al. Sequencing microbial copolymers of 3-hydroxybutyric and 3-mercaptoalkanoic acids by NMR, electrospray ionization mass spectrometry, and size exclusion chromatography NMR. *Biomacromolecules* 2007; 8(3): 985–991.
104. Kelley AS, Srienc F. Production of two phase polyhydroxyalkanoic acid granules in *Ralstonia eutropha*. *Int J Biol Macromol* 1999; 25(1–3): 61–67.
105. Mantzaris NV, Kelley AS, Srienc F, et al. Optimal carbon source switching strategy for the production of PHA copolymers. *Aiche J* 2001; 47(3): 727–743.
106. Kelley AS, Mantzaris NV, Daoutidis P, et al. Controlled synthesis of polyhydroxyalkanoic (PHA) nanostructures in *R. eutropha*. *Nano Lett* 2001; 1(9): 481–485.
107. Pederson EN, McChalicher CW, Srienc F. Bacterial synthesis of PHA block copolymers. *Biomacromolecules* 2006; 7(6): 1904–1911.
108. Li SY, Dong CL, Wang SY, et al. Microbial production of polyhydroxyalkanoate block copolymer by recombinant *Pseudomonas putida*. *Appl Microbiol Biotechnol* 2011; 90(2): 659–669.
109. Tripathi L, Wu LP, Chen J, et al. Synthesis of diblock copolymer poly-3-hydroxybutyrate -*block*-poly-3-hydroxyhexanoate [PHB-*b*-PHHx] by a b-oxidation weakened *Pseudomonas putida* KT2442. *Microb Cell Fact* 2012; 11: 44.
110. Hu D, Chung AL, Wu LP, et al. Biosynthesis and characterization of polyhydroxyalkanoate block copolymer P3HB-*b*-P4HB. *Biomacromolecules* 2011; 12(9): 3166–3173.

111. Tajima K, Han X, Satoh Y, et al. *In vitro* synthesis of polyhydroxyalkanoate (PHA) incorporating lactate (LA) with a block sequence by using a newly engineered thermostable PHA synthase from *Pseudomonas* sp. SG4502 with acquired LA-polymerizing activity. *Appl Microbiol Biotechnol* 2012; 94(2): 365–376.

112. Ochi A, Matsumoto K, Ooba T, et al. Engineering of class I lactate-polymerizing polyhydroxyalkanoate synthases from *Ralstonia eutropha* that synthesize lactate-based polyester with a block nature. *Appl Microbiol Biotechnol* 2013; 97(8): 3441–3447.

113. Matsumoto K, Hori C, Fujii R, et al. Dynamic changes of intracellular monomer levels regulate block sequence of polyhydroxyalkanoates in engineered *Escherichia coli*. *Biomacromolecules* 2018; 19(2): 662–671.

114. Aldor I, Keasling JD. Metabolic engineering of poly(3-hydroxybutyrate-*co*-3-hydroxyvalerate) composition in recombinant *Salmonella enterica* serovar typhimurium. *Biotechnol Bioeng* 2001; 76(2): 108–114.

115. Iadevaia S, Mantzaris NV. Genetic network driven control of PHBV copolymer composition. *J Biotechnol* 2006; 122(1): 99–121.

116. Gardner TS, Cantor CR, Collins JJ. Construction of a genetic toggle switch in *Escherichia coli*. *Nature* 2000; 403(6767): 339–342.

117. Iadevaia S, Mantzaris NV. Synthesis of PHBV block copolymers driven by an oscillatory genetic network. *J Biotechnol* 2007; 128(3): 615–637.

118. Elowitz M, BLeibler S. A synthetic oscillatory network of transcriptional regulators. *Nature* 2000; 403(6767): 335–338.

119. Portle S, Iadevaia S, San KY, et al. Environmentally-modulated changes in fluorescence distribution in cells with oscillatory genetic network dynamics. *J Biotechnol* 2009; 140(3–4): 203–217.

Part II

Feedstocks

8 Inexpensive and Waste Raw Materials for PHA Production

Sebastian L. Riedel and Christopher J. Brigham

CONTENTS

8.1 INTRODUCTION

Worldwide plastic pollution and continuing climate change are among the biggest challenges of today's society [1,2]. The natural polyester family polyhydroxyalkanoates (PHA) have similar properties to synthetic plastics but are fully biodegradable in nature by a variety of microorganisms [3–5]. Additionally, PHA can be produced with a lower CO_2 footprint than fossil-based plastics [6]. PHA are classified based on the length of their side chains as short-chain-length PHA with ≤5 carbon atoms (*scl*-PHA), medium-chain-length PHA with 6–14 carbon atoms (*mcl*-PHA), and long-chain-length PHA (*lcl*-PHA) with >14 carbon atoms [7]. The *scl*-PHA tend to be more crystalline and brittle and less flexible, compared with *mcl-/lcl*-PHA, through the co-crystallization of their sidechains [8,9]. The most common PHA is the *scl*-homopolymer polyhydroxybutyrate (PHB) [10,11]. However, copolymers in and between PHA classes are quite common as the *scl-co-scl* polymer poly(3-hydroxybutyrate-*co*-3-hydroxyvalerate) [P(3HB-*co*-3HV) or (PHBV)] or the *scl-co-mcl* polymer poly(3-hydroxybutyrate-*co*-3-hydroxyhexanoate), P(3HB-*co*-3HHx). The desired properties of the PHA can be tuned by the type and number of different monomers present in the polymer and fine-tuned by the molar ratio of the different monomers [5]. Additionally, the molecular weight of each PHA polymer is critical for its processing [12]. This almost endless possible combinations of all PHA monomers (over 150 known) [7] also creates a great challenge for its processing. PHA production with constant monomer composition and constant molecular weight is mandatory for successful transfer to a reliable industrial processing process [13].

This is challenging since PHA are synthesized by a variety of different microorganisms, and the feedstocks used for production may vary in their composition, which again affects the composition of the synthesized PHA from the microorganisms [14].

For the past few decades, PHA have been touted as a bio-based, biodegradable, and biocompatible alternative to "traditional" chemical-based plastics. However, the worldwide PHA production is still negligible, due to high production costs, partly caused by the substrate supply. In 2019, only 25,000 tons of PHA were entering commercial markets, which represented only 1.2% of the worldwide produced bioplastic [15]. Also, all types of bioplastic (half of which are non-biodegradable) add up to only 2.1 million tons, which is compared with the continually rising petroleum-based plastic production of currently 360 million tons, still a tiny production [15]. To address the issue of worldwide negligible PHA production, biogenic feedstocks, which show little competition to other industries, have the potential to enable low-cost PHA production, boosting further PHA production at the commercial level. An ideal production process should work as "substrate flexible" to avoid dependencies on a single carbon feedstock. Seasonal shortcomings or other (emerging) industrial processes are price drivers in the competitive bioeconomy, which is especially critical for the bioplastic industry, due to the price pressure of synthetic plastics. Also, there is not one ideal single carbon feedstock that can be used to substitute all or a majority of the petroleum-based plastic material. Besides the choice of carbon feedstocks, which should be available in large quantities at a reasonable price, the following must be considered for an efficient PHA production process: (i) selection of a PHA production strain or consortia that accumulate PHA to high amounts per cell dry mass (CDM), (ii) a substrate flexible bioprocess that allows for the control of the PHA composition, (iii) a cultivation strategy by which high cell densities can be reached with a high production titer of >1 g/(L·h), (iv) an efficient PHA downstream process [also depending on (i)], and (v) choice of the proper additives for compounding before the targeted processing. All of these factors will be discussed in this book series.

This chapter focuses on various side and waste streams from the food and agriculture industry, including food waste, plant- and animal-based waste oils and fats, volatile fatty acids (VFA), wastewaters, and sugar or carbohydrate-rich side streams as well as CO_2, syngas, and pyrolysis products of conventional plastics as inexpensive feedstocks for the production of PHA polymers with pure or microbial mixed cultures (MMC). These raw materials can be used directly as carbon sources or after conversion to consumable molecules (e.g., sugars or VFA), as shown in Figure 8.1.

A selected overview of PHA production from these carbon sources with different microorganisms is given in Table 8.1.

8.2 OLEAGINOUS LIPID-BASED FEEDSTOCKS

Plant- or animal-based waste raw materials, such as oil, fat, and/or fatty acids, are favorable feedstocks for PHA production due to their high carbon content and usability as a "100%" feedstock. This enables production processes to reach excellent yields (up to 0.8 g PHA per gram of feedstock), high cell densities of 100–200 g/L, and

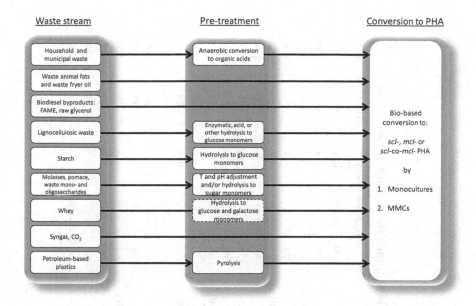

FIGURE 8.1 Overview of different pre-treatment methods to convert various waste and side streams from the food industry and agriculture into fermentable feedstocks for biotechnological production of diverse types of PHA by monoseptic or mixed microbial cultures.

also excellent productivity of 1–2.5 g PHA/(L·h) during fed-batch cultivations with wild-type or recombinant strains of *Ralstonia eutropha* [17,20,31–33]. Additionally, *scl-co-mcl*-PHA with high *mcl*-PHA content of over 10 mol-% is often produced with these raw materials, since the necessary *mcl*-PHA precursor molecules can be produced most efficiently from β-oxidation intermediates [34]. However, using lipids as substrates, especially in bioreactor cultivations, is much more challenging than using carbon sources like sugars because the hydrophobic lipids can create a three-phase system in the aqueous medium between the air, the lipid, and the media. Also, a standard fed-batch protocol with exponential carbon limiting feed, as often applied in industrial processes, to reach high cell densities [35] cannot be applied because the feedstock fed (oil/fat) is not identical to the carbon source used (fatty acids) for growth by the microorganisms. The lipids need to be bioavailable during the cultivation before consumption by the microorganisms. Some microorganisms, such as *R. eutropha,* can emulsify lipids by secreting a lipase, which cleaves the fatty acids from the glycerol backbone of the triacylglycerols from the oil/fat [36]. In the resulting emulsion, the lipid droplets have a high surface area in the aqueous media, which increases their bioavailability for the microorganisms. A side effect of the emulsification, especially in aerated stirred tank reactor systems, is heavy foaming of the culture broth. This can be fought with mechanical foam breakers or the addition of antifoam reagents, whereby the latter decreases the oxygen transfer into the media and can also have negative effects on the cell physiology [37]. Also, some antifoam agents may serve as a carbon source and can be integrated into the polymer chain, which can decrease the molecular weight of the PHA [38].

TABLE 8.1

Examples of Polyhydroxyalkanoate Production by Different Microorganisms from Various Carbon Feedstocks

Carbon Source	Strain	Product	Production /Yield			PHA/CDM	Ref.
			[g/L]	[g/(L·h)]	[g/g]	[wt.-%]	
Coffee oil	R. eutropha H16	PHB	049	1.3	0.8	89	[16]
Plant oil	Recombinant R. eutropha	P(3HB-co-19 mol-% 3HHx)	103	1.1	0.6–0.8	74	[17]
Tallow	R. eutropha H16	PHB	024	0.3	0.4	63	[18]
Waste animal fats	Recombinant R. eutropha	P(3HB-co-19 mol-% 3HHx)	027	0.4	0.5	60	[18]
Waste glycerol	R. eutropha H1	PHB	038	1.1		50	[19]
Waste rapeseed oil + propanol	R. eutropha H16	P(3HB-co-8 mol-% 3HV)	138	1.5	0.8	76	[20]
Hydrolyzed waste cooking oil fatty acids	P. putida KT2440	mcl-co-mcl PHA	58	1.9	0.7–0.8	36	[21]
Acetic, propionic, butyric acid	R. eutropha H16	P(3HB-co-6 mol-% 3HV)	094	2.1		83	[22]
Activated sludge + acetate	Mixed culture	PHB				70	[23]
Palm oil mill effluent	Mixed culture	P(3HB-co-23 mol-% 3HV)			0.6	64	[24]
Oil palm trunk sap	B. megaterium MC1	PHB	3	0.21		30	[25]
Sucrose	A. latus	PHB	068	4.0		50	[26]
Whey	Recom.[a] E. coli	PHB	168	4.6		87	[27]
Whey	H. mediterranei	P(3HB-co-8 mol-% 3HV)	006	0.05	0.3	50	[28]
Waste glycerol (biomass)/CO₂ (PHB)	R. eutropha DSM 545	PHB	28	0.17	0.35	61	[29]
CO₂	C. eutrophus B-10646	PHB	43			85	[30]

Direct use of the most produced plant oils in the world, palm, soybean, and rapeseed oil (160 million tons of 207 million tons in 2019 [39]), has been widely studied as feedstocks for PHA production [17,40,41]. Also, spent (waste) vegetable oils were used to produce various types of PHA [20,21,42–46], with the highest scl-PHA production of 138 g/L P(3HB-co-8 mol-%-3HV) with R. eutropha H1 using waste rapeseed oil with propanol as a precursor for the 3HV fraction [20]. The highest mcl-PHA production from spent vegetable oil was reached with Pseudomonas putida KT2440 with 58 g/L of an mcl-PHA copolymer consisting of five different mcl-HA monomers using hydrolyzed waste cooking oil fatty acids as feedstock [21]. Another source of waste lipids for PHA production in the million-ton scale is spent coffee grounds, which contain around 15 wt.-% coffee oil [47–49]. Using extracted coffee oil, wild-type R. eutropha could produce 49 g/L PHB with a high PHB content of CDM of 89 wt.-% and productivity >1 g PHB/(L·h) [16]. The scl-co-mcl-PHA copolymer P(3HB-co-3HHx) with a high molar content of 22 mol-% 3HHx was synthesized from coffee waste oil using a recombinant R. eutropha strain as the biocatalyst [50]. Also, waste streams from fisheries (annual production in 2016, 171 million tons [51]) can be used to produce PHA. About 20–80% of a fish used for food or feed enters the waste stream, which makes waste fish oil an interesting feedstock for industrial processes. Thuoc et al. isolated the halophilic bacterium, Salinivibrio sp. M318, from fermented fish sauce in Vietnam and used it for PHB production. Using a mixture of fish sauce, waste fish oil from the Basa fish (Pangasius bocourti), and glycerol in fed-batch bioreactor cultivations resulted in a PHB production of 69 g/L after 78 h. Also, the PHB content of the biomass was reasonably high, with 51.5 wt.-% [52].

Nevertheless, all triacylglycerol-containing feedstocks that can be used for PHA production are, in general, competing with other industries, such as animal feed or biodiesel production [53,54], which will affect their prices through increased demand. In 2017, 78% of the global biodiesel was made from virgin plant oils (palm, soya, rapeseed oil), 10% from used cooking oil (UCO), and 7% from waste animal fats (WAF). However, there are large geographical differences in global biodiesel production. In the EU-28, the portion of UCO and WAF for biodiesel production is 15and 4%, respectively, and the UCO and WAF portions used in Germany are 43 and <1%, respectively [55]. The two main by-products of biodiesel production are fatty acid ethyl esters (FAEE) and fatty acid methyl esters (FAME), which are poor quality and raw glycerol; they have also been used as feedstock for scl- and mcl-PHA production [56–61]. With the increasing content of free fatty acids (FFA), the use of an oil or fat becomes uneconomical for biodiesel production due to soap formation from FFA through the alkaline catalysts, which prevents the separation of the FAMEs from the raw glycerol after the transesterification process. With a slower acidic catalyst, an efficient biodiesel production process can tolerate higher FFA contents, up to 30 wt.-% [53,62]. Riedel et al. [18] investigated the use of low-quality WAF with a high content of FFA (>50 wt.-%) for PHA production with a recombinant strain [63] and the wild-type strain of R. eutropha. Because of the high content of saturated FA (>50 wt.-%), the WAF had a high melting temperature of 47°C, which made it at first not consumable for R. eutropha during bioreactor cultivations at the cultivation temperature of 30°C. The WAF formed chunks in the aqueous media, which could not be emulsified by the bacteria. Through optimized pre-culture conditions and the initial use of waste frying oil, liquified WAF could be used in fed-batch bioreactor cultivations for the production of 50 g/L CDM with a PHA content of 75 wt.-% of the *scl-co-mcl* copolymer P(HB-*co*-19 mol-% 3HHx) [18]. Annually 500,000 tons of waste lipids are produced by the animal processing industry in Europe, which shows the potential of this raw material as a low-cost feedstock for PHA production [64].

8.3   MIXED ORGANIC ACID FEEDSTOCKS

Mixed organic acids, often termed VFAs, are produced by anaerobic fermentation of biomass using a microbial consortium. During the acidogenic phase of waste activated sludge (WAS) treatment, compounds like acetic, propionic, *n*-butyric, and *n*-valeric acids (among others) can be produced. The ratios of the types of VFAs synthesized can be altered, depending on the carbon feedstock used to produce the acids; for example, propionic acid production can be increased if the activated sludge is rich in carbohydrates and pH-adjusted to ~8.0 [65]. Also, the types of acidogenic microorganisms present in a microbial consortium have an effect on the VFAs produced. A VFA production scheme using cheese whey and a mixed microbial culture of *Lactobacillus*, *Olsenella*, *Actinomyces* spp. and other, unclassified bacterial strains produced C2–C8 VFAs, with the majority being hexanoic acid (33 wt.-%) and octanoic acid (25 wt.-%) [66–68]. Often, VFAs are the carbon source of choice for nitrogen or phosphorus removal from wastewater [65–69], but these organic acids can be used as a carbon source to produce PHA. After the acidogenic phase, the VFA-containing liquor can be concentrated and/or used directly for a carbon feedstock for organisms with a high potential of producing and accumulating PHA [70]. VFAs can also be recovered and used as a feedstock for independent monocultures of PHA-producing organisms [71].

The types of acids present in a VFA mixture greatly influences the type of PHA produced when using the VFA mixture as a carbon feedstock, depending on which biocatalyst is used. Yang et al. [72] adjusted the organic acid concentrations/ratios and examined PHA synthesis using the wild-type *R. eutropha* strain, H16. The addition of increasing amounts of butyric acid was found to maximize cellular biomass accumulation and intracellular PHA concentrations. The addition of propionic acid was found to increase the concentration of the 3-hydroxyvalerate (3HV) monomer in the resulting PHA. Building on these findings, Huschner et al. devised a bioreactor feeding strategy to produce P(3HB-*co*-3HV) where acetic, propionic, and butyric acids (in a 3.1:1.4:1 mass ratio) were fed via pH control and acid sodium salts were fed to the culture by pO_2 control [22]. Adjustment of the carbon/nitrogen (C/N) ratio allowed for the maximization of PHA accumulation (>80% PHA/cell dry mass, >100 g/L PHA). While propionic acid provides odd-chain-length carbon substrates for PHA synthesis, the amount of propionate fed to the culture is generally not reflected in the mol-% 3HV content in the resulting PHA. Metabolic flux analysis (MFA) has been performed on *R. eutropha* using combinations of acetate, propionate, and butyrate as carbon sources. Notable conclusions from this analysis include: (i) only a maximum of 11.5% of the carbon from propionate was shown to go into the 3HV monomer and (ii) a majority of carbon (69.6–86.6%) from butyrate was converted to 3-hydroxybutyryl-CoA (3HB-CoA) and the 3-hydroxybutyrate

(3HB) monomer [73]. Jeon et al., however, altered the carbon flux in *R. eutropha* to synthesize poly(3-hydroxybutyrate-*co*-3-hydroxyhexanoate) [P(3HB-*co*-3HHx)] from cultures fed with butyrate. By deleting the gene encoding the native PHA polymerase (*phaC1*) in *R. eutropha* and adding heterologous broad-substrate-specificity PHA synthase and (*R*)-specific enoyl-CoA ligase genes *in trans*, P(3HB-*co*-3HHx), 60 wt.-% of CDM, containing ~20 wt.-% 3HHx monomer, was produced by the engineered *R. eutropha* strain using butyrate as the main carbon source [74]. *Escherichia coli* has also been engineered to produce PHA using VFA feedstocks. By expressing a propionate/CoA transferase gene (*pct*) along with a PHA synthesis operon, P(3HB-*co*-3HV), with a 3HV monomer, content of 80 wt.-% was synthesized from cultures containing propionate [75].

Using MMC for PHA synthesis from VFA feedstocks can save operating costs (e.g., open cultivation, no sterilization). Several waste streams have been used as carbon feedstocks for VFA, and ultimately PHA production, including sugar cane molasses, brewery wastewater, olive oil mill effluent, palm oil mill effluent, and many others [70]. However, different microorganisms have different PHA synthesis capabilities, so a stable MMC is ideal for PHA production because a PHA mixture that is constantly changing would present significant processing challenges [56,70]. Since PHA-producing MMCs typically come from activated sludge and other waste biomass, enrichment of these communities for PHA synthesis involves feast-famine cycling where cells are allowed to grow with a VFA substrate for a short duration (i.e., "feast"), and then once the exogenous carbon substrate is consumed, enter into "famine" conditions. By cycling through these conditions, PHA-producing bacteria are enriched over non-PHA producers [76]. Butyrate and acetate were compared as carbon sources for PHA synthesis by MMCs in fed-batch culture. Butyrate was preferred by the cultures over acetate as a substrate for PHA production [77]. VFAs were used for PHA production using MMC from wastewater sludge, and the pH and feeding regime of the culture were tested. The optimal conditions for VFA synthesis were a pH of 9.0 (25.934 ± 1.485 mg chemical oxygen demand (COD)/L), and optimal PHA production conditions consisted of continuous pulse feeding of VFA production fermentation broth [78].

Annually, over 1 billion tons of food is wasted every year, which has the potential to be used as a low-cost feedstock for the production of *scl*-PHA in controlled MMC with non-GMOs after a fermentation step for VFAs [68]. The company, Full Cycle Bioplastics LLC, patented [71] a complex bioprocess that allows the production of VFA mixtures with constant/defined compositions from heterogenous food wastes, which then allows for PHA production with a constant monomer composition. This latter part is mandatory for the production of PHA with consistent properties, which is one of the major barriers to PHA materials entering industrial processing [68]. To date (December 2019), Full Cycle Bioplastics is raising capital to scale-up their production and is so far not selling PHA to commercial markets [79].

8.4 MONO- AND POLYSACCHARIDE FEEDSTOCKS

The most well-studied PHA-producing organisms have been shown to grow and synthesize polymer using sugars as a carbon feedstock. In PHA producers and other bacteria, sugars are broken down by way of central metabolic pathways, for example, glycolysis and pentose phosphate pathways, to provide energy and carbon skeletons for anabolism [80]. In agricultural and industrial waste, there exists a variety of carbohydrate-containing streams that could provide carbon for lower-cost PHA synthesis.

One valuable agro-industrial by-product is molasses, which is the residual syrup from sugar refining. Even though little or no crystalline sugar (sucrose) can be extracted from molasses, this side stream is 50 wt.-% sugars (mostly sucrose, but also containing some glucose and fructose). Molasses is sometimes used for human consumption but generally finds application as a carbon-rich fertilizer, cattle feed supplement, or feedstock in ethanol production [81]. Many works have examined PHA synthesis using molasses as the main carbon source, with yields ranging from 1 to 4.5 g/L and intracellular PHA contents ranging from 17 to 55 wt.-% of CDM [82–85]. Molasses has been used as a carbon feedstock for PHA production by MMC. The cultures were enriched with feast-famine cycling for PHA-producing organisms, and a maximum PHA content of 74.6 wt.-% was achieved [86]. The same group later examined the relationship between the MMC composition and PHA production performance. Notably, the presence of *Azoarcus* was correlated with higher PHA production yields, whereas the presence of *Thauera* was linked with higher 3HV fractions in the resulting PHA [87]. In Malaysia and other palm oil-producing countries, the use of waste products from palm fronds, felled trees, palm fruit husks, and more have been a priority among researchers to produce more value-added products, like PHA. Lokesh et al. examined oil palm trunk sap as a possible carbon source for microbial growth and PHA synthesis using *Bacillus megaterium* as the biocatalyst. Oil palm trunk sap was shown to contain monosaccharides like glucose and fructose, disaccharides like sucrose and maltose, and sugar alcohols like inositol. *B. megaterium* was shown to grow and produce up to 30% of its cell dry mass as PHA when oil palm trunk sap was present as the main carbon source [25]. Fermentable sugars can also be obtained from the oil palm fronds. The fronds can be pressed using sugarcane pressing/processing equipment to obtain sugars, which, like the oil palm trunk sap, could be used as a carbon source for PHA synthesis [88]. The oil palm frond juice, consisting mainly of glucose, fructose, and sucrose, was fed to *R. eutropha* strain CCUG52238^T as the main carbon source and produced over 40 wt.-% of its CDM as PHB [89]. In another study, >20 g/L of fermentable sugars from oil palm frond juice was provided to *R. eutropha*, and this was shown to produce as much PHB as using technical grade sugars, indicating the absence of growth or production inhibitors in the oil palm frond pressings [90]. Vinasse is a waste stream containing high amounts of sugars obtained from molasses-based ethanol production. Due to the low pH value, high temperature, and high ash content of typical vinasse waste streams, pre-treatment is required before using vinasse as a carbon source for microbial growth and production. Raw and pre-treated vinasse was used for PHA synthesis by *Haloarcula marismortui*; up to 30% of cell dry mass PHB was produced [91]. Pre-treated (pH-adjusted and phenolic compounds removed) vinasse was also used as a carbon source for growth and PHA production by the halophilic archaeon, *Haloferax mediterranei*. This organism was able to synthesize up to 70 wt.-% of its CDM as PHA using pre-treated vinasse as the carbon source [92].

Waste polysaccharides have also been used as carbon sources for polymer synthesis. Starch, a glucose polymer, is (like PHA) a carbon storage compound in plants and can be found in a variety of foods consumed by people and other animals. Roughly half the starch from crop plants is used in food applications [81]. Starch has been used as a bio-based plastic itself in a structural capacity in disposable dining utensils and as packing material [93]. A small number of studies have demonstrated direct conversion of starch to PHB using bacteria that are naturally capable of hydrolyzing the polysaccharide [94–96]. Bhatia et al. engineered *E. coli* to both breakdown starch and synthesize PHB by heterologously expressing the *amyl* gene from *Paenibacillus* sp. and a synthetic PHA synthesis operon of *R. eutropha* origin. The engineered strain produced up to 40% of its CDM of PHB when grown in minimal medium in flask cultures with starch as the sole carbon source [97]. Generally, however, for starch to be used effectively as a carbon source for PHA production, it requires pre-treatment in the form of hydrolysis to produce glucose. PHA production from starch hydrolysates has been examined using several different biocatalyst organisms [81,98,99].

In agricultural waste streams, there are plenty of non-starch polysaccharides that can be converted to feedstocks for PHA-producing organisms. A large portion of agricultural waste is lignocellulosic fibrous waste. Pre-treatment is generally necessary for the microbial conversion of lignocellulose into a value-added product. Lignocellulosic biomass consists of lignin, which is a structural component containing large amounts of polyphenolics, as well as cellulose and hemicellulose. When hydrolyzed by enzymatic action, acid hydrolysis, or other methods, cellulose and hemicellulose can provide fermentable sugars like D-glucose, D-xylose, D-arabinose, and D-mannose for use as feedstocks in PHA synthesis. However, there are two major roadblocks in the design of a system for producing PHA from lignocellulose: (i) several lignin hydrolysis products are known to be toxic to cells and act as growth inhibitors [81,100] and (ii) most PHA producers cannot use pentose sugars for growth or PHA synthesis [81]. Pre-treated lignocellulosic biomass hydrolysates have been used as a carbon source for PHA production using many different biocatalysts, including *R. eutropha*, recombinant *E. coli*, *Burkholderia sacchari*, and others (reviewed in [101]). Lignocellulosic biomass from the palm oil industry has also been studied as a carbon source for PHA production [102,103]. Sugarcane bagasse hydrolysate has also been used as a carbon feedstock for PHA production. *Burkholderia cepacia* and *B. sacchari* produce up to 62 wt.-% of their CDM as PHB when cultivated with sugarcane bagasse hydrolysate as the main carbon source in bioreactor cultures [104].

8.5 CARBON DIOXIDE AS A FEEDSTOCK

CO_2 is a very familiar waste gas, as it is one of the main combustion products of carbon-containing compounds. Carbon dioxide is also a greenhouse gas, and the continued increase in CO_2 concentration in Earth's atmosphere and oceans presents a potentially catastrophic environmental problem. Mitigation of CO_2 is a welcome prospect, and many organisms are capable of using CO_2 for growth and synthesizing useful metabolic products. *R. eutropha* is one of the principal microorganisms that has been studied for autotrophic PHA synthesis. *R. eutropha* uses CO_2 for growth and production via the Calvin–Benson–Bassham (CBB) cycle, relying on the enzyme ribulose 1,5-bisphosphate carboxylase/oxygenase (RuBisCO) as the key enzyme for the introduction of CO_2 into cellular metabolic pathways. Culturing non-photosynthetic bacteria like *R. eutropha* using CO_2 as the sole carbon source requires a bioreactor setup that delivers a mixture of H_2, O_2, and CO_2 to cells; the H_2 is supplied for energy and the O_2 acts as the terminal electron acceptor [105,106]. Early studies of autotrophic *R. eutropha* cultures reported a PHB content of ~20% of cell dry mass [107,108]. It should be noted that these early studies were focused not on *R. eutropha* as a PHA biocatalyst, but as a microbial chassis for the production of single-cell proteins [109]. The working group of Tatiana Volova at the Siberian Federal University, Russia, focused on the autotrophic production of PHA using *R. eutropha* [110] and a locally isolated bacterial species, *Cupriavidus eutrophus* B-10646 [30]. The same group has studied the growth and PHA production of *R. eutropha* in the presence of carbon monoxide. With carbon monoxide concentrations up to 20 vol%, *R. eutropha* was shown to produce up to 72.8 wt.-% PHA [111]. The ability of *R. eutropha* to produce PHA from CO_2 in the presence of carbon monoxide suggests its suitability for growth and production using syngas as the carbon feedstock. Syngas, or synthesis gas, is the product of the gasification of organic waste and contains mainly CO_2, CO, and H_2 [81,112]. Other microorganisms have been shown to produce PHA using syngas as a carbon and energy source. *Rhodospirillum rubrum* can grow on syngas in anaerobic conditions. *R. rubrum* has photosynthetic capabilities but can grow and produce PHB in light and darkness, as demonstrated using syngas from microwave-pyrolyzed household waste as the carbon source [112]. An engineered *R. rubrum* strain that produces PHA precursors, not via the traditional PHB synthesis pathway but using metabolic intermediates from the fatty acid synthesis pathway, was produced. This strain was able to synthesize small quantities of medium-chain-length PHA, containing mostly 3-hydroxyoctanoate (3HO) and 3-hydroxydecanoate (3HD) monomers, from syngas [113]. A more detailed discussion of PHA synthesis using syngas is available in a later chapter of this volume.

Many photosynthetic cyanobacteria are capable of producing PHA using CO_2 as the main carbon source in photoautotrophic growth conditions. Cyanobacterial genera like *Synechocystis* and *Synechococcus* have been shown to produce small amounts of PHA (usually PHB) photoautotrophically. In some cases, these organisms have been engineered to increase photoautotrophic PHA yield [114–117], albeit the maximum yields are still some way off those that are seen in the model organisms of PHA synthesis grown heterotrophically. In general, intracellular PHA makes up no more than 30 wt.-% of CDM in cyanobacterial species grown photoautotrophically [118].

8.6   OTHER CARBON FEEDSTOCKS

Whey is a by-product of the dairy industry and consists of water, sugars, fats, proteins, and salts [119]. As such, it is a mixed carbon source for PHA production. Roughly half of all whey is used in human and livestock food production [119,120], leaving the other half of the whey produced to be classified as a pollutant waste stream. Many different organisms, both as mixed cultures and individual pure cultures, have been grown on whey for PHA biosynthesis. One key challenge for using whey as a substrate for microbial PHA production is its protein content, which suggests a low carbon-to-nitrogen (C/N) ratio. However, for some organisms, phosphate starvation could potentially be used to trigger PHA production on whey. Also, lactose is generally not a carbon source that is readily used by most well-known PHA producers. For this challenge, MMC makes an attractive solution. Also, engineered *E. coli* could potentially be used to alleviate both challenges [119]. Koller et al. have examined multiple species of microorganisms for their ability to convert whey to PHA, including *H. mediterranei* [121], *P. hydrogenovora* [122], and *Hydrogenophaga pseudoflava* [123]. Generally, the PHA produced using whey as the carbon feedstock was PHB unless valeric acid was added to the culture, in which case the P(3HB-*co*-3HV) copolymer could be produced. However, *H. mediterranei* was able to produce P(3HB-*co*-3HV) containing a small amount of 3HV monomer without any additional supplementation [121–123].

With interest in biodiesel growing, waste glycerol from its production has become an important carbon source for the synthesis of value-added products. There are a variety of PHA-producing organisms that are capable of using glycerol as the main carbon source, such as *Bacillus thuringiensis* [124], *P. corrugata*, *P. oleovorans* [125], and *Novosphingobium* sp. [126], to name a few. Also, MMC has been used for the production of PHA from crude waste glycerol, and, depending on the genus/species makeup of the MMC, short- or medium-chain-length PHA can be produced [127–129].

Other, more "non-conventional" waste streams have been used for PHA production, including petroleum-based plastics. Plastics like poly(ethylene terephthalate) (PET) or poly(styrene) (PS) have been used as carbon feedstocks. However, in both cases, the polymers must be pyrolyzed at a high temperature before being fed to the microorganisms. *P. putida* is metabolically capable of using the pyrolysis products of both polymers [130].

8.7 CONCLUSIONS AND OUTLOOK

The use of carbon-containing waste streams as feedstocks for the production of renewable, value-added products is a sphere that is being examined closely to create a more sustainable society. Waste streams of many critical industries (e.g., agriculture, food service, energy) contain fermentable carbon compounds that can be harvested and used in bio-based productions. The synthesis of PHA bioplastics is one such production. Many diverse microorganisms with diverse metabolic capabilities have been shown to synthesize PHA. Thus, for a plentiful carbon compound present in a waste stream, a biocatalyst can be selected to metabolize that compound for growth and polymer synthesis.

For large-scale, sustainable PHA production, the outlook is as it always has been. Many research groups have shown that PHA can be produced in a high yield from various carbon feedstocks, with the thermal properties, mechanical properties, and biocompatibility of the resulting polymers sufficiently characterized using a combination of classical and state-of-the-art assays. We, the PHA research community, know that high-quality bio-based polymers can be made. The question remains:

will it matter? Research groups and companies have tried to produce consumer and industrial products made from PHA. However, the hurdle to mainstream acceptance is the fact that PHA and polymers like it are costly as raw materials compared with petroleum-based, chemically synthesized polymers. The use of cheap, plentiful carbon feedstocks is a step in the direction of price competitiveness of PHA, but it is only a part of the remedy for this challenge. Another hurdle would be the development of a cost-effective recovery process to extract and purify PHA from biomass. The topic of polymer extraction is covered in another chapter of this book series. Time and the continued devotion of many world-class researchers and innovative companies will tell if we ever pick up a consumer product in our local supermarkets to read the label and see the phrase, "The packaging of this product is made from bio-based, biodegradable polyhydroxyalkanoate." Until such time, there is plenty of available carbon to keep filling cells with polymer.

REFERENCES

1. Kellogg WW, Schware R. *Climate Change and Society: Consequences of Increasing Atmospheric Carbon Dioxide*. London, New York, NY: Routledge, 2019.
2. Jambeck JR, Geyer R, Wilcox C, et al. Plastic waste inputs from land into the ocean. *Science* 2015; 347(6223): 768–771.
3. Dilkes-Hoffman LS, Lant PA, Laycock B, et al. The rate of biodegradation of PHA bioplastics in the marine environment: A meta-study. *Mar Pollut Bull* 2019; 142: 15–24.
4. Boyandin AN, Prudnikova SV, Filipenko ML, et al. Biodegradation of polyhydroxyalkanoates by soil microbial communities of different structures and detection of PHA degrading microorganisms. *Appl Biochem Microbiol* 2012; 48(1): 28–36.
5. Noda I, Lindsey SB, Caraway D. Nodax™ class PHA copolymers: Their properties and applications. In: Chen GQ, Ed., Steinbüchel A, Series Ed., *Plastics from Bacteria*. Berlin, Heidelberg: Springer, 2010; pp. 237–255.
6. Dietrich K, Dumont MJ, Del Rio LF, et al. Producing PHAs in the bioeconomy—Towards a sustainable bioplastic. *Sustain Prod Consum* 2017; 9: 58–70.
7. Rehm BHA. Polyester synthases: Natural catalysts for plastics. *Biochem J* 2003; 376(1): 15–33.
8. de Koning G. Physical properties of bacterial poly((R)-3-hydroxyalkanoates). *Can J Microbiol* 1995; 41(13): 303–309.
9. Yoshie N, Sakurai M, Inoue Y, et al. Cocrystallization of isothermally crystallized poly(3-hydroxybutyrate-co-3-hydroxyvalerate). *Macromolecules* 1992; 25(7): 2046–2048.
10. Lemoigne M. Produits de dehydration et de polymerisation de l'acide ß-oxobutyrique. *Bull Soc Chim Biol* 1926; 8: 770–782.
11. Lenz RW, Marchessault RH. Bacterial polyesters: Biosynthesis, biodegradable plastics and biotechnology. *Biomacromolecules* 2005; 6(1): 1–8.
12. Renstad R, Karlsson S, Albertsson AC. The influence of processing induced differences in molecular structure on the biological and non-biological degradation of poly (3-hydroxybutyrate-co-3-hydroxyvalerate), P(3-HB-co-3-HV). *Polym Degrad Stab* 1999; 63(2): 201–211.
13. Chen GQ, Chen XY, Wu FQ, et al. Polyhydroxyalkanoates (PHA) toward cost competitiveness and functionality. *Adv Ind Eng Polym Res* 2019; In Press.
14. Koller M. Advances in polyhydroxyalkanoate (PHA) production. *Bioengineering* 2017; 4(4): 88.
15. European Bioplastics. Bioplastics Market Data, 2019. Available from: https://www.european-bioplastics.org/market/. Accessed on: December 31, 2019.

16. Obruca S, Petrik S, Benesova P, et al. Utilization of oil extracted from spent coffee grounds for sustainable production of polyhydroxyalkanoates. *Appl Microbiol Biotechnol* 2014; 98(13): 5883–5890.

17. Riedel SL, Bader J, Brigham CJ, et al. Production of poly(3-hydroxybutyrate-*co*-3-hydroxyhexanoate) by *Ralstonia eutropha* in high cell density palm oil fermentations. *Biotechnol Bioeng* 2012; 109(1): 74–83.

18. Riedel SL, Jahns S, Koenig S, et al. Polyhydroxyalkanoates production with *Ralstonia eutropha* from low quality waste animal fats. *J Biotechnol* 2015; 214: 119–127.

19. Cavalheiro JMBT, de Almeida MCMD, Grandfils C, et al. Poly(3-hydroxybutyrate) production by *Cupriavidus necator* using waste glycerol. *Process Biochem* 2009; 44(5): 509–515.

20. Obruca S, Marova I, Snajdar O, et al. Production of poly(3-hydroxybutyrate-*co*-3-hydroxyvalerate) by *Cupriavidus necator* from waste rapeseed oil using propanol as a precursor of 3-hydroxyvalerate. *Biotechnol Lett* 2010; 32(12): 1925–1932.

21. Ruiz C, Kenny ST, Babu PR, et al. High cell density conversion of hydrolysed waste cooking oil fatty acids into medium chain length polyhydroxyalkanoate using *Pseudomonas putida* KT2440. *Catalysts* 2019; 9(5): 468.

22. Huschner F, Grousseau E, Brigham CJ, et al. Development of a feeding strategy for high cell and PHA density fed-batch fermentation of *Ralstonia eutropha* H16 from organic acids and their salts. *Process Biochem* 2015; 50(2): 165–172.

23. Serafim LS, Lemos PC, Oliveira R, et al. Optimization of polyhydroxybutyrate production by mixed cultures submitted to aerobic dynamic feeding conditions. *Biotechnol Bioeng* 2004; 87(2): 145–160.

24. Lee WS, Chua ASM, Yeoh HK, et al. Strategy for the biotransformation of fermented palm oil mill effluent into biodegradable polyhydroxyalkanoates by activated sludge. *Chem Eng J* 2015; 269: 288–297.

25. Lokesh BE, Hamid ZAA, Arai T, et al. Potential of oil palm trunk sap as a novel inexpensive renewable carbon feedstock for polyhydroxyalkanoate biosynthesis and as a bacterial growth medium. *CLEAN Soil Air Water* 2012; 40(3): 310–317.

26. Yamane T, Fukunaga M, Lee YW. Increased PHB productivity by high-cell-density fed-batch culture of *Alcaligenes latus*, a growth-associated PHB producer. *Biotechnol Bioeng* 1996; 50(2): 197–202.

27. Ahn WS, Park SJ, Lee SY. Production of poly (3-hydroxybutyrate) from whey by cell recycle fed-batch culture of recombinant *Escherichia coli*. *Biotechnol Lett* 2001; 23(12): 235–240.

28. Koller M, Hesse P, Bona R, et al. Biosynthesis of high quality polyhydroxyalkanoate Co- and terpolyesters for potential medical application by the archaeon *haloferax mediterranei*. *Macromol Symp* 2007; 253: 33–39.

29. Garcia-Gonzalez L, Mozumder MSI, Dubreuil M, et al. Sustainable autotrophic production of polyhydroxybutyrate (PHB) from CO_2 using a two-stage cultivation system. *Catal Today* 2015; 257: 237–245.

30. Volova TG, Kiselev EG, Shishatskaya EI, et al. Cell growth and accumulation of polyhydroxyalkanoates from CO_2 and H_2 of a hydrogen-oxidizing bacterium, *Cupriavidus eutrophus* B-10646. *Bioresour Technol* 2013; 146: 215–222.

31. Sato S, Maruyama H, Fujiki T, et al. Regulation of 3-hydroxyhexanoate composition in PHBH synthesized by recombinant *Cupriavidus necator* H16 from plant oil by using butyrate as a co-substrate. *J Biosci Bioeng* 2015; 120(3): 246–251.

32. Arikawa H, Matsumoto K. Evaluation of gene expression cassettes and production of poly(3-hydroxybutyrate-*co*-3-hydroxyhexanoate) with a fine modulated monomer composition by using it in *Cupriavidus necator*. *Microb Cell Fact* 2016; 15(1): 1–11.

33. Gutschmann B, Schiewe T, Weiske MTH, et al. *In-line* monitoring of polyhydroxyalkanoate (PHA) production during high-cell-density plant oil cultivations using photon density wave spectroscopy. *Bioengineering* 2019; 6(3): 85.

34. Riedel SL, Lu J, Stahl U, et al. Lipid and fatty acid metabolism in *Ralstonia eutropha*: Relevance for the biotechnological production of value-added products. *Appl Microbiol Biotechnol* 2014; 98(4): 1469–1483.
35. Ongey EL, Santolin L, Waldburger S, et al. Bioprocess development for lantibiotic ruminococcin-A production in *Escherichia coli* and kinetic insights into LanM enzymes catalysis. *Front Microbiol* 2019; 10(2133): 1–15.
36. Lu J, Brigham CJ, Rha C, et al. Characterization of an extracellular lipase and its chaperone from *Ralstonia eutropha* H16. *Appl Microbiol Biotechnol* 2013; 97(6): 2443–2454.
37. Routledge SJ. Beyond de-foaming: The effects of aantifoams on bioprocess productivity. *Comput Struct Biotechnol J* 2012; 3(4): e201210001.
38. Shi F, Gross RA, Rutherford DR. Microbial polyester synthesis: Effects of poly(ethylene glycol) on product composition, repeat unit sequence, and end group structure. *Macromolecules* 1996; 29(1): 10–17.
39. Brazil Sees Record Soybean Exports in October–November 2019. Available from: https ://apps.fas.usda.gov/psdonline/circulars/oilseeds.pdf. Accessed on: December 31, 2019.
40. Ciesielski S, Mozejko J, Pisutpaisal N. Plant oils as promising substrates for polyhydroxyalkanoates production. *J Clean Prod* 2015; 106: 408–4221.
41. Fadzil FIBM, Tsuge T. Bioproduction of polyhydroxyalkanoate from plant oils. In: Kalia V, Ed., *Microbial Applications*, Vol. 2. Cham: Springer, 2017; pp. 231–260.
42. Kourmentza C, Costa J, Azevedo Z, et al. *Burkholderia thailandensis* as a microbial cell factory for the bioconversion of used cooking oil to polyhydroxyalkanoates and rhamnolipids. *Bioresour Technol* 2018; 247: 829–837.
43. Benesova P, Kucera D, Marova I, et al. Chicken feather hydrolysate as an inexpensive complex nitrogen source for PHA production by *Cupriavidus necator* on waste frying oils. *Lett Appl Microbiol* 2017; 65(2): 182–188.
44. Kamilah H, Al-Gheethi A, Yang TA, et al. The use of palm oil-based waste cooking oil to enhance the production of polyhydroxybutyrate [P(3HB)] by *Cupriavidus necator* H16 strain. *Arab J Sci Eng* 2018; 43(7): 3453–3463.
45. Pernicova I, Kucera D, Nebesarova J, et al. Production of polyhydroxyalkanoates on waste frying oil employing selected Halomonas strains. *Bioresour Technol* 2019; 292: 122028.
46. Ruiz C, Kenny ST, Narancic T, et al. Conversion of waste cooking oil into medium chain polyhydroxyalkanoates in a high cell density fermentation. *J Biotechnol* 2019; 306: 9–15.
47. Massaya J, Prates Pereira A, Mills-Lamptey B, et al. Conceptualization of a spent coffee grounds biorefinery: A review of existing valorisation approaches. *Food Bioprod Process* 2019; 118: 149–166.
48. Obruca S, Benesova P, Kucera D, et al. Biotechnological conversion of spent coffee grounds into polyhydroxyalkanoates and carotenoids. *New Biotechnol* 2015; 32(6): 569–574.
49. Kovalcik A, Obruca S, Marova I. Valorization of spent coffee grounds: A review. *Food Bioprod Process* 2018; 110: 104–119.
50. Bhatia SK, Kim JH, Kim MS, et al. Production of (3-hydroxybutyrate-*co*-3-hydroxyhexanoate) copolymer from coffee waste oil using engineered *Ralstonia eutropha*. *Bioprocess Biosyst Eng* 2018; 41(2): 229–235.
51. FAO. 2018. *The State of World Fisheries and Aquaculture 2018 – Meeting the Sustainable Development Goals*. Rome: FAO. Licence: CC BY-NC-SA 3.0 IGO.
52. Van Thuoc D, My DN, Loan TT, et al. Utilization of waste fish oil and glycerol as carbon sources for polyhydroxyalkanoate production by *Salinivibrio sp.* M318. *Int J Biol Macromol* 2019; 141: 885–892.

53. Canakci M. The potential of restaurant waste lipids as biodiesel feedstocks. *Bioresour Technol* 2007; 98(1): 183–190.
54. Dahiya S, Kumar AN, Shanthi Sravan J, et al. Food waste biorefinery: Sustainable strategy for circular bioeconomy. *Bioresour Technol* 2018; 248: 2–12.
55. UFOP. *UFOP Report on Global Market Supply: 2017/2018*. Union zur Förderung von Oel- und Proteinpflanzen eV, 2017; pp. 51.
56. Koller M, Maršálek L, de Sousa Dias MM, et al. Producing microbial polyhydroxyalkanoate (PHA) biopolyesters in a sustainable manner. *New Biotechnol* 2017; 37: 24–38.
57. Titz M, Kettl KH, Shahzad K, et al. Process optimization for efficient biomediated PHA production from animal-based waste streams. *Clean Technol Environ Policy* 2012; 14(3): 495–503.
58. Muhr A, Rechberger EM, Salerno A, et al. Novel description of *mcl*-PHA biosynthesis by *Pseudomonas chlororaphis* from animal-derived waste. *J Biotechnol* 2013; 165(1): 45–51.
59. Hermann-Krauss C, Koller M, Muhr A, et al. Archaeal production of polyhydroxyalkanoate (PHA) Co- and terpolyesters from biodiesel industry-derived by-products. *Archaea* 2013; 2013: 129268.
60. Cavalheiro JMBT, Raposo RS, de Almeida MCMD, et al. Effect of cultivation parameters on the production of poly(3-hydroxybutyrate-*co*-4-hydroxybutyrate) and poly(3-hydroxybutyrate-4-hydroxybutyrate-3-hydroxyvalerate) by *Cupriavidus necator* using waste glycerol. *Bioresour Technol* 2012; 111: 391–397.
61. Koller M, Shahzad K, Braunegg G. Waste streams of the animal-processing industry as feedstocks to produce polyhydroxyalkanoate biopolyesters. *Appl Food Biotechnol* 2018; 5(4): 193–203.
62. Canakci M, Van Gerpen J. Biodiesel production from oils and fats with high free fatty acids. *Trans ASAE* 2001; 44(6): 1429–1436.
63. Budde CF, Riedel SL, Willis LB, et al. Production of poly(3-hydroxybutyrate-*co*-3-hydroxyhexanoate) from plant oil by engineered *Ralstonia eutropha* strains. *Appl Environ Microbiol* 2011; 77(9): 2847–2854.
64. Koller M, Braunegg G. Advanced approaches to produce polyhydroxyalkanoate (PHA) biopolyesters in a sustainable and economic fashion. *EuroBiotech J* 2018; 2(2): 89–103.
65. Feng L, Chen Y, Zheng X. Enhancement of waste activated sludge protein conversion and volatile fatty acids accumulation during waste activated sludge anaerobic fermentation by carbohydrate substrate addition: The effect of pH. *Environ Sci Technol* 2009; 43(12): 4373–4380.
66. Domingos JMB, Martinez GA, Scoma A, et al. Effect of operational parameters in the continuous anaerobic fermentation of cheese whey on titers, yields, productivities, and microbial community structures. *ACS Sustain Chem Eng* 2017; 5(2): 1400–1407.
67. Domingos JMB, Puccio S, Martinez GA, et al. Cheese whey integrated valorisation: Production, concentration and exploitation of carboxylic acids for the production of polyhydroxyalkanoates by a fed-batch culture. *Chem Eng J* 2018; 336: 47–53.
68. Brigham CJ, Riedel SL. The potential of polyhydroxyalkanoate production from food wastes. *Appl Food Biotechnol* 2019; 6(1): 7–18.
69. Hong C, Haiyun W. Optimization of volatile fatty acid production with co-substrate of food wastes and dewatered excess sludge using response surface methodology. *Bioresour Technol* 2010; 101(14): 5487–5493.
70. Riedel SL, Brigham CJ. Polymers and adsorbents from agricultural waste. In: Simpson BK, Ed., *Byproducts from Agriculture and Fisheries: Adding Value for Food, Feed, Pharma, and Fuels*. NJ: John Wiley & Sons Ltd., 2019; pp. 523–544.
71. Anderson JH, Anderson DH. Producing resins from organic waste products. *U.S. Patent No. 10,465,214*. November 5, 2019.

72. Yang YH, Brigham CJ, Budde CF, et al. Optimization of growth media components for polyhydroxyalkanoate (PHA) production from organic acids by *Ralstonia eutropha*. *Appl Microbiol Biotechnol* 2010; 87(6): 2037–2045.
73. Yu J, Si Y. Metabolic carbon fluxes and biosynthesis of polyhydroxyalkanoates in *Ralstonia eutropha* on short chain fatty acids. *Biotechnol Prog* 2004; 20(4): 1015–1024.
74. Jeon JM, Brigham CJ, Kim YH, et al. Biosynthesis of poly(3-hydroxybutyrate-*co*-3-hydroxyhexanoate) (P(HB-*co*-HHx)) from butyrate using engineered *Ralstonia eutropha*. *Appl Microbiol Biotechnol* 2014; 98(12): 5461–5469.
75. Yang YH, Brigham CJ, Song E, et al. Biosynthesis of poly(3-hydroxybutyrate-*co*-3-hydroxyvalerate) containing a predominant amount of 3-hydroxyvalerate by engineered *Escherichia coli* expressing propionate-CoA transferase. *J Appl Microbiol* 2012; 113(4): 815–823.
76. Coats ER, Watson BS, Brinkman CK. Polyhydroxyalkanoate synthesis by mixed microbial consortia cultured on fermented dairy manure: Effect of aeration on process rates/yields and the associated microbial ecology. *Water Res* 2016; 106: 26–40.
77. Marang L, Jiang Y, van Loosdrecht MCM, et al. Butyrate as preferred substrate for polyhydroxybutyrate production. *Bioresour Technol* 2013; 142: 232–239.
78. Chen H, Meng H, Nie Z, et al. Polyhydroxyalkanoate production from fermented volatile fatty acids: Effect of pH and feeding regimes. *Bioresour Technol* 2013; 128: 533–538.
79. Full Cycle Bioplastics LLC. Available from: http://fullcyclebioplastics.com/. Accessed on: December 31, 2019.
80. Durica-Mitic S, Göpel Y, Görke B. Carbohydrate utilization in bacteria: Making the most out of sugars with the help of small regulatory RNAs. In: Storz G, Ed., *Regulating with RNA in Bacteria and Archaea*. Washington, DC: ASM Press, 2018; pp. 229–248.
81. Nikodinovic-Runic J, Guzik M, Kenny ST, et al. Chapter four – Carbon-rich wastes as feedstocks for biodegradable polymer (polyhydroxyalkanoate) production using bacteria. *Adv Appl Microbiol* 2013; 84: 139–200.
82. Page WJ. Suitability of commercial beet molasses fractions as substrates for polyhydroxyalkanoate production by*Azotobacter vinelandii* UWD. *Biotechnol Lett* 1992; 14(5): 385–390.
83. Solaiman DKY, Ashby RD, Hotchkiss AT, et al. Biosynthesis of medium-chain-length poly(hydroxyalkanoates) from soy molasses. *Biotechnol Lett* 2006; 28(3): 157–162.
84. Omar S, Rayes A, Eqaab A, et al. Optimization of cell growth and poly(3-hydroxybutyrate) accumulation on date syrup by a *Bacillus megaterium* strain. *Biotechnol Lett* 2001; 23(14): 1119–1123.
85. Braunegg G, Genser K, Bona R, et al. Production of PHAs from agricultural waste material. *Macromol Symp* 1999; 144: 375–383.
86. Albuquerque MGE, Torres CAV, Reis MAM. Polyhydroxyalkanoate (PHA) production by a mixed microbial culture using sugar molasses: Effect of the influent substrate concentration on culture selection. *Water Res* 2010; 44(11): 3419–3433.
87. Carvalho G, Oehmen A, Albuquerque MGE, et al. The relationship between mixed microbial culture composition and PHA production performance from fermented molasses. *New Biotechnol* 2014; 31(4): 257–263.
88. Zahari MAKM, Ariffin H, Mokhtar MN, et al. Case study for a palm biomass biorefinery utilizing renewable non-food sugars from oil palm frond for the production of poly(3-hydroxybutyrate) bioplastic. *J Clean Prod* 2015; 87: 284–290.
89. Mohd Zahari MAK, Ariffin H, Mokhtar MN, et al. Factors affecting poly(3-hydroxybutyrate) production from oil palm frond juice by *Cupriavidus necator* (CCUG52238 T). *J Biomed Biotechnol* 2012; 2012: 1–8.
90. Zahari MAKM, Zakaria MR, Ariffin H, et al. Renewable sugars from oil palm frond juice as an alternative novel fermentation feedstock for value-added products. *Bioresour Technol* 2012; 110: 566–571.

91. Pramanik A, Mitra A, Arumugam M, et al. Utilization of vinasse for the production of polyhydroxybutyrate by *Haloarcula marismortui*. *Folia Microbiol* 2012; 57(1): 71–79.

92. Bhattacharyya A, Pramanik A, Maji SK, et al. Utilization of vinasse for production of poly-3-(hydroxybutyrate-*co*-hydroxyvalerate) by *Haloferax mediterranei*. *AMB Expr* 2012; 2(1): 34.

93. Brigham CJ. Biopolymers: Biodegradable alternatives to traditional plastics. In: Török B, Ed., *Green Chemistry – An Inclusive Approach*. New York, NY: Elsevier, 2018; pp. 753–770.

94. González-García Y, Rosales MA, González-Reynoso O, et al. Polyhydroxybutyrate production by *Saccharophagus degradans* using raw starch as carbon source. *Eng Life Sci* 2011; 11(1): 59–64.

95. Halami PM. Production of polyhydroxyalkanoate from starch by the native isolate *Bacillus cereus* CFR06. *World J Microbiol Biotechnol* 2008; 24(6): 805–812.

96. Ramadas NV, Singh SK, Soccol CR, et al. Polyhydroxybutyrate production using agro-industrial residue as substrate by *Bacillus sphaericus* NCIM 5149. *Braz Arch Biol Technol* 2009; 52(1): 17–23.

97. Bhatia SK, Shim YH, Jeon JM, et al. Starch based polyhydroxybutyrate production in engineered *Escherichia coli*. *Bioprocess Biosyst Eng* 2015; 38(8): 1479–1484.

98. Fonseca GG, Fonseca GG, de Arruda-Caulkins JC, et al. Production and characterization of poly-(3-hydroxybutyrate) from recombinant *Escherichia coli* grown on cheap renewable carbon substrates. *Waste Manag Res* 2008; 26(6): 546–552.

99. Haas R, Jin B, Zepf FT. Production of poly(3-hydroxybutyrate) from waste potato starch. *Biosci Biotechnol Biochem* 2008; 72(1): 253–256.

100. Wang W, Yang S, Hunsinger GB, et al. Connecting lignin-degradation pathway with pre-treatment inhibitor sensitivity of *Cupriavidus necator*. *Front Microbiol* 2014; 5: 247.

101. Sawant SS, Salunke BK, Tran TK, et al. Lignocellulosic and marine biomass as resource for production of polyhydroxyalkanoates. *Korean J Chem Eng* 2016; 33(5): 1505–1513.

102. Hassan MA, Yee LN, Yee PL, et al. Sustainable production of polyhydroxyalkanoates from renewable oil-palm biomass. *Biomass Bioenerg* 2013; 50: 1–9.

103. Zakaria MR, Fujimoto S, Hirata S, et al. Ball milling pretreatment of oil palm biomass for enhancing enzymatic hydrolysis. *Appl Biochem Biotechnol* 2014; 173(7): 1778–1789.

104. Silva LF, Taciro MK, Michelin Ramos ME, et al. Poly-3-hydroxybutyrate (P3HB) production by bacteria from xylose, glucose and sugarcane bagasse hydrolysate. *J Ind Microbiol Biotechnol* 2004; 31(6): 245–254.

105. Chakravarty J, Brigham CJ. Solvent production by engineered *Ralstonia eutropha*: channeling carbon to biofuel. *Appl Microbiol Biotechnol* 2018; 102(12): 5021–5031.

106. Brigham CJ, Gai CS, Lu J, et al. Engineering *Ralstonia eutropha* for production of isobutanol from CO_2, H_2, and O_2. In: Lee JW, Ed., *Advanced Biofuels and Bioproducts*. New York, NY: Springer, 2013; pp. 1065–1090.

107. Morinaga Y, Yamanaka S, Ishizaki A, et al. Growth characteristics and cell composition of *Alcaligenes eutrophus* in chemostat culture. *Agric Biol Chem* 1978; 42(2): 439–444.

108. Siegel RS, Ollis DF. Kinetics of growth of the hydrogen-oxidizing bacterium *Alcaligenes eutrophus* (ATCC 17707) in chemostat culture. *Biotechnol Bioeng* 1984; 26(7): 764–770.

109. Ishizaki A, Tanaka K, Taga N. Microbial production of poly-D-3-hydroxybutyrate from CO_2. *Appl Microbiol Biotechnol* 2001; 57(1): 6–12.

110. Volova TG, Kalacheva GS. The synthesis of hydroxybutyrate and hydroxyvalerate copolymers by the bacterium *Ralstonia eutropha*. *Microbiology* 2005; 74(1): 54–59.

111. Volova T, Kalacheva G, Altukhova O. Autotrophic synthesis of polyhydroxyalkanoates by the bacteria *Ralstonia eutropha* in the presence of carbon monoxide. *Appl Microbiol Biotechnol* 2002; 58(5): 675–678.

112. Revelles O, Beneroso D, Menéndez JA, et al. Syngas obtained by microwave pyrolysis of household wastes as feedstock for polyhydroxyalkanoate production in *Rhodospirillum rubrum*. *Microb Biotechnol* 2017; 10(6): 1412–1417.

113. Heinrich D, Raberg M, Fricke P, et al. Synthesis gas (syngas)-derived medium-chain-length polyhydroxyalkanoate synthesis in engineered *Rhodospirillum rubrum*. *Appl Environ Microbiol* 2016; 82(20): 6132–6140.

114. Panda B, Jain P, Sharma L, et al. Optimization of cultural and nutritional conditions for accumulation of poly-β-hydroxybutyrate in *Synechocystis* sp. PCC 6803. *Bioresour Technol* 2006; 97(11): 12296–12301.

115. Kamravamanesh D, Pflügl S, Nischkauer W, et al. Photosynthetic poly-β-hydroxybutyrate accumulation in unicellular cyanobacterium *Synechocystis* sp. PCC 6714. *AMB Expr* 2017; 7(1): 143.

116. Taroncher-Oldenburg G, Nishina K, Stephanopoulos G. Identification and analysis of the polyhydroxyalkanoate-specific β-ketothiolase and acetoacetyl coenzyme A reductase genes in the *Cyanobacterium* sp. strain PCC6803. *Appl Environ Microbiol* 2000; 66(10): 4440–4448.

117. Carpine R, Du W, Olivieri G, et al. Genetic engineering of *Synechocystis* sp. PCC6803 for poly-β-hydroxybutyrate overproduction. *Algal Res* 2017; 25: 117–127.

118. Kamravamanesh D, Lackner M, Herwig C. Bioprocess engineering aspects of sustainable polyhydroxyalkanoate production in cyanobacteria. *Bioengineering* 2018; 5(4): 111.

119. Amaro TMMM, Rosa D, Comi G, et al. Prospects for the use of whey for polyhydroxyalkanoate (PHA) production. *Front Microbiol* 2019; 10: 992.

120. Pantazaki AA, Papaneophytou CP, Pritsa AG, et al. Production of polyhydroxyalkanoates from whey by *Thermus thermophilus* HB8. *Process Biochem* 2009; 44(8): 847–853.

121. Koller M. Recycling of waste streams of the biotechnological poly(hydroxyalkanoate) production by *Haloferax mediterranei* on whey. *Int J Polym Sci* 2015; 2015: 8.

122. Koller M, Bona R, Chiellini E, et al. Polyhydroxyalkanoate production from whey by *Pseudomonas hydrogenovora*. *Bioresour Technol* 2008; 99(11): 4854–4863.

123. Koller M, Hesse P, Bona R, et al. Potential of various archae- and eubacterial strains as industrial polyhydroxyalkanoate producers from whey. *Macromol Biosci* 2007; 7(2): 218–226.

124. Kumar P, Ray S, Patel SKS, et al. Bioconversion of crude glycerol to polyhydroxyalkanoate by *Bacillus thuringiensis* under non-limiting nitrogen conditions. *Int J Biol Macromol* 2015; 78: 9–16.

125. Ashby RD, Solaiman DKY, Foglia TA. Bacterial poly(hydroxyalkanoate) polymer production from the biodiesel co-product stream. *J Polym Environ* 2004; 12(3): 105–112.

126. Teeka J, Imai T, Reungsang A, et al. Characterization of polyhydroxyalkanoates (PHAs) biosynthesis by isolated *Novosphingobium* sp. THA_AIK7 using crude glycerol. *J Ind Microbiol Biotechnol* 2012; 39(5): 749–758.

127. Ashby RD, Solaiman DKY, Foglia TA. Synthesis of short-/medium-chain-length poly(hydroxyalkanoate) blends by mixed culture fermentation of glycerol. *Biomacromolecules* 2005; 6(4): 2106–2112.

128. Moralejo-Gárate H, Kleerebezem R, Mosquera-Corral A, et al. Substrate versatility of polyhydroxyalkanoate producing glycerol grown bacterial enrichment culture. *Water Res* 2014; 66: 190–198.

129. Moita R, Freches A, Lemos PC. Crude glycerol as feedstock for polyhydroxyalkanoates production by mixed microbial cultures. *Water Res* 2014; 58: 9–20.

130. Blank LM, Narancic T, Mampel J, et al. Biotechnological upcycling of plastic waste and other non-conventional feedstocks in a circular economy. *Curr Opin Biotechnol* 2020; 62: 212–219.

9 The Sustainable Production of Polyhydroxyalkanoates from Crude Glycerol

Neha Rani Bhagat, Preeti Kumari,
Arup Giri, and Geeta Gahlawat

CONTENTS

9.1   INTRODUCTION – POLYHYDROXYALKANOATES

Continuous exploitation of non-biodegradable synthetic plastics and the gradual depletion of global petroleum reserves have created a big environmental pollution problem for humankind. Hence, there is an urgent need to look for a sustainable solution to replace petroleum-derived plastics and reduce the dependency on non-renewable fossil fuels [1]. Polyhydroxyalkanoates (PHA) are considered as an interesting alternative for synthetic plastics as they are synthesized from renewable resources and are completely biodegradable in the environment [2]. PHA are bio-based thermoplastics, which are synthesized and accumulated intracellularly by bacteria as energy-reserve inclusions, and their physicomechanical properties are quite similar to petroleum-derived plastics [3]. PHA have found a broad range of applications in packaging, wound management, drug delivery, tissue engineering, prolonged release of fertilizers in agriculture, and cartilage repair.

PHA are biopolyesters that are composed of 3-hydroxyalkanoic acids (HA acids). PHA are generally grouped into three categories based on the number of carbon atoms in their monomeric units [4]. The first category includes short-chain-length (scl-) polymers having 3–5 carbon atoms in a monomer, such as poly(3-hydroxybutyrate) [P(3HB) a.k.a. PHB] and poly(3-hydroxybutyrate-co-3-hydroxyvalerate) P(3HB-co-3HV) copolymer, synthesized by various bacteria such as *Cupriavidus necator* and *Alcaligenes latus* (presently known as *Azohydromonas lata*) [5]. The second category includes medium-chain-length (mcl-) polymers with 6–14 carbon atoms, such as poly(3-hydroxyhexanoate-co-3-hydroxyoctanoate) or P(3HHx-co-3HO), produced by *Pseudomonas* sp. PHA with 14 or more carbon atoms are long chain length (lcl), and they are very rare [6,7]. PHA are biodegradable in nature and undergo either anaerobic or aerobic degradation depending upon the provided conditions, and release CO_2 and water as final products in the environment [8].

Although PHA are excellent biodegradable materials, the high cost of production has restricted their industrial application and successful commercialization [3]. The high-cost issue can be resolved by using inexpensive renewable substrates, high cell density cultivations, optimizing bioprocess fermentation strategies that result in high productivity/yield, and developing simple, efficient recovery and purification strategies [2,9,10]. The most crucial factors contributing to the production cost are expensive raw materials and the low conversion rates of carbon substrates into PHA. Other important factors are low product yield/concentrations and a slow microbial growth rate using such materials.

It has been suggested in the literature that raw feedstock itself contributes to around 40–50% of the overall production cost; therefore, it is indispensable to use inexpensive substrates and optimize their concentration for high production [11]. This is because PHA production takes place in aerobic conditions, which results in a high loss of carbon substrate by the cellular respiration process [12]. This loss is primarily due to the release of the majority of carbon in the form of CO_2 through respiration, and, to a small extent, through excretion of other carbonaceous compounds. Only less than half of the carbon substrate is used for biomass production and PHA accumulation. Thus, continuous sincere efforts are needed in this direction for improvement in the conversion rate of carbon substrates into PHA for the successful utilization of economical and renewable substrates. More focus should be on the use of efficient carbon substrates, such as waste glycerol, molasses, cheese whey, food waste, hemicellulose, and agricultural waste, for the industrial production of PHA [3,13,14].

In recent years, researchers have also adopted different strategies for the enhanced production of PHA, such as bioprocess engineering, statistical optimization of media, high cell density cultivation strategies, multi-stage continuous cultivation, mathematical modeling, and mixed culture approaches [10,15–21]. Thus, emphasis should be on PHA production using inexpensive and sustainable substrates, along with novel, high PHA accumulating strains, high yield, and productivity cultivation strategies for successful commercialization of PHA. Moreover, new technological advancements are currently needed on an industrial scale for using waste raw materials.

As waste materials from different sources are accumulating at increasing rates, crude glycerol (CG) from the biodiesel industry or other petrochemical industries has been considered as a promising feedstock for PHA production and, therefore, represents the core part of this chapter. These glycerol-rich waste streams can be used by PHA-producing microbial strains as inexpensive carbon and energy sources for polymer (PHA) synthesis [12]. Glycerol is a by-product of the biodiesel manufacturing industry and is generated in large quantities as waste (around 1 kg of CG is released per 10 kg of biodiesel), which raises several problems related to its disposal [1,22]. Although glycerol is used in cosmetics, food, and pharmaceutical applications, it is costly to refine crude glycerol into a highly purified form appropriate for commercial use. An alternative is the valorization of waste glycerol by (bio)conversion into high-value products like PHA. This route seems to be an economically viable and sustainable approach. CG has been used for the synthesis of PHA by different bacteria and haloarchaea, such as *Bacillus megaterium* [23], *Cupriavidus necator* [2], *Pseudomonas putida* [24], *Haloferax mediterranei* [25], and mixed microbial cultures [26].

In this chapter, research mainly related to PHA production using CG is summarized. In recent studies, CG originating from different sources has been used successfully for PHA synthesis. However, the composition of the resulting PHA varies depending on the origin of individual CG samples [27]. There are several decisive factors such as catalyst type, the efficiency of the transesterification process, percentage of methanol, recovery yield of biodiesel, and level of impurities in the feedstock that affect the chemical composition of CG [28]. All these variations need to be considered during the valorization of CG into the high-value product, PHA, using microbes. Moreover, metabolic pathways involved in the synthesis of different types of PHA from glycerol have not been sufficiently discussed in the literature. Hence, the present chapter summarizes all the metabolic pathways involving glycerol, including metabolic engineering strategies adopted based on these pathways. Finally, cultivation processes and characterization strategies adopted for the production of different types of PHA from glycerol are discussed in the present chapter.

PHA synthesized from glycerol generally have lower molecular weight than PHA synthesized from other substrates, typically significantly less than 1000 kDa [12]. The molecular weight of a given polymer is an important property for commercialization purposes, as it determines the material characteristics of the polymer to a large extent. Thus, if glycerol is used as a feedstock for PHA production, we need to evaluate the suitability of recovered biopolymers for proposed applications, for example, low molecular weight PHA are useful for food packaging and diverse medical applications [12]. Thus, the mechanical properties of synthesized PHA should be analyzed thoroughly after extraction to meet the expectation of the intended application. This chapter also covers the impact of glycerol on the molecular mass of PHA biopolymers and their physicomechanical properties.

9.2 CRUDE GLYCEROL FROM BIODIESEL MANUFACTURE

Biodiesel production is achieved by using different types of oils extracted from plants or animal fats, such as castor, cottonseed, corn, crambe, peanut, coconut, mustard, sunflower, rapeseed, soybean, tung, and palm [29–31]. In the biodiesel industry, CG is one of the major by-products that is released into the environment as waste. It is estimated that for the production of 10 kg biodiesel, approximately 1 kg of crude glycerol is generated, which constitutes 10 wt.-% of the produced biodiesel [32,27]. Due to the increased demand for biodiesel to replace fossil fuels in recent years, the glycerol production rate has steadily increased, which leads to the excess generation of CG. In the absence of proper disposal procedures, CG may become an environmental pollution problem [31]. Although glycerol has many industrial applications in the medical, cosmetic, and food industries, these applications require highly purified materials, which further increases the cost [27]. Thus, CG can be directly valorized into PHA to eliminate the refining cost and, at the same time, yield a value-added product.

9.2.1 CHEMICAL COMPOSITION OF CRUDE GLYCEROL

CG obtained from biodiesel industries contains aqueous and organic compounds and impurities. The aqueous compounds encompass water, methanol, and glycerol, while the organic fraction contains free fatty acids (FFAs), fatty acid methyl esters (FAMEs), soap, and mono-, di-, and triglycerides [33]. FAMEs mainly include linoleic acid, palmitic acid, linolenic acid, oleic acid, and stearic acid. At the same time, sodium (Na), potassium (K), P, Ca, Mg, and Fe are detected in trace amounts in CG [33]. The composition of CG depends on both the process strategy for biodiesel production and the lipophilic feedstock used during biodiesel synthesis. Table 9.1 lists the oils and fats used as feedstock for biodiesel production and summarizes the composition of generated by-products.

CG also contains 2–5% ash; the high concentration of salt challenges the further processing of CG. A transesterification reaction is implemented for the conversion of the lipophilic feedstock into biodiesel. In the presence of a catalyst, the triglycerides (oil feedstock) react with alcohol (typically methanol) and produce FAMEs or other fatty acid esters (biodiesel) and CG [29]. Figure 9.1 explains the reaction of transesterification used for biodiesel and glycerol production. Methanol (CH_3OH), amyl alcohol (pentanol; $C_5H_{11}OH$), ethanol (C_2H_5OH), propanol (C_3H_7OH), and butanol (C_4H_9OH)

TABLE 9.1

Chemical Compositions of Crude Glycerol from Different Sources

Oils	Glycerol (%)	Methanol (%)	Water (%)	Soap (%)	Salts (%)	FFAs (%)	FAMEs (%)	Glycerides (%)	Ref.
Sunflower	30	50	2	13	2–3	13		2–3	[123]
Soybean	33	12.6	6.5	26.1	5–7	1.4	19.3	1.6	[33]
Waste Vegetable	27.8	8.6	4.1	20.5	2–3	3.0	28.8	7.0	[33]
Soybean and waste vegetable	57.1	11.3	1.0	31.4	2–3	BDL*	0.5	0.4	[33]
Canola	56.45	28.27	–	15.28	–	–	–	–	[124]
Rapeseed	65.7	23.4– 37.5	–	–	1–2	–	–	9.74	[125]
Crambe	62.5	23.4– 37.5	–	–	1-2	–	–	8.08	[125]

* Below detection limit; FFAs, Free fatty acids; FAMEs, Fatty acid methyl esters

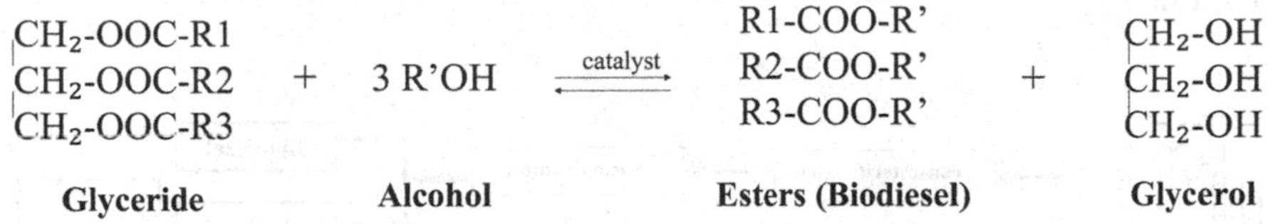

$$
\begin{array}{lcl}
\text{CH}_2\text{-OOC-R1} & & \text{R1-COO-R'} \qquad \text{CH}_2\text{-OH}\\
\text{CH}_2\text{-OOC-R2} \quad + \quad 3\,\text{R'OH} \xrightleftharpoons{\text{catalyst}} & \text{R2-COO-R'} \quad + \quad \text{CH}_2\text{-OH}\\
\text{CH}_2\text{-OOC-R3} & & \text{R3-COO-R'} \qquad \text{CH}_2\text{-OH}
\end{array}
$$

| Glyceride | Alcohol | Esters (Biodiesel) | Glycerol |

FIGURE 9.1 Transesterification reaction of triglycerides with alcohol (Adapted from Ma et al. [29] with permission from Elsevier).

are alcohols typically used in the transesterification process; an alcohol-to-triglycerides ratio of 3:1 is required stoichiometrically in the transesterification process [29]. The stoichiometric ratio should be higher to achieve the equilibrium in the chemical reaction and to obtain the maximum ester yield. These reactions are catalyzed either by enzymes or bases/acids. Catalysts play a major role in the purity of glycerol in the process of transesterification [30]. The alkali catalysts used here are KOH, NaOH, carbonates, Na, and K alkoxides like CH_3ONa, C_2H_5ONa, C_3H_7ONa, and C_4H_9ONa. For acidic catalysis, H_2SO_4, HCl, and sulfonic acids are generally used. Lipase enzymes are biocatalysts that can also be used in the transesterification reaction. Alkali-catalyzed transesterification processes are much faster than acid-catalyzed reactions, which makes the former more acceptable in the industry compared with the latter [29].

9.2.2 CRUDE GLYCEROL BASED PHA PRODUCTION

A large number of PHA-producing microbes can grow on CG in optimized environmental conditions and metabolize it into cell biomass and PHA biopolymer. The types of biodegradable PHA plastics that are commonly synthesized include poly(3-hydroxybutyrate) [P(3HB) or PHB], poly(3-hydroxyvalerate) (PHV), poly(3-hydroxyhexanoate) (PHHx), and PHA containing 3-hydroxy-2-methylbutyrate or

3-hydroxy-2-methylvalerate (3H2MV) [12]. Generally, PHA are produced by pure cultures through microbial fermentation, and because of this, the average production cost of green plastic increases, which is almost double the price of poly(vinylchloride) (PVC) production [34]. Crude glycerol-based PHA production could be done with mixed microbial cultures (MMC) and feast-famine approaches, which have lower operating costs [35,36]. MMC has emerged as a viable process to reduce the production cost of PHA processes. They offer several advantages compared with pure cultures, such as growth on a variety of industrial waste feedstocks, the possibility for axenic cultivation of microbes, and easy assimilation of the carbon substrate.

In conditions of physiological stress, native PHA-producing microbes accumulate lipophilic reserve materials (PHA) inside the cells. Typically, this stress occurs under the excess availability of carbon source (e.g. glycerol) and deficiency of essential nutrients like phosphorus, nitrogen, or oxygen etc. [37]. In these stress conditions, glycerol is converted to P(3HB); this biocatalytic cascade starts with the action of three different enzymes, namely glycerol kinase, glycerol-3-phosphate dehydrogenase, and triosephosphate isomerase. The scheme of biodegradable plastic (PHA) production from CG is shown in Figure 9.2. Bioconversion of waste glycerol to P(3HB) has been studied in detail using *C. necator* DSM 545 [1,2,13].

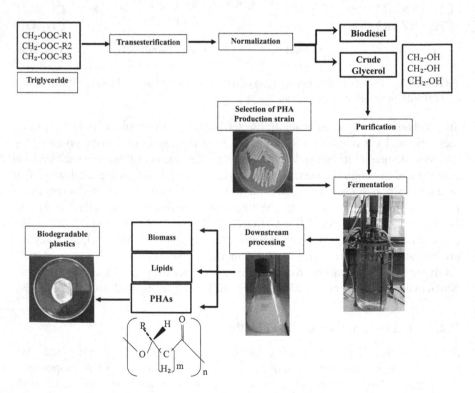

FIGURE 9.2 Production of biodegradable plastic (PHA) from crude glycerol (Adapted from Koller et al. [12] with permission from SBMU).

Based on the polymer length, PHA are categorized into *scl*-, *mcl*-, and *lcl*-PHA. Monomeric units (3-hydroxy fatty acids) have various combinations in PHA, so, based on these combinations, polymers are classified as homopolymers, copolymers, and heteropolymers [38]. *Scl*-PHA are brittle in nature due to their high crystallinity, while *mcl*-PHA are elastic and flexible due to their low crystallinity and low tensile strength [5,39]. *Lcl*-PHA are very rare, and their properties are only scarcely described [40]. Molecular weight (Mw) and monomer composition have huge effects on PHA thermal properties, for example, the Tg (glass transition temperature) and melting point, and mechanical properties like tensile strength, elastic module, elongation, and crystallinity [40]. *Scl*-PHA has high Tg, high melting temperatures, and lower elongation at break when compared with *mcl*-PHA [41]. The PHA polymer's physicochemical properties are dependent on various factors such as carbon substrate, cultivation conditions, and the microbial strain used for production. PHA have quite similar mechanical properties to synthetic polymers; these properties allow PHA to be used in industries directly or by further thermal processing [8].

9.3   METABOLIC PATHWAYS OF PHA SYNTHESIS FROM GLYCEROL

The biosynthetic pathway for PHA synthesis varies among different microbial groups where bacteria have the potential to synthesize PHA in both stationary and exponential growth phases [42]. Many bacteria have the potential to form various PHA biopolyesters like *scl*-PHA, including P(3HP), P(4HB), P(3HV), P(3HB-*co*-3HV), and *mcl*-PHA. In the past few years, PHA biosynthesis has been well investigated from different sources such as glucose, vegetable oils, beet sugar, cane sugar, sucrose, and corn [43]. However, using CG is an economical and environmentally sustainable alternative for PHA synthesis. This approach involves the use of CG as a substrate, which is commonly thrown away as industrial waste. The carbon atoms in glycerol molecules are chemically in a highly reduced state in comparison with glucose or lactose [12]. Hence, cells growing on glycerol as substrate are in a more reduced state, which further favors PHA synthesis. Thus, glycerol can be considered as one of the most suitable substrates for the synthesis of the P(3HB) precursor, acetyl-CoA.

So far, there have been different metabolic pathways described for PHA synthesis, while the pathways involved in PHA synthesis from glycerol have not been discussed in much detail. The glycerol-based pathways are summarized in Figure 9.3, including the tricarboxylic acid (TCA) cycle, P(3HB) synthesis, fatty acid *de novo* synthesis, β-oxidation, and the hydroxypropionate synthesis pathway [12,43]. PHA biosynthesis begins with the metabolism of glycerol to form dihydroxyacetone phosphate (DHAP) and 3-hydroxypropionaldehyde [28]. Then, the first molecule, DHAP, is converted into acetyl-CoA through glycolysis. Acetyl-CoA is a vital molecule in PHA synthesis that leads to the formation of the molecule 3-hydroxyalkanoyl-CoA, either hydroxyacyl-CoA (HA-CoA) or hydroxybutyryl-CoA (HB-CoA), of different lengths after going through several reactions in the PHA biosynthesis pathway. Thus, polymerization of these molecules results in the biosynthesis of PHA molecules of different chain lengths.

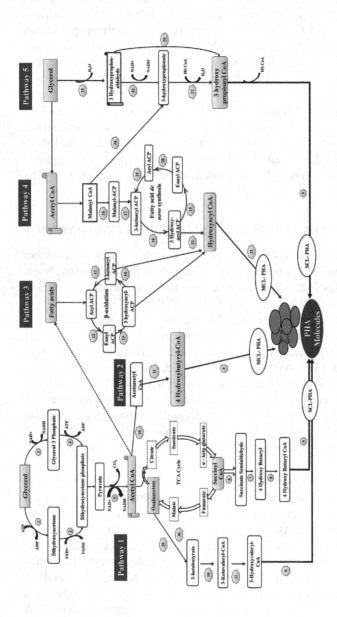

FIGURE 9.3 Schematic representation of various pathways involved in PHA synthesis from glycerol [46,50,129]. Note: Enzymes involved in the biosynthesis are: 1. DhaD, Glycerol dehydrogenase; 2. DhaK, Dihydroxyacetone kinase; 3. GlpK, Glycerol kinase; 4. G3P dehydrogenase; 5. Coenzyme A; 6. SucD, Succinyl semialdehyde dehydrogenase; 7. 4hbD, 4-hydroxybutyrate dehydrogenase; 8. 4-hydroxybutyrate-CoA: CoA transferase (OrfZ); 9. PhaC, PHA synthase; 10. PhaA, β- ketothiolase; 11. PhaB, NADP dependent acetoacetyl-CoA reductase; 12. Acyl CoA Dehydrogenase; 13. FadB, S-3-hydroxyacyl-CoA reductase, YdiO, enoyl-CoA reductase; 14. FadB, hydroxyacyl-CoA dehydrogenase/enoyl-CoA hydratase; 15. Ygef/FadA, thiolase; 16. ACC, acyl CoA Carboxylase; 17. FabD, Malonyl transacylase; 18. β-ketoacyl ACP synthase; 19. β-ketoacyl ACP reductase; 20. β-hydroxyacyl ACP dehydrase; 21. enoyl ACP reductase; 22. PhaG, 3-hydroxyacyl-acyl carrier protein CoA transferase; 23. PhaC1 (STQK), PHA synthase; 24. Malonyl CoA reductase, mcr; 25. DhaB, Glycerol dehydratase; 26. AldD, aldehyde dehydrogenase; 27. PCS', Propanoyl CoA synthatase; 28. PduP, Propionaldehyde dehydrogenase; 29. ThrA, aspartokinase 1, ThrB, homoserine kinase, ThrC, Threonine synthase; 30. Ilv, threonine deaminase.

9.3.1 PATHWAY 1: TCA CYCLE FOR *SCL*-PHA SYNTHESIS

The first pathway for PHA synthesis from glycerol involves the combination of the glycolytic pathway and the TCA cycle. Here, glycerol is broken down into dihydroxyacetone (DHA) and glycerol-3-phosphate (G3P) with the help of the enzymes glycerol dehydrogenase (DhaD) and glycerol kinase (GlpK). These two molecules are then broken down into DHAP by the enzymes dihydroxyacetone kinase (DhaK) and glycerol 3-phosphate dehydrogenase [28]. This DHAP molecule is then converted into pyruvate via the glycolysis pathway. Furthermore, through the pyruvate dehydrogenase reaction, pyruvate is converted into acetyl-CoA in the presence of coenzyme A with the release of a CO_2 molecule and reduction of NAD^+ into NADH. Acetyl-CoA is the major intermediate molecule that further participates in the TCA cycle to form the next substrate, namely succinyl-CoA. Succinyl-CoA is then converted into succinate semialdehyde by the enzyme succinate semialdehyde dehydrogenase (SucD) [44]. Furthermore, succinate semialdehyde is converted into 4-hydroxybutyrate and 4-hydroxybutyryl-CoA via 4-hydroxybutyrate dehydrogenase (4hbD) and 4-hydroxybutyrate-CoA: CoA transferase (OrfZ) [44,45]. This formation of either HA-CoA or HB-CoA is considered as a major step in PHA synthesis where these end molecules get polymerized into *scl*-PHA molecules via PHA synthase enzymes.

9.3.2 PATHWAY 2: *SCL*-PHA SYNTHESIS

In comparison with pathway 1, this metabolic pathway involves DHAP formation through G3P, which is finally condensed into acetyl-CoA through glycolysis and the pyruvate dehydrogenase reaction. This acetyl-CoA molecule is further converted into acetoacetyl-CoA via the enzyme 3-ketothiolase (formerly known as β-ketothiolase) (PhaA), which afterward, gets reduced to 3-hydroxybutyryl-CoA by acetoacetyl-CoA dehydrogenase (PhaB) using the cofactor NADPH [44]. This 3-HB-CoA molecule is then finally polymerized into P(3HB) by P(3HB) polymerase (PhaC). P(3HB) is one of the most common *scl*-PHA molecules formed during PHA synthesis. The enzymes involved in this pathway play a key role in *scl*-PHA biosynthesis.

9.3.3 PATHWAY 3: FATTY ACID β-OXIDATION FOR *MCL*-PHA SYNTHESIS

The central metabolite acetyl-CoA plays a vital role in the formation of fatty acids during metabolism. Moreover, β-oxidation of fatty acids is involved in this pathway, where the substrates formed after the degradation of fatty acids lead to PHA synthesis. During β-oxidation, first, these fatty acids get activated for degradation by conjugating with coenzyme A, forming acyl-acyl carrier protein (acyl-ACP) [46]. This acyl-ACP molecule is then converted into enoyl ACP, which further forms hydroxyacyl ACP and ketoacyl ACP under the influence of enzymes such as FabE, FabB, FabA, YdiCoA, FadB, and FadA, as mentioned in Figure 9.3

[44,46]. These two molecules, hydroxyacyl ACP and ketoacyl ACP, play a major role in the formation of hydroxyacyl-CoA. In the final step, this hydroxyacyl-CoA is polymerized into *mcl*-PHA molecules by the catalytic action of the enzyme PHA synthase [46].

9.3.4 PATHWAY 4: FATTY ACID *DE NOVO* PATHWAY FOR THE SYNTHESIS OF *MCL*-PHA

This pathway involves the intermediate molecule of the metabolism, acetyl-CoA, which undergoes carboxylation to form malonyl CoA. Afterward, it gets converted into malonyl ACP, which is then condensed into ketoacyl ACP by the enzyme β-ketoacyl ACP synthase. Ketoacyl ACP is the first step of fatty acid synthesis, and this is further converted into hydroxyacyl ACP by the enzyme ketoacyl ACP reductase [47]. Hydroxyacyl ACP is an important molecule that conjugates with coenzyme A to form hydroxyacyl-CoA. This final intermediate molecule of PHA synthesis is then polymerized into *mcl*-PHA molecules with the help of the enzyme, PHA synthase, and the hydroxyl acyl carrier protein, CoA transferase [46].

9.3.5 PATHWAY 5: POLY(3-HYDROXYPROPIONATE) SYNTHESIS PATHWAY

Among the *scl*-PHA, poly(3-hydroxypropionate) [P(3HP)] is one of the most interesting candidates. The pathway for P(3HP) synthesis comprises three different steps: dehydration of glycerol, oxidation, and coenzyme A coupling. During this process, 3-hydroxypropionaldehyde (3HPA) is converted to 3-hydroxypropionyl-CoA (3HP-CoA), which is finally polymerized into P(3HP) polymer [47]. These reactions are catalyzed by the enzymes glycerol dehydratase (*DhaB1*), propionaldehyde dehydrogenase (*PduP*), and PHA synthase (*PhaC1*) H16 [48]. Genes encoding these enzymes are heterogeneously expressed in microorganisms; therefore, only recombinant strains are known to synthesize P(3HP).

This polymer can also be formed via the route of malonyl CoA, where malonyl CoA formation from fatty acid *de novo* synthesis generates hydroxypropionate with the help of the enzyme, malonyl CoA reductase [49]. Then, this hydroxypropionate molecule undergoes ligation with coenzyme A, along with the release of water for generating hydroxy propionyl CoA. This final intermediate substrate of hydroxy propionyl CoA biosynthesis is then polymerized to PHA via the PHA synthase enzyme, finally synthesizing poly(3HP). The enzymes involved in this pathway are not harbored in a single microbe; therefore, the genes for malonyl CoA reductase (*mcr*), propionyl-CoA synthetase (*prpE*), acetyl-CoA carboxylase (*accABCD*), and PHA synthase (*phaC1*) were transferred into *E. coli* strains [48,49].

9.4 PRODUCTION OF PHA FROM CRUDE GLYCEROL

PHA production from CG involves several steps such as suitable strain development, shake flask optimization, fermentation studies in bioreactors, industrial scale-up, extraction, and purification. Production of PHA is very expensive as several factors, such as bacterial growth rate, final cell biomass, PHA content in biomass, time taken to achieve maximum PHA production, substrate-to-PHA conversion efficiency, PHA yield on the substrate, and efficient recovery methods to isolate the PHA from biomass, control the production cost. Some PHA have achieved large-scale production, and still, studies are continuing to lower the production cost [50]. Primarily, three steps, (i) inoculation and adaptation in substrate, (ii) fermentation, and (iii) isolation and purification, are included in the PHA production process using glycerol as the main carbon source. A schematic representation of PHA production from CG is provided in Figure 9.4. Some researchers discovered that the direct fermentation method of biodiesel-derived glycerol to PHA using *C. necator* results in expedient PHA yields [51,13]. However, some studies initially purified glycerol from CG using acids and multi-step distillation processes to remove the inhibitors present in it, such as methanol, NaCl, or sodium soaps (see Figure 9.4).

In the adaptation process, suitable strains must be adapted in the waste glycerol medium. For this purpose, different concentrations of glycerol are used varying between 10 and 60 g/L for cell growth, and optimized environmental conditions were provided to the strain for adaptation in particular glycerol media [2,13]. In the next fermentation step, PHA production is normally done in two stages. In the first stage, balanced cell growth is achieved by providing a fresh nutrient medium,

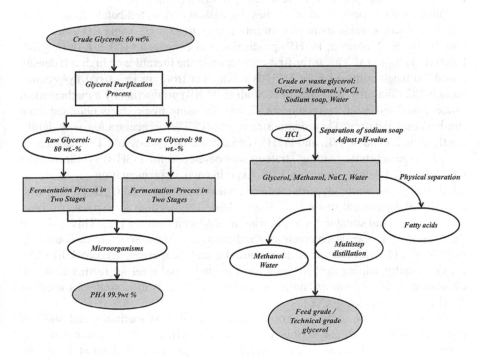

FIGURE 9.4 Glycerol as a carbon source for PHA production along with purification steps [60,128].

which is followed by a second stage of culturing the cells in excess carbon source and nutrient limitation conditions to trigger PHA accumulation [52,53]. In the last step, the polymer is extracted and purified from the cells using different solvents, such as chloroform, sodium hypochlorite, acetone, and 1,2 propylene carbonate [2,7,24,54,55]. Generally, chloroform is used for PHA extraction, but it is very toxic and causes environmental pollution; therefore, some recent studies have used 1,2 propylene carbonate, a non-toxic solvent, for PHA extraction obtained from crude glycerol [2,56].

9.4.1 PRODUCTION OF P(3HB) FROM CRUDE GLYCEROL

Among all *scl*-PHA, P(3HB) is the most widely and commonly studied PHA (see Table 9.2). Many bacteria like *Cupriavidus necator, Bacillus megaterium, Burkholderia cepacia, Pseudomonas oleovorans, Paracoccus* sp., and the haloarchaeon *Haloferax mediterranei* have been used for P(3HB) production from CG as the main carbon source [13,23,25,57– 61]. Shrivastav et al. screened and isolated several bacterial strains for P(3HB) production from *Jatropha*-based CG and reported an accumulation of 4.0 g/L biomass containing 76 wt.-% P(3HB) during shake flask optimization studies [54]. The direct valorization of CG into P(3HB) showed the production of a final biomass concentration of 6.69 g/L with a P(3HB) content of 64% of cell dry mass (CDM) after 42 h of shake flask cultivation of *C. necator* [2]. The maximum polymer productivity reported was 0.15 g/(L·h).

Other process optimization studies have shown that fed-batch fermentation approaches can increase biomass concentration up to 82.5 g/L using CG as a renewable feedstock. Moreover, P(3HB) production was enhanced to 51 g/L during fed-batch cultivation [13]. This is the first-ever report in the literature on high cell density based fed-batch cultivation for P(3HB) production from an industrial by-product, namely CG. Similarly, Špoljarić et al. studied P(3HB) production by a mathematical model based fed-batch cultivation of *C. necator* using by-products obtained from the biodiesel industry. They demonstrated that P(3HB) concentration was significantly high, around 30 g/L, and P(3HB) content was 64.3 wt.-% of CDM [22]. Later, a three-stage substrate feeding strategy was developed for P(3HB) synthesis from CG during fed-batch cultivation [62]. Experimental implementation of the developed strategy for waste glycerol revealed an accumulation of 65.6 g/L P(3HB) with a total P(3HB) content of 62.7 wt.-% of CDM. This feeding strategy was robust, cost-effective, and suitable for application in fed-batch cultures [62]. This indicated that optimized fermentation conditions and efficient cultivation strategies could help improve the PHA production from a renewable and inexpensive feedstock like CG, thus eventually helping in their scale-up at the industrial level for further commercialization. Table 9.2 shows the reports of different cultivation strategies used for P(3HB) production using CG.

Furthermore, CG contains various impurities, such as methanol and sodium/potassium salts, which can seriously affect cell growth and PHA accumulation by bacteria. For example, one study reported that the presence of NaCl at 5.5% concentration in waste glycerol could decrease the yield of PHA due to osmoregulatory challenge [51]. Similar findings were documented by Ashby et al., who demonstrated

TABLE 9.2

Comparison of PHA Production by Different Cultivation Strategies Using Crude Glycerol

Microorganism	Cultivation Mode	PHA Type	PHA^a (g/L)	PHA^b (% CDM)	PHA^c [g/(L·h)]	Ref.
Cupriavidus necator JMP 134	Fed-batch	P(3HB)	10.56	48	0.21	[51]
C. necator DSM 545	Fed-batch	P(3HB)	38.1	50	1.1	[13]
Burkholderia cepacia ATCC 17759	Fed-batch	P(3HB)	7.41	31	0.06	[58]
C. necator DSM 7237	Fed-batch	P(3HB)	20.8	74.5	0.27	[20]
C. necator DSM 545	Fed-batch	P(3HB)	69	65	0.76	[77]
Paracoccus sp. LL1	Fed-batch	P(3HB)	24.2	39.3	0.04	[59]
C. necator IPT 026	Fed-batch	P(3HB)	4.3	65	0.12	[55]
C. necator DSM4058	Fed-batch	P(3HB)	24.75	85.6	0.21	[126]
Pseudomonas oleovorans	Batch	P(3HB)	1.14	38	0.016	[57]
Bacillus megaterium	Batch	P(3HB)	3.4	62	0.08	[23]
C. necator DSM 545	Batch	P(3HB)	5.26	67	0.15	[2]
Mixed microbial culture	Batch	P(3HB)	–	59	–	[21]
Mixed microbial culture	Fed-batch	P(3HB)	–	47	0.24 g/(L·d)	[26]
Pandoraea sp. MA03	Fed-batch	P(3HB-co-3HV)	5.73	63.7	0.14	[75]
C. necator DSM 545	Fed-batch	P(3HB-co-3HV)	10.9	55.6	0.25	[61]
Haloferax mediterranei DSM1411	Fed-batch	P(3HB-co-3HV)	15.2	75.4	0.12	[25]
Yangia sp. ND199	Fed-batch	P(3HB-co-3HV)	8.3	40.6	0.25	[127]
C. necator DSM 545	Batch	P(3HB-co-3HV)	4.84	71	0.12	[2]
C. necator DSM 545	Fed-batch	P(3HB-co-4HB)	10.9	36.1	0.15	[1]
Cupriavidus sp. USMAHM13	Batch	P(3HB-co-4HB)	3.07	49	–	[74]
P. putida GO16	Fed-batch	mcl-PHA*	6.25	33	0.13	[24]

(Continued)

TABLE 9.2 (CONTINUED)
Comparison of PHA Production by Different Cultivation Strategies Using Crude Glycerol

Microorganism	Cultivation Mode	PHA Type	PHA^a (g/L)	PHA^b (% CDM)	PHA^c [g/(L·h)]	Ref.
C. necator IPT 027	Batch	mcl-PHA$^{\#}$	1.52	65.8	0.06	[7]
Recombinant E. coli ABC$_{Ab}$	Batch	P(3HB-co-3HHx)	0.6	40.4	0.025	[68]
Pseudomonas corrugata 388	Batch	mcl-PHA‡	0.67	20	0.01	[73]
Recombinant: E. coli LS5218	Batch	P(3HHx-co-3HO-co-3HD-co-3HDD)	0.392	11.6	–	[67]

^{a} PHA concentration (g/L); ^{b} PHA content in biomass (% CDM); ^{c} PHA productivity [g/(L·h)]; PHB-poly(3-hydroxybutyrate); P(3HB-co-3HV)-poly(3-hydroxybutyrate-co-3-hydroxyvalerate); P(3HB-co-3HHx); P(3HHx-co-3HO-3HD-co-3HDD)-poly(3-hydroxyhexanoate-co-3-hydroxyoctanoate-co-3-hydroxydecanoate-co-3-hydroxydodecanoate); *-Poly(3-hydroxyhexanoate-co-3-hydroxyoctanoate-co-3-hydroxydo decanoate-co-3-hydroxydodecenoate-co-3-hydroxytetradecanoate-co-3-hydroxy-tetradecenoate); # - Poly(3-hydroxybutyrate-co-11-hydroxyhexadecanoate-co-15-hydroxypentadecanoate-co-3-hydroxytetradecanoate); ‡ - Poly(3-hydroxydecanoate-co-3-hydroxydodecenoate).

that the direct use of CG containing 40 wt.-% methanol could affect bacterial growth and P(3HB) synthesis [57]. Thus, optimization of the biodiesel synthesis processes followed by methanol recycling will not only help in minimizing the production cost for the biodiesel industry but will also support the direct conversion of CG into PHA. Nevertheless, a few impurities present in CG, such as FFAs and FAMEs, could enhance the growth of bacteria by acting as supplementary carbon sources. This was clearly observed in *P. oleovorans* cultivation, a strain that metabolized FFAs and FAMEs and used them for cell growth [57]. These findings are similar to another study on CG, wherein an unidentified strain, AIK7, was isolated from wastewater, and when grown on CG containing FFAs and FAMEs, showed improvement in bacterial growth and polymer yield [63].

Various investigators have analyzed the effect of different glycerol concentrations on P(3HB) production [2,13,58,64]. In one study, increasing the concentration of glycerol beyond 3% in the culture medium inhibited the microbial growth due to osmotic pressure on the cells [58]. Thus, proper maintenance of glycerol concentration within the range of 1–3 wt.-% should be done to achieve high cell growth, and P(3HB) yields during the optimization process. Gahlawat and Soni (2017) reported that *C. necator* DSM 545 exhibited a maximum growth rate of 0.26 h^{-1} at 20 g/L glycerol concentration [2]. Increasing the glycerol concentration beyond 20 g/L decreased the maximum growth rate of the culture. Thus, the conversion efficiency of glycerol into PHA depends on its concentration inside the culture medium. The mixed microbial culture approach has also been used as an alternative approach for the successful production of PHA from glycerol [21,26,36]. In contrast with pure culture, mixed culture approaches provide several advantages, such as growth on low-quality substrates, sustained variation in substrate (glycerol) concentration, and no need to maintain the axenic cultivation condition, which ultimately lowers the production cost. Thus, mixed culture cultivation or the microbial community engineering approach are strong trends currently emerging for obtaining improved PHA yields in a cost-effective manner [36].

9.4.2 PRODUCTION OF P(3HB-CO-3HV) FROM CRUDE GLYCEROL

P(3HB-*co*-3HV) copolymer production is typically achieved only when a culture medium is supplied with two feedstocks; the first is glycerol acting as a substrate for the generation of 3HB units, while the second feedstock is a supplementary carbon source for the production of the 3HV moiety [28]. The P(3HB-*co*-3HV) copolymers are synthesized by microbes using CG and odd-numbered alkanoic acids, such as propionic acid, valeric acid, and heptanoic acid, as carbon substrates [65]. These odd carbon number organic acids are directly used as the precursor for the synthesis of 3HV monomeric units through β-oxidation or the fatty acid *de novo* pathway.

The fraction of 3HV monomers in polymer could be controlled by altering the concentration of odd carbon number substrates in the medium. Zhu et al. studied the synthesis of P(3HB-*co*-3HV) copolymers by *B. cepacia* using CG and levulinic acid (4-oxopentanoic acid) as co-substrates [66]. Continuous supplementation of levulinic acid in the culture exhibited accumulation of P(3HB-*co*-3HV) copolymers, wherein the mole fraction of the 3HV units increased from 5 to 32%. However, all P(3HB-*co*-3HV) copolymers synthesized in these conditions showed low molecular weight in comparison with the homopolyester P(3HB). Recently, P(3HB-*co*-3HV) copolymer production was investigated using *C. necator* DSM 545 by feeding different odd carbon number precursors (i.e., valeric acid, propionic acid, and sodium propionate) along with crude glycerol as a by-product from biodiesel manufacturing [2]. The concentration of organic acids was varied from 2 to 6 g/L to study their effect on the molar fraction of 3HV. The maximum concentration of P(3HB-*co*-3HV) copolymer of 3.4 g/L and a 3HV fraction of 30 mol-% was achieved at 4 g/L valeric acid.

Various fed-batch cultivation strategies have also been designed for further optimization in P(3HB-*co*-3HV) production from crude glycerol. P(3HB-*co*-3HV) synthesis by fed-batch culture of *Haloferax mediterranei* yielded PHA concentration and content of 15.2 g/L and 75.4 wt.-% of CDM, respectively [25]. Another investigation on the use of CG for P(3HB-*co*-3HV) production by fed-batch cultivation of *C. necator* DSM 545 showed an accumulation of 10.9 g/L PHA with increased productivity of 0.25 g/(L·h) [61]. The authors of this study concluded that by-products of the biodiesel industry could be used as an inexpensive, renewable feedstock for P(3HB-*co*-3HV) synthesis, an alternative to refined carbon substrates (e.g., glucose, fructose).

9.4.3   PRODUCTION OF *MCL*-PHA FROM CRUDE GLYCEROL

To date, only a limited number of studies have been carried out on *mcl*-PHA production from crude glycerol (Table 9.2). Table 9.2 shows the list of differently composed *mcl*-PHA synthesized from CG using batch and fed-batch cultivation strategies. Wang et al. illustrated *mcl*-PHA copolymer production by a genetically engineered *E. coli* strain using two different carbon sources, namely glycerol and glucose [67]. The *E. coli* strain was engineered by cloning three genes, *phaA*, *phaB*, and *phaC*, involved in P(3HB) synthesis, while two other genes, *phaG* and *alkK*, were involved in *mcl*-PHA synthesis. This engineered strains exhibited the ability to synthesize P(3HB-*co*-*mcl*-PHA) polymers from glycerol, and accumulated cell biomass of 5.73 g/L concentration and a copolymer of 60% of CDM. The synthesized P(3HB-*co*-*mcl*-PHA) copolymers were then extracted from the cells, and their monomeric composition was determined. The composition of two copolymers was P(3HB-*co*-2.7 mol-% 3HO-*co*-2.5 mol-% 3HD) and P(3HB-*co*-1.4 mol-% 3HO-*co*-1.7 mol-% 3HD-*co*-0.1 mol-% 3HDD). The introduction of *mcl*-PHA building blocks into P(3HB) for P(3HB-*co*-*mcl*-PHA) production greatly improves the material properties of PHA.

Furthermore, Phithakrotchanakoon et al. investigated the production of PHA from CG obtained from a palm oil based biodiesel plant, using recombinant *E. coli* ABC*Ah* [68]. This *E. coli* ABC*Ah* strain was designed by cloning three genes, namely 3-ketothiolase (PhaA*Re*), acetoacetyl-CoA reductase (PhaB*Re*) from *R. eutropha*, and PHA synthase (PhaC*Ah*) from *Aeromonas hydrophila*. This strain showed an accumulation of the highest P(3HB) content of 14% of CDM and a polymer concentration of 0.6 g/L when 1% (v/v) glycerol concentration was used for growth. The structural characterization of synthesized PHA was done by ^{1}H-NMR (hydrogen-1 nuclear magnetic resonance) and ^{13}C-NMR (carbon-13 nuclear magnetic resonance), and it showed the presence of an *mcl*-PHA, namely P(3HB-*co*-3HHx). Thermogravimetric analysis (TGA) of *mcl*-PHA showed that the degradation temperature (T_d) was around 299°C, which was higher than that observed for P(3HB) synthesized from CG using *Burkholderia cepacia* $(T_d -281.5°C)$ [58]. Thermal stability is an important physical property of a given polymer, which is beneficial during thermal processing. Moreover, the mechanical properties of *mcl*-PHA, such as elasticity and flexibility, are also generally better than that for *scl*-PHA, which makes them suitable candidates for tissue engineering purposes [69]. However, P(3HB) is highly brittle and fragile, which hinders its use as cartilage tissue material in engineering. Several other studies on *mcl*-PHA production from CG are listed in Table 9.2.

9.5 CHARACTERIZATION OF PHA SYNTHESIZED FROM GLYCEROL

Characterization is the most important step in PHA production for the qualitative analysis of synthesized polymers to assess their suitability for a particular application. PHA characterization has been done by using different techniques such as spectrophotometric analysis, nuclear magnetic resonance (NMR), Fourier transform infrared spectroscopy (FTIR), differential scanning calorimetry (DSC), TGA, gas chromatography (GC), and gel permeation chromatography (GPC). All physicomechanical properties of PHA, such as melting temperature, thermal stability, glass transition temperature, crystallinity, the molecular mass of the polymer, type of monomeric units, structural composition, and their association and conformation could be determined by these techniques.

Quantification of PHA synthesized from glycerol can be done by a simple spectrophotometric method at 235 nm using sulfuric acid as a blank [70]. The principle of this method is that the extracted polymer will be converted into crotonic acid after mixing with concentrated sulfuric acid. This method is also useful in the determination of the effect of different glycerol concentrations on the final yield of PHA. The NMR technique can be used to analyze the structural characterization of PHA, wherein the number of double bonds present in the polymer is calculated. Generally, ^{1}H-NMR and ^{13}C-NMR are used for this purpose, and different resonances are used for the determination of different copolymers [71,72]. In one study, the ^{1}H-NMR analysis confirmed that P(3HB)s produced from crude glycerol feedstock are end-capped with glycerol moiety through covalent bonding [28]. Zhu et al. observed that the methyl group, methylene group, and methine groups were found at the resonance of 1.25, 2.45 and 2.65, and 5.25 ppm, respectively [58]. This study also reported that terminal glycerol groups of the polymer were detected after the expansion of the spectral region between 3 and 4.5 ppm. The end-capping of P(3HB) through glycerol's secondary hydroxyl group was confirmed by resonance at 3.86 ppm. This indicated that glycerol acts as a polymer chain termination agent due to the covalent binding of the secondary hydroxyl group of glycerol to P(3HB). Similar NMR results have been reported in another study [73].

The total quantity of PHA and its monomer composition were analyzed by GC. GC is used to determine the composition of only volatile or semi-volatile compounds; therefore, the PHA has to be first hydrolyzed and acylated into volatile esters. Such a GC technique can be used for screening and isolation of CG-utilizing PHA producers from soil, sludge, and water samples [74]. The GC chromatogram revealed the presence of three peaks, namely 3-hydroxybutyrate methyl ester, 4-hydroxybutyrate methyl ester, and 4-butyrolactone, which further indicated the presence of P(3HB-co-4HB) accumulation. GC can be used in combination with mass spectrometry (MS) to obtain information about the mass-to-charge ratio of monomeric units present in a polymer. In one study, GC-MS was used to determine the composition of the copolymers synthesized by *C. necator* IPT 027 grown on CG [7]. GC-MS confirmed that the composition of PHA was 11-hydroxyhexadec-anoate (53.89–69.05 mol-%), 15-hydroxypentadecanoate (4.29–6.04 mol-%), and 3-hydroxytetradecanoate (5.54–13.10 mol-%). GC-MS was also used to character-ize PHA produced by *Pandoraea* sp. MA03 from CG; 3HB and 3HV units were identified by the peaks occurring at 2.95 min and 4.46 min, respectively, based on mass spectral analysis [75].

FTIR spectroscopy is another important tool that is used for studying the chemi-cal interactions and conformation of polymer molecules. It is also used for deter-mining the crystallinity index and changes in the morphology of the polymer. The crystallinity of a polymer is measured by the FTIR technique after scanning at different spectral ranges [54,76]. Different functional groups of PHA polymers are found at different spectral ranges. The FTIR technique has been used for scanning CG-derived PHA, where the hydroxyl (OH) groups were present near the spectral range of 3440 cm^{-1}, while the most important characteristic band of ester carbonyl was found near the spectral range of 1720–1740 cm^{-1} [2]. P(3HB-co-3HV) exhibited a peak at 1728.25 cm^{-1}, while P(3HB) showed a crystallinity band near 1182.14 cm^{-1}, the methine (–CH) group occurred as a peak at 1278 cm^{-1}, and the CH$_3$ group was identified by the peak at 1379 cm^{-1}. Other literature reports also confirmed the pres-ence of these peaks in glycerol-derived PHA [7,54,55].

DSC is used to determine all thermal properties of PHA. This includes the glass transition temperature (T_g), cooling crystalline temperature (T_{cc}), melting tempera-ture (T_m), and degree of crystallization (X_c) [28]. Thus, the thermal properties of PHA synthesized from CG can be analyzed by DSC, which significantly affects the appli-cability of a polymer. In one study, the DSC thermogram revealed that the melting point of P(3HB) was 175°C, and for a P(3HB-co-3HV) copolymer, two thermal peaks were obtained at 142 and 156°C due to the melting-recrystallization phenomenon [2]. Similar T_m results (173 to 175°C) have been reported for P(3HB) produced from glucose [7,77]. Another study on CG revealed a slight variation in results and showed that the T_m of P(3HB), obtained from different fermentation runs, was in the range of 157.2 to 184.3°C [55]. In the case of P(3HB-co-3HV), the melting temperature decreases with the increase in the 3HV fraction in the polymer [2]. This decrease in melting temperatures within the range of 130 to 140°C has also been observed when the 3HV units were incorporated into P(3HB) matrices to yield P(3HB-co-3HV) copolymers [25].

TGA is generally used to analyze the thermal stability of polymer in terms of decomposition or the degradation temperature (T_d) of PHA. This process includes subsequent heating and cooling steps to understand the degradation phenomenon

[70,78]. The decomposition temperature of P(3HB) was observed to be 281.5°C when CG was used for *B. cepacia* cultivations in bioreactors [58]. It was found that there was no significant change in T_g and T_m when xylose was used, except for T_d, which was 13°C higher for P(3HB) synthesized from glycerol compared with xylose-derived P(3HB). The thermal degradation temperature (T_d) of P(3HB-*co*-3HV) was reported at approximately 283°C when glycerol was used as a carbon source for *C. necator* fermentation [56]. Thus, T_d was almost comparable to a P(3HB) polymer observed in another study [58]. The degradation temperature is an essential parameter for checking the suitability of a polymer for desirable applications. A high degradation temperature indicates that the thermal stability of the polymer is also high, which is desirable for thermal processing at high temperatures.

GPC is used to determine the molecular mass/weight (Mw) and the number average molecular weight (Mn) of a PHA polymer [76]. However, it is essential to note that the production of PHA from glycerol generates polymers with relatively low molecular weight compared with PHA polymers synthesized from refined sugars [28,62]. This is discussed in detail in section 9.7 on the impact of CG on the molecular weight of PHA.

9.6 METABOLIC ENGINEERING FOR GLYCEROL-BASED PHA PRODUCTION

At present, many problems prevent the widespread use of PHA polymers, including the high cost of production and the failure to meet the demand and supply ratio of PHA. This high production cost is due to the high energy demand required for complex downstream processing strategies, the lower conversion rate of carbon substrates into the polymer, and the slow growth of microbes [79]. So far, many efforts have been made to solve these problems by isolating high PHA-producing bacterial strains, using biowaste as a carbon source, and developing efficient fermentation and recovery processes [80]. Based on these strategies, the use of unrefined crude glycerol streams for PHA production has greatly reduced the expenses of biopolymer synthesis. However, still, there is a need to enhance the cell biomass and PHA production by using this waste material as it is already reported that both the growth and polymer properties are affected by the amount and kind of impurities present in this raw material [81,82].

However, several bacterial strains have shown the potential of hydrolyzing complex substrates to use simple carbon sources, but do not have the capacity to synthesize PHA polymers under natural conditions. This highlights that there is a need for suitable host strains for economic and sustainable production of PHA. *E. coli* is one of these bacterial strains that is most widely studied. However, it lacks native metabolic pathways for polyester biosynthesis and requires heterologous gene expression encoding necessary enzymes for PHA synthesis. Moreover, *E. coli* is a *par excellence* engineering model due to its growing potency on cheap substrates, the use of a broad range of carbon sources, faster growth, and cell fragility, allowing easy recovery of the product [83]. So, specific technological improvements are required for the enhancement of PHA productivity and content.

In this context, metabolic engineering represents a suitable tool for enhancing PHA synthesis. *E. coli* is selected as the most common workhorse for large-scale production of PHA biopolymers, and several synthetic biological tools were also examined for fine-tuning the expression of heterologous genes in *E. coli*. Also, genetically engineered *E. coli* has shown more accumulation of biopolymers due to the absence of PHA depolymerase enzymes [83,84]. Meanwhile, many other metabolically engineered bacterial strains have been developed for obtaining high PHA productivity from CG. This includes members of the genera *Pseudomonas*, *Alcaligenes*, *Azotobacter*, or *Halomonas*. There are several engineering strategies that can be exploited for enhancing PHA production from crude glycerol (shown in Table 9.3).

9.6.1 ENGINEERED PHA SYNTHASES

PHA synthase specificity is the most pivotal factor for controlling PHA monomer composition. Its concentration can also have a direct influence on the molecular weight of a biopolymer [79]. Thus, fluctuation in PHA synthase activity can create a variation in the monomer's composition and molecular weight of PHA polymers. So far, many PHA synthases have been sequenced and cloned from different bacteria for creating engineered synthases with higher biocatalytic potential [85]. Thus, depending on the available information, the PHA synthase operon sequences can promote enhanced production of novel PHA from glycerol by the identification and construction of new PHA synthases.

For instance, various metabolically engineered *E. coli* species have been used for the synthesis of P(3HP) from CG, where the resulting biopolymer possesses promising properties in comparison with those of other *scl*-polymers [47,81,86]. Moreover, the 3HP monomers polymerized to PHA molecules showed a beneficial effect in reducing the polymer crystallinity and fragility [86]. This copolymer with promising advantages was synthesized by the introduction of PHA synthase genes from *R. eutropha*, along with glycerol dehydratase genes from *Clostridium butyricum*, and propionaldehyde dehydrogenase genes from *Salmonella enterica* [47].

The production of other copolymers and block polymers (*b*-PHA) from glycerol has also been demonstrated using recombinant *E. coli* strains without adding any precursors. Lately, an engineered *E. coli* strain was reported for achieving the highest PHA concentration from CG and glucose as co-substrates [87]. This recombinant strain efficiently expressed PHA synthase (*phaC1*) genes from *C. necator*, along with propionaldehyde dehydrogenase (*pduP*) genes from *Salmonella enterica* serovar Typhimurium, and glycerol dehydratase (*dhaB123* and *gdrAB*) genes from *Klebsiella pneumoniae*.

9.6.2 AERATED PHA SYNTHESIS

Pseudomonas putida KT217, a mutant strain of *P. putida* KT2442, was designed for increased production of *mcl*-PHA under aerobic conditions [88]. This was achieved by knocking out the *aceA* (isocitrate lyase enzyme) gene, which causes an increased metabolic flux of acetyl-CoA toward fatty acid biosynthesis for the synthesis of *mcl*-PHA. *P. putida* KT217 was then used for PHA production using a corn soluble

TABLE 9.3

Metabolic Engineering Approaches to Enhance PHA Production Using Crude Glycerol

S. No.	Approaches	Benefits	Drawbacks	Ref.
1.	Engineered PHA synthases	High productivity	Plasmid instability, expensive	[47,87]
2.	Phasin proteins	Enhanced PHA synthesis	Slow growth, requires advance technology, plasmid instability, expensive	[106,107]
3.	Increasing flexibility of carbon substrate assimilation	Enhances PHA production	Plasmid instability	[98,99]
4.	Deletion of competitive pathways (β-oxidation or D-lactate synthesis deletion)	Enhances PHA production, high PHA yield	Lower cell density	[100,103]
5.	Bacterial shape engineering	Increased cell size to accumulate more biopolymers	Reduced cell growth ability, difficult to do changes in high cell density	[94–96]
6.	Aerated PHA synthesis	Enhanced cell biomass	PHA content low, high energy maintenance	[89,90,107]
7.	Microaerophilic/anoxic cultivation conditions for PHA synthesis	Enhanced PHA synthesis, convenient, economical	Slow growth, less cell biomass	[91–93]
8.	CRISPR/CRISPRi/ CAS 9	Multiple genes deletion and repression for enhanced PHA synthesis	Genomic instability	[104–105]
9.	Next-generation extremophiles	Economical, no contamination, high cell growth in extreme environment	–	[108–111]

medium containing glycerol in aerated shake flask conditions [89]. A maximum CDM of 22 g/L was obtained in this study after supplementation of the nitrogen source along with corn solubles. In another study, *P. putida* KT217 was grown in a complex medium containing 400 g/L condensed corn solubles (on a wet basis) as the carbon source, and 2.2 g/L NH$_4$OH as a nitrogen source, to check its suitability for *mcl*-PHA production in aerated conditions [90]. After the growth phase, by-products of the biodiesel industry (CG and soapstock) were fed to the culture in a bioreactor for increased PHA production during 100 h of cultivation under continuous aeration. Here, compared with other carbon sources, foaming was of less concern when using glycerol. *P. putida* showed an accumulation of total cell biomass of 30 g/L and a PHA content of 31 wt.-% of CDM when a total of 75 g/L glycerol was added. Hence, a maximum PHA content of 31 wt.-% in cell dry mass was obtained only in long-term incubation for 100 h when using glycerol as the carbon source [90]. This showed that aeration caused efficient cell growth; however, PHA accumulation was very low. Thus, low intracellular PHA content can cause poor extraction economics compared with the current industrial processes having a target demand of 90% PHA biomass for efficient extraction.

9.6.3 MICROAEROPHILIC/ANOXIC CONDITIONS FOR PHA SYNTHESIS

PHA polymers are of great importance due to their sustainable, biodegradable, and biocompatible nature. However, current production processes require fully aerobic conditions for the synthesis of PHA [2,91]. This shows that high energy-consuming processes are needed to fulfill these requirements. Thus, microaerophilic cultivation processes have gained more attention for PHA synthesis due to convenient controllability and operation strategies, its low energy maintenance requirement, and the overall economic, eco-friendly, and sustainable nature of the process. For example, an arcA mutant *Escherichia coli* strain using glycerol as a carbon source was studied for P(3HB) synthesis under microaerobic conditions [92]. The microaerophilic conditions during fed-batch cultivation showed 2.57 times enhancement in PHA productivity in comparison with the batch culture setups. The microaerophilic batch fermentation gave rise to only 42 wt.-% P(3HB) content in biomass, while the microaerobic fed-batch cultivation resulted in 51 wt.-% P(3HB) content in biomass. Due to the increased market value of micro-aerophilic culturing, another approach, including the introduction of anaerobic promoters, can be explored to enhance the PHA production in anoxic conditions [93]. These promoters have also resulted in increased PHA production by a factor of 100 by changing the culture condition from aerobic to anaerobic [93]. Thus, the acetate mutant strain of *E. coli* has the potential for enhanced P(3HB) production without a high aeration demand.

9.6.4 BACTERIAL SHAPE ENGINEERING

Genetic manipulation to change bacterial shape/morphology allows the exploitation of bacterial shapes from rods to small spheres, small to large spheres, and fibers. The advantages of doing morphological changes for PHA production include high cell density, simplified downstream processing, space enlargement to allow more PHA accumulation, and more economical bio-production [94]. To the best of our knowledge, there are no reports to date on bacterial shape engineering based PHA production from glycerol. However, several shape-related genes can be exploited for availing of crude glycerol based process benefits, where the exploitation of relevant genes modulate the bacterial length and diameter limit. Enhancement of these limits can enhance the competitiveness in bio-processing, including improvement in the effectiveness of up- and downstream processing. Thus, enhancing bacterial morphology limits can create a promising enhancement in polymer production from glycerol [94–96]. Moreover, the rigid cell wall plays a limiting role for inclusion bodies accumulation inside the bacterial cells by limiting the space, as the weak cell wall allows easy cell size expansion, thus allowing increased storage of PHA granules [94]. So, the rigidity determining genes can also be exploited for enhancing polymer accumulation and production. Other reports have shown the role of essential genes, *mreB* and *ftsZ*, encoding for the cytoskeleton protein, MreB, and the cell division protein, FtsZ, in dividing the bacterial cytoskeleton [95]. Thus, the inactivation of these genes could result in an increase in cell sizes and lengths. So, a lot of effort has been made in this area for enhancing PHA production.

9.6.5 INCREASING FLEXIBILITY OF CARBON SUBSTRATE ASSIMILATION

PHA polymer accumulation is a multistep pathway, and the over-expression of the PHA synthesis genes and other related genes for increased assimilation of carbon sources could result in enhanced PHA production [97,98]. For example, the PHA synthase gene from *Pseudomonas* sp. LDC-5, *phaC1*, was cloned into the *E. coli* genome to study its effect on polymer synthesis [98]. This genetically engineered *E. coli* strain, containing *phaC1* synthesis genes, exhibited high PHA accumulation of 3.4 g/L when the medium was supplied with 1 wt.-% glycerol along with fish peptone. In another approach, a recombinant strain of *R. eutropha* H16 was developed to enhance the glycerol assimilation efficiency of by-products from the biodiesel industry [99]. This recombinant strain was engineered by cloning genes encoding for aquaglyceroporin (glpF) and glycerol kinase (glpK) from *E. coli* onto the chromosome of *R. eutropha* H16, and the developed recombinant strain showed an improved glycerol utilization ability, and high amounts of P(3HB) were synthesized from crude glycerol. These studies demonstrated the potential of *R. eutropha* and *E. coli* sp. for PHA production using CG as an economical and sustainable substrate.

9.6.6 DELETION OF COMPETITIVE PATHWAYS

PHA biopolymer synthesis competes with many other metabolic pathways for intermediate compounds. Therefore, it is necessary to eliminate or deactivate the competing pathways to channel more substrates toward the PHA biosynthesis pathway [100]. For example, the TCA cycle plays the role of the main competitive pathway for using glucose as a carbon substrate. However, TCA is crucial for the metabolism; therefore, this cycle cannot be deleted but weakened. The weakening of this cycle has shown a slight increase in PHA synthesis [100]. The elimination and weakening of competing pathways have been done by the inactivation of different genes involved in the metabolism, such as gene coding for the synthesis of acetate (*ackA*, *pta*, and *poxB*) or D-lactate (*ldhA*). In the literature, the elimination of the *pta* gene and the *frdA* and *ldhA* genes in *E. coli* have shown an increase in P(3HB) accumulation from glucose [101,102]. In 2010, Nikel et al. investigated the effect of inactivation of the D-lactate synthesis gene (*ldhA*) on P(3HB) production using glycerol as the main carbon substrate [103]. The recombinant *E. coli* K24KL strain, engineered by deleting the *ldhA* gene, showed enhancement in P(3HB) production and reduction in acetate synthesis, which could be due to the increased availability of acetyl-CoA and reduced cofactors for PHA production [103]. *E. coli* K24KL synthesized biomass of 41.9 g/L concentration and accumulated a P(3HB) content of 63 wt.-%, with a total polymer yield of 0.4 g/g on glycerol in 60 h of fed-batch cultivation. This is the highest reported PHA yield reported so far from waste glycerol (Table 9.3). Monitoring the intracellular reduced state revealed that strain K24KL has a high ratio of reducing equivalent cofactors to non-reducing ones; hence, *ldhA* deletion results in cells in a higher redox state.

9.6.7 CRISPR/CRISPRi/CAS9

CRISPRi (clustered regularly interspaced short palindromic repeats interference) is a highly favorable tool for metabolic engineering of multiple genes in prokaryotes. CRISPRi is an efficient approach that is used for selective inhibition and insertion of genes [104]. It has great potential for applications in prokaryotic multiple gene expression regulation during PHA synthesis. This CRISPRi system and small regulatory RNA (sRNA) could regulate the gene expression level in different ways; it can influence the gene coding sequence by blocking transcription initiation or elongation, while the sRNA system can control the target genes' expression. Recently, the application of the CRISPRi toolkit has been used for gene silencing in three *Burkholderia* sp., namely *B. cenocepacia*, *B. thailandensis*, and *B. multivorans*. A gene encoding for P(3HB) synthesis (*phbC*) was targeted and silenced using three guided RNAs (gRNA) to target a 50 bp region before the start codon at the putative promoter to study the effect of CRISPRi on PHA accumulation from glycerol [105].

9.6.8 PHASIN PROTEINS

Apart from the genes that enhance PHA synthesis, several natural PHA producers were also reported for genes translating different proteins for PHA granule formation, known as phasins, which can affect the polymer production [106]. Thus, the effect of a phasin, PhaP, has been investigated for the bacterial cell growth and PHA accumulation using glycerol as a raw material in recombinant *E. coli* culture carrying *phaBAC* genes from *Azotobacter* sp. strain FA8 [107]. The results confirmed that the cells expressing *phaP* showed 1.9 times more biomass and contained 2.6 times more P(3HB) production than the cells without the phasins. The phasins showed enhancement in growth and polymer accumulation resulting in high-density cultures, which can have a positive impact during large-scale polymer production. In addition, the recombinant strain presented a suitable model for P(3HB) production from glycerol by allowing 7.9 g/L of polymer production in 48 h of batch cultures [106,107]. The development of such recombinant bacterial strains can help in making PHA production economical and feasible in large-scale industrial production.

9.6.9 NEXT-GENERATION EXTREMOPHILES

Extremophilic bacteria possess several beneficial properties, such as growth under extreme conditions and resistance to microbial contamination. In this context, halophilic bacteria and haloarchaea are a class of extremophiles that are also able to grow on high salt content ocean water without contamination in open, unsterile reactors, thus can enable economic PHA production by reducing the maintenance charges and freshwater requirements [108]. Therefore, future PHA production strategies, such as starting from CG, should focus on these contamination resistant microbes. Among halophiles, *Halomonas* sp. seems to be an easy molecular manipulation target for enhanced PHA biosynthesis. In one report, a halophilic strain, *Halomonas* sp. KM-1, was used for P(3HB) production using crude glycerol as the sole carbon source [109]. However, the PHA content and yield were low on waste glycerol when compared with pure glycerol. PHA productivity and content can be further enhanced in halophiles by the introduction of several new pathways or new genes through metabolic engineering approaches [110,111]. This approach presents a potential future prospective in enhancing polymer production by discovering unique halophiles, and by exploiting their PHA production ability from CG for economical production on a large scale.

9.7 IMPACT OF CRUDE GLYCEROL ON THE MOLECULAR MASS OF PHA

PHA are generally used in various industries either in pure form or in mixture form; *mcl*-PHA are commonly used in drug delivery systems, implants, and surgical devices, while *scl*-PIIA are used for fiber materials, food packaging, and disposable products [112–114]. During PHA production, the type of substrate, feeding strategies, type of bacterial strain, PHA synthase specificity to a substrate, and physiological conditions have a great impact on the monomer composition and molecular weight (Mw) of PHA [71,115,116]. Mechanical resistance, elastic behavior, crystallization, and the thermoplasticity of PHA mainly depends upon its chemical composition and Mw [117]. PHA produced from CG obtained from the biodiesel industry exhibits significantly lower molecular mass in comparison with other industrial or agricultural waste [12]. PHA obtained from glycerol synthesized by *P. mediterranea* has unsaturated side chains with double bonds, which can potentially improve its applications [118]. Early termination of PHA polymerization due to chain transfer agents, such as glycerol and other polyols, results in a low molecular mass of 3.2 kDa.

According to Ashby et al., *mcl*-PHA was synthesized by *P. corrugata* and P(3HB) was synthesized by *P. oleovorans* using glycerol as a substrate; its concentration affects the molecular mass of *mcl*-PHA and P(3HB). By increasing the glycerol concentration in the cultivation medium from 1 to 5 wt.-%, a decrease in the molecular mass of *mcl*-PHA and P(3HB) was observed [73]. Madden et al. worked on *Ralstonia eutropha*, which produced P(3HB) on glycerol and glucose media [119]. Glycerol acted as a chain termination agent by removing covalent bonds of the carboxyl group from P(3HB), and the end-group structure produced a primary hydroxyl of glycerol. As such, the molecular mass of P(3HB) was decreased when compared with glucose [119]. Another group cultivated the highly osmophilic microorganism *Haloferax mediterranei* for the production of PHA on hydrolyzed lactose and crude glycerol as a primary carbon source [60]. PHA polymer produced by the fermentation from glycerol and carbohydrate showed Mw of 253 kDa and Pi (polydispersity index) 2.7, and 696 kDa and Pi 2.2, respectively. Thus, the molecular mass was very low in the case of glycerol compared with glucose [60]. In a follow-up study, PHA production from *Haloferax mediterranei* using crude glycerol as a substrate showed a molecular mass of copolyesters of 253 kDa and Pi 2.7 [25]. These results further confirmed that the PHA from glycerol has a lower molecular mass compared with other substrates under the same experimental conditions.

In another study, P(3HB) extracted from *Cupriavidus necator* had a molecular mass of 302.5 kDa with Pi 4.72, and *Burkholderia sacchari* produced P(3HB) with a molecular mass of 200 kDa and Pi 2.50 ± 0.43 [77]. Hence, PHA produced from CG using different bacterial strains generated a large range of PHA with low to high molecular masses. A Gram-negative bacterium, *Novosphinobium* sp., was used for the production of endotoxin-free PHA from CG, which can further be used for biomedical applications [76]. P(3HB) homopolyester produced from *Novosphinobium* sp. had a low molecular mass of 23.8 kDa and a high Pi of 5.6. Tanadchangsaeng et al. reported that the average Mw of P(3HB) polymer gradually decreased with time from 380 to 250 kDa, which was accompanied by an increase in polydispersity from 1.9 to 2.3 during late fermentation of *C. necator* [120]. The study explains that it could be due to the degradation of longer chains into shorter ones due to the random breakage of chemical bonds of polymer chains, which lowers the molecular weight. A previous report on PHA synthase functioning proposed that it acts as a controlling factor of Mw in intracellular PHA production by chain transfer reactions and chain propagation at the PHA synthase active site [121].

Tsuge et al. cultivated recombinant *R. eutropha* in media with oleic acid (20 g/L) and glycerol (0–10 g/L) and demonstrated that glycerol, even when present only in low concentrations in culture media, could affect the Mw of PHA [122]. As glycerol concentration was increased in the media, the PHA chain number increased, which in turn decreased the molecular weight [123–129]. Mothes et al. adopted *C. necator* and *P. denitrificans* for isolation of PHA using crude glycerol; the molecular mass varied from 620 to 750 kDa, but their properties were very similar to PHA isolated from glucose media. The molecular mass of PHA was not correlated with the purity of glycerol used in culture media [51]. However, no significant decrease in Mw of P(3HB) was observed when glycerol was used for the synthesis of the polymer by *R. eutropha* H16, and M_w varied between 680 to 920 kDa [99]. So far, the highest molecular weight PHA has been synthesized from *C. necator* DSM 545 using CG; it varied from 550 to 1370 kDa [1]. Various reports in the area of the production of PHA from CG highlights the causes of lower molecular masses of biopolymers, which cannot be ignored, so different techniques could be explored to produce the desirable biopolymer from biodiesel waste.

9.8 CONCLUSIONS AND OUTLOOK

The direct valorization of CG into high value-added products like PHA is very important for the successful and yet cost-effective operation of biodiesel industries. This will not only help in minimizing an environmental load of by-products generated by biodiesel industries but will also support the production of different types of PHA having unique physicomechanical properties for diverse applications. In the long-term, the use of biomass-derived CG will be beneficial for both the development of integrated biorefineries and the sustainable production of PHA. However, some issues need to be addressed for the wide-scale use of CG for PHA synthesis. Firstly, the chemical composition of CG significantly differs with the type of process and feed-stock used for the production of biodiesel. Thus, researchers across the world need to pay more attention when selecting the type of CG for the development of particular PHA production processes. Secondly, impurities present in CG, such as methanol and salts, can directly affect the conversion efficiency of glycerol into PHA polymers. In some PHA production processes, impurities in waste glycerol can inhibit cell growth and metabolism and result in low PHA productivity and yields. Therefore, it is essential to establish a strict specification for CG feed composition. The "standardized" CG will have high importance due to its consistency. Lastly, the concentration of CG in the culture medium has a great influence on cell growth and needs to be optimized.

Furthermore, various process development strategies, such as optimization of process parameters and fermentation conditions, development of metabolically engineered strains, design of efficient bioreactor systems for stable, aseptic operation, and improvement in the activity of catalysts, need to be adopted to enhance PHA production yields. Other promising techniques of microbial community engineering and glycerol fermentation using engineered halophilic strains are now required to further minimize the cost of production and increase the diversity of PHA. These strategies have low energy requirements because of their resistance to microbial contamination during cultivation. Bacterial shape engineering can simplify the downstream recovery process and reduce the production cost further. Another major concern associated with the use of CG, such as the main carbon substrates, is the low molecular weight of PHA polymers. Thus, it is essential to decide their suitability for particular types of applications. In some cases, feeding of supplementary carbon sources along with glycerol could help in preserving their molecular weight. Another possibility is the use of low molecular weight PHA for medical devices and packaging applications. In conclusion, CG has been considered as a renewable and inexpensive substrate. However, there are still more technical issues to be solved for the direct use of CG from the biodiesel industry and their complete valorization into PHA on a large scale.

ACKNOWLEDGMENTS

One of the authors (Geeta Gahlawat) wants to thank the University Grants Commission (UGC), Government of India, for providing Dr. D. S. Kothari a fellowship for the execution of the project on PHA. All authors want to thank the Defence Research & Development Organization Headquarter, New Delhi, India, for providing the facilities.

REFERENCES

1. Cavalheiro, J. M., Raposo, R. S., de Almeida M. C. M., et al. Effect of cultivation parameters on the production of poly(3-hydroxybutryrate-*co*-4-hydroxybutyrate) and poly(3-hydroxybutyrate-4-hydroxybutyrate-3-hydroxyvalerate) by *Cupriavidus necator* using waste glycerol. *Bioresour Technol* 2012; 111: 391–397.

2. Gahlawat, G., Soni, S. K., Valorization of waste glycerol for the production of poly (3-hydroxybutyrate) and poly (3-hydroxybutyrate-*co*-3-hydroxyvalerate) copolymer by *Cupriavidus necator* and extraction in a sustainable manner. *Bioresour Technol* 2017; 243: 492–501.

3. Reddy, M. V., Mawatari, Y., Yajima, Y., et al. Production of poly-3-hydroxybutyrate (P3HB) and poly (3-hydroxybutyrate-*co*-3-hydroxyvalerate) P (3HB-*co*-3HV) from synthetic wastewater using *Hydrogenophaga palleronii*. *Bioresour Technol* 2016; 215: 155–162.

4. Sudesh, K., Abe, H., Doi, Y., Synthesis, structure and properties of polyhydroxyalkanoates: Biological polyesters. *Prog Polym Sci* 2000; 25(10): 1503–1555.

5. Akaraonye, E., Keshavaraz, T., Roy, I., Production of polyhydroxyalkanoates: The future green materials of choice. *J Chem Technol Biotechnol* 2010; 85(6): 732–743.

6. Singh, A. K., Mallick, N., Enhanced production of SCL-LCL-PHA co-polymer by sludge-isolated *Pseudomonas aeruginosa* MTCC 7925. *Lett Appl Microbiol* 2008; 46(3): 350–357.

7. Ribeiro, P. L. L., da Silva, A. C. M. S., Menezes Filho, J. A., et al. Impact of different by-products from the biodiesel industry and bacterial strains on the production, composition, and properties of novel polyhydroxyalkanoates containing achiral building blocks. *Ind Crop Prod* 2015; 69: 212–223.

8. Chen, G. Q., Plastics completely synthesized by bacteria: Polyhydroxyalkanoates. In: Chen, G. Q., Ed., Steinbüchel, A., Series Ed., *Plastics from Bacteria*. Berlin, Heidelberg: Springer, 2010; 17–37.

9. Ng, K. S., Wong, Y. M., Tsuge, T., et al. Biosynthesis and characterization of poly(3-hydroxybutyrate-*co*-3-hydroxyvalerate) and poly(3-hydroxybutyrate-*co*-3-hydroxy hexanoate) copolymers using jatropha oil as the main carbon source. *Process Biochem* 2011; 46: 1572–1578.

10. Gahlawat, G., Srivastava, A., Model-based nutrient feeding strategies for the increased production of polyhydroxybutyrate (PHB) by *Alcaligenes latus*. *Appl Biochem Biotechnol* 2017; 183: 530–542.

11. Koller, M., Maršálek, L., Miranda de Sousa Dias, M., et al. Producing microbial polyhydroxyalkanoate (PHA) biopolyesters in a sustainable manner. *New Biotechnol* 2017; 37: 24–38.

12. Koller M., Marsalek, L., Principles of glycerol-based polyhydroxyalkanoate production. *Appl Food Biotechnol* 2015; 2(4): 3–10.

13. Cavalheiro, J. M., de Almeida, M. C., Grandfils, C., et al. Poly (3-hydroxybutyrate) production by *Cupriavidus necator* using waste glycerol. *Process Biochem* 2009; 44(5): 509–515.

14. Moreno, P., Yañez, C., Cardozo, N. S. M., et al. Influence of nutritional and physicochemical variables on PHB production from raw glycerol obtained from a Colombian biodiesel plant by a wild-type *Bacillus megaterium* strain. *New Biotechnol* 2015; 32(6): 682–689.

15. Yu, S. T., Lin, C. C., Too, J. R., PHBV production by *Ralstonia eutropha* in a continuous stirred tank reactor. *Process Biochem* 2005; 40: 2729–2734.

16. Gahlawat, G., Srivastava, A., Estimation of fundamental kinetic parameters of polyhydroxybutyrate fermentation process of *Azohydromonas australica* using statistical approach of media optimization. *Appl Biochem Biotechnol* 2012; 168(5): 1051–1064.

17. Zafar, M., Kumar, S., Kumar, S., et al. Artificial intelligence based modeling and optimization of poly (3-hydroxybutyrate-*co*-3-hydroxyvalerate) production process by using *Azohydromonas lata* MTCC 2311 from cane molasses supplemented with volatile fatty acids: A genetic algorithm paradigm. *Bioresour Technol* 2012; 104: 631–641.

18. Gahlawat, G., Srivastava, A. K., Development of a mathematical model for the growth associated Polyhydroxybutyrate fermentation by Azohydromonas australica and its use for the design of fed-batch cultivation strategies. *Bioresour Technol* 2013; 137: 98–105.

19. Horvat, P., Špoljarić, I. V., Lopar, M., et al. Mathematical modelling and process optimization of a continuous 5-stage bioreactor cascade for production of poly [-(R)-3-hydroxybutyrate] by *Cupriavidus necator*. *Bioproc Biosyst Eng* 2013; 36(9): 1235–1250.

20. Kachrimanidou, V., Kopsahelis, N., Papanikolaou, S., et al. Sunflower-based biorefinery: Poly(3-hydroxybutyrate) and poly(3-hydroxybutyrate-*co*-3-hydroxyvalerate) production from crude glycerol, sunflower meal and levulinic acid. *Bioresour Technol* 2014; 172: 121–130.

21. Freches, A., Lemos, P. C., Microbial selection strategies for polyhydroxyalkanoates production from crude glycerol: Effect of OLR and cycle length. *New Biotechnol* 2017; 39: 22–28.

22. Špoljarić, I. V., Lopar, M., Koller, M., et al. Mathematical modeling of poly [(R)-3-hydroxyalkanoate] synthesis by *Cupriavidus necator* DSM 545 on substrates stemming from biodiesel production. *Bioresour Technol* 2013; 133: 482–494.

23. Naranjo, J. M., Posada, J. A., Higuita, J. C., et al. Valorization of glycerol through the production of biopolymers: The PHB case using *Bacillus megaterium*. *Bioresour Technol* 2013; 133: 38–44.

24. Kenny, S. T., Runic, J. N., Kaminsky, W., et al. Development of a bioprocess to convert PET derived terephthalic acid and biodiesel derived glycerol to medium chain length polyhydroxyalkanoate. *Appl Microbiol Biotechnol* 2012; 95(3): 623–633.

25. Hermann-Krauss, C., Koller, M., Muhr, A., et al. Archaeal production of polyhydroxyalkanoate (PHA) co-and terpolyesters from biodiesel industry-derived by-products. *Archaea* 2013; 2013: 1–10.

26. Moita, R., Freches, A., Lemos, P. C., Crude glycerol as feedstock for polyhydroxyalkanoates production by mixed microbial cultures. *Water Res* 2014; 58: 9–20.

27. Yang, F., Hanna, M. A., Sun, R., Value-added uses for crude glycerol-a byproduct of biodiesel production. *Biotechnol Biofuels* 2012; 5(13): 1–10.

28. Zhu, C., Chiu, S., Nakas, J. P., et al. Bioplastics from waste glycerol derived from biodiesel industry. *J Appl Polym Sci* 2013; 130(1): 1–13.

29. Ma, F., Milford, A. H., Biodiesel production: A review. *Bioresour Technol* 1999; 7: 1–5.

30. Mu, Y., Teng, H., Zhang, D. J., et al. Microbial production 1,3–propanediol by *Klebsiella pneumonia* using crude glycerol from biodiesel preparation. *Biotechnol Lett* 2006; 28: 1755–1759.

31. da Silva, G. P., Mack, M., Contiero, J., Glycerol: A promising and abundant carbon source for industrial microbiology. *Biotechnol Adv* 2009; 27: 30–39.

32. Johnson, D. T., Taconi, K. A., The glycerine glut: Options for the value-added conversion of crude glycerol resulting from biodiesel production. *Environ Prog* 2007; 26: 338–348.

33. Hu, S., Luo, X., Wan, C., et al. Characterization of crude glycerol from biodiesel plants. *J Agric Food Chem* 2012; 60: 5915–5921.

34. Chanprateep, S., Current trends in biodegradable polyhydroxyalkanoates. *J Biosci Bioeng* 2010; 110(6): 621–632.

35. Albuquerque, M. G. E., Eiroa, M., Torres, C., et al. Strategies for the development of a side stream process for polyhydroxyalkanoate (PHA) production from sugar cane molasses. *J Biotechnol* 2007; 130: 411–421.

36. Moralejo-Gárate, H., Kleerebezem, R., Mosquera-Corral, A., et al. Impact of oxygen limitation on glycerol-based biopolymer production by bacterial enrichments. *Water Res* 2013; 47(3): 1209–1217.
37. Bitar, A., Underhill, S., Effect of ammonium supplementation on production of poly-β-hydroxybutyric acid by *Alcaligenes eutrophus* in batch culture. *Biotechnol Lett* 1990; 12: 563–568.
38. Anjum, A., Zuber, M., Zia, K. M., *et al.* Microbial production of polyhydroxyalkanoates (PHAs) and its copolymers: A review of recent advancements. *Int J Biol Macromol* 2016; 89: 161–174.
39. Zinn, M., Witholt, B., Egli, T., Occurrence, synthesis and medical application of bacterial polyhydroxyalkanoate. *Adv Drug Deliv Rev* 2001; 53(1): 5–21.
40. Singh, M., Kumar, P., Ray, S., et al. Challenges and opportunities for customizing polyhydroxyalkanoates. *Ind J Microbiol* 2015; 55(3): 235–249.
41. Noda, I., Green, P. R., Satkowski, M. M., et al. Preparation and properties of a novel class of polyhydroxyalkanoate copolymers. *Biomacromolecules* 2005; 6: 580–586.
42. Mohapatra, S., Maity, S., Dash, H. R., et al. *Bacillus* and biopolymer: Prospects and challenges. *Biochem Biophys Rep* 2017; 12: 206–213.
43. Jiang, X. R., Chen, G. Q., Morphology engineering of bacteria for bio-production. *Biotechnol Adv* 2016; 34(4): 435–440.
44. Możejko-Ciesielska, J., Kiewisz, R., Bacterial polyhydroxyalkanoates: Still fabulous? *Microbiol Res* 2016; 192: 271–282.
45. Valentin, H. E., Dennis, D., Production of poly (3-hydroxybutyrate-*co*-4-hydroxybutyrate) in recombinant *Escherichia coli* grown on glucose. *J Biotechnol* 1997; 58(1): 33–38.
46. Meng, D. C., Chen, G. Q., Synthetic biology of polyhydroxyalkanoates (PHA). In: *Synthetic Biology – Metabolic Engineering*. Cham: Springer, 2017; 147–174.
47. Andreeßen, B., Lange, A. B., Robenek, H., et al. Conversion of glycerol to poly (3-hydroxypropionate) in recombinant *Escherichia coli*. *Appl Environ Microbiol* 2010; 76(2): 622–626.
48. Andreeßen, B., Taylor, N., Steinbüchel, A., Poly (3-hydroxypropionate): A promising alternative to fossil fuel-based materials. *Appl Environ Microbiol* 2014; 80(21): 6574–6582.
49. Wang, Q., Liu, C., Xian, M., et al. Biosynthetic pathway for poly (3-hydroxypropionate) in recombinant *Escherichia coli*. *J Microbiol* 2012; 50(4): 693–697.
50. Chen, G. Q., Hajnal, I., Wu, H., et al. Engineering biosynthesis mechanisms for diversifying polyhydroxyalkanoates. *Trends Biotechnol* 2015; 33(10): 565–574.
51. Mothes, G., Schnorpfeil, C., Ackermann, J. U., Production of PHB from crude glycerol. *Eng Life Sci* 2007; 7(5): 475–479.
52. Ganduri, V. S. R. K., Ghosh, S., Patnaik, P. R., Mixing control as a device to increase PHB production in batch fermentations with co-cultures of *Lactobacillus delbrueckii* and *Ralstonia eutropha*. *Process Biochem* 2005; 40: 257–264.
53. Khanna, S., Srivastava, A. K., Computer simulated fed-batch cultivation for over production of PHB: A comparison of simultaneous and alternate feeding of carbon and nitrogen. *Biochem Eng J* 2006; 27: 197–203.
54. Shrivastav, A., Mishra, S. K., Shethia, B., et al. Isolation of promising bacterial strains from soil and marine environment for polyhydroxyalkanoates (PHAs) production utilizing Jatropha biodiesel byproduct. *Int J Biol Macromol* 2010; 47(2): 283–287.
55. Campos, M. I., Figueiredo, T. V. B., Sousa, L. S., et al. The influence of crude glycerin and nitrogen concentrations on the production of PHA by *Cupriavidus necator* using a response surface methodology and its characterizations. *Ind Crop Prod* 2014; 52: 338–346.

56. Gahlawat, G., Soni, K. S., Study on sustainable recovery and extraction of polyhydroxyalkanoates (PHAs) produced by *Cupriavidus necator* using waste glycerol for medical applications. *Chem Biochem Eng Q* 2019; 33(1): 99–110.

57. Ashby, R. D., Solaiman, D. K., Strahan, G. D., Efficient utilization of crude glycerol as fermentation substrate in the synthesis of poly (3-hydroxybutyrate) biopolymers. *J Am Oil Chem Soc* 2011; 88(7): 949–959.

58. Zhu, C., Nomura, C. T., Perrotta, J. A., et al. Production and characterization of poly-3-hydroxybutyrate from biodiesel-glycerol by *Burkholderia cepacia* ATCC 17759. *Biotechnol Prog* 2010; 26(2): 424–430.

59. Kumar, P., Jun, H. B., Kim, B. S., Co-production of polyhydroxyalkanoates and carotenoids through bioconversion of glycerol by *Paracoccus* sp. strain LL1. *Int J Biol Macromol* 2018; 107: 2552–2558.

60. Koller, M., Bona, R., Braunegg, G., et al. Production of polyhydroxyalkanoates from agricultural waste and surplus materials. *Biomacromolecules* 2005; 6: 561–565.

61. García, I. L., López, J. A., Dorado, M. P., et al. Evaluation of by-products from the biodiesel industry as fermentation feedstock for poly (3-hydroxybutyrate-*co*-3-hydroxyvalerate) production by *Cupriavidus necator*. *Bioresour Technol* 2013; 130: 16–22.

62. Mozumder, M. S. I., De Wever, H., Volcke, E. I., et al. A robust fed-batch feeding strategy independent of the carbon source for optimal polyhydroxybutyrate production. *Process Biochem* 2014; 49(3): 365–373.

63. Teeka, J., Imai, T., Cheng, X., et al. Screening of PHA-producing bacteria using biodiesel-derived waste glycerol as a sole carbon source. *J Water Environ Technol* 2010; 8(4): 373–381.

64. Ibrahim, M. H. A., Steinbüchel, A., *Zobellella denitrificans* strain MW1, a newly isolated bacterium suitable for poly (3-hydroxybutyrate) production from glycerol. *J Appl Microbiol* 2010; 108(1): 214–225.

65. Khanna, S., Srivastava, A. K., Production of poly (3-hydroxybutyric-*co*-3-hydroxyvaleric acid) having a high hydroxyvalerate content with valeric acid feeding. *J Ind Microbiol Biotechnol* 2007; 34: 457–461.

66. Zhu, C., Nomura, C. T., Perrotta, J. A., et al. The effect of nucleating agents on physical properties of poly-3-hydroxybutyrate (PHB) and poly-3-hydroxybutyrate-*co*-3-hydroxyvalerate (PHB-*co*-HV) produced by *Burkholderia cepacia* ATCC 17759. *Polym Test* 2012; 31(5): 579–585.

67. Wang, Q., Tappel, R. C., Zhu, C., et al. Development of a new strategy for production of medium-chain-length polyhydroxyalkanoates by recombinant *Escherichia coli* via inexpensive non-fatty acid feedstocks. *Appl Environ Microbiol* 2012; 78(2): 519–527.

68. Phithakrotchanakoon, C., Champreda, V., Aiba, S. I., et al. Production of polyhydroxyalkanoates from crude glycerol using recombinant *Escherichia coli*. *J Polym Environ* 2015; 23(1): 38–44.

69. Wang, Y., Bian, Y. Z., Wu, Q., et al. Evaluation of three-dimensional scaffolds prepared from poly(3-hydroxybutyrate-*co*-3-hydroxyhexanoate) for growth of allogeneic chondrocytes for cartilage repair in rabbits. *Biomaterials* 2008; 29(19): 2858–2868.

70. Salgaonkar, B. B., Mani, K., Bragança, J. M., Characterization of polyhydroxyalkanoates accumulated by a moderately halophilic salt pan isolate *Bacillus megaterium* strain H 16. *J Appl Microbiol* 2013; 114(5): 1347–1356.

71. Dai, Y., Lambert, L., Yuan, Z., et al. Characterization of polyhydroxyalkanoate copolymers with controllable four-monomer composition. *J Biotechnol* 2008; 134(1–2): 137–145.

72. Tan, G. Y., Chen, C. L., Ge, L., et al. Enhanced gas chromatography-mass spectrometry method for bacterial polyhydroxyalkanoates analysis. *J Biosci Bioeng* 2014; 117(3): 379–382.

73. Ashby, R. D., Solaiman, D. K., Foglia, T. A., Synthesis of short-medium-chain-length poly(hydroxyalkanoate) blends by mixed culture fermentation of glycerol. *Biomacromolecules* 2005; 6(4): 2106–2112.
74. Ramachandran, H., Amirul, A. A., Yellow-pigmented *Cupriavidus* sp., a novel bacterium capable of utilizing glycerine pitch for the sustainable production of P (3HB-co-4HB). *J Chem Technol Biotechnol* 2013; 88(6): 1030–1038.
75. de Paula, F. C., de Paula, C. B., Gomez, J. G. C., et al. Poly(3-hydroxybutyrate-*co*-3-hydroxyvalerate) production from biodiesel by-product and propionic acid by mutant strains of *Pandoraea* sp. *Biotechnol Prog* 2017; 33(4): 1077–1084.
76. Teeka, J., Imai, T., Reungsang, A., et al. Characterization of polyhydroxyalkanoates (PHAs) biosynthesis by isolated *Novosphingobium* sp. THA_AIK7 using crude glycerol. *J Ind Microbiol Biotechnol* 2012; 39(5): 749–758.
77. Rodríguez-Contreras, A., Koller, M., Miranda-de Sousa, D. M., et al. Influence of glycerol on poly (3-hydroxybutyrate) production by *Cupriavidus necator* and *Burkholderia sacchari*. *Biochem Eng J* 2015; 94: 50–57.
78. Nair, A. M., Annamalai, K., Kannan, S. K., et al. Characterization of polyhydroxyalkanoates produced by *Bacillus subtilis* isolated from soil samples. *Malaya J Biosci* 2014; 1(1): 8–12.
79. Chen, G. Q., Jiang, X. R., Engineering bacteria for enhanced polyhydroxyalkanoates (PHA) biosynthesis. *Synth Syst Biotechnol* 2017; 2(3): 192–197.
80. Vijay, R., Tarika, K., Production of polyhydroxyalkanoates (PHA) using synthetic biology and metabolic engineering approaches. *Res J Biotechnol* 2018; 13(1): 99–109.
81. Favaro, L., Basaglia, M., Casella, S., Improving polyhydroxyalkanoate production from inexpensive carbon sources by genetic approaches: a review. *Biofuel Bioprod Bioref* 2019; 13(1): 208–227.
82. Chatzifragkou, A., Papanikolaou, S., Effect of impurities in biodiesel-derived waste glycerol on the performance and feasibility of biotechnological processes. *Appl Microbiol Biotechnol* 2012; 95(1): 13–27.
83. Horng, Y. T., Chang, K. C., Chien, C. C., et al. Enhanced polyhydroxybutyrate (PHB) production via the coexpressed phaCAB and vgb genes controlled by arabinose PBAD promoter in *Escherichia coli*. *Lett Appl Microbiol* 2010; 50(2): 158–167.
84. Leong, Y. K., Show, P. L., Ooi, C. W., et al. Current trends in polyhydroxyalkanoates (PHAs) biosynthesis: Insights from the recombinant *Escherichia coli*. *J Biotechnol* 2014; 180: 52–65.
85. Meng, D. C., Shen, R., Yao, H., et al. Engineering the diversity of polyesters. *Curr Opin Biotechnol* 2014; 29: 24–33.
86. Park, S. J., Kim T. W., Kim, M. K., et al. Advanced bacterial polyhydroxyalkanoates: Towards a versatile and sustainable platform for unnatural tailor-made polyesters. *Biotechnol Adv* 2012; 30: 1196–1206.
87. Wang, Q., Yang, P., Xian, M., et al. Production of block copolymer poly (3-hydroxybutyrate)-block-poly (3-hydroxypropionate) with adjustable structure from an inexpensive carbon source. *ACS Macro Lett* 2013; 2(11): 996–1000.
88. Klinke, S., Dauner, M., Scott, G., et al. Inactivation of isocitrate lyase leads to increased production of medium-chain-length poly(3-hydroxyalkanoates) in *Pseudomonas putida*. *Appl Environ Microbiol* 2000; 66(3): 909–913.
89. Javers, J., Gibbons, W., Karunanithy, C., Optimizing a nitrogen-supplemented, condensed corn soluble medium for growth of the polyhydroxyalkanoate producer *Pseudomonas putida* KT217. *Int J Agric Biol Eng* 2012a; 5(4): 62–67.
90. Javers, J., Karunanithy, C., Polyhydroxyalkanoate production by *Pseudomonas putida* KT217 on a condensed corn solubles based medium fed with glycerol water or sunflower soapstock. *Adv Microbiol* 2012b; 2(3): 241.

91. Nikel, P. I., Pettinari, M. J., Galvagno, M. A., et al. Poly (3-hydroxybutyrate) synthesis by recombinant *Escherichia coli* arcA mutants in microaerobiosis. *Appl Environ Microbiol* 2006; 72(4): 2614–2620.

92. Nikel, P. I., Pettinari, M. J., Galvagno, M. A., et al. Poly (3-hydroxybutyrate) synthesis from glycerol by a recombinant *Escherichia coli* arcA mutant in fed-batch microaerobic cultures. *Appl Microbiol Biotechnol* 2008; 77(6): 1337–1343.

93. Wei, X. X., Shi, Z. Y., Yuan, M. Q., et al. Effect of anaerobic promoters on the microaerobic production of polyhydroxybutyrate (PHB) in recombinant *Escherichia coli*. *Appl Microbiol Biotechnol* 2009; 82(4): 703.

94. Jiang, X. R., Chen, G. Q., Morphology engineering of bacteria for bio-production. *Biotechnol Adv* 2016; 34(4): 435–440.

95. Jiang, X. R., Wang, H., Shen, R., et al. Engineering the bacterial shapes for enhanced inclusion bodies accumulation. *Metab Eng* 2015; 29: 227–237.

96. Wang, Y., Wu, H., Jiang, X., et al. Engineering *Escherichia coli* for enhanced production of poly (3-hydroxybutyrate-*co*-4-hydroxybutyrate) in larger cellular space. *Metab Eng* 2014; 25: 183–193.

97. Zheng, Y., Chen, J. C., Ma, Y. M., et al. Engineering biosynthesis of polyhydroxyalkanoates (PHA) for diversity and cost reduction. *Metab Eng* 2019. doi:10.1016/j.ymben.2019.07.004.

98. Sujatha, K., Shenbagarathai, R., A study on medium chain length-polyhydroxyalkanoate accumulation in *Escherichia coli* harbouring phaC1 gene of indigenous *Pseudomonas* sp. LDC-5. *Lett Appl Microbiol* 2006; 43: 607–614.

99. Fukui, T., Mukoyama, M., Orita, I., et al. Enhancement of glycerol utilization ability of *Ralstonia eutropha* H16 for production of polyhydroxyalkanoates. *Appl Microbiol Biotechnol* 2014; 98(17): 7559–7568.

100. Wang, Y., Chung, A., Chen, G. Q., Synthesis of medium-chain-length polyhydroxyalkanoate homopolymers, random copolymers, and block copolymers by an engineered strain of *Pseudomonas entomophila*. *Adv Healthc Mater* 2017; 6(7): 1601017.

101. Chang, D. E., Shin, S., Rhee, J. S., et al. Acetate metabolism in a pta mutant of *Escherichia coli* W3110: Importance of maintaining acetyl coenzyme A flux for growth and survival. *J Bacteriol* 1999; 181: 6656–6663.

102. Wlaschin, A. P., Trinh, C. T., Carlson, R., et al. The fractional contributions of elementary modes to the metabolism of *Escherichia coli* and their estimation from reaction entropies. *Metab Eng* 2006; 8: 338–352.

103. Nikel, P. I., Giordano, A. M., de Almeida, A., et al. Elimination of D-lactate synthesis increases poly (3-hydroxybutyrate) and ethanol synthesis from glycerol and affects cofactor distribution in recombinant *Escherichia coli*. *Appl Environ Microbiol* 2010; 76(22): 7400–7406.

104. Tao, W., Lv, L., Chen, G. Q., Engineering *Halomonas* species TD01 for enhanced polyhydroxyalkanoates synthesis via CRISPRi. *Microb Cell Fact* 2017; 16(1): 48.

105. Hogan, A. M., Rahman, A. Z., Lightly, T. J., et al. A broad-host range CRISPRi toolkit for silencing gene expression in *Burkholderia*. *ACS Synth Biol* 2019; 8: 2372–2384.

106. de Almeida, A., Nikel, P. I., Giordano, A. M., et al. Effects of granule-associated protein PhaP on glycerol-dependent growth and polymer production in poly (3-hydroxybutyrate)-producing *Escherichia coli*. *Appl Environ Microbiol* 2007; 73(24): 7912–7916.

107. de Almeida, A., Giordano, A. M., Nikel, P. I., et al. Effects of aeration on the synthesis of poly(3-hydroxybutyrate) from glycerol and glucose in recombinant *Escherichia coli*. *Appl Environ Microbiol* 2010; 76(6): 2036–2040.

108. Yue, H., Ling, C., Yang, T., et al. A seawater-based open and continuous process for polyhydroxyalkanoates production by recombinant *Halomonas campaniensis* LS21 grown in mixed substrates. *Biotechnol Biofuels* 2014; 7(1): 1–12.

109. Kawata, Y., Aiba, S. I., Poly (3-hydroxybutyrate) production by isolated *Halomonas* sp. KM-1 using waste glycerol. *Biosci Biotechnol Biochem* 2010; 74 (1): 175–177.
110. Fu, X. Z., Tan, D., Aibaidula, G., et al. Development of *Halomonas* TD01 as a host for open production of chemicals. *Metab Eng* 2014; 23: 78–91.
111. Zhao, H., Zhang, H. M., Chen, X., et al. Novel T7-like expression systems used for *Halomonas*. *Metab Eng* 2017; 39: 128–140.
112. Johnston, B., Jiang, G., Hill, D., et al. The molecular level characterization of biodegradable polymers originated from polyethylene using non-oxygenated polyethylene wax as a carbon source for polyhydroxyalkanoate production. *Bioengineering* 2017; 4: 1–14.
113. Kumar, P., Mehariya, S., Ray, S., et al. Biodiesel industry waste: A potential source of bioenergy and biopolymers. *Ind J Microbiol* 2015; 55: 1–7.
114. Gonzalez-Ausejo, J., Rydz, J., Musioł, M., et al. A comparative study of three-dimensional printing directions: The degradation and toxicological profile of a PLA/PHA blend. *Polym Degrad Stab* 2018; 152: 191–207.
115. Albuquerque, M. G. E., Martino, V., Pollet, E., et al. Mixed culture polyhydroxyalkanoate (PHA) production from volatile fatty acid (VFA)-rich streams: Effect of substrate composition and feeding regime on PHA productivity, composition and properties. *J Biotechnol* 2011; 151: 66–76.
116. Bengtsson, S., Pisco, A. R., Johansson, P., et al. Molecular weight and thermal properties of polyhydroxyalkanoates produced from fermented sugar molasses by open mixed cultures. *J Biotechnol* 2010; 147: 172–179.
117. Iwata, T., Strong fibers and films of microbial polyesters. *Macromol Biosci* 2005; 5: 689–701.
118. Pappalardo, F., Fragala, M., Mineo, P. G., et al. Production of filmable medium-chain-length polyhydroxyalkanoates produced from glycerol by *Pseudomonas mediterranea*. *Int J Biol Macromol* 2014; 65: 89–96.
119. Madden, L. A., Anderson, A. J., Shah, D. T., et al. Chain termination in polyhydroxyalkanoate synthesis: Involvement of exogenous hydroxy-compound as chain transfer agents. *Int J Biol Macromol* 1999; 25(1): 43–53.
120. Tanadchangsaeng, N., Yu, J., Microbial synthesis of polyhydroxybutyrate from glycerol: Gluconeogenesis, molecular weight and material properties of biopolyester. *Biotechnol Bioeng* 2012; 109(11): 2808–2818.
121. Kawaguchi, Y., Doi, Y., Kinetics and mechanism of synthesis and degradation of poly(3-hydroxybutyrate) in *Alcaligenes eutrophus*. *Macromolecules* 1992; 25(9): 2324–2329.
122. Tsuge, T., Ko, T., Tago, M., et al. Effect of glycerol and its analogs on polyhydroxyalkanoate biosynthesis by recombinant *Ralstonia eutropha*: A quantitative structure-activity relationship study of chain transfer agents. *Polym Degrad Stab* 2013; 98(9): 1586–1590.
123. Asad-ur-Rehman, Saman, W. R. G., Nomura, N., et al. Pre-treatment and utilization of raw glycerol from sunflower oil biodiesel for growth and 1, 3-propanediol production by *Clostridium butyricum*. *J Chem Technol Biotechnol* 2008; 83: 1072–1080.
124. Pyle, D. J., Garcia, R. A., Wen, Z., Producing docosahexaenoic acid (DHA)-rich algae from biodiesel-derived crude glycerol: Effects of impurities on DHA production and algal biomass composition. *J Agric Food Chem* 2008; 56(11): 3933–3939.
125. Thompson J. C., He, B. B., Characterization of crude glycerol from biodiesel production from multiple feedstocks. *Appl Eng Agric* 2006; 22(2): 261–265.
126. Salakkam, A., Webb, C., Production of poly (3-hydroxybutyrate) from a complete feedstock derived from biodiesel by-products (crude glycerol and rapeseed meal). *Biochem Eng J* 2018; 137: 358–364.

127. Van-Thuoc, D., Huu-Phong, T., Minh-Khuong, D., et al. Poly (3-hydroxybutyrate-*co*-3-hydroxyvalerate) production by a moderate halophile *Yangia* sp. ND199 using glycerol as a carbon source. *Appl Biochem Biotechnol* 2015; 175(6): 3120–3132.
128. de Castro, J. S., Nguyen, L. D., Seppala, J., Bioconversion of commercial and waste glycerol into value-added polyhydroxyalkanoates by bacterial strains. *J Microb Biochem Technol* 2014; 6(6): 337–345.
129. da Silva, G. P., Mack, M., Contiero, J., Glycerol: A promising and abundant carbon source for industrial microbiology. *Biotechnol Adv* 2009; 27(1): 30–39.

10 Biosynthesis of Polyhydroxyalkanoates (PHA) from Vegetable Oils and Their By-Products by Wild-Type and Recombinant Microbes

*Manoj Lakshmanan, Idris Zainab-L,
Jiun Yee Chee, and Kumar Sudesh*

CONTENTS

10.1 INTRODUCTION

10.1.1 CURRENT POLLUTION ISSUE: THE DEVELOPMENT OF BIOPOLYMER

Plastics are indispensable materials in various human activities owing to their wide range of applications and excellent properties. These petroleum-based plastics are durable, lightweight, and low cost. Advancements in the field of science and technology have been increasing the global production and use of synthetic polymers, such that 280 million tonnes of plastic production in the year 2011 [1] increased to 311 million tonnes in 2014 [2] to the current 350 million tonnes [3]. The major areas of plastic applications include the development of medical appliances, building materials, aerospace, automobiles, food packaging, carry bags, coffee cups, or plastic bottles. Single-use plastics constitute about 8% by weight of solid waste generated daily [4]. The used plastics are either dumped on land or buried in landfills. Regrettably, the detrimental effects of such plastics on our environment was only realized within the last 30 years. Petroleum-based plastics are recalcitrant to physical, chemical, and biological degradation. Over 500 billion pieces of plastic bags and 35 million bottles used worldwide end up on the land, beaches, and in oceans [5]. Synthetic plastics only photodegrade into smaller pieces and contaminate water bodies, particularly the Pacific, Atlantic, and Indian Ocean [6]. The United Nations reported the presence of about 5–10 million tonnes of plastics in the North Pacific Ocean between Japan and California. Years later, Gill projected 200,000 plastic pieces per square kilometre in the North Atlantic Ocean and the Caribbean Sea. Since then, the degree of plastic litter carried to the sea has become problematic [7]. Often, marine biota are endangered due to entanglement with or ingestion of plastic debris found in the environment. Over 270 species of turtle, mammals, fish, and seabirds have experienced restricted movement, starvation, or death [2]. Microplastics are mistaken for phytoplankton and eaten by zooplankton, fish, and cetaceans such as whales and dolphins. Upon consumption, plastic debris causes internal injuries, intestinal blockage, reduces stomach capacity and deters growth [8]. Worldwide, no less than 86% of sea turtle species, 23% of marine mammal species and 36% of seabird species are affected by plastic debris [9]. Debris, such as ropes, fishing line, nets, plastic bags and Styrofoam, has been removed from the digestive tracts of turtles. The turtles ingest floating plastic bags because they resemble jellyfish, which consequently block the digestive passage and eggs in female genital canals [8,10]. There has been a growing concern that plastics are acting as a faster medium for the introduction of foreign species (molluscs, barnacles, and algae) compared with the natural means known for centuries. The hard surfaces of floating plastics provide a suitable substrate for several opportunistic colonizers to attach. As the quantities of these non-biodegradable materials upsurge in marine debris, dispersal will be faster, and prospects for invasion by alien and possibly aggressive invasive species could be enhanced [11].

Considerable efforts have been directed towards reducing environmental plastic pollution through incineration and recycling as the standard plastic waste management process. Although incineration deflects a significant fraction (20%–25%) of the plastic waste away from landfill, the major drawback of this method is high cost and the emission of harmful/toxic molecules, such as dioxin, that are more destructive than the plastic itself. Recycling processes were proposed as a replacement to landfill and incineration. During recycling, plastic materials, such as poly(ethylene) (PE), poly(styrene) (PS) or poly(propylene) (PP), are thermally degraded at 25°C to 430°C and recycled or converted into liquid hydrocarbon fuel. However, recycling is laborious as the waste consists of mixed plastic types; therefore, only about 10% of household plastic items collected are fully recycled back into plastics, while the larger percentage of the plastic is disposed of in landfills or incinerated. Furthermore, beneficial material properties may be lost after a few cycles of recycling [12].

Despite new emerging technologies and research, plastic does not vanish but remains in our environment [5]; the presence of plastic is increasing at an alarming rate. Thus, alternative packaging materials are needed to reduce our dependence on petroleum-based plastics that are non-renewable. Renewable resources from living organisms, such as plants and animals, can reliably provide raw materials for the continuous production of bio-based plastics, such as polyhydroxyalkanoates (PHA). PHA does not take up landfill space because it is biodegradable and compostable after use. The degradation rate of a piece of PHA varies between a few months (under aerobic or anaerobic conditions) to years (in seawater), depending on the environmental conditions and polymer composition [13]. PHA are hydrolyzed by the action of microorganisms secreting PHA depolymerase enzymes. These enzymes produce low molecular weight oligomers, which are further metabolized by microbes to H_2O and CO_2 under aerobic conditions or to H_2O, CH_4 and CO_2 in an anaerobic environment [14]. The CO_2 released from PHA hydrolysis does not contribute to atmospheric CO_2 concentration because the PHA was produced from CO_2 fixed by plants during photosynthesis [15]. Therefore, PHA is 100% bio-based and carbon neutral.

10.1.2 INTRODUCTION AND BACKGROUND OF POLYHYDROXYALKANOATES

PHA, a lipid-like, water-insoluble inclusion body in microbes was initially observed in *Azotobacter chroococcum* [16]. PHA are polyesters of 3-, 4-, 5- and 6-hydroxyalkanoate (HA) units synthesized as carbon or energy storage materials in many microbial cells, such as *Bacillus megaterium, Cupriavidus necator, Alcaligenes latus, Pseudomonas putida*, as well as some archaea and cyanobacteria. The primary functions of PHA in cells are regulation of intracellular energy flow, protection against ultraviolet irradiation, osmotic shock, oxidative stress, desiccation and channelling carbon compounds to metabolic pathways [17]. PHA occur in spherical granular form within the cell cytoplasm measuring between 0.2 and 0.8 μm in diameter. It can be directly visualised using a phase-contrast light microscope (see Figure 10.1A) or stained with Nile blue, Nile red and Sudan black B. The finer details of the distinct PHA granules can be visualised through transmission electron microscopy (TEM) as shown in Figure 10.1B.

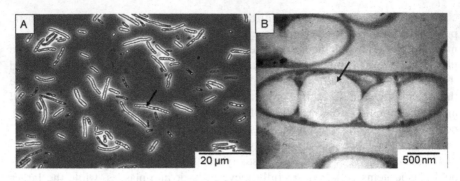

FIGURE 10.1 Observation of PHA granules in bacterial cells under (A) phase-contrast and (B) transmission electron microscope. Black arrows indicate PHA granules inside the cells.

Depending on the length of the monomer side chains, PHA are mainly divided into short-chain-length (*scl*) monomers of 3–5 carbon atoms and medium-chain-length (*mcl*) monomers containing 6–14 carbon atoms [18]. The most common representative of PHA is the homopolyester poly(3-hydroxybutyric acid) [P(3HB); a.k.a. PHB]. P(3HB) is the simplest homopolyester of 3-hydroxybutyrate isolated from *Bacillus megaterium* and characterised by Maurice Lemoigne in 1926. P(3HB) is a high molecular weight polyester with material properties similar to PP [19]. However, the industrial application of P(3HB) as a thermoplastic is limited due to its brittleness, high melting temperature, and only 5% extension to break compared with 400% for PP [20]. Other HA units such as 3-hydroxyvalerate (3HV), 3-hydroxyhexanoate (3HHx), 3-hydroxyheptanoate (3HHp) and 3-hydroxyoctanoate (3HO) were identified in the 1970s [21,22,23]. The occurrence of 3HV and 3HHx monomer units was identified from chloroform extracts of activated sewage sludge samples [21]. PHA copolymers consisting of both *scl*- and *mcl*-monomers had better properties than P(3HB) homopolymer and therefore received commercial interest. Then, the first industrial production of poly(3-hydroxybutyrate-*co*-3-hydroxyvalerate) [P(3HB-*co*-3HV); a.k.a. PHBV] copolymer took place. Among the important properties of copolymers are its lower melting temperatures, higher flexibility, elasticity and suitability for the development of scaffolds for tissue engineering matrices and surgical devices [17]. In the meantime, the high production costs alongside the low price of conventional plastics have prevented the wide use of PHA. The cost of carbon feedstocks accounts for about 50% of the total production costs [24]. PHA can be synthesized via a fermentative process using a wide range of substrates as carbon sources [25]. The common renewable resources mostly used for PHA production are sugars (glucose, molasses, sucrose), whey and plant/vegetable oil [26]. Amongst these, vegetable oils are the most attractive feedstock due to their availability, low toxicity, comparatively low price and biodegradability [27]. Vegetable oils have shown great potential as a source of carbon for PHA production. Figure 10.2 shows the lifecycle of PHA from vegetable oils.

Vegetable oils are important renewable resources obtained from numerous plants and are usually named for their biological sources, such as groundnut oil, soybean oil and palm oil. The oils are generally obtained by mechanically pressing the plant seeds or fruit through solvent or mechanical extraction methods. Seed oils include

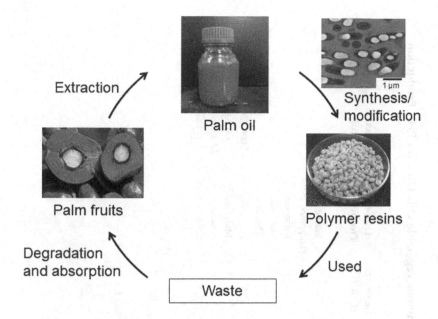

FIGURE 10.2 Lifecycle of PHA synthesized from vegetable oil.

soybean, canola, sunflower, groundnuts, cotton and palm kernel, while palm and olive oils are pressed out of the soft fruit [28]. Oil consists of mainly triglycerides formed between various fatty acids and glycerol. The triacylglycerol (TAG) composition of the oil is responsible for its high carbon content; therefore, it is preferred over sugar for PHA production. Theoretically, 1 g of oil produces about 0.8 g of PHA compared with only 0.4 g of PHA/g of sugar substrates [20,29]. Bacteria use oil by the action of lipases that are secreted to hydrolyze the ester bonds between fatty acids and the glycerol backbone. The free fatty acids (FFA) are then transported into the cell for catabolism through the β-oxidation pathway [30]. Fatty acids are long, straight-chain compounds of varying degrees of unsaturation. Saturated fatty acids contain only single bonds in the carbon backbone, while the unsaturated fatty acids contain two or more double bonds. The number of double bonds per triglyceride differs by oil, and vegetable oils containing more double bonds remain liquid at room temperature. For example, palm oil, having roughly 1.7 double bonds per molecule, solidifies more than soybean oil, which contains 10.6 double bonds [31]. The chemical and physical properties of oil depend on the chain length and degree of unsaturation of its fatty acids. The presence of multiple carbon-carbon double bonds makes vegetable oil a perfect and natural building block for the preparation of a range of useful polymeric materials. The common vegetable oils that have been found to be suitable for PHA production include palm oil, [32] soybean oil [33] and canola oil [34]. For instance, the use of soybean oil [35] and canola oil [34] as carbon sources yielded high PHA contents of 83 wt.-% and 92 wt.-% of the cell dry mass (CDM), respectively. Generally, the fatty acid chain length of vegetable oil is from C12 to C20 and dominated by palmitic (C16:0), stearic (C18:0), oleic (C18:1), linoleic (C18:2) and linolenic acid (C18:3), as shown in Table 10.1.

TABLE 10.1

Major Fatty Acid Compositions and Countries of the Common and Uncommon Vegetable Oils Used for PHA Production

Vegetable oils	Major Fatty Acids (%)							Leading Producers	Production [tons]	Production Cost [US$/ton] [36]
	Lauric (C12)	Myristic (C14)	Palmitic (C16)	Stearic (C18)	Oleic (C18:1)	Linoleic (C18:2)	Linolenic (C18:3)			
Common										
Coconut*	55.8	18.7	7.9	0.3	3.3	0.1	–	Indonesia, Philippines, India	–	–
Crude palm kernel**	48.3	15.5	8.0	2.1	15.4	2.6	–	Indonesia, Malaysia, Thailand	–	–
Palm mesocarp**	0.1	0.9	43.8	4.0	42.1	8.9	0.2	Indonesia, Malaysia, Thailand	35,000,000 20,000,000 2,600,000	165.20 239.40
Soybean***	–	0.5	9.0	4.0	28.5	49.5	8.0	China, US Brazil	– –	400.60, 459.90
Uncommon										
Jatropha	–	0.1	17.1	4.3	42.0	34.8	0.1	–		
Date seed****	19.1	9.3	8.3	1.7	55.3	5.3	–			
African elemi**	–	–	34.5	1.6	34.7	26.2	1.3			
Bitter apple**	–	–	9.7	14.0	15.4	60.0	0.1			
Desert date**	–	–	14.1	9.3	31.3	44.3	–			
*Amygdalus pedunculata**	–	–	2.0	0.6	69.1	26.5	0.9			

*[37]; **[38]; ***[39] ****[40];
—: Not reported

In general, the countries producing the highest quantities of vegetable oil are Indonesia, Malaysia, Thailand, China and the U.S. The production cost of palm oil in either Indonesia or Malaysia is cheaper compared with soybean oil. This is due to the relatively cheaper inputs to the production process and mainly because of the higher productivity of palm compared with other vegetable oil crops. Indonesia, for example, achieved palm oil production at US$ 165/tonne while Malaysia's production cost is about US$ 239/tonne, as shown in Table 10.1. China, USA and Brazil contribute about 90% of the global soybean oil production, while 5% production comes from other Asian countries and Africa.

10.1.3 GLOBAL PRODUCTION OF VEGETABLE OIL PER YEAR

The average global annual production of vegetable oil is increasing, such that its production escalated from 117.1 million tonnes in the year 2000 to 125 million tonnes between the year 2001 and 2005 [27]. Currently, vegetable oil production is rising at 4% per year and production of 187 million tonnes was reported in 2016/2017 (see Figure 10.3). The USDA-FSA (U.S. Department of Agriculture – Farm Service Agency), in regard of increasing food consumption in emerging countries and biofuel applications, has predicted 195 million tonnes of vegetable oil production in the future with the highest (37.6%) supply from palm oil and palm kernel, followed by soybean at 30%; where the remaining 32.5% will be supplied by other vegetable oils [41].

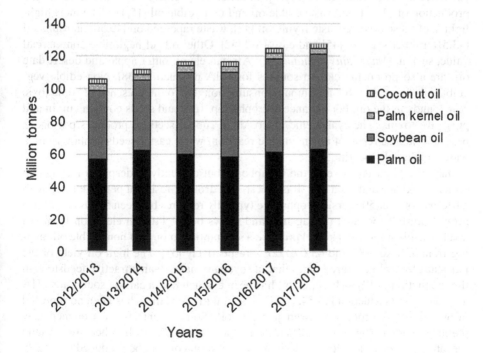

FIGURE 10.3 Global production of vegetable oil that has been used for PHA production.

Palm oil has been one of the world's most traded vegetable oils. It represents about 37.6% of total vegetable oil production, followed by 30% supply from soybean; other oils account for the remaining 32.5%. Vegetable oils are now in the spotlight for both laboratory and industrial polymeric material research due to their universal availability and relatively low price [42]. Vegetable oils are also important carbon feedstock for PHA production. However, the primary use of vegetable oil is in the food sector. The global escalating vegetable oil consumption resulting from rapid population growth is believed to create a global imbalance in the food supply and demand market in the near future. It is necessary to replace an edible feedstock with an inedible one for the profitable and sustainable industrial-scale production of PHA in order to solve the problem.

10.1.4 Underused Vegetable Oils That Can Be Potentially Used for PHA Production

In addition to the vegetable oils that are produced for food applications, there are also vegetable oils with no nutritional value, such as jatropha and rubber seed oil. There are other attractive oils from desert dates, African elemi and bitter apple that do not compete with commonly used edible oils. The suitability of these oils for cell growth and PHA production is less known; therefore, several studies were conducted [43,44,45]. The use of vegetable oils of no nutritional value as a feedstock for PHA production is attractive and becoming a promising approach in lowering the overdependence on food grade oils for PHA production [46]. Researchers have reviewed the production of PHA from waste edible oil and non-edible oil [15,47]. The most highlighted of these oils are waste frying oil [48], waste rapeseed oil [49,50], jatropha oil [43,51], rubber seed oil [45] and castor oil [52]. Other oils of negligible commercial value, such as *Amygdalus pedunculata*, African elemi, bitter apple and desert date oil, are also promising carbon sources for PHA production [38]. Non-edible vegetable oils are unsuitable for human consumption due to the presence of some toxic compounds in the oil. For instance, jatropha sap, fruit and seeds contain curcin and purgative toxins. The cyanogenic glucoside in rubber seed oil produces poisonous prussic acid as a result of an enzymatic reaction, while castor seeds contain ricin, a water-soluble phytotoxin [39].

Jatropha plant is native to the Caribbean, but currently widespread throughout America, Africa and Asia [43]. It is a pest- and drought-resistant perennial tree with a lifespan of over 50 years. Jatropha tree typically requires between 50 mm and 1200 mm of rainfall for rapid propagation and grows below 1400-m elevation from sea level. The plant seed and kernel produce a substantial amount of non-edible oil ranging from 35% to 50% and 45% to 60%, respectively [53]. The high oil yield of the jatropha plant (almost three times that of soybean) makes it an attractive feedstock in the production of PHA. Jatropha oil has been evaluated as a carbon source for PHA and biodiesel production [43,54,55]. However, until now, there has been no reported pilot-scale PHA production from jatropha oil. Some countries have opened new plantations for jatropha. India has allocated a total of 1.72 million hectares of land for jatropha cultivation. It is yet to be seen if jatropha oil can be produced profitably and sustainably for various new applications including the production of PHA on a commercial scale. Another attractive non-edible oil is castor oil. The castor oil plant is native to Africa and tropical Asia. It was grown in Egypt about 6000 years ago and used for medicinal purposes. Currently, castor plants are cultivated globally for their oil content. Because castor oil is non-edible, it is mainly used in the formulation of bases for lubricants, feedstock for oleo-chemicals and fuels, reactive components for paints, inks and coatings as well as foams and polymers. Due to its numerous applications, the world market for castor oil has been steadily increasing, and India supplies about 70% of the world's total exports of castor seed oil; China, Brazil and Thailand cover the remaining 30% [56]. The rubber tree is renowned for its popular products, rubber latex and rubberwood. According to Njwe et al. [57], the often-neglected rubber seed contains about 44% oil, 50% cake and 6% waste. According to Natural Rubber Statistics [58], rubber plantation cover about 11.5 million hectares of land globally and the major contributing countries are Indonesia, Thailand, Malaysia and China at 3.45, 2.90, 1.03 and 1.00 million hectares, respectively [59]. The combined rubber plantation area of these four countries in the global rubber agricultural production is placed at 76%. The rubber seed availability of major rubber-producing Southeast Asian countries and their probable oil and biomass products are shown in Figure 10.4. Depending on the rubber clones, the quantity of rubber seed production ranges between 150 and 1,000 kg/(hectare·year) at a selling price of US$ 350 to US$ 1,000 per tonne [56]. The price analysis of rubber seed oil for biodiesel shows the possibility of achieving lower cost biodiesel production from rubber seed oil.

The selection of non-edible vegetable oils as feedstocks for PHA production requires careful consideration of several factors such as oil suitability, cost, yield

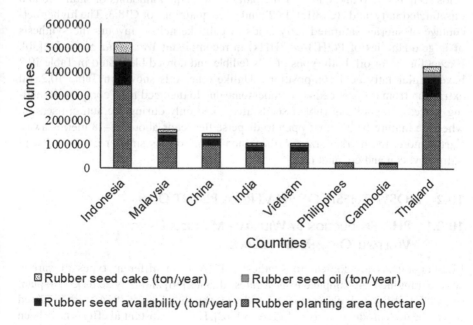

FIGURE 10.4 Rubber seed availability of major rubber-producing Southeast Asian countries and their probable oil and biomass products [60].

TABLE 10.2
Plantation Cost, Oil Yield and Composition of Common Non-Edible and Edible Oil [39]

		Non-Edible Oil			Edible Oil	
		Jatropha	Rubber Seed	Castor	Soybean	Palm
Plantation	US$/hectare	620	–	140–160	615	950
cost	US$/kg oil	0.39	–	0.12–0.14	1.64	0.19
Oil yield	wt.-%	Seeds (35–40) Kernel (50–60)	40–50	53	20	20
	kg oil/hectare	1590	80–120	1188	375	5000
Fatty acid	Palmitic	14.2	10.2	1.0	10.0	45.0
composition	Stearic	6.9	8.7	1.0	4.0	5.0
(%)	Oleic	43.1	24.6	3.0	23.0	40.0
	Linoleic	34.3	39.6	4.2	51.0	10.0
	Linolenic	–	16.3	0.3	7.0	–
	Ricinoleic	–	–	89.5	–	–

–: Not reported

and composition. Oil composition has a profound effect on the quantity and properties of the PHA obtained. For instance, palm kernel oil is rich in saturated fatty acids such as C12:0 and C14:0, while palm oil has equal amounts of saturated and unsaturated fatty acids (C18:1, C18:2) and trace quantities of C18:3. The higher percentage of shorter saturated fatty acids in palm kernel oil favours the synthesis of large quantities of P(3HB-*co*-3HHx) in recombinant *Wautersia eutropha* [30]. Except for castor oil, both types of oils (edible and non-edible) listed in Table 10.2 have similar fatty acid compositions. Unlike other oils shown in Table 10.2, oil extraction from rubber seeds is cumbersome due to the need for a complex deshelling process. In addition, rubber seeds are picked only during the autumn season, when the mature fruits burst open to disperse the seeds about 15–18 metres away. Furthermore, the market size for these non-edible oils is smaller compared with palm, soybean and coconut oil.

10.2 BIOSYNTHESIS OF PHA FROM PLANT OILS

10.2.1 PHA PRODUCTION BY WILD-TYPE MICROBES USING VEGETABLE OILS AND BY-PRODUCTS

Microorganisms are known to synthesize PHA from different types of carbon sources ranging from simple carbohydrates, alkanes [56] and fatty acids [57] to plant oils [58]. Until now, bulk production and wider applications of PHA were hindered due to the high production cost of PHA [60–62]. Hence, substantial efforts have been devoted to decrease the cost of production through bioengineered bacterial strains, use of low-value by-products and efficient fermentation and recovery processes

[18,63]. Nevertheless, the main cost impact in the upstream process is the cost of the substrates [61]. Selection of suitable carbon substrates is a critical factor that determines the overall output of the fermentation process and significantly influences the cost of the final product. The ideal approach is to choose sustainable, economical and most readily available carbon substrates that can support not only microbial growth but also PHA production efficiently. By-products from the agricultural sector, such as beet molasses and alphéchin (wastewater from olive oil mills) [64], waste cooking oil, palm sludge oil and glycerol from plants [33,65–68], are among the various attractive renewable resources that can be used in fermentation processes to produce PHA in high-cell-density cultures.

Vegetable oils, such as soybean oil, palm oil and rapeseed oil, are desirable carbon sources for PHA production because it is cheaper than most of the sugars on the marketplace. Although optimization works have been done to achieve high productivity through the use of sugars, when discussing the relationship of production cost to yield, it is still higher than the 'acceptable' level [69]. Approximately 0.3–0.4 g P(3HB)/g glucose is reported to be the highest product yield from glucose. In contrast, plant oils are expected to give better cell biomass and PHA yields (0.6–0.8 g of PHA per 1 g of oil) as they contain a higher carbon content per weight [29]. It has been reported that famine conditions favour biopolymer synthesis, with some bacteria capable of producing up to 80% (w/w) PHA [33,66]. In this context, Mravec et al. observed that bacteria extend the cell size from its original length to suit the cytoplasmic storage as well as control the PHA volume portion so that the bacterial cell volume will stay below 40% [70].

10.2.1.1 Cupriavidus necator H16

Chronologically, *Cupriavidus necator* H16 was previously declared as *Hydrogenomonas eutropha*, *Alcaligenes eutrophus*, *Ralstonia eutropha* and *Wautersia eutropha*. In 2004, it was renamed after confirmation of DNA–DNA hybridization experiments and phenotypic characteristics evaluation, DNA base ratios and 16S rRNA and placed under the β-subclass of the Proteobacteria [71,72].

Cupriavidus necator could heterotrophically produce PHA up to 80 wt.-% from its CDM in nitrogen limitation conditions with *scl*-PHA as the monomeric compounds. With a wide substrate specificity to enable the intake of various carbon sources, this metabolically versatile bacterium has become an efficient cell factory for PHA production. Kahar et al. studied *C. necator* H16 (wild-type) and a transformant strain derived from the transformation of a PHA synthase gene from *Aeromonas caviae* using soybean oil as the sole carbon source to synthesize P(3HB) homopolymer and P(3HB-*co*-5 mol-% 3HHx) copolymer, respectively. In both the wild-type and transformant, the production of PHA from soybean oil was up to 80 wt.-% CDM [33,66]. The incorporation of a minor molar fraction of 3HHx altered the undesirable physical and thermal properties of P(3HB) homopolymer.

Another wild-type glucose-utilizing strain of *C. necator* was used by Imperial Chemical Industries (UK) to produce P(3HB-*co*-3HV) copolymer under the trade name Biopol™ [73]. Propionate was added together with glucose in order to produce the copolymer. Instead of glucose, *C. necator* cells can also grow on oils and fatty acids. Besides propionate, 1-pentanol can also be added to produce the 3HV monomer. In a laboratory-scale study, the latter was fed together with oleic acid to *Cupriavidus* sp. USMAA2-4 that yielded 66 wt.-% of P(3HB-*co*-3HV) of CDM [74].

Waste frying rapeseed oil is one of the waste cooking oils used for PHA production by *C. necator* H16 [48,49]. Upon cooking, the properties of the pure oil were affected bceause of the presence of O_2 and H_2O. In a 72-hour one-stage batch fermentation, the highest P(3HB) concentration obtained was 1.2 g/L using waste frying rapeseed oil, compared with 0.62 g/L achieved when using fresh rapeseed oil [48]. The residual food remaining in the oil after frying probably supplied more nutrients to the bacterium and indirectly enhanced the production of PHA. Another possible reason for this could be the preference for saturated fatty acid uptake by bacteria to transform them into acetyl-CoA [48,75] via the β-oxidation pathway.

Based on Kamilah et al. [68], *C. necator* H16 was able to produce high CDM of 25.4 g/L with 71 wt.-% of P(3HB) when grown in a minimal medium with waste cooking oil (palm oil) as the carbon source. This result was comparable to fresh cooking oil, where 24.8 g/L of CDM with 72 wt.-% P(3HB) was achieved. In another study, it was shown that waste cooking oil could yield more P(3HB) than fresh cooking oil with similar molecular weights [48]. The reason for the better yield was because the waste cooking oil contained dissolved residual nutrients that enriched the medium. Usually, the composition of triglycerides in fresh and pure oil is more than 95%; the amount of triglycerides will eventually decrease after the frying process which will produce more FFA [75].

10.2.1.2 *Burkholderia*

Besides *C. necator* H16, several other bacteria have been extensively employed as model platforms for obtaining PHA from plant oils, namely *Burkholderia cepacia*, *Comamonas testosteroni* and *Pseudomonas putida*. Alias and Tan [76] isolated *B. cepacia* from palm oil mill effluent (POME); a P(3HB) content of more than 50 wt.-% CDM was produced from palm oil. This bacterium, however, was only able to synthesize P(3HB) homopolymer from plant oil as the sole carbon source. The ability to synthesize P(3HB-*co*-3HV) copolymer was investigated by Chee et al. [66,77] by co-feeding crude palm kernel oil (CPKO) and sodium valerate to *Burkholderia* sp. USM (JCM15050), and also by Zhu et al. [78,79] via continuous feeding of levulinic acid as a co-substrate together with xylose, which led to the production of P(3HB-*co*-3HV) by *B. cepacia* ATCC 17759.

Burkholderia thailandensis E264, an isolate from a rice field soil sample in central Thailand, was investigated for its ability to produce PHA and rhamnolipid using waste cooking oil derived from sunflower as the carbon source. After entering the stationary phase, the cell biomass and PHA concentration reached 12.2 and 7.5 g/L respectively. Assimilation of waste cooking oil was channelled to the cell biomass, PHA and rhamnolipids production [80]. The PHA produced was found to be a P(3HB) homopolymer. The authors demonstrated a high average molecular weight, M_w, of 5.11×10^5 g/mol and a polydispersity index (PDI) of 2.86 for the P(3HB) produced by *B. thailandensis*. This result is comparable with P(3HB) accumulated by *C. necator* supplemented with waste rapeseed oil; this polymer had an M_w of 5.77×10^5 g/mol and a PDI of 2.66 [81].

10.2.1.3 Comamonas

In 2003, *Comamonas testosteroni* was found to produce P(3HB) from naphthalene [82]. Its ability to synthesize *mcl*-PHA from vegetable oils, for example, castor seed oil, coconut oil, mustard oil, cottonseed oil, groundnut oil, olive oil and sesame oil, for growth and PHA accumulation was also discovered [83]. Approximately 78.5–87.5 wt.-% of *mcl*-PHA in CDM was synthesized by this bacterium, and it mainly contained 3-hydroxyoctanoate (3HO, C_8) and 3-hydroxydecanoate (3HD, C_{10}) monomers among C_6–C_{14} 3-hydroxyalkanoate monomers in a polymer chain. The amount of PHA accumulated was comparable with different varieties of vegetable oils [83].

10.2.1.4 Pseudomonas

Bacteria from the genus *Pseudomonas* are generally associated with *mcl*-PHA production. However, an exception is the strain *Pseudomonas* sp. 61-3, which can incorporate both *scl*- and *mcl*-PHA [84] with 3HB typically a minor constituent from renewable carbon feedstocks [14]. The monomer composition of *mcl*-PHA is controlled by three factors: the chemical structures of fatty acids, the substrate specificity of the PHA synthase and the β-oxidation pathway for long chain fatty acids [65]. PHA consisting mainly of 3HO was synthesized by *P. oleovorans* as the first example reported to comprise *mcl*-monomers [85].

The soil-isolated strain *Pseudomonas* sp. 61-3 turned out to be capable of producing a blend P(3HB) homopolymer and a copolymer consisting of monomers ranging from C_4 to C_{12} from plant oils and alkanoic acids of even carbon numbers, while P(3HB-co-3HV) copolymer was produced using alkanoic acids of odd carbon numbers [86]. The two different types of PHA were stored in separate inclusions in the same bacterial cell [86], suggesting that there are two types of PHA synthases with different substrate specificities, whereby one of the synthases can incorporate a wide compositional range of *mcl*-PHA [84].

Ashby et al. reported on *P. oleovorans* B-14682 and *P. corrugata* 388 for PHA production from a co-product stream of soy-based biodiesel production containing glycerol, FFA and residual fatty acid methyl ester [87]. The study showed that *P. oleovorans* B-14682 preferred glycerol over fatty acid derivatives and was able to produce 13–27 wt.-% P(3HB) with 1.3 g/L CDM. Glycerol is considered a favourable substrate for the formation of acetyl-CoA, an intermediate substrate in the PHA synthesis pathway. In fact, *P. corrugata* 388 consumed both glycerol and fatty acid derivatives at approximately the same rate and β-oxidized the fatty acids to form *mcl*-PHA. Ashby et al. showed that mixed culture fermentations of glycerol as a carbon substrate could be used in the formation of polymer blends of P(3HB) and *mcl*-PHA by both *P. oleovorans* B-14682 and *P. corrugata* 388. However, the proton nuclear magnetic resonance (¹H-NMR) analysis revealed that the number average molecular weight (M_n) of P(3HB) homopolymer decreased because of the esterification of glycerol and growing PHA chains, which leads to polymer chain termination [88].

Sometimes there are exceptional cases where a lipase enzyme is absent in certain strains such as *P. putida* PGA1 and *P. putida* KT2440. Carbon sources in the form of triglycerides cannot support both bacterial growth and PHA production [89].

Therefore, an additional saponification step is required to hydrolyze the triglycerides into FFA and glycerol, which can be assimilated by *P. putida* PGA1 for growth and PHA production. An example is the use of saponified palm kernel oil. As the saponified palm kernel oil concentration was varied from 0.5% to 1.0% (w/v), the CDM increased from 3.0 to 6.8 g/L. A PHA content of up to 37 wt.-% was reported [89]. In aerobic conditions, fatty acids are metabolized through the β-oxidation pathway; thus adequate oxygen supply to the bacterial culture is crucial to maximize the PHA yield and productivity. Annuar et al. showed that an increase in saponified palm kernel oil concentration from 5 to 10 g/L in the culture medium caused a significant reduction in the volumetric oxygen transfer coefficient ($k_L a$) value by 50% [90]. Therefore, proper aeration of the culture is necessary when using oil and fatty acids as the carbon sources.

Pseudomonas putida KT2440, when grown on lauric acid, could synthesize high amounts of PHA, which was not affected by nitrogen limitation [91,92]. In this context, a study reported that PHA production (17.9 wt.-%) in 1.7 g/L CDM, with 3HO being the major monomer (60 mol-%), was almost 10-fold higher than those from cells grown on other carbon sources [91]. An improved feeding strategy was developed to delay the bacterial stationary phase to achieve higher biomass and PHA productivity prior to the harvest stage. In fed-batch feeding, *P. putida* KT2440 achieved a PHA yield of 58 g/L and CDM of 159 g/L with PHA volumetric productivity of 1.93 g/(L·h) using hydrolyzed waste cooking oil fatty acids as the sole carbon substrate [93].

In earlier studies, *Pseudomonas* sp. has been demonstrated to use saponified palm kernel oil and saponified rapeseed oil for PHA production [94,95]. In 2017, the ability of wild-type *P. chlororaphis* PA23, a plant growth promoting rhizobacterium (PGPR), to use sole waste canola frying oil was only discovered by Sharma et al. to generate 4.88 g/L CDM and 10.6 wt.-% of *mcl*-PHA, which contained mainly 3HO and 3HD monomers [96]. The addition of octanoic and non-anoic acid as co-substrates, together with fresh canola oil increased the PHA content by 10–13 wt.-% [96]. Another group of researchers reported that *P. chlororaphis* HS21, when previously fed with palm kernel oil, could produce 45% *mcl*-PHA in CDM, which consisted mainly of 3HO and 3HD monomers with up to 3.3 g/L CDM. The ability of this bacterium to synthesize PHA solely from alkanoic acids showed that this metabolic pathway was closely linked to the β-oxidation pathway, but neither *de novo* fatty acid synthesis nor chain elongation reaction was involved [97].

10.2.1.5 *Bacillus*

Bacillus species is the first extensively studied bacterium since the exploration of *in vivo* P(3HB) in *B. megaterium* by Lemoigne in 1926 [98]. A preliminary study was carried out to determine the effects of soybean oil when added to the preculture medium of *B. subtilis* to induce the production of a lipase enzyme to accelerate lipase activity at the initial stage of cultivation. In the first 18 hours of bacterial growth, the lipase activity rose by 88% in comparison with the culture without the induction of oil. The results showed that the *B. subtilis* isolated from soil samples contaminated with edible oil had increased P(3HB) accumulation and CDM; no residual soybean oil was observed at the end of the culture when the initial oil concentration was 10 g/L. The CDM and P(3HB) content were found to be the highest with 13.1 g/L and 87 wt.-%, respectively when 5 g/L soybean oil was used [99].

The potential of *Bacillus cereus* FA11 to use olive oil for PHA terpolymer production in two-stage fermentation was evaluated. The presence of olive oil (1% v/v) and volatile fatty acids (propionic, butyric, valeric, hexanoic acids) in the medium resulted in the biosynthesis of terpolymer that amounted to 60.3% (w/w) after 48 hours of cultivation. The resulting PHA was characterized by Fourier Transform InfraRed spectroscopy and ^{1}H-NMR. The latter analysis further confirmed that the terpolymer was comprised of three different monomers: 3HB, 3HV, and 6HHx [100].

10.2.1.6 *Aeromonas*

Aeromonas caviae is a wild-type strain that can produce P(3HB-*co*-3HHx) from different vegetable oils and fatty acids. *Aeromonas caviae* FA440, a soil-isolated strain, possesses ability to synthesize PHA containing both *scl* 3HB and *mcl* 3HHx monomers. This random *scl–mcl* copolymer was produced under two-stage cultivation from olive oil and alkanoic acids with even carbon numbers [101] and its ability to synthesize 3HV monomer was also discovered using alkanoic acids with odd carbon numbers [102]. The highest PHA content produced from lauric acid (fatty acid with 12 carbons) was 36 wt.-%, and the 3HHx fraction in the copolymer was 20 mol-%. Olive oil on the other hand resulted in the production of PHA having about 13 mol-% 3HHx [102]. The addition of 3HHx monomer into P(3HB) decreased the melting temperature (T_m) of the resulting P(3HB-*co*-3HHx) samples from 177°C to 130°C when the 3HHx fraction increased from 0 to 17 mol-% [101]. In fact, the P(3HB-*co*-3HHx) with 10–17 mol-% of 3HHx composition had an elongation to break of up to 850%, which exhibited better performance than P(3HB-*co*-20 mol-% 3HV) [103].

Because of its attractive PHA synthase gene (*phaC*), various efforts have been made to clone and express the *A. caviae phaC* to enhance the productivity of the *scl–mcl* copolymer [104–106]. Various strategies were developed for this purpose, such as the co-expression of the (*R*)-specific enoyl-CoA hydratase gene (*phaJ*) with *phaC* [107], polymerase chain reaction (PCR)-mediated or *in vivo* random mutagenesis [104,108], double mutation of N149S/D171G (NSDG) in *phaC* [106,109,110], construction of a chimeric enzyme inheriting both *phaC* of *A. caviae* and *C. necator* [111], and co-expression of a phasin as an activator for PhaC [112]. Table 10.3 summarizes all the wild-type bacterial strains discussed in this section that are capable of producing PHA using vegetable oils.

10.2.2 PHA PRODUCTION BY RECOMBINANT MICROBE STRAINS WITH NEW GENES USING VEGETABLE OILS AND BY-PRODUCTS

Section 10.2.1 showed that many wild-type microbes are able to produce PHA by metabolizing vegetable oils and its by-product wastes as the sole carbon source. In addition to that, there are numerous studies by various research groups worldwide that used genetic engineering tools to modify these PHA or non-PHA producers to either improve the PHA yield by increasing the usage efficiency of the carbon substrates, simplify the PHA production process or tailor the strains to produce PHA copolymers with desired properties. While plant oils stand as one of the most

TABLE 10.3

Summary of Wild-Type and Recombinant Bacterial Strains that Are Capable of Using Vegetable Oils for PHA Production

Bacterial Strain	Carbon Substrate	Type of PHA	Cell Dry Mass (CDM) [g/L]	PHA Content [Wt.-%]	PHA Concentration [g/L]	Ref.
Wild type						
C. necator H16	Soybean oil	P(3HB)	118–126	72–76		[33]
	Waste frying Rapeseed oil	P(3HB)			1.2	[48]
Burkholderia cepacia FLP1	Waste cooking oil	P(3HB)	25.4	71		[68]
Burkholderia sp. USM (JCM15050)	Crude palm oil	P(3HB)	4.2	50		[76]
	Crude palm kernel oil	P(3HB)	2.2	70		[67]
	Crude palm kernel oil + Sodium valerate (added at 36 hour)	P(3HB-co-3HV)	2.3	72	1.7	[77]
B. thailandensis	Used cooking oil	P(3HB)	12.2	7.5		[80]
Pseudomonas sp. 61-3	Palm oil	P(3HB), mcl-PHA	5.82	15	0.87	[86]
P. putida PGA1	Saponified palm kernel oil	mcl-PHA	3.0	37	1.1	[89]
P. putida KT2440	Hydrolysed waste cooking oil fatty acids	mcl-PHA	159.4	36.4	58	[93]
P. chlororaphis PA23	Waste canola frying oil	mcl-PHA	4.88	10.6		[96]
P. chlororaphis HS21	Palm kernel oil	mcl-PHA	3.3	45		[97]
P. mosselli TO7	Palm kernel oil	mcl-PHA	4.31	47.1		[146]

(Continued)

TABLE 10.3 (CONTINUED)
Summary of Wild-Type and Recombinant Bacterial Strains that Are Capable of Using Vegetable Oils for PHA Production

Bacterial Strain	Carbon Substrate	Type of PHA	Cell Dry Mass (CDM) [g/L]	PHA Content [Wt.-%]	PHA Concentration [g/L]	Ref.
	Soybean oil	mcl-PHA	3.76	49.8		[146]
Recombinant strains						
C. necator PHB⁻4/pJRDEE32d13	Olive oil, corn oil and palm oil	3HB-co-3HHx	3.5–3.6	76–81		[66]
C. necator PHB⁻4/pJRDEE32d13	Soya bean oil	3HB-co-3HHx	128–132	71–74		[33]
C. necator PHB⁻4/pBBREE32d13	Soya bean oil	3HB-co-3HHx	1.3–3.4	54–83		[116]
C. necator PHB⁻4/pBBREE32d13	PKO, PO, CPKO, PAO	3HB-co-3HHx	3.1–4.3	40–87		[32]
C. necator H16C_AC/pJRDEE32d13	Soya bean oil	3HB-co-3HHx	0.9–6.8	26–89	0.1–4.9	[119]
C. necator Re2058/pCB113	Palm oil	3HB-co-3HHx	3.24–3.6	68–78		[120]
C. necator Re2160/pCB113	Palm oil	3HB-co-3HHx	2.00–2.74	56–63		[120]

attractive carbon substrates for PHA production due to the high carbon content and high conversion rates to PHA, it is essential to establish bacterial strains that can efficiently use them to produce PHA for targeted applications. This section highlights developments in the biosynthesis of PHA using vegetable oils and its by-products via recombinant bacterial strains. The contents in this section are laid out according to the type of recombinant bacterial strains that have been developed.

The bacterium *C. necator* still remains one of the most studied bacterial strains for PHA production. The wild-type strain has been widely used as a model organism for PHA biosynthesis due to its capability to accumulate high PHA contents (90 wt.-% of its CDM) as P(3HB) [113,114]. It has also been shown that it can grow well using plant oil as the sole carbon source [112]. However, P(3HB), which is an *scl*-PHA, is less favourable for industrial applications due to its brittle nature. A PHA with a combination of *scl*- and *mcl*-PHA monomers is preferred as a replacement for petroleum-based plastics due to enhancements in properties like flexibility and processability. As such, *C. necator* has been expansively used in genetic engineering to produce recombinant strains that are capable of producing PHA that have a combination of *scl*- and *mcl*-PHA monomers within them. Several research teams have explored the possibility of developing recombinant *C. necator* strains. Fukui and Doi cloned the PHA synthase gene from *A. caviae* into a PHA-negative mutant, *C. necator* PHB⁻4, to enable the transformant strain to produce P(3HB-*co*-3HHx) copolymer with 4 mol-% 3HHx by using various plant oils, such as palm, olive and corn [66]. Also, the PHA content in the recombinant strain was about 80 wt.-% of its CDM, which was comparable to the P(3HB) produced by the wild-type strain using the same carbon sources. Kahar et al. used the same transformant strain to produce P(3HB-*co*-3HHx) copolymer using soya bean oil [33]. It was reported that the cell biomass obtained from their study was between 128 and 132 g/L with a PHA content between 71 and 74 wt.-%. The 3HHx copolymer content was found to be around 5 mol-% [33]. Similarly, Loo et al. demonstrated the use of recombinant *C. necator* harbouring the PHA synthase gene from *A. caviae* using palm kernel oil, palm olein, CPKO and palm acid oil as the sole carbon sources [32]. In this study, a plasmid harbouring the mutated $phaC_{Ac}$ gene was used instead of the plasmid carrying the wild-type $phaC_{Ac}$ synthase gene. Of all the carbon sources tested, palm kernel oil was found to be the best in giving a high yield of PHA, whereby 87 wt.-% CDM of PHA was obtained. Apart from that, the 3HHx mol fraction remained constant at 5 mol-% across all oils tested [32]. A mutated PHA synthase gene of *A. caviae* was developed in a prior study by Tsuge et al. This mutated PHA synthase showed an increase in the 3HHx mol fraction up to 5.1 mol-% when the transformant was cultivated using soybean oil compared to the wild-type synthase that could produce up to 3.5 mol-% 3HHx using the same carbon source [116]. The synthase from *A. caviae* was used in these studies for the development of recombinant strains because this bacterial strain was among the first to have been found to produce P(3HB-*co*-3HHx) [101]. The PHA synthase from *A. caviae* was found to efficiently polymerize 3HB-CoA and 3HHx-CoA [117]. Besides that, it was also found to possess a gene encoding enoyl-CoA hydratase, $phaJ_{Ac}$, which enables the conversion of fatty acid β-oxidation intermediates into PHA precursors [66,118]. Mifune et al., further engineered a *C. necator* mutant strain for enhanced 3HHx production. In their work, they used the

PHA synthase gene of *A. caviae* that was mutated by changing two amino acids at positions 149 and 171 and designated it as N149S/D171G mutant. In addition, the β-ketothiolase gene of *C. necator* (*phaA$_{Cn}$*) was modified, resulting in a slight reduction in 3HB molar fraction and a concomitant small increase in the 3HHx molar fraction. Further deletion of *phaB1$_{Cn}$* caused a further reduction of 3HB monomers in P(3HB-*co*-3HHx). Finally, the, insertion of the *phaJ$_{Ac}$* gene from *A. caviae* into the *pha* operon resulted in a significant increase in the 3HHx fraction without adversely affecting the growth of cells. Biosynthesis using soya bean oil as the sole carbon source with this enhanced mutant strain gave improved 3HHx fractions in P(3HB-*co*-3HHx) between 5.7 and 9.9 mol-% [119].

In almost all engineered bacterial strains, antibiotics are added to the growth medium for plasmid maintenance. However, using antibiotics for a large-scale PHA production process will result in the elevation of production costs at an unimaginable scale. Thus, an engineered strain that does not require antibiotics for plasmid maintenance is ideally preferred, especially for large-scale productions. Budde et al. attempted to solve this problem and they successfully developed a mutant *C. necator* strain that does not require the addition of antibiotics [120]. Many mutant strains were developed by them. However, only two strains, *C. necator* Re2058/pCB113 and Re2160/pCB113, were used for subsequent fermentation studies using plant oil as these strains were developed to be antibiotic-independent [120]. The development of these strains first started with the insertion of a new PHA synthase gene (*phaC2$_{Ra}$*) from *Rhodococcus aetherivorans* I24 along with *phaJ1* from *Pseudomonas aeruginosa* PAO1 (*phaJ1$_{Pa}$*) in a *C. necator* mutant (Re2152), where the native *phaC1* and *phaB* genes were deleted. The PHA operon from this mutant strain was amplified via PCR and cloned into pBBR1MCS-2 plasmid, creating the plasmid pCB81. For antibiotic independence and plasmid maintenance, strains Re2058 and Re2160 had the *proC* gene deleted. The *proC* gene encodes the enzyme pyrolline-5-carboxylate synthase, which is responsible for proline biosynthesis. Proline is an essential amino acid that functions as one of the key organic compatible solutes to maintain the osmolarity of cells for bacterial growth [121]. Alternatively, this deleted gene was cloned into the pCB81 plasmid generating pCB113 plasmid. This plasmid was inserted to the Re2058 and Re2160 strains generating mutants that are antibiotic independent [(Re2058/pCB113) and (Re2160/pCB113)]. This is a common strategy to maintain the stability of plasmid without the use of antibiotics, whereby an auxotrophic mutant is developed by genome mutation. Then the mutation is complemented by the expression of the mutated gene in a plasmid. An auxotrophic mutant expressing successful complementation of uracil and proline along with heterologous protein expression was reported in *Pseudomonas fluorescens* [122]. Biosynthesis using palm oil as the sole carbon source revealed that these strains could produce up to 71 wt.-% PHA with 17 mol-% 3HHx (Re2058/pCB113) and 66 wt.-% PHA with 30 mol-% 3HHx (Re2160/pCB113) by the end of a 96-h fermentation period [120]. Following the successful development of this *C. necator* mutant strain, Riedel et al. explored the optimal production of P(3HB-*co*-3HHx) with one these strains (Re2058/pCB113) using palm oil as the sole carbon source. In this study, various fermentation strategies (batch, extended batch and fed-batch) were employed to produce high-cell-density cultures. PHA accumulation was triggered by nitrogen limitation. Compared with all the strategies tested, fed-batch fermentations were found to result in the highest levels of cell density and PHA with 139 g/L CDM, which contains about 74 wt.-% PHA with 19 mol-% 3HHx monomer fractions [123]. In a separate study, Wong et al. used the C. necator Re2160/pCB113 mutant strain to produce P(3HB-co-3HHx) copolymers using various types of plant oils in the first stage and CPKO with varied concentrations as the sole carbon source in the second stage. The first stage that was to screen for the best carbon source for PHA production found that all plant oils except for CPKO and coconut oil showed a similar range of 3HHx monomer contents, which is between 41 and 46 mol-%. Biosynthesis using CPKO and coconut oils showed elevated levels of 3HHx monomers of about 56 and 62 mol-%, respectively [124]. Based on the local availability plus sustainability and encouraging results for PHA production, CPKO was chosen for subsequent optimization studies. When 2.5 g/L CPKO was used for biosynthesis, the bacterial strain could produce up to 70 mol-% 3HHx monomers at 12 h; this decreased to 60 mol-% after 24 h and remained mostly constant up to the end of a 72-h fermentation period [124]. In a study by Obruca et al., random chemical mutagenesis of C. necator H16 resulted in the improvement of P(3HB) accumulation (35 wt.-% CDM) compared with the wild-type strain when waste rapeseed oil was used as the sole carbon source. The developed mutant exhibited high levels of specific enzymatic activity that is responsible for the oxidative stress response, such as malic enzyme, NADP-dependent isocitrate, glucose-6-phosphate dehydrogenase and glutamate dehydrogenase [125]. Due to the increase in the NADPH/NADP$^+$ ratio, the authors deduced that the activity of the P(3HB) biosynthetic pathway was improved. Apart from that, the mutant strain was also able to incorporate improved levels of 3HV monomers compared with the wild-type strain. This phenomenon was postulated to have occurred due to the lower availability of oxaloacetate in the 2-methylcitrate cycle for the use of propionyl-CoA. This was as an effect of the increased malic enzyme activity in the mutant [125]. In a separate study, Murugan et al. discovered that the compositions of 3HB and 3HHx monomers in P(3HB-co-3HHx) produced by the engineered C. necator strain Re2058/pCB113 could be controlled by supplementing sugar and oil as carbon sources simultaneously [126]. In this study, CPKO and oil palm trunk sap (OPTS), which contains high amounts of sugar, were used. When high concentrations of CPKO with low concentrations of OPTS (based on total sugar content) were fed, high molar fractions of 3HHx were incorporated. Conversely, when the concentration of OPTS was increased while lowering the concentration of CPKO, more incorporation of 3HB monomers was observed [126]. Based on these findings, Murugan et al. concluded that this simple co-feeding strategy could be applied using inexpensive waste carbon sources to tailor the composition of the desired copolymers for the economical production of PHA. Similar studies were also conducted by Murugan et al. using palm olein (PO) and fructose as carbon feedstocks with the same recombinant C. necator strain used in their previous study. When the bacterial culture was supplemented with 5 g/L palm olein and 7 g/L fructose, a copolymer of P(3HB-co-3HHx) with 17 mol-% 3HHx fraction was produced. The cells were also found to have accumulated about 80 wt.-% PHA in 7.1 g/L CDM [127]. The molecular weights of the polymers varied with varying 3HHx compositions at different fermentation time points. Copolymers with molecular weights in the range of 547–685 kDa were produced in this study. These values were at least two-fold higher than any reported values in the literature [127]. A recent discovery by Foong et al. reported a unique novel PHA synthase gene isolated from a mangrove soil metagenome [128]. Moreover, a mutant strain of *C. necator* PHB$^-$4 expressing the plasmid pBBRMCS-2 harbouring the novel PHA synthase gene was developed. This PHA synthase was found to be able to include myriad PHA monomers during biosynthesis due to its broad substrate specificity. This includes the *scl-co-mcl*-PHA P(3HB-*co*-3HHx) using CPKO as the sole carbon source [128]. Following the discovery of this new PHA synthase, Lakshmanan et al. successfully produced an interesting combination of co- and terpolymers by feeding CPKO as the sole carbon source while co-supplementing precursor compounds to the mutant strain for the inclusion of other monomers along with 3HHx in the polymer chain in a subsequent study. As a result, a copolymer with a combination of four different monomers was produced in this study P(3HB-*co*-4HB-*co*-5HV-*co*-3HHx) [129]. Studies by Zainab et al. have shown that the mutant strain *C. necator* Re2058/pCB113 was capable of producing P(3HB-*co*-3HHx) with up to 31 mol-% of 3HHx in PHA using underused oils such as desert date, African elemi, bitter apple and *Amygdalus pedunculata* oils that have no food value [38]. The molecular weights (M_w) of the copolymers produced in this study were reported to be relatively high, 510–630 kDa [38]. This finding suggested that the underused plant oils stand as a good potential alternative to the food grade plant oils that are commonly used for PHA biosynthesis. Thinagaran and Sudesh, however, have reported the use of emulsified sludge palm oil (SPO) for producing P(3HB-*co*-3HHx) via the recombinant strain *C. necator* Re2058/pCB113 [130]. Initially, it was reported that feeding a total of 10 g/L SPO yielded around 9.7 g/L CDM with a PHA content of 74 wt.-% that had about 22 mol-% 3HHx monomers. In an attempt to improve the PHA yield in this study, a fed-batch fermentation strategy was employed in a 13-L bioreactor setting. It was observed that the biomass productivity markedly increased up to 1.9 g/(L·h), while the PHA productivity improved to 1.1 g/(L·h) [130]. Table 10.3 summarizes all the recombinant bacterial strains discussed in this section.

10.3 CHALLENGES IN USING DIFFERENT TYPES OF MICROORGANISMS IN LARGE-SCALE PHA PRODUCTION

Sustainable production of PHA by suitable microorganisms with a low cost but high substrate-to-PHA conversion efficiency are critical matters that global researchers are currently trying to overcome. Apart from well-known species involved in industrial PHA production, such as *C. necator*, an industrial production bacterial strain requires several features to be selected as promising PHA producers, for example, it must be non-pathogenic, easy to manipulate genomically, produce no toxins and grow fast [131,132]. However, among the oil-using PHA producers, strains like *Burkholderia*, *Pseudomonas* and *Aeromonas* are regarded as pathogenic bacteria (risk group II), which might hamper the upscaling of PHA production. *Burkholderia* and *Pseudomonas* species are known as animal, plant and human pathogens; *B. cepacia* and *P. aeruginosa* specifically are dangerous pathogens for patients suffering from cystic fibrosis [133]. *Aeromonas* species are mainly pathogenic to poikilothermic animals and aquaculture. In certain cases, *A. caviae* and *A. hydrophila* cause extraintestinal and gastrointestinal infections in humans [134]. One of the main virulence factors associated with bacterial pathogenicity is surface polysaccharides such as capsular polysaccharides, lipopolysaccharide and glucans [134]. Moreover, lipopolysaccharides exist in the outer membrane of Gram-negative bacteria; they are heat stable endotoxins and are believed to cause septicaemia in humans [135]. They provoke a strong immune response because of the polysaccharide chain linked to a lipid moiety known as lipid A [136]. Lipid A is an endotoxin primarily responsible for the toxicity property of Gram-negative bacteria. Therefore, the elimination of microbial components from raw PHA, including endotoxins, is necessary [15]. Purification of PHA will co-purify this endotoxic substance, and hence an additional oxidizing step is essential to reduce the endotoxin content for medical applications [137,138].

However, the selection of the *Bacillus* species as a model for a PHA producer is beneficial because of the absence of the toxic lipopolysaccharides external layer and simplification of the PHA extraction process [99,138]. Nevertheless, the major disadvantage of using *Bacillus* is due to their sporulation. In *Bacillus*, both sporulation and formation of PHA granules in the cytoplasm are initiated due to stress conditions [139]. However, Valappil et al. [140] investigated how intracellular mobilization of PHA and spore formation in *B. cereus* SPV were inhibited in low pH conditions. Research done by Wakisaka et al. [141] proved that potassium plays a crucial role not only by affecting *Bacillus* growth and endotoxin formation but also for sporulation. The sporulation of *Bacillus* strains was suppressed when they grew in potassium-deficient media, which, in parallel, enhanced the formation of P(3HB). By controlling the pH and potassium concentration in the culture media, the target of pilot-scale PHA production by *Bacillus* sp. can be attained [141].

Initially, the initial interest of *C. necator* H16 was aimed at the chemolithoautotrophic production of single-cell protein for food consumption but not for its PHA synthesizing ability [142]. The accumulation of undesirable PHA in *C. necator* has lowered its nutritive value. Because of this issue, a *C. necator* PHB$^-$4 mutant was constructed to avoid the synthesis of PHA in bacterial cells. As discussed earlier, *A. hydrophila* is a pathogenic bacterium that can produce endotoxin. The access of *A. hydrophila* as a single-cell protein was reported by Bajpai [143]; however, there is no detailed information on the effect of its immunogenicity level to humans or animals. A study on the immunogenicity of *Methylococcus capsulatus* as a single-cell protein showed that the whole cell preparation samples caused immune responses in mice. In contrast, cell-free preparation samples where the cell wall was removed were safe to consume [144,145].

In summary, the safety requirements associated with the use of a particular type of microorganism for large-scale PHA production followed by biological recovery of PHA polymers through animals would be a crucial element to consider for future studies if single-cell proteins or insect proteins are used for human and animal consumption. The details of the biological recovery method of PHA using animals/insects are covered in the next section.

10.4 APPLICATION OF WASTE VEGETABLE OILS AND NON-FOOD-GRADE PLANT OILS FOR LARGE-SCALE PRODUCTION OF PHA

The biggest challenge in scaling up PHA for commercial use is the production cost. Although numerous studies have been carried out in the field of PHA, it is still a big challenge to produce PHA economically and sustainably. Major bottlenecks in scaling up PHA production exist at the selection of a suitable carbon source and downstream processing (PHA recovery). The recent advancements and discoveries of various waste carbon feedstocks have brought us a step closer to achieving the successful scale-up and promising commercialization of PHA. Vegetable and plant oil waste stand as attractive, inexpensive carbon feedstocks due to their high carbon content and good conversion to PHA. Their high carbon content also allows low flow-rate feeding of the substrates to the cultures, thus reducing the dilution of the fermentation broth [29,32,33,43,66]. Palm oil by far has the largest yield per hectare compared with other plant oils. This also translates to the large amount of oily waste that is generated alongside mainstream palm oil production that can be a waste 'goldmine' for PHA production. These waste oils are usually converted to animal feed and other low-grade non-food applications [147]. Using these inexpensive waste materials for PHA production automatically converts them to value-added materials, thus enabling the generation of wealth while promoting zero-waste technology. Waste oils such as SPO, POME and waste frying oil also have high FFA content that is readily available for assimilation by bacterial strains for PHA production.

Besides having inexpensive carbon feedstock for PHA production, it is also essential to ensure that the appropriate bacterial strains are used for large-scale production. *C. necator* still stands as the best candidate for the use of oily waste in PHA production, as mentioned in Section 10.2.2. With the successful genetic alteration to enable this strain to produce industrially useful PHA copolymers, such as P(3HB-co-3HHx), and a few published studies that have already shown that this engineered strain can convert oily waste to the desired PHA in a bioreactor setting, it is practical to use engineered *C. necator* for the large-scale production of PHA. In fact, some companies, such as KANEKA (Japan), have been routinely using an engineered *C. necator* strain for their pilot-scale PHA production from palm oil.

In the PHA production flow, the upstream process that involves fermentation and recovery of cells from the culture medium is well optimized. However, the downstream process where the polymer needs to be extracted and recovered from cells remains a big challenge. Conventional methods, such as solvent extraction, is impractical for large-scale recovery of PHA since it involves eco-unfriendly and costly solvents. The use of mechanical disrupting methods or enzymes may be costly and add to the production cost. In addition, the residual cellular debris has to be treated, which further adds to the cost of wastewater treatment. Recovery methods that are environmentally friendly and sustainable are preferred for large-scale processing and recovery of PHA. The biological recovery method is a new avenue where a few research groups are exploring the large-scale production of PHA. The biological recovery method involves the feeding of PHA-laden dried bacterial cells to living organisms and recovering the PHA in the form of faecal excrement. The faecal pellets are further subjected to simple and environmentally friendly purification methods that do not involve the use of harsh chemicals to bring the purity of the recovered PHA close to that of solvent extracted ones. An accidental discovery reported by Kunasundari et al. that Sprague-Dawley rats can biologically recover PHA has initiated a subsequent stream of studies in other living organism models such as insects [148,149]. Mealworm-based biological recovery has proven to be a successful method in recovering PHA from bacterial cells at the laboratory scale [150–153]. Despite rat and mealworm models being equally successful in recovering PHA of high purity without compromising the physicochemical properties, the mealworms would be a better candidate for large-scale production of PHA since handling small animals, such as rats, at a large scale would be expensive and unsuitable for the downstream process. Mealworms are much easier to handle at a large scale and are convertible to value-added food products for animals and humans too in the long run. The dried bacterial cells act as a single-cell protein for the mealworms. Mealworms thriving on the dried cell diet were proven to significantly increase its total protein content up to 79 wt.-% and decrease the lipid content up to 8.3-% of its CDM [153]. This implies that protein-enriched mealworms are produced via this biological recovery method, which can be processed into high-protein food for animals. However, as discussed in Section 10.3, it is imperative to carefully assess the immunological and toxicity effects of these protein-enriched worms if they are going to be used as food supplements for humans. Our team showed that there are no negative effects in terms of immunological and toxicity generated in animals or insects when they were fed with *C. necator* as a single-cell protein. Also, because of the pathogenicity of other reported PHA-producing strains in Section 10.3, it will be critical to evaluate the immunological and toxicity responses in animals or insects if the PHA biological recovery method is to be adapted in PHA production using these pathogenic strains. The combination of cheap waste carbon feedstocks from the palm oil and food industries, engineered bacterial strains for the optimal utilization of these waste oils plus the biological recovery method for downstream processing paves a sustainable and economical way of producing PHA. This model can be adopted for large- and subsequently commercial-scale production of PHA. Figure 10.5 shows the overall process beginning from the oil palm plantation up to obtaining purified PHA using a sustainable and cost-effective system. Setting up the PHA production facilities close to the source of the carbon feedstock supply helps to further reduce the cost in terms of transportation of waste. At the same time, it could also promote a circular economy whereby the production of waste is kept at significantly minimum levels. All processes within the PHA production system complement each other, thus avoiding the generation of waste and pollutants. This will be a sustainable system, whereby it does not have any negative environmental impact and, at the same time, helps society in terms of socioeconomic benefit by creating new jobs.

10.5 CONCLUSIONS AND OUTLOOK

In conclusion, vegetable oils and waste by-products are shown to have huge potential for high PHA production. Theoretical calculations in published literature have

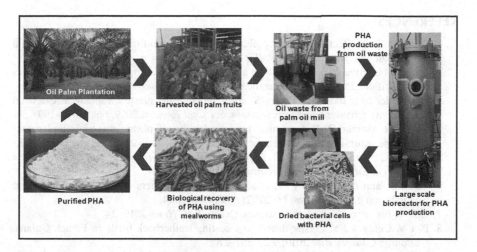

FIGURE 10.5 Model for sustainable and cost-effective production of PHA using palm oil waste.

also supported this observation whereby for every kilogram of oil fed, one kilogram of PHA is expected to be obtained. Many studies highlighted in Sections 10.2.1 and 10.2.2 have shown the conversion of vegetable oil to PHA close to the theoretical yield. However, for the sustainable production of PHA, it is important to source feedstocks that are also available in a sustainable manner. The production process of feedstocks should ideally have minimal impact on the environment while maintaining maximum production capacity. Malaysia, for example, is one of the largest producers of palm oil in the world. As of 2019, about 5 million hectares of land is needed to produce about 20 million tonnes of palm oil per year. Based on the theoretical conversion of feedstock to PHA at 1:1 ratio, about 1 million tonnes of palm oil would be required to produce 1 million tonnes of PHA. About 0.25 million hectares of land is needed to produce 1 million tonnes of palm oil. This is equal to a landmass that is 3.5 times larger than Singapore. Acquiring such a large landmass for PHA production is indeed unsustainable because it also has direct competition with the food production industry. Palm oil productions around the world cater mostly for food applications. Therefore, a better way is to source the waste by-products of the palm oil industry for PHA production. However, another challenge that needs to be addressed here is how to gather the waste oil feedstocks at a particular location for PHA production. Setting up PHA production plants close to waste feedstock sources is a better choice but constructing PHA production plants at every plantation site scattered throughout the country is also impractical. Therefore, a proper balance has to be reached in determining the most strategic locations to set up PHA production facilities that does not incur huge costs in transporting and pooling up the waste feedstock, and, at the same time, is easily accessible by plantation developers to consistently transfer waste feedstocks. Once determined, the PHA production model outlined in section 10.4 can be adapted for setting up an economical and sustainable PHA production process.

REFERENCES

1. Sigler M. The effects of plastic pollution on aquatic wildlife: current situations and future solutions. *Water Air Soil Pollut* 2014; 225(11): 2184.
2. Schmidt N, Thibault D, Galgani F, et al. Occurrence of microplastics in surface waters of the Gulf of Lion (NW Mediterranean Sea). *Prog Oceanogr* 2018; 163: 214–220.
3. Heidbreder LM, Bablok I, Drews S, et al. Tackling the plastic problem: a review on perceptions, behaviors, and interventions. *Sci Total Environ* 2019; 668: 1077–1093.
4. Mohee R, Unmar G. Determining biodegradability of plastic materials under controlled and natural composting environments. *Waste Manag* 2007; 27(11): 1486–1493.
5. Shaw DK, Sahni P. Plastic to oil. *Paper Presented at the International Conference on Advances in Engineering and Technology Singapore*, 2014.
6. Vert M, Santos ID, Ponsart SP, et al. Degradable polymers in a living environment: where do you end up? *Polym Int* 2002; 51(10): 840–844.
7. Gill, V. Plastic rubbish blights Atlantic Ocean. *BBC News* 2010; 24.
8. Plot V, Georges J-Y. Plastic debris in a nesting leatherback turtle in French Guiana. *Chelonian Conserv Biol* 2010; 9(2): 267–270.
9. Stamper MA, Spicer CW, Neiffer DL, et al. Morbidity in a juvenile green sea turtle (*Chelonia mydas*) due to ocean-borne plastic. *J Zoo Wildl Med* 2009; 40(1): 196–198.
10. Mascarenhas R, Santos R, Zeppelini D. Plastic debris ingestion by sea turtle in Paraíba, Brazil. *Mar Pollut Bull* 2004; 49(4): 354–355.
11. Gregory MR. Environmental implications of plastic debris in marine settings-entanglement, ingestion, smothering, hangers-on, hitch-hiking and alien invasions. *Philos Trans R Soc B Biol Sci* 2009; 364(1526): 2013–2025.
12. Sarker M, Mamunor Rashid M, Molla M, et al. Thermal conversion of waste plastics (HDPE, PP and PS) to produce mixture of hydrocarbons. *Am J Environ* 2012; 2(5): 128–136.
13. Madison LL, Huisman GW. Metabolic engineering of poly(3-hydroxyalkanoates): From DNA to plastic. *Microbiol Mol Biol Rev* 1999; 63(1): 21–53.
14. Lee EY, Jendrossek D, Schirmer A, et al. Biosynthesis of copolyesters consisting of 3-hydroxybutyric acid and medium-chain-length 3-hydroxyalkanoic acids from 1,3-butanediol or from 3-hydroxybutyrate by *Pseudomonas* sp. A33. *Appl Microbiol Biotechnol* 1995; 42(6): 901–909.
15. Koller M. Biodegradable and biocompatible polyhydroxy-alkanoates (PHA): auspicious microbial macromolecules for pharmaceutical and therapeutic applications. *Molecules* 2018; 23: 362.
16. Meyer A. *Practicum der Botanischen Bakterienkunde*, vol. 2. Gustav-Fischer, 1903.
17. Sudesh K, Abe H, Doi Y. Synthesis, structure and properties of polyhydroxyalkanoates: biological polyesters. *Prog Polym Sci* 2000; 25(10): 1503–1555.
18. Lee SY. Plastic bacteria? Progress and prospects for polyhydroxyalkanoate production in bacteria. *Trends Biotechnol* 1996; 14(11): 431–438.
19. Braunegg G, Lefebvre G, Genser KF. Polyhydroxyalkanoates, biopolyesters from renewable resources: physiological and engineering aspects. *J Biotechnol* 1998; 65(2–3): 127–161.
20. Tsuge T. Metabolic improvements and use of inexpensive carbon sources in microbial production of polyhydroxyalkanoates. *J Biosci Bioeng* 2002; 94(6): 579–584.
21. Wallen LL, Rohwedder WK. Poly-β-hydroxyalkanoate from activated sludge. *Environ Sci Technol* 1974; 8(6): 576–579.
22. Kunioka M, Nakamura Y, Doi Y. New bacterial copolyesters produced in *Alcaligenes eutrophus* from organic acids. *Polym Commun* 1988; 29: 174–176.
23. Doi Y. *Microbial polyesters*. VCH Publishers New York, 1990.

24. Riedel SL, Jahns S, Koenig S, et al. Polyhydroxyalkanoates production with *Ralstonia eutropha* from low quality waste animal fats. *J Biotechnol* 2015; 214: 119–127.

25. Anderson AJ, Dawes EA. Occurrence, metabolism, metabolic role, and industrial uses of bacterial polyhydroxyalkanoates. *Microbiol Rev* 1990; 54(4): 450–472.

26. Sudesh K. *Polyhydroxyalkanoates from Palm Oil: Biodegradable Plastics*, 1 edn. Springer Verlag Berlin Heidelberg, 2013.

27. Lu Y, Larock RC. Novel polymeric materials from vegetable oils and vinyl monomers: preparation, properties, and applications. *ChemSusChem* 2009; 2(2): 136–147.

28. Gunstone FD. *Vegetable Oils in Food Technology: Composition, Properties and Uses*, 2 edn. Blackwell Publishing Ltd., 2011.

29. Akiyama M, Tsuge T, Doi Y. Environmental life cycle comparison of polyhydroxyalkanoates produced from renewable carbon resources by bacterial fermentation. *Polym Degrad Stab* 2003; 80(1): 183–194.

30. Riedel SL, Lu J, Stahl U, et al. Lipid and fatty acid metabolism in *Ralstonia eutropha*: relevance for the biotechnological production of value-added products. *Appl Microbiol Biotechnol* 2013; 98(4): 1469–1483.

31. Xia Y, Larock RC. Vegetable oil-based polymeric materials: synthesis, properties, and applications. *Green Chem* 2010; 12(11): 1893.

32. Loo C-Y, Lee W-H, Tsuge T, et al. Biosynthesis and characterization of poly(3-hydroxybutyrate-*co*-3-hydroxyhexanoate) from palm oil products in a *Wautersia eutropha* mutant. *Biotechnol Lett* 2005; 27(18): 1405–1410.

33. Kahar P, Tsuge T, Taguchi K, et al. High yield production of polyhydroxyalkanoates from soybean oil by *Ralstonia eutropha* and its recombinant strain. *Polym Degrad Stab* 2004; 83(1): 79–86.

34. López-Cuellar MR, Alba-Flores J, Rodríguez JNG, et al. Production of polyhydroxyalkanoates (PHAs) with canola oil as carbon source. *Int J Biol Macromol* 2011; 48(1): 74–80.

35. Park DH, Kim BS. Production of poly(3-hydroxybutyrate) and poly(3-hydroxybutyrate co 4 hydroxybutyrate) by *Ralstonia eutropha* from soybean oil. *New Biotechnol* 2011; 28(6): 719–724.

36. Basiron Y, Simeh MA. Vision 2020 – the palm oil phenomenon. *OPIEJ* 2005; 5(2): 1–10.

37. Kumar P, Krishna A. Physicochemical characteristics of commercial coconut oils produced in India. *Grasas Aceites* 2015; 66(1): 062.

38. Zainab LI, Uyama H, Li C, et al. Production of polyhydroxyalkanoates from underutilized plant oils by *Cupriavidus necator*. *Clean (Weinh)* 2018; 46(11): 1700542.

39. Gui MM, Lee K, Bhatia S. Feasibility of edible oil vs. non-edible oil vs. waste edible oil as biodiesel feedstock. *Energy* 2008; 33(11): 1646–1653.

40. Purama R, Al-Sabahi J, Sudesh K. Evaluation of date seed oil and date molasses as novel carbon sources for the production of poly (3-hydroxybutyrate-*co*-3-hydroxyhexanoate) by *Cupriavidus necator* H16 Re 2058/pCB113. *Ind Crops Prod* 2018; 119: 83–92.

41. Colombo CA, Chorfi Berton LH, Diaz BG, et al. Macauba: a promising tropical palm for the production of vegetable oil. *OCL* 2017; 25(1): D108.

42. Miao S, Wang P, Su Z, et al. Vegetable-oil-based polymers as future polymeric biomaterials. *Acta Biomater* 2014; 10(4): 1692–1704.

43. Ng K-S, Ooi W-Y, Goh L-K, et al. Evaluation of jatropha oil to produce poly(3-hydroxybutyrate) by *Cupriavidus necator* H16. *Polym Degrad Stab* 2010; 95(8): 1365–1369.

44. Batcha AFM, Prasad DR, Khan MR, et al. Biosynthesis of poly (3-hydroxybutyrate) (PHB) by *Cupriavidus necator* H16 from jatropha oil as carbon source. *Bioprocess Biosyst Eng* 2014; 37(5): 943–951.

45. Kynadi AS, Suchithra TV. Formulation and optimization of a novel media comprising rubber seed oil for PHA production. *Ind Crops Prod* 2017; 105: 156–163.
46. Koller M, Atlić A, Dias M, et al. Microbial PHA production from waste raw materials. In: *Plastics from Bacteria*, vol. 14. Springer, 2010: pp. 85–119.
47. Tian P-Y, Shang L, Ren H, et al. Biosynthesis of polyhydroxyalkanoates: current research and development. *Afr J Biotechnol* 2009; 8(5): 709–714.
48. Verlinden RAJ, Hill DJ, Kenward MA, et al. Production of polyhydroxyalkanoates from waste frying oil by *Cupriavidus necator*. *AMB Expr* 2011; 1(1): 11.
49. Obruca S, Marova I, Snajdar O, et al. Production of poly(3-hydroxybutyrate-*co*-3-hydroxyvalerate) by *Cupriavidus necator* from waste rapeseed oil using propanol as a precursor of 3-hydroxyvalerate. *Biotechnol Lett* 2010; 32(12): 1925–1932.
50. Możejko J, Ciesielski S. Pulsed feeding strategy is more favorable to medium-chain-length polyhydroxyalkanoates production from waste rapeseed oil. *Biotechnol Prog* 2014; 30(5): 1243–1246.
51. Ng K-S, Wong Y-M, Tsuge T, et al. Biosynthesis and characterization of poly(3-hydroxybutyrate-*co*-3-hydroxyvalerate) and poly(3-hydroxybutyrate-*co*-3-hydroxyhexanoate) copolymers using jatropha oil as the main carbon source. *Process Biochem* 2011; 46(8): 1572–1578.
52. Singh AK, Bhati R, Mallick N. *Pseudomonas aeruginosa* MTCC 7925 as a biofactory for production of the novel SCL-LCL-PHA thermoplastic from non-edible oils. *Curr Biotechnol* 2015; 4: 65–74.
53. Chhetri A, Tango M, Budge S, et al. Non-edible plant oils as new sources for biodiesel production. *Int J Mol Sci* 2008; 9(2): 169–180.
54. Allen AD, Anderson WA, Ayorinde FO, et al. Biosynthesis and characterization of copolymer poly(3HB-*co*-3HV) from saponified *Jatropha curcas* oil by *Pseudomonas oleovorans*. *J Ind Microbiol Biotechnol* 2010; 37(8): 849–856.
55. Chitra P, Venkatachalam P, Sampathrajan A. Optimisation of experimental conditions for biodiesel production from alkali-catalysed transesterification of *Jatropha curcas* oil. *Energy Sustain Dev* 2005; 9(3): 13–18.
56. Mutlu H, Meier MA. Castor oil as a renewable resource for the chemical industry. *Eur J Lipid Sci Technol* 2010; 112(1): 10–30.
57. Njwe R, Chifon M, Ntep R. Potential of rubber seed as protein concentrate supplement for dwarf sheep of Cameroon. In: *Utilization of Research Results on Forage and Agricultural Byproduct Materials as Animal Feed Resources in Africa. Proceedings of the First Joint Workshop Held in Lilongwe*, Malawi, 1988: pp. 5–9.
58. *Report NRS*. Malaysia Rubber Board, 2015.
59. Ng WPQ, Lim MT, Lam HL, et al. Overview on economics and technology development of rubber seed utilisation in Southeast Asia. *Clean Technol Environ Policy* 2014; 16(3): 439–453.
60. Chen G-Q, Chen X-Y, Wu F-Q, et al. Polyhydroxyalkanoates (PHA) toward cost competitiveness and functionality. *Adv Ind Eng Polym Res* 2020; 3: 1–7.
61. Choi GG, Kim HW, Rhee YH. Enzymatic and non-enzymatic degradation of poly(3-hydroxybutyrate-*co*-3-hydroxyvalerate) copolyesters produced by *Alcaligenes* sp. MT-16. *J Microbiol* 2004; 42(4): 346–352.
62. Timm A, Byrom D, Steinbüchel A. Formation of blends of various poly(3-hydroxyalkanoic acids) by a recombinant strain of *Pseudomonas oleovorans*. *Appl Microbiol Biotechnol* 1990; 33(3): 296–301.
63. Grothe E, Moo-Young M, Chisti Y. Fermentation optimization for the production of poly(β-hydroxybutyrate) microbial thermoplastic. *Enzyme Microb Technol* 1999; 25: 132–141.

64. Pozo C, Toledo M, Rodelas B, et al. Effects of culture conditions on the production of polyhydroxyalkanoates by *Azotobacter chroococcum* H23 in media containing a high concentration of alpechin (wastewater from olive oil mills) as primary carbon source. *J Biotechnol* 2002; 97: 125–131.

65. Eggink G, van der Wal H, Huijberts GNM, et al. Oleic acid as a substrate for poly-3-hydroxyalkanoate formation in *Alcaligenes eutrophus* and *Pseudomonas putida*. *Ind Crops Prod* 1992; 1(2–4): 157–163.

66. Fukui T, Doi Y. Efficient production of polyhydroxyalkanoates from plant oils by *Alcaligenes eutrophus* and its recombinant strain. *Appl Microbiol Biotechnol* 1998; 49(3): 333–336.

67. Chee J-Y, Tan Y, Samian MR, et al. Isolation and characterization of a *Burkholderia* sp. USM (JCM15050) capable of producing polyhydroxyalkanoate (PHA) from triglycerides, fatty acids and glycerols. *J Polym Environ* 2010; 18(4): 584–592.

68. Kamilah H, Tsuge T, Yang TA, et al. Waste cooking oil as substrate for biosynthesis of poly(3-hydroxybutyrate) and poly(3-hydroxybutyrate-*co*-3-hydroxyhexanoate): turning waste into a value-added product. *Malays J Microbiol* 2013; 9(1): 51–59.

69. Lee S, Choi J, Han K, et al. Removal of endotoxin during the purification of poly(3-hydroxybutyrate) from Gram-negative bacteria. *Appl Environ Microbiol* 1999; 65: 2762–2764.

70. Mravec F, Obruca S, Krzyzanek V, et al. Accumulation of PHA granules in *Cupriavidus necator* as seen by confocal fluorescence microscopy. *FEMS Microbiol Lett* 2016; 363(10): fnw094.

71. Vandamme P, Coenye T. Taxonomy of the genus *Cupriavidus*: a tale of lost and found. *Int J Syst Evol Microbiol* 2004; 54(6): 2285–2289.

72. Poehlein A, Kusian B, Friedrich B, et al. Complete genome sequence of the type strain *Cupriavidus necator* N-1. *J Bacteriol* 2011; 193: 5017.

73. Holmes PA. Applications of PHB – a microbially produced biodegradable thermoplastic. *Phys Technol* 1985; 16(1): 32–36.

74. Shantini K, Yahya ARM, Amirul AA. Influence of feeding and controlled dissolved oxygen level on the production of poly(3-hydroxybutyrate-co-3-hydroxyvalerate) copolymer by *Cupriavidus* sp. USMAA2-4 and its characterization. *Appl Biochem Biotechnol* 2015; 176(5): 1315–1334.

75. Martino L, Cruz M, Scoma A, et al. Recovery of amorphous polyhydroxybutyrate granules from *Cupriavidus necator* cells grown on used cooking oil. *Int J Biol Macromol* 2014; 71: 117–123.

76. Alias Z, Tan IKP. Isolation of palm oil-utilising, polyhydroxyalkanoate (PHA)-producing bacteria by an enrichment technique. *Bioresour Technol* 2005; 96(11): 1229–1234.

77. Chee JY, Lau NS, Samian MR, et al. Expression of *Aeromonas caviae* polyhydroxyalkanoate synthase gene in *Burkholderia* sp. USM (JCM15050) enables the biosynthesis of SCL-MCL PHA from palm oil products. *J Appl Microbiol* 2011; 112: 45–54.

78. Zhu C, Chiu S, Nakas JP, et al. Bioplastics from waste glycerol derived from biodiesel industry. *J Appl Polym Sci* 2013; 130(1): 1–13.

79. Zhu C, Nomura CT, Perrotta JA, et al. Production and characterization of poly-3-hydroxybutyrate from biodiesel-glycerol by *Burkholderia cepacia* ATCC 17759. *Biotechnol Prog* 2009; 26(2): 424–430.

80. Kourmentza C, Costa J, Azevedo Z, et al. *Burkholderia thailandensis* as a microbial cell factory for the bioconversion of used cooking oil to polyhydroxyalkanoates and rhamnolipids. *Bioresour Technol* 2018; 247: 829–837.

81. Obruca S, Benesova P, Oborna J, et al. Application of protease-hydrolyzed whey as a complex nitrogen source to increase poly(3-hydroxybutyrate) production from oils by *Cupriavidus necator*. *Biotechnol Lett* 2013; 36(4): 775–781.

82. Thakor NS, Patel MA, Trivedi UB, et al. Production of poly(β-hydroxybutyrate) by *Comamonas testosteroni* during growth on naphthalene. *World J Microbiol Biotechnol* 2003; 19(2): 185–189.

83. Thakor N, Trivedi U, Patel KC. Biosynthesis of medium chain length poly(3-hydroxyalkanoates) (*mcl*-PHAs) by *Comamonas testosteroni* during cultivation on vegetable oils. *Bioresour Technol* 2005; 96(17): 1843–1850.

84. Matsusaki H, Manji S, Taguchi K, et al. Cloning and molecular analysis of the poly(3-hydroxybutyrate) and poly(3-hydroxybutyrate-*co*-3-hydroxyalkanoate) biosynthesis genes in *Pseudomonas* sp. strain 61-3. *J Bacteriol* 1998; 180(24): 6459–6467.

85. De Smet MJ, Eggink G, Witholt B, et al. Characterization of intracellular inclusions formed by *Pseudomonas oleovorans* during growth on octane. *J Bacteriol* 1983; 154(2): 870–878.

86. Kato M, Fukui T, Doi Y. Biosynthesis of polyester blends by *Pseudomonas* sp. 61-3 from alkanoic acids. *Bull Chem Soc Jpn* 1996; 69(3): 515–520.

87. Ashby RD, Solaiman DKY, Foglia TA. Bacterial poly(hydroxyalkanoate) polymer production from the biodiesel co-product stream. *J Polym Environ* 2004; 12(3): 105–112.

88. Ashby RD, Solaiman DKY, Foglia TA. Synthesis of short-/medium-chain-length poly(hydroxyalkanoate) blends by mixed culture fermentation of glycerol. *Biomacromolecules* 2005; 6(4): 2106–2112.

89. Tan IKP, Kumar KS, Theanmalar M, et al. Saponified palm kernel oil and its major free fatty acids as carbon substrates for the production of polyhydroxyalkanoates in *Pseudomonas putida* PGA1. *Appl Microbiol Biotechnol* 1997; 47(3): 207–211.

90. Annuar M, Tan I, Ibrahim S, et al. Production of medium-chain-length poly(3-hydroxyalkanoates) from saponified palm kernel oil by *Pseudomonas putida*: kinetics of batch and fed-batch fermentations. *Malays J Microbiol* 2006; 2(2): 1–9.

91. Wang Q, Nomura CT. Monitoring differences in gene expression levels and polyhydroxyalkanoate (PHA) production in *Pseudomonas putida* KT2440 grown on different carbon sources. *J Biosci Bioeng* 2010; 110(6): 653–659.

92. Mozejko-Ciesielska J, Dabrowska D, Szalewska-Palasz A, et al. Medium-chain-length polyhydroxyalkanoates synthesis by *Pseudomonas putida* KT2440 relA/spoT mutant: bioprocess characterization and transcriptome analysis. *AMB Expr* 2017; 7: 92.

93. Ruiz C, Kenny ST, Babu PR, et al. High cell density conversion of hydrolysed waste cooking oil fatty acids into medium chain length polyhydroxyalkanoate using *Pseudomonas putida* KT2440. *Catalysts* 2019; 9: 468.

94. Możejko J, Przybyłek G, Ciesielski S. Waste rapeseed oil as a substrate for medium-chain-length polyhydroxyalkanoates production. *Eur J Lipid Sci Technol* 2011; 113(12): 1550–1557.

95. Sun YH, Kim DY, Chung CW, et al. Characterization of a tacky poly(3-hydroxyalkanoate) produced by *Pseudomonas chlororaphis* HS21 from palm kernel oil. *J Microbiol Biotechnol* 2003; 13(1): 64–69.

96. Sharma PK, Munir RI, de Kievit T, et al. Synthesis of polyhydroxyalkanoates (PHAs) from vegetable oils and free fatty acids by wild-type and mutant strains of *Pseudomonas chlororaphis*. *Can J Microbiol* 2017; 63(12): 1009–1024.

97. Yun H, Kim D, Chung C, et al. Characterization of a tacky poly(3-hydroxyalkanoate) produced by *Pseudomonas chlororaphis* HS21 from palm kernel oil. *J Microbiol Biotechnol* 2003; 13(1): 64–69.

98. Lemoigne M. Produits de dehydration et de polymerisation de l acide ßoxobutyrique. *Bull Soc Chim Biol* 1926; 8: 770–782.

99. Marjadi D, Dharaiya N. Microbial production of poly-3-hydroxybutyric acid from soybean oil by *Bacillus subtillis*. *Eur J Exp Biol* 2013; 3(5): 141–147.

100. Masood F, Abdul-Salam M, Yasin T, et al. Effect of glucose and olive oil as potential carbon sources on production of PHAs copolymer and tercopolymer by *Bacillus cereus* FA11. *3 Biotech* 2017; 7: 87.

101. Shimamura E, Kasuya K, Kobayashi G, et al. Physical properties and biodegradability of microbial poly(3-hydroxybutyrate-*co*-3-hydroxyhexanoate). *Macromolecules* 1994; 27(3): 878–880.

102. Doi Y, Kitamura S, Abe H. Microbial synthesis and characterization of poly(3-hydroxybutyrate-*co*-3-hydroxyhexanoate). *Macromolecules* 1995; 28(14): 4822–4828.

103. Chen G-Q, Zhang G, Park S, et al. Industrial scale production of poly(3-hydroxybutyrate-*co*-3-hydroxyhexanoate). *Appl Microbiol Biotechnol* 2001; 57(1–2): 50–55.

104. Amara A, Steinbüchel A, Rehm B. *In vivo* evolution of the *Aeromonas punctata* polyhydroxyalkanoate (PHA) synthase: isolation and characterization of modified PHA synthases with enhanced activity. *Appl Microbiol Biotechnol* 2002; 59: 477–482.

105. Han J, Qiu Y-Z, Liu D-C, et al. Engineered *Aeromonas hydrophila* for enhanced production of poly(3-hydroxybutyrate-*co*-3-hydroxyhexanoate) with alterable monomers composition. *FEMS Microbiol Lett* 2004; 239: 195–201.

106. Tsuge T, Watanabe S, Shimada D, et al. Combination of N149S and D171G mutations in *Aeromonas caviae* polyhydroxyalkanoate synthase and impact on polyhydroxyalkanoate biosynthesis. *FEMS Microbiol Lett* 2007; 277: 217–222.

107. Fukui T, Yokomizo S, Kobayashi G, et al. Co-expression of polyhydroxyalkanoate synthase and (*R*)-enoyl-CoA hydratase genes of *Aeromonas caviae* establishes copolyester biosynthesis pathway in *Escherichia coli*. *FEMS Microbiol Lett* 1999; 170: 69–75.

108. Kichise T, Taguchi S, Doi Y. Enhanced accumulation and changed monomer composition in polyhydroxyalkanoate (PHA) copolyester by in vitro evolution of *Aeromonas caviae* PHA synthase. *Appl Environ Microbiol* 2002; 68(5): 2411–2419.

109. Kawashima Y, Orita I, Nakamura S, et al. Compositional regulation of poly(3-hydroxybutyrate-*co*-3-hydroxyhexanoate) by replacement of granule-associated protein in *Ralstonia eutropha*. *Microb Cell Fact* 2015; 14: 187.

110. Zhang M, Kurita S, Orita I, et al. Modifcation of acetoacetyl-CoA reduction step in *Ralstonia eutropha* for biosynthesis of poly(3-hydroxybutyrate-*co*-3-hydroxyhexanoate) from structurally unrelated compounds. *Microb Cell Fact* 2019; 18: 147.

111. Matsumoto K, Takase K, Yamamoto Y, et al. Chimeric enzyme composed of polyhydroxyalkanoate (PHA) synthases from *Ralstonia eutropha* and *Aeromonas caviae* enhances production of PHAs in recombinant *Escherichia coli*. *Biomacromolecules* 2009; 10(4): 682–685.

112. Ushimaru K, Motoda Y, Numata K, et al. Phasin proteins activate *Aeromonas caviae* polyhydroxyalkanoate (PHA) synthase but not *Ralstonia eutropha* PHA synthase. *Appl Environ Microbiol* 2014; 80(9): 2867–2873.

113. Hanisch J, Waltermann M, Robenek H, et al. The *Ralstonia eutropha* H16 phasin PhaP1 is targeted to intracellular triacylglycerol inclusions in *Rhodococcus opacus* PD630 and *Mycobacterium smegmatis* MC2155, and provides an anchor to target other proteins. *Microbiology* 2006; 152(Pt 11): 3271–3280.

114. Uchino K, Saito T. Thiolysis of poly(3-hydroxybutyrate) with polyhydroxyalkanoate synthase from *Ralstonia eutropha*. *J Biochem* 2006; 139(3): 615–621.

115. Brigham CJ, Budde CF, Holder JW, et al. Elucidation of β-oxidation pathways in *Ralstonia eutropha* H16 by examination of global gene expression. *J Bacteriol* 2010; 192(20): 5454–5464.

116. Tsuge T, Saito Y, Kikkawa Y, et al. Biosynthesis and compositional regulation of poly[(3-hydroxybutyrate)-*co*-(3-hydroxyhexanoate)] in recombinant *Ralstonia eutropha* expressing mutated polyhydroxyalkanoate synthase genes. *Macromol Biosci* 2004; 4(3): 238–242.

117. Fukui T, Doi Y. Cloning and analysis of the poly(3-hydroxybutyrate-*co*-3-hydroxyhexanoate) biosynthesis genes of *Aeromonas caviae*. *J Bacteriol* 1997; 179(15): 4821–4830.

118. Hisano T, Tsuge T, Fukui T, et al. Crystal structure of the (R)-specific enoyl-CoA hydratase from *Aeromonas caviae* involved in polyhydroxyalkanoate biosynthesis. *J Biol Chem* 2003; 278(1): 617–624.

119. Mifune J, Nakamura S, Fukui T. Engineering of pha operon on *Cupriavidus necator* chromosome for efficient biosynthesis of poly(3-hydroxybutyrate-*co*-3-hydroxyhexanoate) from vegetable oil. *Polym Degrad Stab* 2010; 95(8): 1305–1312.

120. Budde CF, Riedel SL, Willis LB, et al. Production of poly(3-hydroxybutyrate-*co*-3-hydroxyhexanoate) from plant oil by engineered *Ralstonia eutropha* strains. *Appl Environ Microbiol* 2011; 77(9): 2847–2854.

121. Csonka LN. Physiological and genetic responses of bacteria to osmotic stress. *Microb Rev* 1989; 53: 121–127.

122. Schneider JC, Jenings AF, Mun DM, et al. Auxotrophic markers *pyrF* and *proC* can replace antibiotic markers on protein production plasmids in high-cell-density *Pseudomonas fluorescens* fermentation. *Biotechnol Prog* 2005; 21: 343–348.

123. Riedel SL, Bader J, Brigham CJ, et al. Production of poly (3-hydroxybutyrate-*co*-3-hydroxyhexanoate) by *Ralstonia eutropha* in high cell density palm oil fermentations. *Biotechnol Bioeng* 2012; 109(1): 74–83.

124. Wong YM, Brigham CJ, Rha C, et al. Biosynthesis and characterization of polyhydroxyalkanoate containing high 3-hydroxyhexanoate monomer fraction from crude palm kernel oil by recombinant *Cupriavidus necator*. *Bioresour Technol* 2012; 121: 320–327.

125. Obruca S, Snajdar O, Svoboda Z, et al. Application of random mutagenesis to enhance the production of polyhydroxyalkanoates by *Cupriavidus necator* H16 on waste frying oil. *World J Microbiol Biotechnol* 2013; 29(12): 2417–2428.

126. Murugan P, Chhajer P, Kosugi A, et al. Production of P(3HB-*co*-3HHx) with controlled compositions by recombinant *Cupriavidus necator* Re2058/pCB113 from renewable resources. *Clean (Weinh)* 2016; 44(9): 1234–1241.

127. Murugan P, Gan C-Y, Sudesh K. Biosynthesis of P (3HB-*co*-3HHx) with improved molecular weights from a mixture of palm olein and fructose by *Cupriavidus necator* Re2058/pCB113. *Int J Biol Macromol* 2017; 102: 1112–1119.

128. Foong CP, Lakshmanan M, Abe H, et al. A novel and wide substrate specific polyhydroxyalkanoate (PHA) synthase from unculturable bacteria found in mangrove soil. *J Polym Res* 2017; 25(1): 23.

129. Lakshmanan M, Foong CP, Abe H, et al. Biosynthesis and characterization of co and ter-polyesters of polyhydroxyalkanoates containing high monomeric fractions of 4-hydroxybutyrate and 5-hydroxyvalerate via a novel PHA synthase. *Polym Degrad Stab* 2019; 163: 122–135.

130. Thinagaran L, Sudesh K. Evaluation of sludge palm oil as feedstock and development of efficient method for its utilization to produce polyhydroxyalkanoate. *Waste Biomass Valorization* 2019; 10(3): 709–720.

131. Wang Y, Yin J, Chen G-Q. Polyhydroxyalkanoates, challenges and opportunities. *Curr Opin Biotechnol* 2014; 30: 59–65.

132. Kourmentza C, Plácido J, Venetsaneas N, et al. Recent advances and challenges towards sustainable polyhydroxyalkanoate (PHA) production. *Bioengineering* 2017; 4: 55.

133. Eberl L, Vandamme P. Members of the genus *Burkholderia*: good and bad guys. *F1000Research* 2016; 5(F1000 Faculty Rev): 1007.

134. Tomás JM. The main *Aeromonas* pathogenic factors. *ISRN Microbiol* 2012; 2012: Article ID 256261.
135. Rietschel E, Teruo Kirikae T, Shade F, et al. Bacterial endotoxin: molecular relationships of structure to activity and function. *FASEB J* 1994; 8: 217–225.
136. Wang X, Quinn P. Endotoxins: lipopolysaccharides of Gram-negative bacteria. In: Wang X, Quinn P (eds.), *Endotoxins: Structure, Function and Recognition.* Springer, Dordrecht, New York, NY, 2010.
137. Chen G-Q, Wu Q. The application of polyhydroxyalkanoates as tissue engineering materials. *Biomaterials* 2005; 26(33): 6565–6578.
138. Valappil SP, Peiris D, Langley GJ, et al. Polyhydroxyalkanoate (PHA) biosynthesis from structurally unrelated carbon sources by a newly characterized *Bacillus* spp. *J Biotechnol* 2007; 127(3): 475–487.
139. Valappil SP, Rai R, Bucke C, et al. Polyhydroxyalkanoate biosynthesis in *Bacillus cereus* SPV under varied limiting conditions and an insight into the biosynthetic genes involved. *J Appl Microbiol* 2008; 104(6): 1624–1635.
140. Valappil SP, Boccaccini AR, Bucke C, et al. Polyhydroxyalkanoates in Gram-positive bacteria: insights from the genera *Bacillus* and *Streptomyces. Anton Leeuw* 2006; 91(1): 1–17.
141. Wakisaka Y, Masaki E, Nishimoto Y. Formation of crystalline delta-endotoxin or poly-beta-hydroxybutyric acid granules by *Asporogenous* mutants of *Bacillus thuringiensis. Appl Environ Microbiol* 1982; 43(6): 1473–1480.
142. Raberg M, Volodina E, Lin K, et al. *Ralstonia eutropha* H16 in progress: applications beside PHAs and establishment as production platform by advanced genetic tools. *Crit Rev Biotechnol* 2017; 38(4): 494–510.
143. Bajpai P. *Single Cell Protein Production from Lignocellulosic Biomass.* Springer, Singapore, 2017.
144. Steinmann J, Wottge H, Müller-Ruchholtz W. Immunogenicity testing of food proteins: *in vitro* and *in vivo* trials in rats. *Int Arch Allergy Immunol* 1990; 91: 62–65.
145. Ritala Λ, Häkkinen S, Toivari M, et al. Single cell protein—state-of-the-art, industrial landscape and patents 2001–2016. *Front Microbiol* 2017; 8: 2009.
146. Chen Y-J, Huang Y-C, Lee C-Y. Production and characterization of medium-chain-length polyhydroxyalkanoates by *Pseudomonas mosselii* TO7. *J Biosci Bioeng* 2014; 118(2): 145–152.
147. Ainie K, Siew WL, Tan YA, et al. Characterization of a by-product of palm oil milling. *Elaeis* 1995; 7(2): 165–173.
148. Kunasundari B, Murugaiyah V, Kaur G, et al. Revisiting the single cell protein application of *Cupriavidus necator* H16 and recovering bioplastic granules simultaneously. *PLoS One* 2013; 8(10): e78528.
149. Kunasundari B, Arza CR, Maurer FHJ, et al. Biological recovery and properties of poly(3-hydroxybutyrate) from *Cupriavidus necator* H16. *Sep Purif Technol* 2017; 172: 1–6.
150. Murugan P, Han L, Gan CY, et al. A new biological recovery approach for PHA using mealworm, *Tenebrio molitor. J Biotechnol* 2016; 239: 98–105.
151. Ong SY, Zainab LI, Pyary S, et al. A novel biological recovery approach for PHA employing selective digestion of bacterial biomass in animals. *Appl Microbiol Biotechnol* 2018; 102: 2117–2127.
152. Ong SY, Kho HP, Riedel SL, et al. An integrative study on biologically recovered polyhydroxyalkanoates (PHAs) and simultaneous assessment of gut microbiome in yellow mealworm. *J Biotechnol* 2018; 265: 31–39.
153. Zainab LI, Sudesh K. High cell density culture of *Cupriavidus necator* H16 and improved biological recovery of polyhydroxyalkanoates using mealworms. *J Biotechnol* 2019; 305: 35–42.

11 Production and Modification of PHA Polymers Produced from Long-Chain Fatty Acids

Christopher Dartiailh, Nazim Cicek, John L. Sorensen, and David B. Levin

CONTENTS

11.1 INTRODUCTION

Polyhydroxyalkanoates (PHA) are microbial storage polyesters produced to sequester carbon in response to nutrient-limiting environments. Upon depletion of an external carbon source, the polymer may later be consumed as a source of energy (ATP) and reducing power (NADH/NADPH). Some PHA have characteristics similar to a range of petroleum-derived polymers [1] and can be produced using renewable resources. PHA are biodegradable due to a breadth of microorganisms that produce external PHA depolymerases or non-specific lipases, which result in environmental decomposition [2–6]. Another important characteristic of PHA is their biocompatibility, making them suitable for medical applications [7]. Taken together, PHA provide sustainable alternatives for current plastic materials. Formed by the enzymatic (*PhaC*-encoded) condensation of hydroxyalkanoate-CoA monomers, PHA polymers are polyesters consisting of a linear backbone, minimally composed of 3-carbon repeating units, with side chains of varying lengths [8]. Short-chain-length (*scl*-) PHA range in length from three to five carbons (C3–C5), and medium-chain-length (*mcl*-)PHA contain monomers from six to fourteen carbons (C6–C14) [9,10].

Long-chain-length (*lcl-*)PHA are longer than C14 but have only been produced in trace quantities during *mcl*-PHA production [11,12].

The production cost of PHA remains a limitation to broader commercial applications. While research has focused on reducing costs in all facets of the production, extraction, and purification processes, the costs and effects of various substrates have drawn the most attention. The poor substrate yield of *in vivo* systems drives up the associated production cost. Although the estimated substrate cost for *scl*-PHA production was reduced to 22% of the total cost when using methane at thermophilic temperatures [13], the estimates for elastomeric *mcl*-PHA production can exceed 50% of the total cost [14–16].

Long-chain fatty acids (LCFAs) are the main components of vegetable oils and a promising source of cheap, renewable substrates for PHA production [17]. LCFAs are highly reduced and provide ATP as well as reducing equivalents to the cell when metabolized. Worldwide vegetable oil production in 2012 eclipsed 150 million tons [18], and food processing has been estimated to produce over one million tons of waste vegetable oil [19]. LCFAs can be used as inexpensive substrates for *scl*-PHA production with high yield and intracellular content. However, the polymers remain saturated and do not contain any functional moieties [20,21]. Saturated PHA can be modified through the reaction of the polymer ends or free radical mechanisms [10], regardless of the substrate used for their synthesis. The literature has emphasized the functionalization of *mcl*-PHA from LCFAs, as the unsaturated moieties of some LCFAs can be retained to obtain olefinic *mcl*-PHA. The theoretical yield of *mcl*-PHA from LCFAs is relatively high. The estimated theoretical substrate yield from the LCFAs of canola oil was 0.72 g *mcl*-PHA per g substrate [22].

LCFAs can be composed of saturated fatty acids (SFAs) and unsaturated fatty acids (USFAs). The USFAs can further be classified as monounsaturated (MUFAs) or polyunsaturated (PUFAs) [23,24]. The fatty acid composition of vegetable oils varies with crops and cultivars [25,26]. However, since fatty acids are largely incorporated into *mcl*-PHA from fatty acid degradation while conserving the olefin position [27,28], they can be classified by their unsaturation level for the sake of predicting *mcl*-PHA composition. Vegetable oils can be highly comprised of SFAs (coconut, palm), MUFAs (olive, canola), or PUFAs (soybean, flax) [29]. The predominant monomers in *mcl*-PHA are C8 and C10, and, as a result, low incorporation of olefin moieties occur from predominant Δ9-monounsaturated fatty acids (oleic acid); the ω-3 unsaturation of linolenic acid found in polyunsaturated fatty acids can be expected in all monomers C8 and longer [30].

Mcl-PHA produced from octanoic, nonanoic, or decanoic acids have been described as elastomeric materials similar to polyethylene [31]. The incorporation of vinyl moieties reduces the crystallinity of these polymers, such that they become completely amorphous with sufficient unsaturation [32]. LCFAs cannot be considered a cheap substrate replacement if the polymer has drastically different properties. However, these novel polymers can be reassessed for alternative applications, and olefin moieties provide opportunities for polymer modification.

The objective of PHA production is to produce sustainable and renewable plastic products capable of replacing current petroleum-derived plastics that are nonbiodegradable and a major source of environmental pollution. The trends in *mcl*-PHA production from LCFAs along with the implications of olefin moieties and their modifications on thermomechanical properties will be reviewed here.

11.2 STRATEGIES FOR PRODUCTION OF *MCL*-PHA

PHA are stored as intracellular granules, and high production rates of polymer require high cell titers. Strategies to improve cell titer include high mixing rates while maintaining high dissolved oxygen (DO) concentrations, often by sparging with pure oxygen or pressurizing the bioreactor [33,34]. The induction of PHA accumulation has been linked to increased intracellular NADH and acetyl-CoA when growth is limited [35]. Nitrogen, phosphate, sulfur, magnesium, and oxygen limitation have all been used to successfully promote PHA synthesis [36,37]. Further improvement of intracellular PHA content has been observed in the dual limitation of nitrogen and oxygen [38]. Contrary to two-phase growth and PHA accumulation models, growth-associated *mcl*-PHA production has been demonstrated at controlled specific growth rates, suggesting the requirement for nutrient limitation is strain- and substrate-dependent [39]. Approaches for reducing PHA production costs focus on maximizing PHA productivity and substrate yield.

11.3 STRATEGIES FOR MAXIMUM VOLUMETRIC PRODUCTIVITY

Stirred-tank bioreactors of various discontinuous or continuous operational configurations have been applied to maximize volumetric productivity [g PHA/(L·h)], thereby improving the cost of PHA production. High biomass titers are required for optimum production rates since PHA are intracellular products, and inhibitory concentrations of medium components limit the productivity of simple batch reactors [40]. Therefore, strategies for fed-batch and continuous PHA production have been developed, which will be briefly summarized here, with a focus on LCFA PHA production, as more comprehensive reviews of PHA bioreactor operations have recently been published [15,36,41,42].

Optimized fed-batch bioreactors have used various feeding regimes to provide the carbon substrate and other nutrients to maximize volumetric productivity, and have provided the highest volumetric productivities [15]. Fed-batch reactors have been more frequently operated using a two-phase feeding regime to first maximize the cell biomass titer before an induction phase for PHA accumulation. A significant challenge in fed-batch processes is determining nutrient delivery to maintain concentrations between limiting and inhibitory levels, as optimized feeding may reduce the impact of other fed-batch challenges (i.e., heat and mass transfer, foaming). Substrates have been delivered for biomass production based on predicted specific growth rates or calculated cumulative substrate consumption using predetermined yield coefficients [43,44]. Alternatively, response-based substrate delivery has been implemented for pH value [37], dissolved oxygen concentration [37,45,46], or carbon dioxide evolution [43,47]. Ultimately, due to the low solubility and diffusional limitations of oxygen into the medium, all these regimes will approach a maximum biomass titer for any reactor configuration due to limited dissolved oxygen [48].

Substrate and nutrient delivery modifications are required for the PHA induction phase as the maximum biomass titer is approached.

Pseudomonas species with growth-associated *mcl*-PHA biosynthesis have been optimized through carbon-limited control of the specific growth rate. In processes using medium-chain fatty acids (MCFAs), such as nonanoic acid, a reduced specific growth rate lowered the oxygen uptake rate such that higher biomass was achieved before the onset of oxygen-limiting conditions when nonanoic acid buildup became toxic [39]. Carbon-limited growth using a quadratic-decaying exponential feed strategy that switched to a linear feed rate, experimentally modeled to optimize growth rate while avoiding oxygen limitation, resulted in the highest reported volumetric productivity of *mcl*-PHA [2.3 g PHA/(L·h)] [44].

High cell density *mcl*-PHA production from LCFAs in fed-batch is limited to a few reports, all of which have distinct growth and PHA production phases. Concomitantly, two fed-batch systems were reported using oleic acid. The highest reported volumetric productivity of *mcl*-PHA from LCFAs was achieved using a combination of oleic acid delivery methods to maintain the maximum specific growth rate, first by monitoring optical density, then by switching to DO control followed by pH control. Following the depletion of phosphate, PHA accumulation rates increased sharply, leading to a final PHA content of 51.4 wt.-% in 141 g/L of biomass for final volumetric productivity of 1.91 g PHA/(L·h) [37]. A similar DO-control fed-batch reactor using oleic acid, but employing nitrogen limitation, resulted in a lower *mcl*-PHA volumetric productivity of 0.57 g PHA/(L·h), which is partly due to the lower biomass obtained, but also because of the much lower PHA content. Curiously, the PHA content reached its maximum early into the cultivation through growth-associated PHA synthesis, long before nitrogen limitation prevented further cell division. Importantly, oleic acid cultivation could be scaled-up into a 30-L reactor with results similar to those in a 2-L reactor [49]. A DO-control fed-batch system for high cell density using phosphorous limitation was used to promote PHA accumulation on corn oil hydrolysate. This resulted in a volumetric *mcl*-PHA accumulation rate of 0.68 g PHA/(L·h) [45]. Recently, a fed-batch approach with oleic acid controlled through nitrogen-limitation to maintain a low growth rate and couple growth to PHA accumulation was reported to improve the PHA content and carbon yield. In this system, biomass of 125.6 g/L containing 54.4 wt.-% *mcl*-PHA was obtained, but the productivity was lower than that reported by Lee et al. [37] due to the length of cultivation [47]. Finally, the monomer composition and thermal properties of *mcl*-PHA using waste rapeseed oil differed with the method of substrate delivery. Pulse-feeding resulted in higher intracellular polymer concentration but lower C12 monomer content, crystallinity, and molecular weights than a continuous feed [50].

Continuous-feed PHA production has been developed using a single bioreactor (single-stage continuous) or with the addition of subsequent reactors to separate the growth and PHA accumulation phases (dual-stage or multi-stage continuous) [41]. A continuous-feed bioreactor may not achieve the same maximum biomass titer or PHA content as a fed-batch reactor. However, by maintaining a high steady-state concentration of PHA long term, continuous PHA production can theoretically result in the highest PHA productivities with simpler feeding control and lower operating costs [51,52]. Dilution rates between 0.1 and 0.3 1/h are optimal for maximizing volumetric productivity. Lower dilution rates increased the PHA content, while higher dilution rates could result in cell washout [40,51,53–56]. Nitrogen limitation is the most prevalent condition for controlling the growth rate and promoting PHA synthesis [54,55]. However, oxygen limitation and dual nutrient-limiting conditions have been reported [53,57,58]. Early single-stage, continuous-feed *mcl*-PHA cultivation resulted in volumetric productivity of 0.17 g PHA/(L·h) [51], and this was subsequently improved up to 0.76 g PHA/(L·h) by increasing the concentration of nitrogen and improving oxygen transfer rates to maintain higher cell density [55].

A two-stage continuous-feed process for *mcl*-PHA production from octane optimized high cell density production in the first stage, which fed into a second reactor with conditions suitable for PHA production. In this manner, the highest continuous volumetric productivity was reported at 1.06 g PHA/(L·h) [54]. From LCFAs, single-stage cultivation using oleic acid was optimized to 0.69 g PHA/(L·h) under oxygen limitation [53]. While the monomer composition of PHA was not affected by the growth rate from octanoate [38,55], the use of oleic acid resulted in a mild shift toward longer monomer composition at low dilution rates, without affecting the molecular weights of the PHA products [53]. While the theoretical potential of continuous cultivation has yet to be realized, continuous-feed PHA production has proven invaluable for its ability to study the effects of growth rate, substrate, and nutrient limitation on factors such as monomer compositions, PHA content, yields, molecular weight, and thermal properties.

11.4 STRATEGIES FOR IMPROVED SUBSTRATE YIELDS FROM MCFAS AND LCFAS

The substrate costs are estimated to account for the largest proportion of the overall techno-economic assessment [57,58]. High volumetric productivities must be balanced with a high substrate yield to minimize production cost. The maximal *mcl*-PHA yield from LCFAs varies depending on the substrate and monomer composition. However, estimates have been reported between 0.58 and 0.72 g/g, which compares to 0.98 g/g from octanoic acid [22,45,53]. The overall PHA yields are typically much lower when considering other process yields, such as biomass production and maintenance [32,38].

Optimizations for high intracellular PHA content were necessary to improve the overall *mcl*-PHA substrate yield, and reduce the PHA extraction and purification costs [53,59]. This was confirmed in a continuous-feed bioreactor operation as nitrogen-limiting conditions with lower dilution rates resulted in an improved *mcl*-PHA substrate yield [57]. The experimental carbon yield was improved by a stepwise decrease in nitrogen feed during the nitrogen-limitation phase, to couple growth with *mcl*-PHA production [47]. Furthermore, strategies to fulfill the substrate requirement for non-PHA biomass, maintenance, and respiration using cheaper substrates have been developed, greatly improving the overall *mcl*-PHA yield from the more expensive substrate [60,61]. The overall *mcl*-PHA yield from octanoate was improved (0.4 g/g) by providing glucose during the exponential growth and adding octanoate during the nutrient limitation of the PHA accumulation phase, however, the PHA content was suboptimal [38].

A similar approach was scaled-up with glucose and nonanoic acid, which resulted in a higher PHA content to yield *mcl*-PHA at 0.56 g/g from nonanoic acid. PHA accumulation began without lag upon the addition of nonanoic acid, supporting the concept of phase-fed fatty acids [62]. Acrylic acid can be provided to inhibit the fatty acid degradation pathway preventing the use of fatty acids for biomass production and further improve on this approach. A continuous reactor containing glucose and nonanoic acid reported an increase in PHA yield from nonanoic acid from 0.15 to 0.90 g/g with the addition of acrylic acid [63]. Similarly, in a carbon-limited fed-batch, volumetric productivity of 1.8 g PHA/(L·h) was reported with an overall *mcl*-PHA yield from nonanoic acid of 0.78 g/g [64]. Since *mcl*-PHA production has a high affinity for C8, C9, and C10 monomers, the use of acrylic acid with fatty acids of the same length can effectively produce *mcl*-PHA [65]. Partial knockout of the fatty acid degradation pathway was effective at producing high C14 monomer content from tetradecanoic acid [66]. Ultimately, blocking the fatty acid degradation pathway may not be effective for increasing the *mcl*-PHA yield from LCFAs, certainly not without drastically changing the monomer composition and production kinetics. Two-reactor systems promise cost-savings through improved substrate yields since no growth occurs in the second PHA-accumulating reactor [52]. This was demonstrated by an overall *mcl*-PHA yield of 0.63 g/g from octane in the second stage, nearly reaching the maximum theoretical yield of 0.66 g/g [55].

11.5 EXTRACELLULAR LIPASE FOR TRIACYLGLYCERIDE CONSUMPTION

Many *Pseudomonads* do not have the extracellular lipase enzymes required for the metabolism of vegetable oils. The growth of these strains from vegetable or animal sources of triacylglycerides (TAGs) required chemical (hydrolysis) or enzymatic (lipases) pretreatment [60,67]. However, *P. aeruginosa*, *P. resinovorans*, *P. chlororaphis*, and other isolated *Pseudomonas* sp. strains have been reported to grow directly from TAGs [17,67–70]. Growth and PHA synthesis directly from TAGs is important because it reduces preprocessing steps. It has been reported that the lipase activity is relatively low in *P. resinovorans* [71]. Fed-batch cultivation of *P. resinovorans* using olive oil deodorizer distillate had a maximum specific growth rate of 0.19 1/h [72], compared to higher rates of 0.55 1/h reported by *P. putida* KT2440 [34]. The apparent decrease in the growth rate for *P. resinovorans* is not necessarily detrimental for PHA productivity when considering high productivities at set growth rates of 0.2 1/h [39] if the cost benefits of growth from TAGs offset productivity loss. Genes encoding the lipase precursor protein (LipA) and lipase chaperone protein (LimA), which together confer the ability to grow and synthesize PHA directly from TAG substrates, have been cloned and expressed in *P. putida* KT2442. The total biomass of the recombinant bacteria was the same, whether cultured with free fatty acids or TAGs, indicating that the lipase activity enabled direct catabolism of the TAGs [73].

11.6 BIOSYNTHESIS AND MONOMER COMPOSITION

Medium-chain-length polyhydroxyalkanoates can be synthesized from a wide range of substrates primarily by species of *Pseudomonas* [74]. *Mcl*-PHA synthesis relies on the microorganism's central fatty acid metabolism, in which fatty acid degradation is the major pathway for *mcl*-PHA production from LCFAs, although it has also been shown to work in concert with fatty acid biosynthesis pathways [75,76]. LCFAs are longer than the monomers typically incorporated into *mcl*-PHA, and, as such, undergo several rounds of fatty acid degradation. Each round of fatty acid degradation produces an $FADH_2$ and NADH and releases acetyl-CoA, which shortens the fatty acid by two carbons (see Figure 11.1). Therefore, continued fatty acid degradation is preferred for energy production while the cells are actively dividing, but PHA production rates increased upon onset of conditions limiting to cell growth [77].

Intermediates of fatty acid degradation can be converted to (R)-3-hydroxyacyl-CoA for polymerization, a process that requires no expenditure of ATP or NADH [47,78]. Thereby, LCFAs can be more directly converted into *mcl*-PHA than substrates that are unrelated to the 3-hydroxyalkanoate subunits that make up PHA polymers (i.e., carbohydrates), resulting in higher yields [79]. Table 11.1 summarizes the monomer compositions of *mcl*-PHA produced from various fatty acids and demonstrates that the activities and specificities of enzymes involved in PHA production vary considerably among *Pseudomonads* given similar substrate and culture conditions.

Three classes of monomer-supplying enzymes are hypothesized to convert fatty acid degradation intermediates to (R)-3-hydroxyacyl-CoA for PHA synthesis: hydratases, reductases, and epimerases. Hydratases, encoding at least four different *phaJ* genes, convert enoyl-CoA to (R)-3-hydroxyalkanoate, which have been confirmed to supply monomers in recombinant hosts (see Figure 11.1) [80–82]. Reductases, encoded by *fabG*, were confirmed to convert 3-ketoacyl-CoA to (R)-3-hydroxyacyl-CoA in recombinant hosts [83,84]. Epimerases, encoded by *fadB* in *E. coli*, as part of the multi-enzyme complex of fatty acid degradation, convert (S)-3-hydroxyacyl-CoA to (R)-3-hydroxyacyl-CoA [85]. The putative epimerase was proposed to become more active in response to a knockout of the (S)-3-hydroxyacyl-CoA dehydrogenase, which prevents further fatty acid degradation [86], and the existence of granule-bound epimerase activity is a possible explanation for *in vitro* polymerization of PHA from (S)-3-hydroxyacyl-CoA [87]. However, PHA monomer-supplying epimerase activity has not been observed in *Pseudomonas* spp. [78].

The activities and specificities of these monomer-supplying enzymes vary with acyl length [81], and since the expression levels of these monomers can be expected to change with microbial strain and stress response, the monomer composition of *mcl*-PHA may depend on the monomer-supplying enzymes to a greater extent than the specificity of the PHA synthase [88]. However, the PHA synthase genes, *phaC1* and *phaC2*, have been shown to have varying monomer specificities [89]. A knockout of the native PHA synthases of *Pseudomonas* sp. resulted in a combination of *scl*-PHA and *mcl*-PHA monomers when provided an alternative PHA synthase [90]. *E. coli* was provided with the *fabG* monomer-providing enzyme, which resulted in

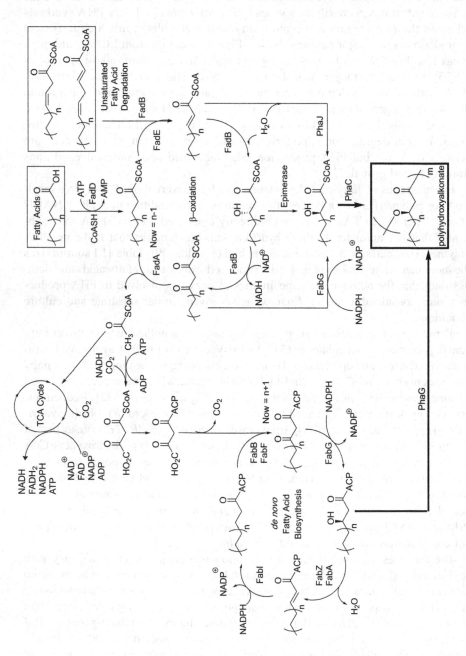

FIGURE 11.1 Biosynthetic pathway for *mcl*-PHA synthesis from fatty acid substrates.

TABLE 11.1

Monomer Composition of *mcl*-PHA Produced by *Pseudomonas* Spp. from Fatty Acids

Substrate	Microorganism [Ref.]	Monomer Composition (mol-%)												
		C_4	C_6	C_8	$C_{8:1}$	C_{10}	$C_{10:1}$	C_{12}	$C_{12:1}$	$C_{12:2}$	C_{14}	$C_{14:1}$	$C_{14:2}$	$C_{14:3}$
Octanoic Acid (C_8)	*P. putida* KT2442 [75]	–	6	92	–	2	–	–	–	–	–	–	–	–
Decanoic Acid (C_{10})	*P. putida* KT2442 [27]	–	5.3	52.3	–	42.3	–	–	–	–	–	–	–	–
Palmitic Acid (C_{16})	*Pseudomonas* sp. DR2 [17]	–	3.4	18.7	–	35.6	–	37.0	–	–	–	–	–	–
Petroselinic ($C_{18:1}$)	*P. putida* KT2442 [24]	–	5.1	45.2	–	33.1	–	12.7	Tr	–	3.9	Tr	–	–
Oleic Acid ($C_{18:1}$)	*P. putida* KT2442 [24]	–	4.4	33.5	–	32.2	–	14.4	Tr	–	–	15.5	–	–
Oleic Acid ($C_{18:1}$)	*P. aeruginosa* 27853 [64]	–	4	55	–	27	–	8	–	–	6	–	–	–
Oleic Acid ($C_{18:1}$)	*P. aeruginosa* 42A2 (30°C) [124]	–	–	24.3	–	30.8	–	3.8	24.2	–	–	3.7	13.1	–
Oleic Acid ($C_{18:1}$)	*P. aeruginosa* 42A2 (37°C) [124]	–	–	2.2	–	39	–	4.7	25.6	–	–	2.1	26.5	–
Oleic Acid ($C_{18:1}$)	*P. resinovorans* B-2649 [71]	3.5	7	37	–	33	–	10	–	–	1	8	–	–
Linoleic Acid ($C_{18:2}$)	*P. putida* KT2442 [26]	–	5.6	38.9	–	22.7	–	–	16.9	–	–	–	15.9	–
Erucic Acid ($C_{22:1}$)	*P. aeruginosa* 27853 [68]	–	3	43	–	36	–	10	–	–	–	8	–	–
Nervonic Acid ($C_{24:1}$)	*P. aeruginosa* 27853 [68]	–	4	28	–	43	–	14	–	–	–	11	–	–
Coconut Oil	*P. resinovorans* B-2649 [69]	–	8	37	–	35	–	17	–	–	3	–	–	–
Canola Fatty Acids	*P. putida* LS46 [22]	–	5.4	41.4	–	26.7	–	9.1	3.6	–	9.3	8.6	–	–
Soybean Oil	*P. resinovorans* B-2649 [69]	Tr	8	29	Tr	30	2	5	9	2	2	2	8	–
Flax Fatty Acids*	*P. putida* KT2442 [30]	–	4.7	23.3	11.4	16.9	9.3	3.9	3.2	~5.5	Tr	2.7	~5.5	~13.6
Linseed Oil	*P. aeruginosa* 42A2 [125]	–	0.4	33.7	5.6	24.3	7.0	4.9	1.4	4.8	0.4	6.4	2.8	8.3

* Values have been assigned as approximate as $C_{12:2}$ and $C_{14:2}$ were combined as 11%, and the 13.6% of $C_{14:3}$ included some $C_{16:3}$.

mcl-PHA when provided an *mcl-phaC* but produced PHB-*co*-PHHx given an *scl-phaC* [84]. Both these examples indicate that the monomer-supplying enzymes provided a variety of monomer lengths, and the selectivity was due to the PHA synthase. Ultimately, the monomer composition is dependent on a variety of factors, but *mcl*-PHA is produced with a preference for C8 and C10 monomers due to enzyme specificities, as shown in Table 11.1.

Despite the aforementioned monomer preferences, the low substrate specificity of *mcl*-PHA synthases accounts for the incorporation of over 150 monomer types [12,91]. The monomer composition of *mcl*-PHA will differ by species and is determined both by the substrate and the culturing conditions. The length of the fatty acid influences the PHA monomer length [68], and functional groups may be incorporated into *mcl*-PHA from the substrate [12]. The functional moieties in LCFAs can be retained during *mcl*-PHA production, depending on their position in the substrate [28,68,73].

The double bonds of MUFAs and PUFAs are removed as the chain is shortened using the fatty acid degradation pathway. The enzyme cis-3,trans-2,enoyl-CoA isomerase is responsible for removing the odd-carbon double bonds, while even-numbered double bonds (such as the $\Delta 12$ olefin group of linoleic acid) are removed by 2,4-dienoyl-CoA reductase [28]. In this way, one may predict the position of double bonds in the PHA polymer based on the substrate provided and the monomer length. For instance, 15.5 mol-% of *mcl*-PHA synthesized by *P. putida* KT2442 grown with oleic acid ($\Delta 9$) was monounsaturated, as the olefin was maintained in the C14 monomers, compared with no retained olefins from petroselinic acid ($\Delta 6$). Growth on linoleic acid ($\Delta 9$, $\Delta 12$) tripled the unsaturation of PHA as both olefins were maintained in the C14, and one olefin remained in C12 [28]. Moreover, PHA produced from linolenic acid ($\Delta 9$, $\Delta 12$, $\Delta 15$) produced highly unsaturated PHA containing mono-(C8, C10), di-(C12), and poly-(C14) unsaturated monomers [30]. Table 11.1 illustrates that the average monomer length and mol-% of unsaturated monomers increased proportionally with the length and unsaturation of the substrate. Hydroxyl and epoxy moieties have also been observed in *mcl*-PHA when *P. aeruginosa* 44T1 was provided with castor or euphorbia oil [92]. *Mcl*-PHA have incorporated halogen, hydroxyl, carboxyl, thiol, epoxy, aromatic, and branched moieties, to name a few, using the appropriate fatty acids [11,12,91]. These functional properties provide the basis for modification of unsaturated *mcl*-PHA, but their inclusion also broadens the applicable chemical modifications.

Culture conditions affect polymer composition due to a shift in the central fatty acid metabolism. As the fatty acids in the bacterial membrane change in length and unsaturation in response to incubation temperature, so too does the *mcl*-PHA composition [76]. PHA production using LCFAs at a lower incubation temperature resulted in a shift toward longer monomers with higher unsaturation. As the unsaturated monomers were not consistent with the LCFAs, it was inferred that the lower temperature increased the incorporation of *mcl*-PHA monomers from *de novo* fatty acid synthesis. The same study also demonstrated a drastic shift in monomer composition based on changing from nitrogen limitation to phosphate limitation [93]. All the above factors ultimately affect the thermal and mechanical properties of *mcl*-PHA.

11.7 FUNCTIONAL MODIFICATIONS OF *MCL*-PHA

The incorporation of vinyl moieties into *mcl*-PHA reduces their crystallinity, ultimately weakening their mechanical properties, but also imparting functionality to the polymer. Chemical modifications have imbued unsaturated *mcl*-PHA with strength or with new properties targeted at niche biomedical applications. The remainder of this chapter discusses the modification of *mcl*-PHA to attain novel properties.

Mcl-PHA with vinyl moieties inherited from MUFAs and PUFAs can be tailored by choice of microorganism and culture conditions, but most simply by the substrate delivery (Table 11.1). Vegetable oils vary in their fatty acid composition, and co-feeding strategies can be applied to control the relative unsaturation [22]. LCFAs appear to increase chain termination during *mcl*-PHA polymerization resulting in lower molecular weights with increasing unsaturation [69]. The glass transition temperatures (T_g) and melting point (T_m) decrease with higher unsaturation in the LCFA substrate [94]. Commonly, 10-undecenoic acid is provided for *mcl*-PHA (PHU) production with terminal vinylic carbons in the side chains [95,96]. Varying the feed ratio of octanoic acid and 10-undecenoic acid resulted in polyhydroxyoctanoate-*co*-undecenoate (PHOU) with an unsaturated monomer concentration equal to the substrate ratio and no effect on molecular weight. When co-feeding octanoate with 10-undecenoate in a continuous steady-state bioreactor, lower dilution rates increased the relative concentration of aliphatic monomers while molecular weights were not affected by the growth rate [56]. Increased feed ratios of 10-undecenoic acid resulted in a decreased glass transition temperature and melting temperature of the *mcl*-PHA and became amorphous with high unsaturation [56]. The same trends were observed with octane:octene ratios, having a decreasing melt endotherm until completely amorphous at 15 mol-% unsaturation [32].

The modification of *mcl*-PHA containing side-chain vinyl groups has produced polymers with new properties (Table 11.2). Chlorination across the double bond drastically elevated the glass transition temperature from −50°C to +58°C with melting temperatures more consistent with *scl*-PHA than *mcl*-PHA, despite hydrolysis resulting in lower molecular weights. The observed changes in these polymers were from sticky to a soft, elastic polymer with moderate chlorine addition. The high chlorine content in the polymers was described as crystalline and brittle. While crystallinity was not measured in that study, these highly chlorinated polymers were observed well below their glass transition temperatures [97].

Carboxylation and hydroxylation at the terminal vinyl position of PHOU have both been demonstrated to change the solubility of the polymer with no reduction in molecular weight [98,99]. Increasing unsaturation content in the PHOU resulted in polymers with higher polarity after modification until the polymers were no longer soluble in the organic solvent [96,99,100]. Epoxidation of PHOU was performed without a reduction in molecular weight or cross-linking. The glass transition temperature was lower with increased epoxidation, but the polymer remained amorphous. The addition of epoxides provided an avenue for cross-linking or producing amphiphilic polymers [94,100,101]. Copolymer grafting has been achieved via free radical polymerization with the side-chain vinyl moieties of unsaturated *mcl*-PHA from soybean fatty acids.

TABLE 11.2

Functional Modification of Unsaturated *mcl*-PHA

Reaction	Described Effect
Carboxylation	Increasing carboxylation results in higher hydrophilicity. This functional moiety is also a precursor for other modifications (see "click-ready") [97,99].

PHOU — KMnO$_4$, 18-crown-6, CH$_2$Cl$_2$, CH$_3$COOH

PHOU — OsO$_4$ in *t*-BuOH, Oxone, DMF 60 °C

Hydroxylation

PHOU — KMnO$_4$, Acetone, NaHCO$_3$, i)H$_2$O/Na$_2$CO$_3$, ii)H$_2$SO$_4$

Hydroxylation increased the hydrophilicity of the polymer, becoming insoluble in organic solvents [100].

Epoxidation

PHOU — MCPBA, CH$_2$Cl$_2$

Reduced glass transition temperature. Precursor to cross-linking, or for further modification (see "transamination") [96,102,103].

Halogenation

LCFA mcl-PHA — Cl$_2$, CCl$_4$

Higher polymer chlorine content resulted in elevated glass transition and melting temperatures [99]. Halogenation provides the leaving group for further substitutions (see "RAFT").

Amination

PHOU — MCPBA, CH$_2$Cl$_2$; Diethanolamine, THF

Complete change in solubility from hydrophobic to hydrophilic. Transamination resulted in a significant decrease in molecular weight due to PHA chain scission [126].

(Continued)

TABLE 11.2 (CONTNIUED)
Functional Modification of Unsaturated *mcl*-PHA

Reaction	Described Effect
RAFT Derivatization	Xanthate substitution converts unsaturated PHA into macro RAFT agents. RAFT polymerization using *N*-isopropyl acrylamide produced thermo-responsive, amphiphilic polymers with glass transition temperatures between 58°C and 100°C [106].
Click-Ready Terminal Alkyne Derivatization	CuAAC-ready *mcl*-PHA for diverse applications. Demonstrated improved PEG grafting compared with previous methodology, resulting in improved crystallinity [127].
Thiol-ene Addition	Thiol-ene reaction with PHOU eliminates polymer premodification for "clickable" functionalization. Significantly broadens types of functional groups and grafting while avoiding chain scission of cross-linking [110,112–115].

Methyl methacrylate (MMA) initiated by benzoyl peroxide formed a copolymer graft with PHA (PHA-*g*-PMMA) in which cross-linking could be prevented with the addition of hydroquinone. Soybean *mcl*-PHA is not in a glassy state and exhibits no crystallinity. However, the grafted polymer became hard and brittle, a property inherited by the glass transition temperature of poly(MMA) (PMMA). Grafted copolymers were produced with higher tensile strength and elongation at break than either homopolymer [102]. Graft polymers were alternatively produced by first activating the MMA or styrene using an oligoperoxide to produce activated PMMA or polystyrene (PS), respectively. Mixing the activated polymers into unsaturated *mcl*-PHA produced PHA-*g*-PMMA and PHA-*g*-PS; however, the PHA-*g*-PS cross-linked [103]. Bromination across the vinyl moieties of unsaturated *mcl*-PHA followed by xanthate substitution, which resulted in macro reversible addition-fragmentation chain transfer (RAFT) agents. RAFT polymerization using N-isopropyl acrylamide produced thermo-responsive, amphiphilic polymers with glass transition temperatures between 58°C and 100°C [104].

Click chemistry can be applied to the vinyl groups of *mcl*-PHA derived from unsaturated fatty acid substrates, which drastically increases the number of modification permutations. Click-ready *mcl*-PHA were first produced by converting PHOU to contain terminal *R*-group carboxylic acids, then by esterification with propargyl alcohol yielding a "clickable" terminal alkyne. The objective was to produce amphiphilic polymers through polyethylene glycol (PEG) grafting. PHA-*g*-PEG was produced using PEG-azide producing polymers using longer PEG oligomers than could be grafted with previous direct esterification methods, ultimately increasing the molecular weight and crystallinity of this copolymer [105]. The terminal alkyne produced enables copper-catalyzed azide-alkyne [3+2] cycloaddition (CuAAC) with molecules containing azide groups.

Instead of modifying the vinyl groups of *mcl*-PHA to introduce click moieties, *mcl*-PHA were cultured with terminal azide groups. The resulting *mcl*-PHA are CuAAC-ready without the need for polymer modification. This was achieved with ω-azido fatty acid substrates delivered to engineered *E. coli*, producing *mcl*-PHA with a yield, molecular weights, and thermal properties consistent with PHOU production [106]. *P. oleovorans* co-fed nonanoate and 11-bromoundecanoic acid produced polymers with terminal bromine groups, which allowed azide substitution. Instead of a terminal azide, the azide-alkyne reaction could be achieved with terminal alkyne *mcl*-PHA produced when co-fed nonanoate and 10-undecynoic acid [107]. Furthermore, a strain-promoted cycloaddition was demonstrated, eliminating the copper requirement of CuAAC [104,105].

Another approach that uses click chemistry without metallic catalysts is thiol-ene addition. In this case, the desired functional molecules contain pendant thiol moieties, which undergo anti-Markovnikov additions to the side-chain vinyl groups of unsaturated *mcl*-PHA. An increase in hydrophilicity without a reduction in the molecular weight of PHOU and unsaturated *mcl*-PHA from soybean LCFAs was achieved when hydroxylated and carboxylated in this manner [108]. Poly(3-hydroxybutyrate)-*co*-undecenoate (PHBU) was cross-linked using a polythiol [pentaerythritol tetrakis (3-mercaptopropionate)] for the thiol-ene click reaction to increase both the elongation and tensile strength compared with native PHBU [109].

A pendant sulfonate addition to PHOU resulted in a dramatic change in solubility, becoming insoluble in organic solvents above 5 mol-% sulfonated monomers and self-aggregated into nanoparticle micelles with sizes dependent on sulfonate concentration [110]. Furthermore, the use of thiol-ene reactions has produced various grafted copolymers from PHOU. Amphiphilic polymers have been designed for medical applications, such as drug delivery by grafting, using thiol-ene click reactions. Jeffamine® grafting onto PHOU increased the hydrophilicity of the polymer [111]. The sequential grafting of fluorinated chains and PEG onto PHOU produced multi-compartment micelles [112]. The same procedure substituting fluorinated chains with sulfonated chains produced amphiphilic polymers coated onto nanometal organic frameworks to produce stable hybrid nanoparticles that displayed no cytotoxicity [113].

11.8 CROSS-LINKING

Cross-linking of *mcl*-PHA involves radical propagation along olefinic moieties, forming a polymer network. Unsaturated *mcl*-PHA have been reported to auto-oxidize and have had cross-linking initiated by reactive peroxides or radiation [32,114–116]. Cross-linking is the result of either C-C, ether, or peroxy bonds, although ether bonds appear to be dominant in cross-linking during auto-oxidation [117,118]. Hydrogen atom abstraction is most likely to occur in the allylic position, at which time the radical reacts with oxygen resulting in a peroxyl radical. The peroxyl radical can abstract another allylic hydrogen until the reaction is finally terminated [119]. The degradation of the hydroperoxide into an epoxide precedes ether bond linkages [118]. The rate of cross-linking can be increased with heat, irradiation, supplementation of oxygen, or with increasing unsaturation. Chain scission and cross-linking will occur simultaneously, but the ester cleavage requires much higher activation energy (almost three-fold), therefore, cross-linking is promoted before chain scission [116,120].

Cross-linking of *mcl*-PHA polymers resulted in markedly different physical and thermal properties. When the octane-based polymer was irradiated, only chain scission occurred and caused a reduction in molecular weight. Irradiation of a polymer with 15 mol-% unsaturated monomers (amorphous) caused cross-linking, and the polymer became solid and less sticky. The cross-linked polymers then showed a constant dynamic modulus from the glass transition temperatures (–15°C to –30°C) to the onset of thermal degradation (170°C), whereas the octane-based polymer showed a significant amount of variation around ambient temperature and softened at 40°C due to melting. The tear resistance of the cross-linked, amorphous polymer was poor, with much less tensile strength than the crystalline, saturated polymer. However, the cross-linked PHA was still biodegradable [32]. Irradiation of *mcl*-PHA with 11 mol-% monounsaturated monomers synthesized by *P. resinovorans* (NRRL B-2649) from tallow made it slightly stronger and more rigid than its crystalline counterpart. However, this further increased the cross-linking density resulting in a decrease in tear resistance, and the soluble fraction showed molecular weight reductions up to 70% [114]. *Mcl*-PHA synthesized from coconut oil (95% saturated), tallow (37% unsaturated oleic acid), and soybean (86% unsaturated or polyunsaturated) fatty acids all displayed increased tensile strength after irradiation.

Mcl-PHA polymers synthesized from soybean oil produced a solid film with a higher Young's modulus than the irradiated polymers produced from coconut or tallow feedstocks [120]. Linseed oil derived *mcl*-PHA were cross-linked by two methods: chemically induced cross-links with meta-chloroperoxybenzoic acid (m-CPBA) and naturally induced cross-links by exposure to air (auto-oxidation). Chemical treatment with m-CPBA resulted in cross-linking in less than 25 days, and by that time, 98% of the polymer was solid and insoluble. Cross-linking by auto-oxidation took between 50 and 75 days under ambient conditions. However, the auto-oxidized PHA had higher tensile strength and was more brittle than the chemically cross-linked polymers. It was suggested that m-CPBA treatment results in ether cross-links, whereas auto-oxidation resulted in carbon-carbon cross-links conferring more strength to the polymer [101].

PHA films were synthesized by *P. oleovorans* grown with various ratios of octanoic, 10-undecenoic acid, and soybean acids followed by cross-linking treatment. Large variations in tensile strength and elongation at break were observed, such that the film characteristics could be tailored by substrate feeding. In all cases, the films were biocompatible, but each elicited different magnitudes of inflammation. They concluded that variation in film properties and degradation rates allow for diverse medical applications [121,122]. A cross-linked network of PHOU-*g*-PEG was produced by UV irradiation to determine the effect on swelling for drug delivery. The addition of PEG into the cross-linked polymer reduced the tensile strength and elongation to break with increasing PEG concentration and increased the degree of cross-linking and hydrophilicity. PEG further improved the biocompatibility of *mcl*-PHA by reducing platelet and protein interactions [123–127].

11.9 CONCLUSIONS AND OUTLOOK

The abundant supply of LCFAs from renewable and waste sources, coupled with high biomass production rates and substrate yields, improves the cost efficiency of sustainable PHA production. Furthermore, the use of unsaturated LCFAs for *mcl*-PHA production inserted functional moieties into the polymer. The content of these moieties can be controlled through cultivation conditions to tailor the composition of the *mcl*-PHA. These unsaturated *mcl*-PHA provide the opportunity to be modified at their vinyl groups (i.e., halogenation, hydroxylation, carboxylation, grafting, cross-linking), and platforms for the modification of these unsaturated *mcl*-PHA continue to emerge. Further exploration and characterization will enable tailored-made PHA for stronger films and amphiphilic polymers for medical applications.

REFERENCES

1. Madison LL, Huisman GW. Metabolic engineering of poly(3-hydroxyalkanoates): From DNA to plastic. *Microbiol. Mol. Biol. Rev.* 1999; 63(1): 21–53.
2. Jendrossek D, Handrick R. Microbial degradation of polyhydroxyalkanoates. *Annu. Rev. Microbiol.* 2002; 56(1): 403–432.
3. Knoll M, Hamm TM, Wagner F, et al. The PHA depolymerase engineering database: A systematic analysis tool for the diverse family of polyhydroxyalkanoate (PHA) depolymerases. *BMC Bioinform.* 2009; 10(89): 1–8.
4. Chuah JA, Yamada M, Taguchi A, et al. Biosynthesis and characterization of poly-hydroxyalkanoate containing 5-hydroxyvalerate units: Effects of 5HV units on biodegradability, cytotoxicity, mechanical and thermal properties. *Polym. Degrad. Stab.* 2013; 98(1): 331–338.
5. Doi Y, Kanesawa Y, Kunioka M, et al. Biodegradation of microbial copolyesters: Poly(3-hydroxybutyrate-co-3-hydroxyvalerate) and poly(3-hydroxybutyrate-co-4-hydroxy butyrate). *Macromolecules* 1990; 23(1): 26–31.
6. Jendrossek D, Knoke I, Habibian RB, et al. Degradation of poly(3-hydroxybutyrate), PHB, by bacteria and purification of a novel PHB depolymerase from *Comamonas* sp. *J. Environ. Polym. Degrad.* 1993; 1(1): 53–63.
7. Bonartsev AP, Myshkina VL, Nikolaeva DA, et al. Biosynthesis, biodegradation, and application of poly (3-hydroxybutyrate) and its copolymers – Natural polyesters produced by diazotrophic bacteria. *Commun. Curr. Res. Educ. Top. Trends Appl. Microbiol.* 2007; 1: 295–307.
8. Lageveen RG, Huisman GW, Preusting H, et al. Formation of polyesters by *Pseudomonas oleovorans*: Effect of substrates on formation and composition of poly-(*R*)-3-hydroxyalkanoates and poly-(*R*)-3-hydroxyalkenoates. *Appl. Environ. Microbiol.* 1989; 54(12): 2924–2932.
9. Lee SY. Bacterial polyhydroxyalkanoates. *Biotechnol. Bioeng.* 1996; 49: 1–14.

10. Hazer B, Steinbüchel A. Increased diversification of polyhydroxyalkanoates by modification reactions for industrial and medical applications. *Appl. Microbiol. Biotechnol.* 2007; 74(1): 1–12.
11. Steinbüchel A, Valentin HE. Diversity of bacterial polyhydroxyalkanoic acids. *FEMS Microbiol. Lett.* 1995; 128(3): 219–228.
12. Kim DY, Kim HW, Chung MG, et al. Biosynthesis, modification, and biodegradation of bacterial medium chain length polyhydroxyalkanoates. *J. Microbiol.* 2007; 45(2): 87–97.
13. Levett I, Birkett G, Davies N, et al. Techno-economic assessment of poly-3-hydroxybutyrate (PHB) production from methane – The case for thermophilic bioprocessing. *J. Environ. Chem. Eng.* 2016; 4(4): 3724–3733.
14. Choi JI, Lee SY. Process analysis and economic evaluation for poly(3-hydroxybutyrate) production by fermentation. *Bioprocess Eng.* 1997; 17(6): 335–342.
15. Blunt W, Levin DB, Cicek N. Bioreactor operating strategies for improved polyhydroxyalkanoate (PHA) productivity. *Polymers* 2018; 10(11): 1197.
16. Koller M, Maršálek L, Miranda de Sousa Dias M, et al. Producing microbial polyhydroxyalkanoate (PHA) biopolyesters in a sustainable manner. *New Biotechnol.* 2017; 37: 24–38.
17. Song JH, Jeon CO, Choi MH, et al. Polyhydroxyalkanoate (PHA) production using waste vegetable oil by *Pseudomonas* sp. strain DR2. *J. Microbiol. Biotechnol.* 2008; 18(8): 1408–1415.
18. Kerenkan AE, Béland F, Do TO. Chemically catalyzed oxidative cleavage of unsaturated fatty acids and their derivatives into valuable products for industrial applications: A review and perspective. *Catal. Sci. Technol.* 2016; 6(4): 971–987.
19. Chhetri AB, Watts KC, Islam MR. Waste cooking oil as an alternate feedstock for biodiesel production. *Energies* 2008; 1: 3–18.
20. Ciesielski S, Mozejko J, Pisutpaisal N. Plant oils as promising substrates for polyhydroxyalkanoates production. *J. Clean. Prod.* 2015; 106: 408–421.
21. Koçer H, Borcakli M, Demirel S, et al. Production of bacterial polyesters from some various new substrates by *Alcaligenes eutrophus* and *Pseudomonas oleovorans*. *Turk. J. Chem.* 2003; 27(3): 365–373.
22. Blunt W, Dartiailh C, Sparling R, et al. Carbon flux to growth or polyhydroxyalkanoate synthesis under microaerophilic conditions is affected by fatty acid chain-length in *Pseudomonas putida* LS46. *Appl. Microbiol. Biotechnol.* 2018; 102(15): 6437–6449.
23. Chowdhury K, Banu LA, Khan S, et al. Studies on the fatty acid composition of edible oils. *Bangladesh J. Sci. Ind. Res.* 2007; 42(3): 311–316.
24. Awogbemi O, Onuh EI, Inambao FL. Comparative study of properties and fatty acid composition of some neat vegetable oils and waste cooking oils. *IJLCT* 2019; 14: 417–425.
25. Gruzdienė D, Anelauskaitė E. Chemical composition and stability of rapeseed oil produced from various cultivars grown in Lithuania. *JFAE* 2010; 39(61): 10–13.
26. Ramos MJ, Fernández CM, Casas A, et al. Influence of fatty acid composition of raw materials on biodiesel properties. *Bioresour. Technol.* 2009; 100: 261–268.
27. Eggink G, de Waard PE, Huijberts GNM. The role of fatty acid biosynthesis and degradation in the supply of substrates for poly (3-hydroxyalkanoate) formation in *Pseudomonas putida*. *FEMS Microbiol. Rev.* 1992; 103(2): 159–163.
28. de Waard P, van der Wal H, Huijberts GN, et al. Heteronuclear NMR analysis of unsaturated fatty acids in poly(3-hydroxyalkanoates). Study of beta-oxidation in *Pseudomonas putida*. *J. Biol. Chem.* 1993; 268(1): 315–319.
29. Bart J, Palmeri N, Cavallero S. Feedstocks for biodiesel production. In: *Biodiesel Science and Technology*. Woodhead Publishing Series in Energy, 2010: 130–225.
30. Casini E, de Rijk TC, de Waard P, et al. Synthesis of poly(hydroxyalkanoate) from hydrolyzed linseed oil. *J. Environ. Polym. Degrad.* 1997; 5(3): 153–158.

31. Antonio RV, Steinbüchel A, Rehm BHA. Analysis of *in vivo* substrate specificity of the PHA synthase from *Ralstonia eutropha*: Formation of novel copolyesters in recombinant *Escherichia coli*. *FEMS Microbiol. Lett.* 2000; 182(1): 111–117.

32. de Koning GJM, van Bilsen HMM, Lemstra PJ, et al. A biodegradable rubber by crosslinking poly(hydroxyalkanoate) from *Pseudomonas oleovorans*. *Polymer* 1994; 35(10): 2090–2097.

33. Sun Z, Ramsay JA, Guay M, et al. Fermentation process development for the production of medium-chain-length poly-3-hyroxyalkanoates. *Appl. Microbiol. Biotechnol.* 2007; 75(3): 475–485.

34. Follonier S, Henes B, Panke S, et al. Putting cells under pressure: A simple and efficient way to enhance the productivity of medium-chain-length polyhydroxyalkanoate in processes with *Pseudomonas putida* KT2440. *Biotechnol. Bioeng.* 2012; 109(2): 451–461.

35. Ren Q, de Roo G, Ruth K, et al. Simultaneous accumulation and degradation of polyhydroxyalkanoates: Futile cycle or clever regulation. *Biomacromolecules* 2009; 10(4): 916–922.

36. Koller M, Braunegg G. Potential and prospects of continuous polyhydroxyalkanoate (PHA) production. *Bioengineering* 2015; 2(2): 94–121.

37. Lee SY, Wong HH, Choi JI, et al. Production of medium-chain-length polyhydroxyalkanoates by high-cell-density cultivation *Pseudomonas putida* under phosphorus limitation. *Biotechnol. Bioeng.* 2000; 68(4): 466–470.

38. Kim GJ, Lee IY, Yoon SC, et al. Enhanced yield and a high production of medium-chainlength poly(3-hydroxyalkanoates) in a two-step fed-batch cultivation of *Pseudomonas putida* by combined use of glucose and octanoate. *Enzyme Microb. Technol.* 1997; 20(7): 500–505.

39. Sun Z, Ramsay JA, Guay M, et al. Carbon-limited fed-batch production of medium-chain-length polyhydroxyalkanoates from nonanoic acid by *Pseudomonas putida* KT2440. *Appl. Microbiol. Biotechnol.* 2009; 82(4): 657–662.

40. Ramsay BA, Saracovan I, Ramsay JA, et al. Continuous production of long-side-chain poly-3-hydroxyalkanoates by *Pseudomonas oleovorans*. *Appl. Environ. Microbiol.* 1991; 57(3): 625–629.

41. Koller M. A review on established and emerging fermentation schemes for microbial production of polyhydroxyalkanoate (PHA) biopolyesters. *Fermentation* 2018; 4(2): 30.

42. Kaur G, Roy I. Strategies for large-scale production of polyhydroxyalkanoates. *Chem. Biochem. Eng. Q.* 2015; 29(2): 157–172.

43. Sun Z, Ramsay JA, Guay M, et al. Automated feeding strategies for high-cell-density fed-batch cultivation of *Pseudomonas putida* KT2440. *Appl. Microbiol. Biotechnol.* 2006; 71(4): 423–431.

44. Maclean H, Sun Z, Ramsay JA, et al. Decaying exponential feeding of nonanoic acid for the production of medium-chain-length poly (3-hydroxyalkanoates) by *Pseudomonas putida*. *Can. J. Chem.* 2008; 86(1): 564–569.

45. Shang L, Jiang M, Yun Z, et al. Mass production of medium-chain-length poly(3-hydroxyalkanoates) from hydrolyzed corn oil by fed-batch culture of *Pseudomonas putida*. *World J. Microbiol. Biotechnol.* 2008; 24(12): 2783–2787.

46. Kellerhals MB, Hazenberg W, Witholt B. High cell density fermentations of *Pseudomonas oleovorans* for the production of *mcl*-PHAs in two-liquid phase media. *Enzyme Microb. Technol.* 1999; 24(1): 111–116.

47. Andin N, Longieras A, Veronese T, et al. Improving carbon and energy distribution by coupling growth and medium chain length polyhydroxyalkanoate production from fatty acids by *Pseudomonas putida* KT2440. *Biotechnol. Bioprocess Eng.* 2017; 22(3): 308–318.

48. Vendruscolo F, Rossi MJ, Schmidell W, et al. Determination of oxygen solubility in liquid media. ISRN Chem. Eng. 2012; 2012: 1–5.

49. Kellerhals MB, Kessler B, Witholt B, et al. Renewable long-chain fatty acids for production of biodegradable medium-chain-length polyhydroxyalkanoates (*mcl*-PHAs) at laboratory and pilot plant scales. *Macromolecules* 2000; 33(13): 4690–4698.

50. Mozejko J, Ciesielski S. Pulsed feeding strategy is more favorable to medium-chain-length polyhydroxyalkanoates production from waste rapeseed oil. *Biotechnol. Prog.* 2014; 30(5): 1243–1246.

51. Preusting H, Hazenberg W, Witholt B. Continuous production of poly(3-hydroxyalkanoates) by *Pseudomonas oleovorans* in a high-cell-density, two-liquid-phase chemostat. *Enzyme Microb. Technol.* 1993; 15(4): 311–316.

52. Hartmann R, Hany R, Witholt B, et al. Simultaneous biosynthesis of two copolymers in *pseudomonas putida* GPO1 using a two-stage continuous culture system. *Biomacromolecules* 2010; 11(6): 1488–1493.

53. Huijberts GNM, Eggink G. Production of poly(3-hydroxyalkanoates) by *Pseudomonas putida* KT2442 in continuous cultures. *Appl. Microbiol. Biotechnol.* 1996; 46(3): 233–239.

54. Jung K, Hazenberg W, Prieto M, et al. Two-stage continuous process development for the production of medium-chain-length poly(3-hydroxyalkanoates). *Biotechnol. Bioeng.* 2001; 72(1): 19–24.

55. Hazenberg W, Witholt B. Efficient production of medium-chain-length poly(3-hydroxyalkanoates) from octane by *Pseudomonas oleovorans*: Economic considerations. *Appl. Microbiol. Biotechnol.* 1997; 48(5): 588–596.

56. Hartmann R, Hany R, Pletscher E, et al. Tailor-made olefinic medium-chain-length poly[(R)-3-hydroxyalkanoates] by *Pseudomonas putida* GPo1: Batch versus chemostat production. *Biotechnol. Bioeng.* 2006; 93(4): 737–746.

57. Durner R, Witholt B, Egli T. Accumulation of poly[(R)-3-hydroxyalkanoates] in *Pseudomonas oleovorans* during growth with octanoate in continuous culture at different dilution rates. *Appl. Environ. Microbiol.* 2000; 66(8): 3408–3414.

58. Durner R, Zinn M, Witholt B, Egli T. Accumulation of poly[(R)]-3-hydroxyalkanoates] in *Pseudomonas oleovorans* during growth in batch and chemostat culture with different carbon sources. *Biotechnol. Bioeng.* 2001; 72(3): 278–288.

59. Chanprateep S. Current trends in biodegradable polyhydroxyalkanoates. J. Biosci. Bioeng. 2010; 110(6): 621–632.

60. Jiang G, Hill D, Kowalczuk M, et al. Carbon sources for polyhydroxyalkanoates and an integrated biorefinery. *Int. J. Mol. Sci.* 2016; 17(7): 1157.

61. Choi J, Lee SY. Factors affecting the economics of polyhydroxyalkanoate production by bacterial fermentation. *Appl. Microbiol. Biotechnol.* 1999; 51(1): 13–21.

62. Davis R, Duane G, Kenny S, et al. High cell density cultivation of *Pseudomonas putida* KT2440 using glucose without the need for oxygen enriched air supply. *Biotechnol. Bioeng.* 2015; 112(4): 725–733.

63. Jiang X, Sun Z, Marchessault RH, et al. Biosynthesis and properties of medium-chain-length polyhydroxyalkanoates with enriched content of the dominant monomer. *Biomacromolecules* 2012; 13(9): 2926–2932.

64. Jiang XJ, Sun Z, Ramsay JA, et al. Fed-batch production of MCL-PHA with elevated 3-hydroxynonanoate content. *AMB Expr.* 2013; 3(1): 50.

65. Qi Q, Steinbüchel A, Rehm BH. Metabolic routing towards polyhydroxyalkanoic acid synthesis in recombinant *Escherichia coli* (fadR): Inhibition of fatty acid {β}-oxidation by acrylic acid. *FEMS Microbiol. Lett.* 1998; 167: 89–94.

66. Liu W, Chen GQ. Production and characterization of medium-chain-length polyhydroxyalkanoate with high 3-hydroxytetradecanoate monomer content by *fadB* and *fadA* knockout mutant of *Pseudomonas putida* KT2442. *Appl. Microbiol. Biotechnol.* 2007; 76(5): 1153–1159.

67. Povolo S, Romanelli MG, Fontana F, et al. Production of polyhydroxyalkanoates from fatty wastes. *J. Polym. Environ.* 2012; 20(4): 944–949.
68. Impallomeni G, Ballistreri A, Carnemolla GM, et al. Synthesis and characterization of poly(3-hydroxyalkanoates) from *Brassica carinata* oil with high content of erucic acid and from very long chain fatty acids. *Int. J. Biol. Macromol.* 2011; 48(1): 137–145.
69. Ashby RD, Foglia TA. Poly(hydroxyalkanoate) biosynthesis from triglyceride substrates. *Appl. Microbiol. Biotechnol.* 1998; 49(4): 431–437.
70. Sharma PK, Munir RI, de Kievit T, et al. Synthesis of polyhydroxyalkanoates (PHAs) from vegetable oils and free fatty acids by wild-type and mutant strains of *pseudomonas chlororaphis. Can. J. Microbiol.* 2017; 63(12): 1009–1024.
71. Lee JH, Ashby RD, Needleman DS, et al. Cloning, sequencing, and characterization of lipase genes from polyhydroxyalkanoate (PHA)-synthesizing *Pseudomonas resinovorans. Appl. Microbiol. Biotechnol.* 2012; 96: 993–1005.
72. Cruz MV, Araújo D, Alves VD, et al. Characterization of medium chain length polyhydroxyalkanoate produced from olive oil deodorizer distillate. *Int. J. Biol. Macromol.* 2016; 82: 243–248.
73. Solaiman DKY, Ashby RD, Foglia TA. Production of polyhydroxyalkanoates from intact triacylglycerols by genetically engineered *Pseudomonas. Appl. Microbiol. Biotechnol.* 2001; 56(5): 664–669.
74. López NI, Pettinari MJ, Nikel PI, et al. Polyhydroxyalkanoates: much more than biodegradable plastics. *Adv. Appl. Microbiol.* 2015; 93: 73–106.
75. Huijberts GNM, Eggink G, de Waard P, et al. 13C Nuclear magnetic resonance studies of *Pseudomonas putida* fatty acid metabolic routes involved in poly(3-hydroxyalkanoate) synthesis. *Synthesis* 1994; 176(6): 1661–1666.
76. Huijberts GNM, Eggink G, de Waard P, et al. *Pseudomonas putida* KT2442 cultivated on glucose accumulates poly(3-hydroxyalkanoates) consisting of saturated and unsaturated monomers. *Appl. Environ. Microbiol.* 1992; 58(2): 536–544.
77. Blunt W, Dartiailh C, Sparling R, et al. Microaerophilic environments improve the productivity of medium chain length polyhydroxyalkanoate biosynthesis from fatty acids in *Pseudomonas putida* LS46. *Process Biochem.* 2017; 59: 18–25.
78. Fiedler S, Steinbüchel A, Rehm BH. The role of the fatty acid β-oxidation multienzyme complex from *Pseudomonas oleovorans* in polyhydroxyalkanoate biosynthesis: Molecular characterization of the *fadBA* operon from *P. oleovorans* and of the enoyl-CoA hydratase genes *phaJ* from *P. oleovorans* and *P. putida. Arch. Microbiol.* 2002; 178(2): 149–160.
79. Blunt W, Lagasse A, Jin, Z. Efficacy of medium chain-length polyhydroxyalkanoate biosynthesis from different biochemical pathways under oxygen-limited conditions using *Pseudomonas putida* LS46. *Process Biochem.* 2019; 82: 19–31.
80. Davis R, Chandrashekar A, Shamala TR. Role of (R)-specific enoyl coenzyme A hydratases of *Pseudomonas* sp in the production of polyhydroxyalkanoates. *Anton. Leeuw.* 2008; 93(3): 285–296.
81. Tsuge T, Taguchi K, Taguchi S, et al. Molecular characterization and properties of (R)-specific enoyl-CoA hydratases from *Pseudomonas aeruginosa*: Metabolic tools for synthesis of polyhydroxyalkanoates via fatty acid B-oxidation. *Int. J. Biol. Macromol.* 2003; 31(4): 195–205.
82. Fukui T, Doi Y. Cloning and analysis of the poly(3-hydroxybutyrate-co-3-hydroxy hexanoate) biosynthesis genes of *Aeromonas caviae. J. Bacteriol.* 1997; 179(15): 4821–4830.
83. Ren Q, Sierro N, Witholt B, et al. FabG, an NADPH-dependent 3-ketoacyl reductase of *Pseudomonas aeruginosa*, provides precursors for medium-chain-length poly-3-hydroxyalkanoate biosynthesis in *Escherichia coli. J. Bacteriol.* 2000; 182(10): 2978–2981.

84. Taguchi K, Aoyagi Y, Matsusaki H, et al. Co-expression of 3-ketoacyl-ACP reductase and polyhydroxyalkanoate synthase genes induces PHA production in *Escherichia coli* HB101 strain. *FEMS Microbiol. Lett.* 1999; 176(1): 183–190.

85. DiRusso C.C. Primary sequence of the *Escherichia coli fadBA* operon, encoding the fatty acid-oxidizing multienzyme complex, indicates a high degree of homology to eucaryotic enzymes. *J. Bacteriol.* 1990; 172(11): 6459–6468.

86. Steinbüchel A. Perspectives for biotechnological production and utilization of biopolymers: Metabolic engineering of polyhydroxyalkanoate biosynthesis pathways as a successful example. *Macromol. Biosci.* 2001; 1(1): 1–24.

87. Kraak MN, Smits THM, Kessler B, et al. Polymerase C1 levels and poly (R-3-hydroxyalkanoate) synthesis in wild-type and recombinant *Pseudomonas* strains. *J. Bacteriol.* 1997; 179(16): 4985–4991.

88. Lageveen RG, Huisman G, Preusting H, et al. Formation of polyesters by *Pseudomonas oleovorans*: Effect of substrates on formation and composition of poly-(R)-3-hydroxy-alkanoates and poly-(R)-3-hydroxyalkenoates. *Appl. Environ. Microbiol.* 1989; 54: 2924–2932.

89. Qi Q, Rehm BHA, Steinbüchel A. Synthesis of poly(3-hydroxyalkanoates) in *Escherichia coli* expressing the PHA synthase gene *phaC2* from *Pseudomonas aeruginosa*: Comparison of PhaC1 and PhaC2. *FEMS Microbiol. Lett.* 1997; 157: 155–162.

90. Sharma P, Munir R, Blunt W, et al. Synthesis and physical properties of polyhydroxyalkanoate polymers with different monomer compositions by recombinant *Pseudomonas putida* LS46 expressing a novel PHA synthase (PhaC1$_{16}$) enzyme. *Appl. Sci.* 2017; 7(3): 242.

91. Steinbüchel A, Lütke-Eversloh T. Metabolic engineering and pathway construction for biotechnological production of relevant polyhydroxyalkanoates in microorganisms. *Biochem. Eng. J.* 2003; 16: 81–96.

92. Eggink G, de Waard P, Huijberts GNM. Formation of novel poly(hydroxyalkanoates) from long-chain fatty acids. *Can. J. Microbiol.* 1995; 41: 14–21.

93. Haba E, Vidal-Mas J, Bassas M, et al. Poly 3-(hydroxyalkanoates) produced from oily substrates by *Pseudomonas aeruginosa* 47T2 (NCBIM 40044): Effect of nutrients and incubation temperature on polymer composition. *Biochem. Eng. J.* 2007; 35(2): 99–106.

94. Park WHO, Lenz RW, Goodwin S. Epoxidation of bacterial polyesters with unsaturated side chains. II. Rate of epoxidation and polymer properties. *J. Polym. Sci. Pt. A Polym. Chem.* 1998; 36(13): 2381–2387.

95. Kurth N, Brachet F, Robic D, et al. Poly(3-hydroxyoctanoate) containing pendant carboxylic groups for the preparation of nanoparticles aimed at drug transport and release. *Polymer* 2001; 43(4): 1095–1101.

96. Follonier S, Riesen R, Zinn M. Pilot-scale production of functionalized *mcl*-PHA from grape pomace supplemented with fatty acids. *Chem. Biochem. Eng. Q.* 2015; 29(2): 113–121.

97. Arkin AH, Hazer B, Borcakli M. Chlorination of poly(3-hydroxyalkanoates) containing unsaturated side chains. *Macromolecules* 2000; 33(9): 3219–3223.

98. Lee MY, Park WH, Lenz RW. Hydrophilic bacterial polyesters modified with pendant hydroxyl groups. *Polymer* 2000; 41(5): 1703–1709.

99. Stigers DJ, Tew GN. Poly(3-hydroxyalkanoate)s functionalized with carboxylic acid groups in the side chain. *Biomacromolecules* 2003; 4(2): 193–195.

100. Bear MM, Leboucher-Durand MA, Langlois V, et al. Bacterial poly-3-hydroxyalkeno-ates with epoxy groups in the side chains. *React. Funct. Polym.* 1997; 34(1): 65–77.

101. Ashby RD, Foglia TA, Solaiman DKY, et al. Viscoelastic properties of linseed oil-based medium chain length poly (hydroxyalkanoate) films: Effects of epoxidation and curing. *Int. J. Biol. Macromol.* 2000; 27: 355–361.

102. Ilter S, Hazer B, Borcakli M, et al. Graft copolymerisation of methyl methacrylate onto a bacterial polyester containing unsaturated side chains. *Macromol. Chem. Phys.* 2001; 202(11): 2281–2286.

103. Cakmakli B, Hazer B, Borcakli M. Poly(styrene peroxide) and poly(methyl methacrylate peroxide) for grafting on unsaturated bacterial polyesters. *Macromol. Biosci.* 2001; 1(8): 348–354.

104. Hazer B. Synthesis and characterization of the novel thermoresponsive conjugates based on poly (3-hydroxyalkanoates). *J. Polym. Environ.* 2014; 22: 159–166.

105. Babinot J, Renard E, Langlois V. Preparation of clickable poly(3-hydroxyalkanoate) (PHA): Application to poly(ethylene glycol) (PEG) graft copolymers synthesis. *Macromol. Rapid Commun.* 2010; 31(7): 619–624.

106. Pinto A, Ciesla JH, Palucci A, et al. Chemically intractable no more: *In vivo* incorporation of 'click'-ready fatty acids into poly-[(R)-3-hydroxyalkanoates] in *Escherichia coli. ACS Macro Lett.* 2016; 5(2): 215–219.

107. Nkrumah-Agyeefi S, Scholz C. Chemical modification of functionalized polyhydroxyalkanoates via 'Click' chemistry: A proof of concept. *Int. J. Biol. Macromol.* 2017; 95: 796–808.

108. Hazer B. Simple synthesis of amphiphilic poly (3-hydroxy alkanoate)s with pendant hydroxyl and carboxylic groups via thiol-ene photo click. *Polym. Degrad. Stab.* 2015; 119: 159–166.

109. Levine AC, Heberlig GQ, Nomura CT. Use of thiol-ene click chemistry to modify mechanical and thermal properties of polyhydroxyalkanoates (PHAs). *Int. J. Biol. Macromol.* 2016; 83: 358–365.

110. Modjinou T, Lemechko P, Babinot J, et al. Poly(3-hydroxyalkanoate) sulfonate: From nanoparticles toward water soluble polyesters. *Eur. Polym. J.* 2015; 68: 471–479.

111. Le Fer G, Babinot J, Versace DL, et al. An efficient thiol-ene chemistry for the preparation of amphiphilic PHA-based graft copolymers. *Macromol. Rapid Commun.* 2012; 33(23): 2041–2045.

112. Babinot J, Renard E, Droumaguet L, et al. Facile synthesis of multicompartment micelles based on biocompatible poly(3-hydroxyalkanoate). *Macromol. Rapid Commun.* 2013; 34(4): 362–368.

113. Jain-Beuguel C, Li X, Houel-Renault L, et al. Water-soluble poly(3-hydroxyalkanoate) sulfonate: Versatile biomaterials used as coatings for highly porous nano-metal organic framework. *Biomacromolecules* 2019; 20(9): 3324–3332.

114. Ashby RD, Cromwick AM, Foglia TA. Radiation crosslinking of a bacterial medium-chain-length poly(hydroxyalkanoate) elastomer from tallow. *Int. J. Biol. Macromol.* 1998; 23(1): 61–72.

115. Hazer B, Demirel SI, Borcakli M, et al. Free radical crosslinking of unsaturated bacterial polyesters obtained from soybean oily acids. *Polym. Bull.* 2001; 46(5): 389–394.

116. Schmid M, Ritter A, Grubelnik A. Autoxidation of medium chain length polyhydroxyalkanoate. *Biomacromolecules* 2007; 8(2): 579–584.

117. Muizebelt WJ, Hubert JC, Venderbosch RAM. Mechanistic study of drying of alkyd resins using ethyl linoleate as a model substance. *Prog. Org. Coatings* 1994; 24(1): 263–279.

118. Mallegol J, Gardette JL, Lemaire J. Long-term behaviour of oil-based varnishes and paints. Fate of hydroperoxides in drying oils. *JACS* 2000; 77(3): 249–255.

119. Simic MG. Free radical mechanisms in autoxidation processes. *J. Chem. Educ.* 1981; 58(2): 125.

120. Ashby RD, Foglia TA, Liu C, et al. Improved film properties of radiation-treated medium-chain-length poly(hydroxyalkanoates). *Biotechnol. Lett.* 1998; 20(11): 1047–1052.

121. Hazer DB, Hazer B, Kaymaz F. Synthesis of microbial elastomers based on soybean oily acids. Biocompatibility studies. Biomed. Mater. 2009; 4(3): 1–9.

122. Hazer B, Hazer DB, Çoban B. Synthesis of microbial elastomers based on soybean oil. Autoxidation kinetics, thermal and mechanical properties. J. Polym. Res. 2010; 17(4): 567–577.

123. Chung CW, Kim HW, Kim YB, et al. Poly(ethylene glycol)-grafted poly(3-hydroxyundecenoate) networks for enhanced blood compatibility. Int. J. Biol. Macromol. 2003; 32(1): 17–22.

124. Fernández D, Rodriguez E, Bassas M, et al. Agro-industrial oily wastes as substrates for PHA production by the new strain Pseudomonas aeruginosa NCIB 40045: Effect of culture conditions. Biochem. Eng. J. 2005; 26(2): 159–167.

125. Bassas M, Diaz J, Rodriguez E, et al. Microscopic examination in vivo and in vitro of natural and cross-linked polyunsaturated mclPHA. Appl. Microbiol. Biotechnol. 2008; 78(4): 587–596.

126. Sparks J, Scholz C. Synthesis and characterization of a cationic poly(β-hydroxyalkanoate). Biomacromolecules 2008; 9(8): 2091–2096.

127. Babinot J, Guigner JM, Renard E, et al. A micellization study of medium chain length poly(3-hydroxyalkanoate)-based amphiphilic diblock copolymers. J. Colloid Interface Sci. 2012; 375(1): 88–93.

12 Converting Petrochemical Plastic to Biodegradable Plastic

Tanja Narancic, Nick Wierckx,
Si Liu, and Kevin E. O'Connor

CONTENTS

12.1 INTRODUCTION: THE PLASTIC WASTE ISSUE

The economies of developed countries so far were based on a continuous growth model that relies on a steady supply of inexpensive resources into a linear supply chain. Versatile, cheap, strong, flexible, and light, plastics are dominant materials in this linear economy. Plastics are fundamental in our everyday lives and well placed to promote new developments in transportation, packaging, agriculture, medical devices, and smart and efficient buildings. So, it is not surprising that global plastic production reached 348 million metric tons in 2017, representing a 13-million metric ton increase compared with 2016 production [1,2]. This powerful industry employed 1.5 million people in the EU and generated a turnover of €355 billion in 2017 [1,3]. The mass production of plastics began after World War II and was influential in the shift toward a single-use culture for commodity items. Some plastic products have a shelf-life of less than a year, while others can be used for decades. Therefore, different plastics will have different lifecycles, and some plastic products will contribute more to waste generation than others. The packaging sector is the largest market for plastics, contributing nearly 40% to total plastics demand [4]. These single-use commodity materials greatly contribute to modern lifestyle, but also create a great environmental concern. For example, the annual demand for polyethylene tere-phthalate (PET) is approximately 33 million tons [2], approximately 7.5% of the total annual plastic demand. While PET bottles are used as an example of success-ful recycling, globally, only 7% of PET produced annually is recycled [5], with a large majority of plastic waste still going to landfill [4,6,7]. Poly(propylene) (PP), poly(ethylene) (PE) (including high, medium, and low density), and poly(styrene) (PS) heavily used in short shelf-life packaging represented 19.3%, 29.8%, and 6.6% of the plastics demands in 2018, respectively [1].

The options to reduce plastic waste include prevention, reuse, recycling, and energy recovery [8,9]. Recycling is viewed as the first option for post-consumer plas-tic packaging, and while plastic waste recycling has seen an increase of almost 75% in the last 10 years, only 30% of the collected waste gets recycled in the EU [1]. This is likely because recycled plastics are often more expensive or of lower quality than virgin plastics. Currently, energy recovery through incineration, gasification, and pyrolysis have the highest impact on plastic waste elimination [2,10]. However, these technologies do not create circularity in the plastics economy.

The circular economy of plastics relies on recycling and reuse as the main pillars to retain the value in the material cycle. However, globally, 80% of all plastic pro-duced has not been reused by society due to the lack of recycling and reuse culture for these materials [2]. The gap in the circular economy thinking for plastics could be filled by focusing on the role of biodegradable plastics and the recycling of plastic waste into a material such as biodegradable plastic. This creates an opportunity to improve both resource efficiency and contribute to a circular economy [11] where biodegradable plastics can be composted, and the compost returned to the soil where plants are grown or used to make biobased and biobased biodegradable plastics, so the cycle is repeated.

Biodegradability depends on the bonds in a polymer but also the material prop-erties of the plastic (e.g., crystallinity), as well as on the environmental condi-tions (temperature, pH, humidity) where these plastics may occur post-consumer. Biodegradable plastics such as thermoplastic starch (TPS), polylactic acid (PLA), and polyhydroxyalkanoate (PHA) can be derived from renewable resources and are thus biobased and biodegradable plastics. However, it is important to mention that there are other biodegradables derived from fossil carbon, for example, polycapro-lactone (PCL) and polybutylene adipate terephthalate (PBAT) [12]. TPS makes up 50% of all bioplastics on the market, with PLA also a major biodegradable bioplastic on the market, with 10.3% of global bioplastic production capacities in 2018 [13]. This is followed by PBAT (7.2%), PBS (4.6%), and PHA (1.4%). Among these three, PHA is the only biopolymer that is completely the product of microbial metabolism, including monomer production and polymerization.

12.2 STRATEGIES FOR UPCYCLING PLASTIC WASTE

Plastic waste has a low or negative value to society, and using this waste as a resource to produce biodegradable plastic addresses two very important aspects: the cost of the substrate for the production of biodegradable plastic and waste management.

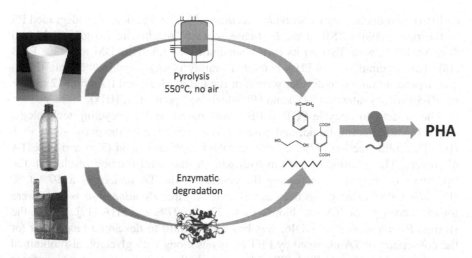

FIGURE 12.1 Strategies for plastic waste upcycling. Polystyrene (PS), polyethylene terephthalate (PET), and polyethylene can be broken down via pyrolysis or enzymatic degradation to monomer constituents: styrene, terephthalic acid (TA), ethylene glycol (EG), and alkanes, and subsequently converted to polyhydroxyalkanoates (PHA) using *Pseudomonas* strains.

Due to the extreme recalcitrance of petrochemical plastics to biodegradation processes [14], the biotechnological strategy to upcycle plastic waste to biodegradable plastics is a two-step process in which petrochemical plastic is firstly depolymerized and then used as a feedstock for the biodegradable plastic producing microorganisms (see Figure 12.1).

PS was the first plastic waste material for which a chemo-biotechnological process was developed for its conversion into the biodegradable plastic, PHA [15]. In this process, polystyrene was pyrolyzed at 520°C to yield oil that contained a high fraction of styrene (83%) and low levels of other aromatic substrates [15]. The pyrolysis oil was used as a feedstock for *Pseudomonas putida* CA-3 strain, and the conversion of styrene to medium-chain-length (*mcl*)-PHA was demonstrated [15]. In a stirred tank bioreactor, 16 g of styrene oil was converted to 2.8 g/L cell dry mass (CDM) with 40% PHA [15]. The PHA had the highest molar fraction of (*R*)-3-hydroxydecanoate, followed by (*R*)-hydroxyoctanoate and (*R*)-hydroxyhexanoate, which is a typical monomer distribution when an unrelated carbon substrate is used for the production of PHA. The conversion of PS to PHA was further improved by varying the nitrogen feed, which resulted in a three-fold increased yield of PHA per g of supplied PS [16]. In these experiments, styrene oil was supplied to the stirred tank reactor through the gaseous phase by using an additional airflow, controlled by a mass flow controller, which was passed through styrene oil into the bioreactor vessel [15]. By changing the feed of gaseous styrene to liquid styrene through the air sparger, further four-fold improvement in PHA production was achieved, yielding 3.36 g/L of PHA and biomass of 10.6 g/L [17].

Recently, a study by Johnston et al. reported the use of a prodegraded PS, PS flakes subjected to thermal oxidation, for the growth and short-chain-length PHA (scl-PHA) production by *Cupriavidus necator* [18]. The addition of prodegraded PS to the rich medium TSB in the best-case scenario resulted in 3.6 g/L CDM and 48% scl-PHA, with TSB on its own contributing to 1.6 g/L CDM and 17% PHB [18]. The accumulated scl-PHA consisted mainly of (R)-3-hydroxybutyrate (3HB) and depending on the molecular weight of the fed prodegraded PS, up to 12 mol-% of (R)-3-hydroxyvalerate (3HV) and (R)-3-hydroxyhexanoate (3HHx).

The process to upcycle PET to PHA was based on PS upcycling technology. Three fractions, solid, liquid, and gaseous, were obtained by the pyrolysis of PET [19]. The solid fraction (77%) contained 51% terephthalic acid (TA) and 20% TA oligomers. The addition of sodium hydroxide to the solid fraction resulted in the hydrolysis of oligomers, increasing the content of the TA molecules to 97 wt.-% [19]. When this solution was used as a feedstock, three *Pseudomonas* isolates were found to convert the TA into biomass and 23% to 27% mcl-PHA [19]. One of the strains, *Pseudomonas* sp. GO16, was brought forward to develop a bioprocess for the conversion of TA obtained by PET pyrolysis along with glycerol, also obtained from waste streams, into PHA [20]. The main challenge in developing a bioprocess for the conversion of TA into PHA is limited TA solubility. The co-feeding strategy allowed for better utilization of TA and overall improved biomass and mcl-PHA accumulation. When TA was used as a sole source of carbon and energy in a bioreactor, *Pseudomonas* sp. GO16 reached 8.7 g/L CDM, with 30% of it being mcl-PHA [20]. The highest biomass and PHA yield (g/L) and productivity [g/(L·h)] result was obtained with a strategy in which waste glycerol (WG) was used as a substrate for biomass accumulation during the growth phase, and TA as a substrate for PHA accumulation when nitrogen became limited. However, in this strategy, TA use was lower compared to the strategy in which both WG and TA were used for biomass and PHA accumulation [20]. Finally, the monomer composition of the accumulated PHA was dependent on the co-feeding strategy, and thus the polymer properties were affected by the applied feeding strategy [20]. Therefore, the variation of substrate feeding strategies allows for the tailoring of polymer properties. For example, when WG is used along with TA in the polymer accumulation phase, the resulting polymer is tacky. In contrast, the polymer produced using feed strategies where only TA is supplied in the polymer accumulation phase exhibits little or no tackiness [20].

PE is the most produced polymer globally; it has a broad range of physicochemical properties, and it is used in a range of commodities, including plastic bags, bubble wrap, farm plastics, water pipes, toys, and many others. Consequently, post-consumer plastic waste contains PE products that pose an environmental pollution risk. We have developed a two-step chemo-biotechnological process for the conversion of post-consumer PE into PHA [21]. In this process, agricultural PE waste was firstly converted to PE pyrolysis wax, comprised of a range of low molecular weight paraffins [21]. Aliphatic hydrocarbons represented 90% of this mixture, while 10% were alkenes, and less than 1% were found to be aromatic hydrocarbons. Among the strains tested for growth with PE wax, *P. oleovorans* B-14682 and *P. aeruginosa* GL-1 were found to use PE wax as a sole carbon and energy source and produced low levels of mcl-PHA [21]. Further improvement in PHA accumulation was achieved by increasing the concentration of PE wax and by switching from NH_4Cl as a nitrogen source to NH_4NO_3 as NO_3^- is known to increase the production of rhamnolipids, such as surfactants. Finally, the addition of exogenous rhamnolipids produced by *P. aeruginosa* GL-1 yielded 0.4 g/L CDM with 19 wt.-% *mcl*-PHA [21].

Recently, the conversion of low-density PE (LDPE) into *scl*-PHA by *C. necator* and *mcl*-PHA by *P. putida* LS46 and *Acinetobacter pittii* IRN19 was demonstrated [22]. LDPE was supplied as a powder to the medium, and it was directly converted to PHA without the need for pretreatment of PE. This report is a great steppingstone for process development in which LDPE could be upcycled to biodegradable PHA.

12.3 ENZYMATIC DEGRADATION OF PETROCHEMICAL PLASTICS

The features of plastics that make them so desirable, namely durability and stability, are at the same time the main challenges associated with their potential for biodegradation. These materials were introduced into the natural environment 60 years ago, and microorganisms have not had enough time to evolve highly efficient enzymes capable of degrading petrochemical plastic. Only limited mineralization, the conversion into biomass, CO_2, H_2O, or CH_4, of petrochemical plastics has been reported [23]. Enzymes identified in the breakage of polymers belong to cutinases, laccases, lipases, and enzymes involved in lignin metabolism [23–27]. These enzymes are emerging as candidates for the development of biocatalytic plastic recycling processes.

In 2016, Yoshida et al. reported the isolation of *Ideonella sakaiensis* 201-F6 with the unusual ability to degrade PET and assimilate its monomers [28]. The capacity of this bacterium for PET degradation is limited; it exhibits very slow degradation of a PET material with significantly lower crystallinity than PET used in packaging. The higher the crystallinity of PET, the more difficult it will be to degrade enzymatically. The key enzymes of *I. sakaiensis* involved in PET degradation, designated as PETase and MHETase [29–31], were characterized in detail and have great potential for further engineering for improved biocatalytic activity in plastic degradation [32].

The improvements to enzymatic hydrolysis of plastics are ongoing [31,32], and one can envisage these enzyme technologies being used in combination with bacterial metabolism of the enzyme-generated monomers to have a complete biotechnological upcycling of plastic waste to PHA and other valuable products.

12.4 METABOLISM OF PLASTIC'S MONOMERS AND THE CONNECTION WITH PHA

12.4.1 METABOLISM

Anthropogenic plastics have only been disseminated into the natural environment in the last century, which is an extremely short period on an evolutionary timescale. This is in contrast to cellulose, which is the most abundant polymer in nature, the metabolism of which has over a billion years of evolutionary history behind it [33]. Despite this, some microbes and enzymes that can degrade plastics, such as PET and polyurethane (PU), have recently been discovered [28,34]. It is highly likely that this enzymatic degradation of plastic is a moonlighting activity of older enzymes, as illustrated by the PET degrading activity of cutinases [26,35]. The same goes for the associated monomer metabolic pathways.

All strategies described here concern a single type of plastic. However, plastic waste will contain a mixture of various polymers, and the capacity of bacteria to convert this mixed plastic waste into PHA will heavily depend on their metabolic capacity to catabolize plastic's monomers. We, therefore, need to have an in-depth understanding of the metabolism of plastic's monomers, which will allow us to engineer it further and improve the process of plastic waste upcycling.

12.4.2 METABOLISM OF STYRENE

Styrene is a simple alkylbenzene used as a starting material for the synthesis of PS and as a solvent in the polymer processing industry. It naturally occurs as a product of fungal decarboxylation of cinnamic acid [36], and this pathway has also been used to enable its biobased production [37]. Microbial pathways for styrene degradation can be found in numerous bacterial species, including *Pseudomonas*, *Rhodococcus*, *Nocardia*, *Xanthobacter*, and *Enterobacter* [38]. Both aerobic and anaerobic degradation of styrene have been described; however, anaerobic degradation is less understood.

In aerobic degradation, styrene is either oxidized to phenylacetic acid via styrene oxide or undergoes direct ring cleavage via 3-vinylcatechol [38]. The oxidation of the side-chain is a hallmark of styrene metabolism in *Pseudomonas*, and this pathway is considered as a styrene-specific pathway [39] (see Figure 12.2). The styrene catabolic operon in *Pseudomonas* contains *styA* and *styB* that encode a two-subunit styrene mono-oxygenase responsible for the transformation of styrene to styrene oxide. In contrast, *styC* encodes an epoxystyrene isomerase that converts styrene oxide to phenylacetylaldehyde [38]. Finally, *styD* encodes dehydrogenase that converts phenylacetylaldehyde into phenylacetic acid. This so-called upper styrene degradation pathway is regulated by a two-component signal transduction system and shows typical carbon source dependent repression of the catabolism of aromatic substrates [40,41]. The lower pathway of styrene degradation, which involves oxidation of the aromatic nucleus of phenylacetyl-CoA followed by ring cleavage, is independently regulated and produces acetyl-CoA [42]. Acetyl-CoA is then funneled into the tricarboxylic acid (TCA) cycle allowing for biomass formation and the *de novo* fatty acid synthesis pathway for PHA accumulation [42]. This is the "indirect" PHA synthesis route that links the catabolism of carbon sources that are structurally unrelated to (*R*)-3-hydroxyalkanoic acids to fatty acid and PHA metabolism [43]. The monomer composition of PHA, in this case, will depend on the strain and PhaC's properties.

The direct ring cleavage is styrene non-specific, and it is found in some *Rhodococcus* and *Pseudomonas* strains [44]. Styrene is oxidized by the activity of an NAD$^+$-dependent dehydrogenase, likely also active with toluene [38]. The product of this oxidation, 3-vinyl-catechol, is further *meta*-cleaved to pyruvate and acetaldehyde [38].

12.4.3 METABOLISM OF TA

TA is converted to protocatechuate (PCA) by the activity of terephthalate dioxygenase and a dehydrogenase (see Figure 12.3). This pathway was identified in *Comamonas*

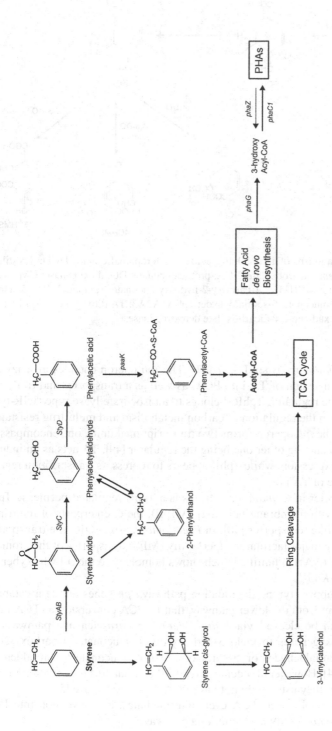

FIGURE 12.2 Aerobic metabolism of styrene: side-chain oxidation and direct ring cleavage. Side-chain oxidation. *styAB*: two-component styrene mono-oxygenase gene; *styC*: styrene oxide isomerase gene; *styD*: phenylacetaldehyde dehydrogenase gene; *paaK*: phenylacetyl-CoA ligase gene. Direct ring cleavage. NAD⁺-dependent dehydrogenase, catechol 2,3–dioxygenase; *phaG*: 3-hydroxyacyl-ACP-CoA transacylase gene; *phaC1*: class II *mcl*-PHA synthase gene; *phaZ*: PHA depolymerase gene. TCA: tricarboxylic acid cycle; PHA: polyhydroxyalkanoates.

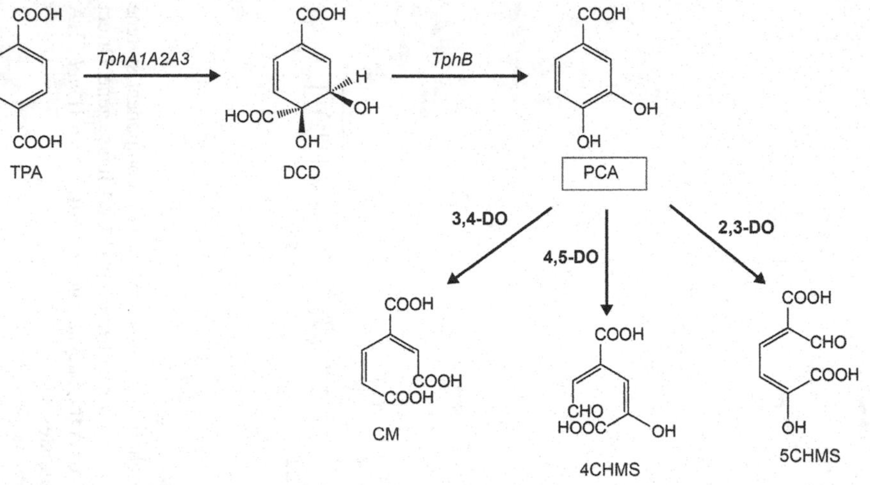

FIGURE 12.3 Catabolism of terephthalic acid. TA: terephthalic acid; DCD: 1,6-dihydr oxycyclohexa-2,4-diene-dicarboxylate; PCA: protocatechuate; DO: dioxygenase; CM: 3-car-boxy-*cis,cis*-muconate; 4CHMS: 4-carboxy-2-hydroxymuconate semialdehyde; 5CHMS: 5-carboxy-2-hydroxymuconate-6-semialdehyde; TphA1A2A3: TA dioxygenase; TphB: 1,2-d ihydroxy-3,5-cyclohexadiene-1,4-dicarboxylate dehydrogenase.

species [45,46], *Rhodococcus* sp. DK17 [47]. The *tph* operon encodes all enzymes required for the conversion of TA into PCA, as well as a transporter and transcriptional regulator. The regulator, TphR, belongs to an isocitrate lyase type (IclR-type) generally involved in the regulation of carbon metabolism and multidrug resistance, for example [48]. The *tph* operon forms two transcriptional units, one encompassing the catabolic genes and the other one being the regulator [49]. TA acts as an inducer of the *tph* operon expression, while TphR appears to repress its transcription regardless of the presence of TA [49].

The TA transporter in *Comamonas* strains is a TphR regulated permease TphC that acts together with membrane-bound TpiA-TpiB, the counterparts of the tripartite tricarboxylate-like transporter [50]. In *Pseudomonas* sp. GO16, the transport of TA is mediated by a major facilitator superfamily (MFS) transporter of the aromatic acid:H+ symporter (AAHS) family, which shows homology to the *p*-hydroxybenzoate transporter *pcaK* [51].

Similar to the upper styrene degradation pathway, *tph* genes act as a catabolic regulon independent from the lower pathway, that is, PCA conversion to TCA intermediates. PCA can be cleaved via *ortho-*, *meta-*, or *para*-cleavage pathways by corresponding dioxygenases to yield 3-carboxy-*cis,cis*-muconate, 4-carboxy-2-hydroxymuconate semialdehyde, or 5-carboxy-2-hydroxymuconate-6-semialdehyde, respectively [52–54]. However, a recent survey of the organisms that contain the *tph* operon showed that they usually do not have PCA-2,3-dioxygenase [51].

Since TA is metabolized to TCA cycle intermediates, its conversion into PHA proceeds via the *de novo* fatty acid synthesis pathway.

12.4.4 Metabolism of Ethylene Glycol

As already mentioned, the enzymatic degradation of plastic and the metabolism of the resulting monomers is likely a moonlighting activity of older enzymes. Glycolate, for instance, is an overflow metabolite of phytoplankton, which is secreted in significant amounts in aquatic environments [55], and its oxidation product glyoxylate is a central metabolite. The microbial utilization of these "old" chemicals can also be adapted to encompass ethylene glycol (EG), a PET degradation product [56]. However, the associated pathways are not optimally induced or balanced, leading to an inability of wildtype *P. putida* KT2440 to use this substrate as a sole carbon source, even though it has the genetic inventory to do so [57].

EG can undergo a series of oxidations to glycolate, catalyzed by broad-substrate-range alcohol and aldehyde dehydrogenases, among which periplasmic pyrrolo-quinoline quinone (PQQ)-dependent dehydrogenases play a prominent role [58,59]. Further oxidation to glyoxylate also occurs [56], catalyzed by a specific membrane-bound glycolate oxidase [57]. A complicating factor of this oxidative pathway is the fact that the aldehyde intermediates are extremely toxic. This also goes for other common polyester-derived α,ω-alcohols like 1,4-butanediol. An efficient upcycling chassis thus requires a delicate balancing of alcohol- and aldehyde-dehydrogenase activities to prevent the accumulation of these highly toxic intermediates, as well as a high tolerance to the associated aldehyde stress.

Even though the glyoxylate resulting from these oxidation reactions is a central metabolite in the glyoxylate shunt, this pathway does not enable its utilization as sole carbon source but converts it into two molecules of CO_2 instead. This conversion can be used to yield a total of five reducing equivalents (NADH, $PQQH_2$, or directly transduced to the electron transport chain) per mole of ethylene glycol. This helps to boost the biomass yield provided that a co-substrate, such as acetate, is provided as a carbon source [60]. In the context of a plastic upcycling process, such co-metabolism may be sufficient because the depolymerization of PET or PU will yield mixtures of ethylene glycol and other monomers.

In *P. putida*, efficient use of ethylene glycol as a sole carbon source requires the alternative Gcl pathway, involving the conversion of two glyoxylates to tartronate semialdehyde and CO_2 through a glyoxylate carboligase and subsequent conversion to 2-phosphoglycerate, which is fed into the central metabolism at the level of the C3 pool [57] (see Figure 12.4). In *P. putida* KT2440, this pathway is not induced by ethylene glycol or its oxidation products. Instead, it is part of a larger metabolic context of purine metabolism induced by xanthine or allantoin as upstream metabolites [60]. Both activation of the Gcl pathway and streamlining of the upstream oxidation steps toward glyoxylate can be achieved by metabolic engineering [57]. The same can also be achieved by adaptive laboratory evolution [60]. Interestingly, these two approaches yielded completely different optimization strategies, with metabolic engineering focused on the overexpression of pathway-encoding genes, and laboratory evolution affecting the associated regulators and other nonintuitive targets such as a porin.

The capacity of *P. putida* KT2440 to upcycle EG to PHA has been demonstrated [57]. Since it was shown that EG is metabolized to TCA cycle intermediates, the

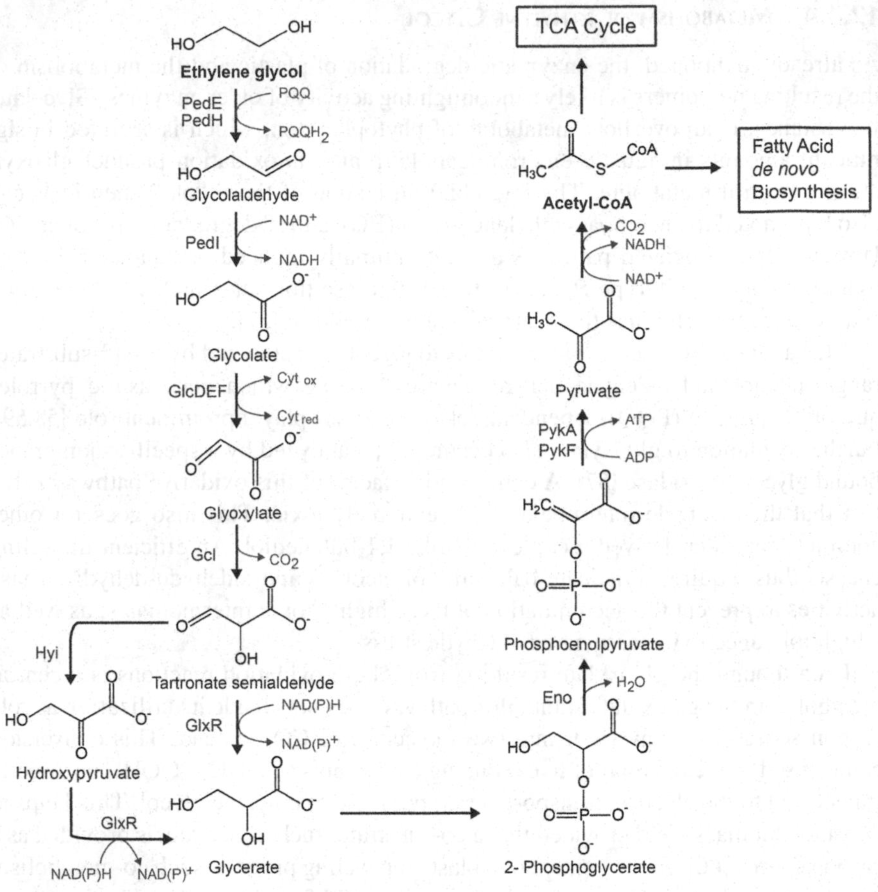

FIGURE 12.4 The metabolism of ethylene glycol. PedE and PedH: quinoprotein ethanol dehydrogenase; PedI: aldehyde-dehydrogenase; GlcDEF: glycolate oxidase; Gcl: glyoxylate carboligase; Hyi: hydroxypyruvate isomerase; GlxR: tartronate semialdehyde reductase; Eno: enolase (phosphopyruvate hydratase); PykA and PykF: pyruvate kinase; TCA: tricarboxylic acids cycle.

connection between the EG catabolism and PHA accumulation is *de novo* fatty acid synthesis.

12.4.5 METABOLISM OF ALKANES

Bacteria, fungi, and yeasts can degrade alkanes. Alkane degraders among bacteria are usually the organisms with versatile metabolisms and can use many other carbon sources in addition to alkanes. However, there are also less numerous specified alkane degraders such as *Alcanivorax borkumensis* [61,62]. The hydrophobic nature of alkanes poses the main challenge for their degradation. The uptake of alkanes is dependent on the molecular weight (M_w) and the characteristics of the environment. Low M_w alkanes (up to hexane) are soluble enough to allow direct uptake from water. However, medium- and long-chain alkanes have to be emulsified, and bacteria that can use these as a carbon and energy sources usually produce surfactants [63,64]. These biosurfactants increase the surface area and lower the surface tension between the carbon substrate and water phase, thereby facilitating their uptake [64]. The uptake itself seems to proceed via a mechanism similar to pinocytosis, as inclusion bodies containing alkanes were observed in alkane degrading bacteria [64,65].

Aerobic degraders of alkanes use monoterminal, diterminal, or subterminal pathways for the degradation of alkanes or all three, as is the case in *Rhodococcus* species [65] (see Figure 12.5). The monoterminal degradation of alkanes starts by oxidation of the terminal carbon, which is catalyzed by mono-oxygenases such as alkane hydroxylases. Additional alkane hydroxylating systems, such as alkane hydroxylase systems belonging to the cytochrome P450 family, have been found in *R. rhodochrous* ATCC 19067 and *Acinetobacter calcoaceticus* EB104 [66,67]. Alkane degraders typically contain multiple alkane hydroxylases with overlapping substrate ranges [65]. Upon the formation of a primary alcohol, an alcohol dehydrogenase and an aldehyde-dehydrogenase oxidize this alcohol in two sequential steps to a fatty acid [62]. Alcohol dehydrogenases were shown to have lower activity when compared with the activity of alkane hydroxylases and, therefore, the step catalyzed by them is considered rate-limiting for the degradation of alkanes [65,68]. The resulting fatty acid is further activated by acyl-CoA ligase to yield fatty acyl-CoA, which is then shunted into β-oxidation [62]. *P. synxantha* LSH-7' seems to have a different

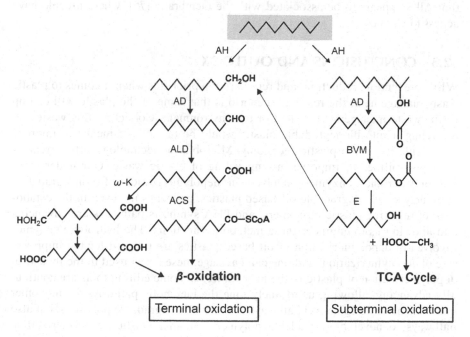

FIGURE 12.5 The metabolism of alkanes. AH: alkane hydroxylase; AD: alcohol dehydrogenase; ALD: aldehyde-dehydrogenase; ACS: acyl-CoA synthetase; ω-H: ω-hydroxylase; BVM: Baeyer–Villiger mono-oxygenase; E: esterase; TCA: tricarboxylic acids cycle.

strategy when grown with n-alkanes. This strain possesses an extracellular alkane hydroxylase and alcohol dehydrogenase, which seems to convert the substrate to a fatty acid prior to its uptake [69].

In cases of diterminal oxidation, a dicarboxylic acid is formed and further processed by β-oxidation [62]. Subterminal oxidation of alkanes yields secondary alcohols, which are then converted to ketones and oxidized to esters. These esters are further hydrolyzed by lipases to yield alcohols and fatty acids.

The organization of alkane degrading genes differs among the degraders. They can be found on a plasmid in an operon [70] or can be dispersed in a chromosome [71], and the regulators may or may not be in the proximity of the genes they regulate [62]. However, the expression of the genes encoding the initial oxidation of alkanes is tightly regulated. Alkane hydroxylases are activated in the presence of an appropriate substrate, such as short-, medium-, or long-chain alkanes [72]. For example, in *P. putida* GPo1 with the presence of C6–C12 n-alkanes, the LuxR family activator AlkS induces the expression of the *alkB* gene cluster encoding the enzymes for the conversion of these alkanes to the corresponding fatty acids. In contrast, when n-alkanes are absent, AlkS binds to its own promoter and represses its expression [73]. The regulation of the expression of CYP153 mono-oxygenase in *Actinobacteria* proceeds via the activation of the CYP153 gene promoter by an AraC family regulator in the presence of a substrate [74]. However, since alkanes are hydrophobic, it remains unclear how they interact with the regulators located in the cytoplasm. A proteomic study of *A. borkumensis* grown with alkanes showed that AlkS appears to be associated with the membrane [75], where it could have access to alkanes.

12.5   CONCLUSIONS AND OUTLOOK

While prevention, reduction, and reuse are top priorities when it comes to plastic waste management, the realistic scenario is that some of the plastic will end up in the waste management system or the environment. Collecting this waste and upcycling it into biodegradable plastic would be one of the means to improve the circularity of the plastics economy. Microbial biotechnology offers tremendous possibilities to improve this upcycling of plastic waste. Our understanding of microbial enzymes involved in depolymerization of nondegradable/extremely slowly degradable oil-based plastics, pathways involved in the catabolism of resulting plastic monomers, and PHA synthesis pathways are advancing and allowing us to engineer these metabolic functions. The hydrolytic enzymes involved in depolymerization of oil-based plastics are tailored for the improved rate of depolymerization, and one can envisage these being used in the industrial depolymerization of plastic in the near future. Genome editing tools are continually advancing, allowing us to finely tune the metabolic pathways for monomer metabolism, to design novel pathways to known biodegradable polymers, but also pathways to novel biodegradable polymers. Finally, plastic depolymerization, monomer metabolism, and PHA production could be integrated to design a custom microbial platform capable of converting plastic into biodegradable counterparts in a single cell [76–78].

ACKNOWLEDGMENTS

TN, SL, and KO'C are supported by Science Foundation Ireland research centre grant number 16/RC/3889. TN is funded by an Ad Astra Fellowship at UCD. TN, KO'C, and NW acknowledge funding from the European Union's Horizon 2020 Research and Innovation Programme under Grant Agreement No. 633962 for the project P4SB and No. 863922 for Mix-UP.

NW acknowledges that the scientific activities of the Bioeconomy Science Center, in the project PlastiCycle, were financially supported by the Ministry of Culture and Science within the framework of the NRW Strategieprojekt BioSC (No. 313/323-400-002 13).

REFERENCES

1. Plastics Europe, *Plastics – The Facts 2018*, 2018.
2. Geyer, R., J.R. Jambeck, and K.L. Law. Production, use, and fate of all plastics ever made. *Sci Adv* 2017; 3(7): e1700782.
3. European Commission. *Questions & Answers*: A European Strategy for Plastics. European Commission: Strasbourg, 2018.
4. Plastics Europe. *Plastics – The Facts 2016*, 2016.
5. Forum, W.E. *Project MainStream – A Global Collaboration to Accelerate the Transition Towards the Circular Economy*. Status Update, 2015.
6. E.B. Recycling America: In the bin. *The Economist*, 2015.
7. Kasper, M. *Energy from Waste Can Help Curb Greenhouse Gas Emissions*. Center for American Progress, 2013.
8. European Commission. *The Seventh Environment Action Programme to 2020 – 'Living Well, Within Limits of our Planet'*, 2013.
9. US EPA. *Fiscal Year 2014–2018 EPA Strategic Plan*, 2014.
10. Ocean Conservancy. *Stemming the Tide: Land-based Strategies for a Plastic-Free Ocean*, 2015.
11. European Commission. *Directive of the European Parliament and of The Council Amending Directive 2008/98/EC on Waste*, 2015.
12. European Bioplastics. *What are Bioplastics? 2018* [cited 2018]. Available from: https://www.european-bioplastics.org/bioplastics/.
13. European Bioplastics. 2018 [cited 2019]. Available from: http://www.european-bioplastics.org/bioplastics/materials/.
14. Narancic, T. and K.E. O'Connor. Plastic waste as a global challenge: Are biodegradable plastics the answer to the plastic waste problem? *Microbiol-SGM* 2019; 165(2): 129–137.
15. Ward, P.G., M. Goff, M. Donner, et al. A two step chemo-biotechnological conversion of polystyrene to a biodegradable thermoplastic. *Environ Sci Technol* 2006; 40(7): 2433–2437.
16. Goff, M., P.G. Ward, and K.E. O'Connor. Improvement of the conversion of polystyrene to polyhydroxyalkanoate through the manipulation of the microbial aspect of the process: A nitrogen feeding strategy for bacterial cells in a stirred tank reactor. *J Biotechnol* 2007; 132(3): 283–286.
17. Nikodinovic-Runic, J., E. Casey, G.F. Duane, et al. Process analysis of the conversion of styrene to biomass and medium chain length polyhydroxyalkanoate in a two-phase bioreactor. *Biotechnol Bioeng* 2011; 108(10): 2447–2455.
18. Johnston, B., I. Radecka, D. Hill, et al. The microbial production of polyhydroxyalkanoates from waste polystyrene fragments attained using oxidative degradation. *Polymers (Basel)* 2018; 10(9): 957.

19. Kenny, S.T., J.N. Runic, W. Kaminsky, et al. Up-cycling of PET (polyethylene tere-phthalate) to the biodegradable plastic PHA (polyhydroxyalkanoate). *Environ Sci Technol* 2008; 42(20): 7696–7701.
20. Kenny, S., J.N. Runic, W. Kaminsky, et al. Development of a bioprocess to convert PET derived terephthalic acid and biodiesel derived glycerol to medium chain length polyhydroxyalkanoate. *Appl Microbiol Biotechnol* 2012; 95: 623–633.
21. Guzik, M.W., S.T. Kenny, G.F. Duane, et al. Conversion of post consumer polyethylene to the biodegradable polymer polyhydroxyalkanoate. *Appl Microbiol Biot* 2014; 98(9): 4223–4232.
22. Montazer, Z., M.B.H. Najafi, and D.B. Levin. Microbial degradation of low-density polyethylene and synthesis of polyhydroxyalkanoate polymers. *Can J Microbiol* 2019; 65(3): 224–234.
23. Sivan, A. New perspectives in plastic biodegradation. *Curr Opin Biotechnol* 2011; 22(3): 422–426.
24. Krueger, M.C., H. Harms, and D. Schlosser. Prospects for microbiological solutions to environmental pollution with plastics. *Appl Microbiol Biot* 2015; 99(21): 8857–8874.
25. Wei, R., T. Oeser, and W. Zimmermann. Synthetic polyester-hydrolyzing enzymes from thermophilic actinomycetes. *Adv Appl Microbiol* 2014; 89: 267–305.
26. Wei, R. and W. Zimmermann. Microbial enzymes for the recycling of recalcitrant petroleum-based plastics: How far are we? *Microb Biotechnol* 2017; 10: 1308–1322.
27. Restrepo-Florez, J.M., A. Bassi, and M.R. Thompson. Microbial degradation and dete-rioration of polyethylene – A review. *Int Biodeter Biodegr* 2014; 88: 83–90.
28. Yoshida, S., K. Hiraga, T. Takehana, et al. A bacterium that degrades and assimilates poly(ethylene terephthalate). *Science* 2016; 351(6278): 1196–1199.
29. Austin, H.P., M.D. Allen, B.S. Donohoe, et al. Characterization and engineering of a plastic-degrading aromatic polyesterase. *Proc Natl Acad Sci USA* 2018; 115(19): E4350–E4357.
30. Palm, G.J., L. Reisky, D. Bottcher, et al. Structure of the plastic-degrading Ideonella sakaiensis MHETase bound to a substrate. *Nat Commun* 2019; 10: 1717.
31. Son, H.F., I.J. Cho, S. Joo, et al. Rational protein engineering of thermo-stable PETase from *Ideonella sakaiensis* for highly efficient PET degradation. *ACS Catal* 2019; 9(4): 3519–3526.
32. Wei, R., T. Oeser, J. Schmidt, et al. Engineered bacterial polyester hydrolases efficiently degrade polyethylene terephthalate due to relieved product inhibition. *Biotechnol Bioeng* 2016; 113(8): 1658–1665.
33. McNamara, J.T., J.L.W. Morgan, and J. Zimmer. A molecular description of cellulose biosynthesis. *Annu Rev Biochem* 2015; 84: 895–921.
34. Howard, G.T. Biodegradation of polyurethane: A review. *Int Biodeter Biodegr* 2002; 49(4): 245–252.
35. Danso, D., J. Chow, and W.R. Streit. Plastics: Environmental and biotechnological per-spectives on microbial degradation. *Appl Environ Microbiol* 2019; 85(19).
36. Shimada, K., E. Kimura, Y. Yasui, et al. Styrene formation by the decomposition by *Pichia carsonii* of trans-cinnamic acid added to a ground fish product. *Appl Environ Microbiol* 1992; 58(5): 1577–1582.
37. Liu, C., X. Men, H. Chen, et al. A systematic optimization of styrene biosynthesis in *Escherichia coli* BL21(DE3). *Biotechnol Biofuels* 2018; 11: 14.
38. O'Leary, N.D., K.E. O'Connor, and A.D.W. Dobson. Biochemistry, genetics and physiology of microbial styrene degradation. *FEMS Microbiol Rev* 2002; 26(4): 403–417.
39. Tischler, D. Pathways for the degradation of styrene. In: *Microbial Styrene Degradation*. Springer, Cham, 2015, pp. 7–22.

40. O'Connor, K., C.M. Buckley, S. Hartmans, et al. Possible regulatory role for non-aromatic carbon-sources in styrene degradation by *Pseudomonas putida* CA-3. *Appl Environ Microbiol* 1995; 61(2): 544–548.

41. O'Leary, N.D., K.E. O'Connor, W. Duetz, et al. Transcriptional regulation of styrene degradation in *Pseudomonas putida* CA-3. *Microbiology (UK)* 2001; 147: 973–979.

42. O'Leary, N.D., K.E. O'Connor, P. Ward, et al. Genetic characterization of accumulation of polyhydroxyalkanoate from styrene in *Pseudomonas putida* CA-3. *Appl Environ Microbiol* 2005; 71(8): 4380–4387.

43. Prieto, A., I.F. Escapa, V. Martinez, et al. A holistic view of polyhydroxyalkanoate metabolism in *Pseudomonas putida*. *Environ Microbiol* 2016; 18(2): 341–357.

44. Warhurst, A.M. and C.A. Fewson. Microbial metabolism and biotransformations of styrene. *J Appl Bacteriol* 1994; 77(6): 597–606.

45. Sasoh, M., E. Masai, S. Ishibashi, et al. Characterization of the terephthalate degradation genes of *Comamonas* sp. strain E6. *Appl Environ Microbiol* 2006; 72(3): 1825–1832.

46. Wang, Y.Z., Y.M. Zhou, and G.J. Zystra. Molecular analysis of isophthalate and terephthalate degradation by *Comamonas testosteroni* YZW-D. *Environ Health Persp* 1995; 103: 9–12.

47. Choi, K.Y., D. Kim, W.J. Sul, et al. Molecular and biochemical analysis of phthalate and terephthalate degradation by *Rhodococcus* sp. strain DK17. *FEMS Microbiol Lett* 2005; 252(2): 207–213.

48. Molina-Henares, A.J., T. Krell, M.E. Guazzaroni, et al. Members of the IclR family of bacterial transcriptional regulators function as activators and/or repressors. *FEMS Microbiol Rev* 2006; 30(2): 157–186.

49. Kasai, D., M. Kitajima, M. Fukuda, et al. Transcriptional regulation of the terephthalate catabolism operon in *Comamonas* sp. strain E6. *Appl Environ Microbiol* 2010; 76(18): 6047–6055.

50. Hosaka, M., N. Kamimura, S. Toribami, et al. Novel tripartite aromatic acid transporter essential for terephthalate uptake in *Comamonas* sp. strain E6. *Appl Environ Microbiol* 2013; 79(19): 6148–6155.

51. Salvador, M., U. Abdulmutalib, J. Gonzalez, et al. Microbial genes for a circular and sustainable bio-PET economy. *Genes (Basel)* 2019; 10(5): 373.

52. Kasai, D., T. Fujinami, T. Abe, et al. Uncovering the protocatechuate 2,3-cleavage pathway genes. *J Bacteriol* 2009; 191(21): 6758–6768.

53. Frazee, R.W., D.M. Livingston, D.C. Laporte, et al. Cloning, sequencing, and expression of the *Pseudomonas putida* protocatechuate 3,4-dioxygenase genes. *J Bacteriol* 1993; 175(19): 6194–6202.

54. Noda, Y., S. Nishikawa, K.I. Shiozuka, et al. Molecular cloning of the protocatechuate 4,5-dioxygenase genes of *Pseudomonas paucimobilis*. *J Bacteriol* 1990; 172(5): 2704–2709.

55. Lau, W.W.Y. and E.V. Armbrust. Detection of glycolate oxidase gene glcD diversity among cultured and environmental marine bacteria. *Environ Microbiol* 2006; 8(10): 1688–1702.

56. Mückschel, B., O. Simon, J. Klebensberger, et al. Ethylene glycol metabolism by *Pseudomonas putida*. *Appl Environ Microbiol* 2012; 78(24): 8531–8539.

57. Franden, M.A., L.N. Jayakody, W.J. Li, et al. Engineering *Pseudomonas putida* KT2440 for efficient ethylene glycol utilization. *Metab Eng* 2018; 48: 197–207.

58. Wehrmann, M., P. Billard, A. Martin-Meriadec, et al. Functional role of lanthanides in enzymatic activity and transcriptional regulation of pyrroloquinoline quinone-dependent alcohol dehydrogenases in *Pseudomonas putida* KT2440. *mBio* 2017; 8(3): e00570–e00617.

59. Wehrmann, M., C. Berthelot, P. Billard, et al. The PedS2/PedR2 two-component system is crucial for the rare earth element switch in *Pseudomonas putida* KT2440. *mSphere* 2018; 3(4): e00376–e00418.

60. Li, W.J., L.N. Jayakody, M.A. Franden, et al. Laboratory evolution reveals the metabolic and regulatory basis of ethylene glycol metabolism by *Pseudomonas putida* KT2440. *Environ Microbiol* 2019; 21(10): 3669–3682.

61. Schneiker, S., V.A.P.M. dos Santos, D. Bartels, et al. Genome sequence of the ubiquitous hydrocarbon-degrading marine bacterium *Alcanivorax borkumensis*. *Nat Biotechnol* 2006; 24(8): 997–1004.

62. Rojo, F. Degradation of alkanes by bacteria. *Environ Microbiol* 2009; 11(10): 2477–2490.

63. Santos, D.K.F., R.D. Rufino, J.M. Luna, et al. Biosurfactants: Multifunctional biomolecules of the 21st century. *Int J Mol Sci* 2016; 17(3): 401.

64. Cameotra, S.S. and P. Singh. Synthesis of rhamnolipid biosurfactant and mode of hexadecane uptake by *Pseudomonas* species. *Microbial Cell Factor* 2009; 8: 16.

65. Mishra, S. and S.N. Singh. Microbial degradation of n-hexadecane in mineral salt medium as mediated by degradative enzymes. *Bioresour Technol* 2012; 111: 148–154.

66. Cardini, G. and P. Jurtshuk. Cytochrome P-450 involvement in the oxidation of n-octane b cell-free extracts of *Corynebacterium* sp. strain 7E1C. *J Biol Chem* 1968; 243(22): 6070–6072.

67. Muller, R., O. Asperger, and H.P. Kleber. Purification of cytochrome P-450 from n-hexadecane grown *Acinetobacter calcoaceticus*. *Biomed Biochim Acta* 1989; 48(4): 243–54.

68. Pirog, T.P., T.A. Shevchuk, and Y.A. Klimenko. Intensification of surfactant synthesis in *Rhodococcus erythropolis* EK-1 cultivated on hexadecane. *Appl Biochem Microbiol* 2010; 46(6): 599–606.

69. Meng, L., H.S. Li, M.T. Bao, et al. Metabolic pathway for a new strain *Pseudomonas synxantha* LSH-7′: From chemotaxis to uptake of n-hexadecane. *Sci Rep (UK)* 2017; 7: 39068.

70. van Beilen, J.B., S. Panke, S. Lucchini, et al. Analysis of *Pseudomonas putida* alkane-degradation gene clusters and flanking insertion sequences: Evolution and regulation of the alk genes. *Microbiology-SGM* 2001; 147: 1621–1630.

71. Lo Piccolo, L., C. De Pasquale, R. Fodale, et al. Involvement of an alkane hydroxylase system of *Gordonia* sp. strain SoCg in degradation of solid n-alkanes. *Appl Environ Microbiol* 2011; 77(4): 1204–1213.

72. Tani, A., T. Ishige, Y. Sakai, et al. Gene structures and regulation of the alkane hydroxylase complex in *Acinetobacter* sp. strain M-1. *J Bacteriol* 2001; 183(5): 1819–1823.

73. Canosa, I., J.M. Sanchez-Romero, L. Yuste, et al. A positive feedback mechanism controls expression of AlkS, the transcriptional regulator of the *Pseudomonas oleovorans* alkane degradation pathway. *Mol Microbiol* 2000; 35(4): 791–799.

74. Liang, J.L., J.H. JiangYang, Y. Nie, et al. Regulation of the alkane hydroxylase CYP153 gene in a Gram-positive alkane-degrading bacterium, *Dietzia* sp. strain DQ12-45–1b. *Appl Environ Microbiol* 2016; 82(2): 608–619.

75. Sabirova, J.S., M. Ferrer, D. Regenhardt, et al. Proteomic insights into metabolic adaptations in *Alcanivorax borkumensis* induced by alkane utilization. *J Bacteriol* 2006; 188(11): 3763–3773.

76. Wierckx, N., T. Narancic, C. Eberlein, et al. Plastic biodegradation: Challenges and opportunities. In: *Consequences of Microbial Interactions with Hydrocarbons, Oils, and Lipids: Biodegradation and Bioremediation*, R. Steffan, Editor. Springer International Publishing, Cham, 2018, pp. 1–29.

77. Wierckx, N., M.A. Prieto, P. Pomposiello, et al. Plastic waste as a novel substrate for industrial biotechnology. *Microb Biotechnol* 2015; 8(6): 900–903.

78. Hatti-Kaul, R., L.J. Nilsson, B. Zhang, et al. Designing biobased recyclable polymers for plastics. *Trends Biotechnol* 2019; 38(1): 50–67.

13 Comparing Heterotrophic with Phototrophic PHA Production

Concurring or Complementing Strategies?

Ines Fritz, Katharina Meixner, Markus Neureiter, and Bernhard Drosg

CONTENTS

13.1 INTRODUCTION – THE *STATUS QUO* OF PHA PRODUCTION

Poly(3hydroxybutyrate) (PHB) was discovered as natural content of bacteria cells [60] at a time when plastic (artificial recalcitrant polymers) were not used on an industrial scale and long before environmental pollution with plastic residues was a fact. It was in the 1960s when ICI (Imperial Chemical Industries) started a visionary

331

approach: biotechnological production of the biodegradable plastic PHA (the copolymer of PHB) and poly(3hydroxyvalerate) (PHV) copolymer, synthesized by and isolated from *Alcaligenes* sp. (meanwhile renamed *Cupriavidus necator*), which was grown on monosaccharides. Although never highly economically feasible, the PHB/PHV polymer was used for cosmetics, and food packaging and the biotechnological production system became the standard for all mid- to large-scale amounts [18], and it remained the standard over several decades.

In the early 1990s, packaging for consumer products made from PHA was on the market but with steadily declining abundance. They disappeared during the 1990s – not because of unsuitable functionality, not because they were biodegradable, not because they were produced from monomeric sugars – because of the high price of the material. ICI stopped production and sold the patent. PHA, although still in existence, fell back into the status of a very small market niche.

Many attempts had been made to reduce production cost and, most importantly, to reduce the effort of polymer extraction from the bacterial cell mass. The PHA price decreased, but it did not fall low enough to compete with the market-dominating polyethylene and polypropylene. PHA were dammed to a shadowy existence.

In 2019, the global annual production of conventional fossil-based plastic reached an estimated amount of close to 400 million tonnes [7]. Parallel with the production, plastic waste increased in amount and in its impact on the environment, mostly due to intentional release, losses and littering. This pollution aspect of conventional plastics made biodegradable plastics, such as PHA, once again, interesting for packaging and other consumer products. While price is still a dominating factor, the lack of available quantities – resulting from a lack of production capacities – is currently limiting its broad use.

PHA are, besides poly(lactic acid) (PLA), among the most promising biobased and biodegradable polymers with interesting material properties and with highly beneficial use in certain applications. Biotechnological production with prokaryotes is, undoubtedly, the most promising way to go and may remain so for the near future. The prospect of significantly increased market demand did not only stimulate manufacturers, it also led to an intensified ethical discussion about the use of agricultural resources for (single-use) consumer products [54].

Regarding the use of substrates, three general production strategies can be identified, similar to biorefinery concepts: (1) using sugar or starch – derived from food crops – as input, (2) using lignocellulosic biomass or other liquid and solid organic residues (waste streams) as input and (3) biotechnology with photo- and autotrophic bacteria – using CO_2 and sunlight as input. All three have specific advantages and disadvantages. While options 1 and 2 are discussed in other chapters, this chapter takes a closer look at the details of phototrophic PHA production using cyanobacteria.

13.2 HETEROTROPHIC PHA PRODUCTION FOR COMPARISON

Commercial processes for PHA production are currently based on first generation biomass like glucose derived from starch and sucrose derived from sugar cane or sugar beet. Co-substrates, such as propionic acid and higher chain length fatty acids, for the production of the respective copolymers are usually derived from petrochemical sources. Typical wild-type production strains are *Cupriavidus necator* (formerly *Ralstonia eutropha*), *Alcaligenes eutrophus* (formerly *Alcaligenes latus*) for P(3HB), P(3HB-*co*-4HB) and P(3HB-*co*-3HB) and *Aeromonas hydrophila* for P(3HB-*co*-3HHx) types of polymer [26,27]. Apart from wild-type strains, also, genetically modified *Escherichia coli* (with the PHA expression system from *Cupriavidus necator*) is used in industry – and also to a great extent in research – for the production of mainly P(3HB) or P(3HB-*co*-3HV) on diverse types of carbon sources. Wild-type as well as genetically modified strains of *Pseudomonas* sp. are used for the production of *mcl*-PHA from different raw materials [74]; however, until now, these processes were only realized at a small scale and due to the differences in substrate as well as polymer composition they do not represent a suitable benchmark for comparison with phototrophic PHA producers.

Production is in most cases realized as a fed-batch process, with unlimited biomass formation in the first stage and the second stage for polymer formation under limitation of an essential nutrient (usually nitrogen or phosphorous) [58] For commercial processes with *C. necator* or *E. coli*, biomass concentrations of more than 100 g L^{-1} (up to 200 g L^{-1}) with polymer concentrations typically exceeding 75% in cell dry mass (CDM) and a process duration of 48–72 h have been reported [26]. This corresponds to a volumetric productivity between approximately 1.4–4.0 g L^{-1} h^{-1}. The maximum theoretical yield based on glucose is 0.48 g P(3HB) per g glucose [120], which is because the process requires oxygen and, according to the biochemical pathway, two carbons of each sugar molecule are converted to CO_2. Actual product yields are therefore in the range of 0.3–0.4 g P(3HB) per g metabolized sugar [37]. However, this means that 60–70% of the carbon source is lost due to respiration. Especially for first generation substrates, which are also in competition with food and feed in terms of crops and arable land, this raises issues regarding sustainability and ecological benefits of the final products.

With this background, it is understandable that researchers – both in terms of substrate costs and ecologically sustainable processes – have been investigating alternative substrates like crude glycerol, cheese whey, waste plant and animal lipids and by-products from the sugar industry over the past 20 years. The results have been summarized in several reviews [25,48,76,90]. Many variations of heterotrophic PHA production strategies had been evaluated – on the substrate side as well as on the use of alternative organisms, like *Bacillus* sp., *Burkholderia* sp., *Halomonas* sp. and *Haloferax* sp., which can efficiently metabolize some of the rarer substrate streams. In the case of using second- and third-generation substrates, it is very important to keep in mind that the residues have to be available in sufficient quantities to satisfy even an increasing demand for the products. Recent publications cover, amongst others, glycerol [115], waste animal fats [88] and desugarized molasses [95]. There is, in general, a high variation in the published results regarding scale and the availability of process parameters and chemical and mechanical material characterization data, which makes comparisons quite challenging.

Lately, increasing efforts have been put into the development of processes based on the carboxylate platform, which uses wastes and wastewaters to produce volatile fatty acids [3]. This has the advantage that large quantities are available without competition to food production and theoretically different input streams can be brought onto one common platform. Traditionally there are mixed culture approaches that have been improved during the past few years that are able to provide polymers of a reasonable consistent quality [52,97]. In addition, there are approaches to process volatile fatty acids from dark fermentation into PHA using pure single strains to obtain a defined polymer [59].

The use of biomass residues or even wastes or wastewaters requires new approaches regarding process design. In these types of substrates, the concentration of the carbon source is often low due to previous processing or pre-treatment steps. Conventional fed-batch approaches are therefore not applicable. Processes using biomass retention, like membrane bioreactors, can help to maintain high cell densities [148 g L^{-1} at 86% P(3HB) in cell dry mass] and high volumetric productivities for PHB (>3 g L^{-1} h^{-1}) despite low substrate concentrations [37,40]. Other improvements in process design include continuous two- or multi-stage processes [53] and the development of non-sterile processes with extremophilic strains [51,56,121] that have the potential to significantly reduce energy and manipulation costs during the fermentation process.

13.3 PHB SYNTHESIS IN CYANOBACTERIA

Many, but not all, cyanobacteria are known for their natural ability to synthesize PHB [8,22,32,103]. Frequently cited genera are: *Anabaena, Anacystis, Aphanicomenon, Aphanocapsa, Arthrospira, Aulosira, Calothrix, Chlorogloea, Chlorogloeopsis, Gloeothece, Gloeotrichia, Halospirulina, Lyngbya, Microcoleus, Microcystis, Nodularia, Nostoc, Oscillatoria, Phormidium, Plectonema, Schizothrix, Scytonema, Spirulina, Stanieria, Synechococcus, Synechocystis, Tolypothrix* and *Trichodesmium*.

As for any other biotechnological process, PHB production with photo-autotrophic microorganisms demands short fermentation times, high volumetric productivity, simple (or at least non-complex) handling, low risk of contaminations and low operating cost.

Cyanobacteria cope with most of these criteria, making them an interesting topic for many research groups worldwide. Strains and cultivation conditions were improved for significantly shorter fermentation times and for increased PHA contents [32]. More specifically, all authors report PHB to be the dominant, if not the only PHA in cyanobacteria [9], while PHB/PHV copolymers are, among others, the most common hydroxyalkanoates observed in heterotrophic bacteria. The different composition can be partly explained by the appearance of cyanobacteria typical synthesis enzymes (Figure 13.1). Those enzymes are phaA (ß-ketothiolase) (today: 3-ketothiolase), phaB (acetoacetyl-CoA reductase), phaC (poly(3-hydroxyalkanoate) synthase) [38] and phaE with unknown function. The respective genes are located in two separate clusters in the genome (Figure 13.2) and seem to be widely similar for all PHB-producing cyanobacteria. In contrast, the phaB and phaC synthesis genes in the heterotrophic bacteria *Ralstonia* sp. appear in multiple copies, which can lead to an increased synthesis speed. The phaE gene does not appear in aerobic heterotrophic bacteria, although it was identified in some species of sulphur bacteria as part of a bigger protein associated with PHB granules [61]. The remaining PhaE in

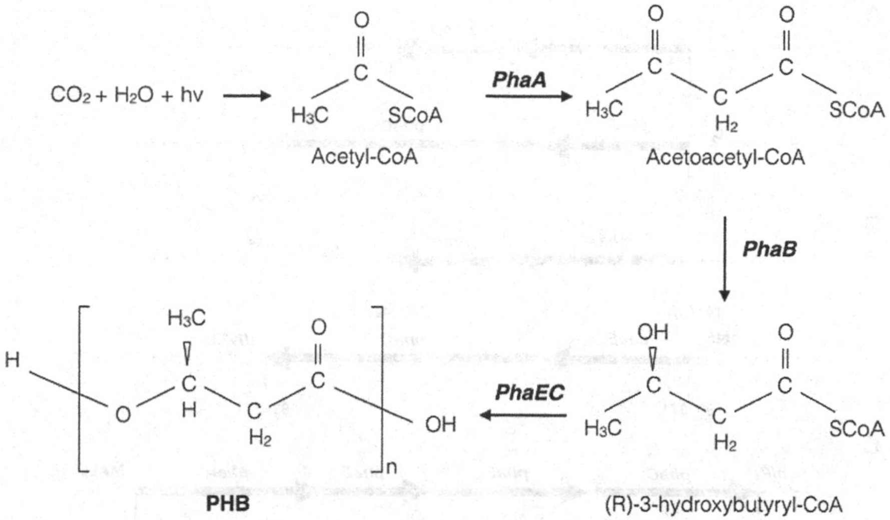

FIGURE 13.1 Biosynthesis of PHB in *Synechocystis* PCC 6803 from acetyl-coenzyme A. Scheme redrawn from [32], based on the general pathway described in [109] and [116] with modifications as suggested by [47].

cyanobacteria is a mostly linear helix protein with no dedicated reaction centre and no catalytic function associated but with a sterical fitting to the PhaC synthase [99] and a potential role as a coenzyme or regulating protein [105].

Another important storage product for cyanobacteria is the polymeric carbohydrate glycogen, which is produced via growth-associated synthesis, and is, in theory, in competition for the acetate pool with the PHB synthesis path [118]. However, the regulation mechanisms are different for both these polymers and knock-out of the glycogen path does not necessarily increase PHB accumulation in the cell [29]. Other than glycogen, PHB in cyanobacteria seems to act as a buffering pathway catching excess acetyl-CoA and reduction equivalents (NADH) during the light phase (energy surplus). However, because of an interrupted tricarboxylic acid cycle, PHB cannot be used as fast as glycogen [30]. As a result and beneficial for biotechnological production, cyanobacteria will consume glycogen first in dark conditions and in the case of temporally unfavourable fermentation conditions. This behaviour can be exploited during the ripening phase, when cyanobacteria have accumulated higher amounts of glycogen and are kept in the dark to convert the carbohydrate into PHB [113].

Within the bigger drawbacks of PHB production with cyanobacteria are their comparably slow autotrophic growth and their cellular PHB content, rarely exceeding 25% in the biomass [47]. This is balanced by the fact that no organic carbon source is needed as it is for PHA production with heterotrophic bacteria.

13.4 LIGHT AS AN ENERGY SOURCE FOR CYANOBACTERIA

Chlorophyll is the most common pigment used by microorganisms for light absorption. Smaller variations in the molecule lead to several types of chlorophyll and

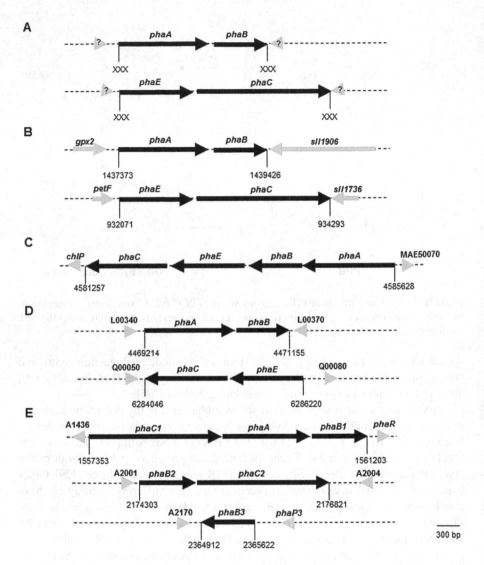

FIGURE 13.2 Location of the pha genes in the four cyanobacteria: *Synechocystis* CCALA192 (A), *Synechocystis* PCC 6803 (B), *Microcystis aeruginosa* (C) and *Arthrospira platensis* (D) and in the heterotrophic bacterium, *Ralstonia eutropha* H16 (E); redrawn from [99].

bacteriochlorophyll, each with a small deviation in the absorbance spectrum as well. Secondary or accessorial pigments, such as carotenoids, phycobiliproteins and rhodopsin, complement the energy absorbance by using the centre of the light spectrum [94,117]. What nature developed by evolution, namely specific light demands, favourable pH-value and salinity as well as low concentrations of nutrients can be set for almost selective conditions in biotechnological production with cyanobacteria.

Water usually acts as the electron donor in the oxygenic photo-autotrophic metabolism, releasing oxygen during the light phases. Instead of water, H_2S acts as the electron donor for purple bacteria and sulphur bacteria in anoxygenic photosynthesis, releasing sulphur. In dark periods, all photo-autotrophic microorganisms switch to a heterotrophic metabolism, aerobically consuming oxygen and releasing carbon dioxide. Any biotechnological production must consider these general differences between light and dark phases. Practically speaking, an oxygenic photo-autotrophic fermentation needs a constant flow of carbon dioxide and constant dissipation of oxygen during the day (light phase), while CO_2 must be removed and oxygen provided during night (dark phase). However, cyanobacteria are able to survive in the dark under limited oxygen availability, which can be exploited to suppress potential contaminants or pathogens in a non-sterile high-volume production environment [113].

Cyanobacteria generally need light within photosynthetically active radiation (PAR) for photosynthesis, which converts light energy to chemical energy. PAR is considered as the wavelength range between 400 and 700 nm and is commonly quantified as photosynthetic photon flux (μmol photon m^{-2} s^{-1}) or expressed as energy unit irradiance I (W m^{-2}). By extending the PAR band, *Acaryochloris marina* can even exploit near-infrared light [12]. Given solar radiation, not more than about 9% of the full spectrum energy can be converted into biomass [15]. However, considering only the PAR region, the maximal efficiency of photosynthesis is estimated to be 11.3% [19].

The light intensity has to be in a certain range to satisfy biomass production. If the intensity is too low, light becomes a limiting factor, which is of course undesirable. If the light intensity is too high, however, it can lead to photoinhibition [57]. This generally means the cyanobacteria are no longer able to repair photosystem II (PSII), which further leads to a loss in the activity of the oxygen-evolving complex [23,91].

The biochemistry of cyanobacteria is based on circadian metabolic cycles, which can be triggered either by light (like the native day-night cycles) or by rhythmic feeding with organic substrate for a stimulated heterotrophic cycling [84]. However, not all cyanobacteria can use sugar (glucose or fructose), especially not when illuminated at the same time [79].

In a production environment, the light-dark cycle is not only driven by day and night but also due to the change in light exposure during flow and mixing in the photobioreactor. The reactor design has, therefore, a crucial impact on the total light energy available to the cells. The light/dark quotient is defined as the sum of the time in the light versus the time in the dark zone. This practical aspect of reactor operation not only prevents light starvation of the organisms, but also allows the dark catalytic reactions of photosynthesis to be completed in order to restore the full capacity of the photosynthetic apparatus [34,35]. Kroon reports that photosynthetic efficiency is highest when the turnover rate of electrons in PSII is equal to the frequency of the change between light and dark [55].

According to variations in the pigment composition, different wavelengths are absorbed with different efficiencies. Besides other factors, the pigment composition depends on the nutrient status [102], which is essential for PHA production typically starting under nutrient limitations. While red light can promote the synthesis of phycobiliproteins as well as biomass production, as reported for *Gloeothece membranacea* [73], excessive amounts of blue light results in an imbalance of photosystems I and II by disturbed electron flow. The phenomenon seems to be exclusive for cyanobacteria and was not observed in green algae growth [21,62]. This finding is partly in contrast with older literature, where the spectral influence on growth and biomass composition in cyanobacteria is reported to be specific to the organism used [119].

From an economical point of view, sufficient biomass has to be produced per batch and the growth rate (λ) should be as high as possible. However, there is a theoretical limit to the productivity of a mass culture, which is heavily determined by the average irradiance per cell (light intensity), the mixing (quotient light/dark time), gas exchange (degassing of surplus oxygen) and temperature (general thermodynamics). In optimized setups, the theoretical maximum growth rate should only be limited by the rate of photosynthesis – resulting in remarkable efforts in photobioreactor design and operation [87]. Overall, for a mid-priced product, such as PHB, artificial illumination is not recommended [83].

13.5 CO_2 AS A CARBON SOURCE FOR CYANOBACTERIA

The inorganic carbon is actively transported through the cytoplasm membrane as hydrogen-carbonate (HCO_3^-), the dissociated form of dissolved CO_2 at neutral or slightly alkaline pH value [9,30,103]. Once in the cell, cyanobacteria bind carbon via the Calvin cycle (also known as reductive pentose phosphate cycle or Calvin–Benson–Bassham cycle) driven by the energy gained from photosynthesis [100].

As the enzyme ribulose-1,5-bisphosphate-carboxylase/oxygenase (RuBisCO) happens to have a low affinity for CO_2 (50% saturation at the normal atmospheric level of currently 400 ppm [41]), cyanobacteria have mechanisms to increase the concentration of this enzyme, so-called carbon-concentrating mechanisms (CCMs) [57], which gives them the ability to maintain photosynthesis under different carbon concentrations [42].

For biotechnological growth of photo-autotrophic organisms, different carbon dioxide sources can be used. In nature, the organisms use the CO_2 in ambient air. Although steadily rising [41], CO_2 concentrations are still very low, which makes air an unattractive carbon source for industrial processes. More interesting are typical exhaust- or off-gases with elevated CO_2 concentrations. These can be flue gas from combustion processes or raw industrial off-gases deriving from biotechnological processes, such as ethanol or biogas production, with no incineration involved.

The main driver for using flue gases is the chance to reduce CO_2 emissions. However, the demand for its biological carbon capture and utilization may not always be synchronous to the emission. In temperate climates, combustion for energy and heat is more intense during winter while the daylength and light intensity is the lowest, resulting in the slowest phototrophic growth.

Flue gases have typical CO_2 concentrations in the range from 3 to 15%. Their advantages are very low price, usually high amounts emitted and location of the sources all over the world. To give an example, 13,000 Mt yr^{-1} of CO_2 are emitted as flue gases from stationary sources (> 0.1 Mt yr^{-1} of CO_2) according to the Intergovernmental Panel on Climate Change (IPCC) [69]. As a disadvantage, direct flue gas utilization will need some conditioning like cooling, dust removal or removal of other harmful incineration residues. In many countries, legal issues can become challenging, as products derived from flue gases might be considered waste and may not be declared as a product.

Raw industrial CO_2-rich off-gases are a very promising alternative. Such carbon dioxide can be used for sparkling beverages, as the typical impurities are not harmful or can be removed with simple measures. The potential is very high, as about 507 Mt yr^{-1} of such off-gases are available worldwide, whereas only about 110 Mt yr^{-1} are already used as raw materials in the chemical industry [11], resulting in an untapped fraction of 78%. However, in large-scale photo-autotrophic PHA production facilities, the transport of such off-gases can become an issue, as the emitters are less abundant than those for flue gases. Optimal photobioreactor operation sites may currently be far from the next off-gas emitter.

The CO_2 dosage has to be maintained to balance a minimum of acidification (pH decreases with surplus CO_2) with carbon loss and growth limitation due to carbon limitation and alkalinization. This leads directly to a regulated CO_2 (or off-gas) inlet flow directly coupled to the pH monitoring. Small bubbles and a long enough retention time, as easily achieved in tubular photobioreactors, are beneficial for dissolving the gaseous CO_2 completely. Measuring the pH and maintaining it at a defined level through CO_2 injection would be a proven strategy to control carbon supply [91]. It is reported that this system only works well as long as there is nitrate uptake due to biomass increase [13]. This can become a problem in N-limited cultures, as the CO_2 demand can stop during the PHA-synthesis phase. While some authors report algorithms and methods to predict the upcoming CO_2 demand [91,92], practical experience shows that the ripening phase (PHB formation decoupled from biomass growth) can be operated in a separate vessel in the dark and without any gas flow [113].

Finally, it should be mentioned that many cyanobacteria could be cultivated under organo-heterotrophic conditions in the dark with similar growth rates compared with autotrophic conditions [65] and with very high PHB contents [106]. Acetate is among the best-known transcription inducer and can be a carbon source for the metabolic chain of PHA biosynthesis. However, organo-heterotrophic cyanobacteria growth has no advantages over organo-heterotrophic growth of other bacteria and is not investigated any deeper.

13.6 NUTRIENTS FOR CYANOBACTERIAL GROWTH

In many freshwater environments, phosphorus (P) is often the limiting nutrient and therefore controls the abundance of natural cyanobacterial populations [96]. For mass development of cyanobacteria, typically less than 0.03 mg L^{-1} P are required [93]. Typical cultivation media for algae and cyanobacteria, often recommended by culture collections, show higher P concentrations than necessary and much higher than commonly found in nature [5]. The BG-11 medium, for instance, contains 7 mg L^{-1} phosphorus. Cyanobacterial blooms, which regularly occur in eutrophic waters, lead to the assumption that they need high phosphorus and nitrogen (N) concentrations. In reality, cyanobacterial blooms often appear when concentrations of dissolved phosphate are low. Because of their high affinity to both nutrients, cyanobacteria can overgrow other photosynthetic organisms and excel over their competitors. Moreover, cyanobacteria can store sufficient amounts of phosphorus (mostly as polyphosphate) which is enough for several cell divisions [77]. In production practice, less than 1 mg L^{-1} P is required for sufficient biomass growth and for achieving a certain degree of P limitation to stimulate elevated PHB accumulation in the cyanobacteria cells [113].

The nitrogen content in cyanobacteria can reach up to 10% of dry matter [104] as a component of peptides and proteins in enzymes, in light harvesting phycobiliproteins [4] and in cyanophycin, a specialized storage protein [77]. Many cyanobacteria species have the ability to fix atmospheric N_2 when dissolved nitrogen concentrations are low [117], but can lose this ability after long-term exposure to dissolved nitrogen in the medium [89]. Cell uptake is preferably in the form of ammonium as long as no other ionic form of N is available, and they assimilate nitrate before fixing N_2 [39]. Any nitrogen uptake is directly connected with photosynthesis and therefore also closely related to CO_2 fixation and biomass growth. These processes compete for electron donors like ferredoxin and energy provided by photosynthesis [77]. Carbohydrates and lipids (including PHB) are synthesized under nitrogen starvation [104,118].

Data from our experiments (unpublished) showed similar stimulation of PHB synthesis in 34 different cyanobacteria strains, of which 31 were formerly not investigated wild-type strains, under nitrogen starvation conditions. Nitrogen surplus but phosphorus starvation induced the PHB accumulation to a lesser degree. However, a combined N and P starvation led to the highest PHB contents in the cells but led also to a slower biomass growth. Minimum intracellular nutrient levels have to be kept sufficient for overall metabolic activity [82]. This is similar to many other reports about PHB synthesis stimulation in well-known cyanobacteria strains from culture collections [32].

Many cyanobacteria species can fix atmospheric nitrogen [17,94], among those some can naturally synthesize PHB. Due to the requirement of an oxygen-free environment, single cell cyanobacteria can fix nitrogen only in dark conditions (no oxygen production). This will make an otherwise beneficial biotechnological process ineffective.. However, the necessary energy is 16 adenosine triphosphate (ATP) per fixed molecule of N_2 [17]. Each ATP invested in nitrogen fixation is missing for PHB synthesis. Nitrogen addition to the cultivation medium is, therefore, the preferred option.

13.7   OTHER GROWTH CONDITIONS FOR CYANOBACTERIA

Cyanobacteria prefer pH values higher than 5 [20] to keep the ATP flux high [44]; most are alkalophiles having their growth optima between pH 7.5 and 10. However, the pH value does not only influence the available energy but also the available dissolved HCO_3^- and influences all other membrane transport of ions [64]. Overall, the optimal pH for the maximal growth rate cannot be generalized as it varies from strain to strain and depends on the situation in the natural environment.

The requirement for a certain total ion concentration (osmotic pressure and salinity) is somehow combined with the pH value, although it can be a limiting factor on its own [78,98]. Cyanobacteria may be best adapted to freshwater (e.g. *Anabaena* sp., *Microcystis* sp., *Synechocystis* PCC 6803) or to brackish water (e.g. *Synechocystis salina*), to sea water (e.g. *Aphanicomenon* sp., *Nodularia* sp.) or they may be halophilic (e.g. *Aphanothece* sp., *Halospirulina* sp.). According to typical tolerance ranges, again, no general selection recommendation may be given.

Cyanobacteria growth is reported from cryophilic (+4°C) up to thermophilic conditions (e.g. *Synechococcus lividus*, 75°C) [24]. Photosynthetic activity, without observable growth, was reported to happen at astonishing −30°C [31]. Elevated temperature increases the metabolic turnover, and thermophilic conditions can significantly reduce the contamination risk. However, thermophilic cyanobacteria able to produce PHA are rare, and thermophilic production in a large-scale photobioreactor will cause very high effort for thermal insulation. The majority of cyanobacterial production is therefore done in the range between 20 and 35°C with the lowest air-condition effort necessary and with respect to the surrounding situation [34].

Combining typical needs, tolerances and environmental conditions makes cyanobacteria the primary candidates for the utilization of nutrients from wastewater, binding of exhaust CO_2 and production of a valuable biopolymer, such as PHB [67].

13.8 CURRENT STATUS OF PHOTOTROPHIC PHA PRODUCTION

Although many research and laboratory scale data for PHB synthesis with cyanobacteria are published, production data on a bigger scale, such as pilot or industrial scale, are rare [112]. This may be because of the large effort required to fight contaminations and because of special precautions necessary when working with genetically modified cyanobacteria in big volumes. At least two success stories were found in the literature, where one describes a single experiment with a randomly mutated *Synechocystis* PCC 6714 strain in two stages [45] and the other provides a comprehensive production concept for non-sterile conditions with the native *Synechocystis* CCALA 192 strain as a single-stage growth and PHB synthesis procedure with an additional ripening phase [113]. These reports summarize the major strategies with their specific advantages and drawbacks.

Increased PHB production was achieved by all researchers by managing nitrogen or nitrogen and phosphorus starvation, in some cases with, in others without, the addition of PHB precursors [32]. Although the practical execution varies, the principle was the same in all cases: fast biomass growth under optimal conditions until a certain cell density was reached, followed by a nutrient starvation phase for rapid PHB synthesis. In the majority of reports, this was achieved by separating the cells from the remaining medium and transferring them into a fresh medium that was free from N and P sources and maybe contained a precursor.

Repeating this principle in a large-scale production fermenter will require enormous technological effort for the cell transfer into a new medium and will cause some type of mechanical and chemical stress to the cells. Shear forces from the pump and the centrifugation as well as an acyclic dark period with oxygen limitation are the most significant. In addition, the separation may not be that precise under technical conditions as it is possible in small volumes in the laboratory, resulting in a new lag phase and in a slowly continued growth at low nutrient concentrations.

Another strategy is the use of a carefully composed medium that contains the pre-calculated amount of nutrients to reach a certain cell density (e.g. 2 g biomass per litre). The continuous nutrient depletion directly shifts the culture into the starvation phase without additional stress and without a new lag phase. Such a shifting cultivation may last approximately one week for a growth phase and one additional

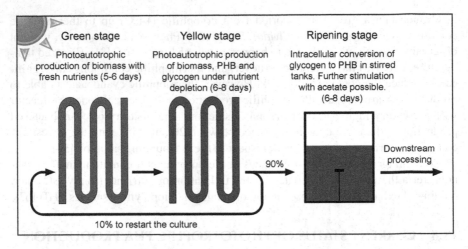

FIGURE 13.3 Operation mode for producing cyanobacterial PHB on a pilot scale in a non-sterile environment and without medium transfer [113].

week for PHB production. Within the ca. 2-day shift, the culture's colour changes from intense blue-green to an orange-yellow, indicating the nutrient starvation. At this time, no new biomass is produced, the cell density remains almost constant and the CO_2 consumption decreases slowly to zero. The remaining glycogen is partly consumed for energy generation and partly slowly transformed into additional PHB. This last step is called the ripening stage and can be operated outside the photobioreactor in a simple vessel without illumination (Figure 13.3). After 13–14 days in the photobioreactor and an additional 7 days in the ripening tank, a concentration of 1 g DM biomass L^{-1} was achieved with a PHB content of ca. 11–13% m/m in the dry matter. *Synechocystis* CCALA 192 was used for this prototype production in a 200-litre scale and with four growth cycles in sequence without cleaning or maintenance of the reactor [113]. The production at a laboratory scale elsewhere under the same conditions resulted in 2 g DM biomass L^{-1} with a PHB content of 12% in the cell dry matter within 14 days [33], indicating sub-optimal growing conditions in the system with 60-mm glass tubes.

However, the approximate 12–15% PHB in the cell dry matter and maximum 250 mg PHB L^{-1} is way too low for a breakthrough in commercial production [83], as are the ca. 30% PHB in the DM of a genetically modified cyanobacteria requiring a closed reactor with increased safety precautions [45]. Besides focused productivity on biomass and PHB content, other aspects and conditions can make the cyanobacterial PHB production more feasible, such as water and nutrient recycling, nutrient utilization from wastewater, side products (phycocyanin), excess biomass conversion into biogas or organic fertilizer and excess biomass use as animal feed, to name a few [67,83]. Although still not an economic breakthrough, such a combined process increases the sustainability of the PHB production [36].

Auto-selectivity, a combination of cultivation conditions favourable for the intended strain and unfavourable for all potential contaminants, is a serious goal for all large-scale processes. For cyanobacteria, this selectivity can be achieved by setting several parameters simultaneously: the lack of dissolved organic carbon, limiting concentrations of nitrogen and phosphorous and a pH value at or above 8.5. However, as the culture reaches its stationary phase, increased numbers of cells will die and release their content. This may act as a carbon and energy source for heterotrophic contaminants, making running batch processes under starvation conditions for long time critical [32].

13.9 PHOTOTROPHIC CULTIVATION SYSTEMS

Generally, cultivation systems for cyanobacteria can be distinguished in open and closed systems. Depending on the cyanobacterial strain, the environmental conditions, the necessary cultivation conditions, the final purpose and application of the biomass, the cultivation system is chosen [16]. The most important restriction is due to safety restrictions and the necessity to use a closed photobioreactor when cultivating a genetically modified strain. Nevertheless, closed as well as open systems can be built to a certain size indoors and outdoors, respectively. The volumes range from a few litres at laboratory scale to several cubic metres at production scale [85].

Raceway ponds, circular ponds, shallow big ponds and tanks are the main types of open systems but here cascade systems are also used. The main advantages of open systems are the low investment and low operating expenditure. However, large areas (which can be industrial sites or building roofs) and high water amounts (respiration) are required. Since large systems are constructed outside, they are mainly built in areas with little precipitation and many sunshine hours. Furthermore, it is difficult to control the cultivation conditions; for this reason, only species requiring highly selective environments (high salt pH values, concentrations, etc.) or occurring as naturally dominant species in this area are beneficial to cultivate in open systems [66]. Useful information about suitable species can be found in the review literature [50].

Biomass concentrations and productivities are rather low in most systems (0.3–0.5 g L^{-1}, 0.05–0.1 g L^{-1} d^{-1}) [85]. Remarkable exceptions are cascade systems with a low thickness layer (about 6 mm), in which high biomass concentrations (40–50 g L^{-1}) and high productivities (32 g L^{-1} d^{-1}) can be achieved [66]. Open systems have been used since the 1970s, manly for cultivating *Spirulina* sp. for food and feed purposes. Long-term experimental data is available to allow estimates for productivities in temperate climate as well as to estimate the water needed to replace evaporation [66].

Closed systems are designed in various shapes. The most common ones are tubular and flat plate photobioreactors [16], which can be adjusted vertically, horizontally or helically or, in the case of flat plate reactors, in vertical orientation with a certain angle of inclination. The reactor content can be pumped or moved by the airlift principle [1]. Stirred tank reactors with artificial illumination and sleeve bags are technically more simple reactor types but have specific drawbacks; stirred tank reactors require artificial illumination and sleeve bags are very difficult to stir, resulting in bad reproducibility. All closed systems are less prone to contamination, need less space than open systems and can be built in areas with moderate climatic conditions. However, their investment and operative expenditures are higher, and the systems are more complex to run and maintain [16]. Nevertheless, in 2018 a vertical tubular photobioreactor system with approximately 40,000 tubes, 6 m in height (estimated total volume of 600 m³) started operation in Bruck/Leitha (Austria) with the aim to be economically competitive [6].

Currently, a cascade system seems to be a preferable option [83] because it fully meets the required criteria combined with high growth rates and the option to operate up to ten reactors in parallel with one day delay between each [66]. Another option to increase the production efficiency and reduce costs (by a factor of eight) is to transfer the cultivation system from central to south Europe. This has not been realized yet; the estimated PHB production costs is €24 kg⁻¹ from an optimized cultivation system under the assumption of elevated PHB concentrations in the biomass [83].

PHA production by cyanobacteria has yet to reach the commercial production scale. No praxis data for the effort and cost of downstream processing in the cubic metre scale is available. As cyanobacteria are Gram-negative bacteria, it can be assumed that the PHB isolation and purification will not be much different from the isolation and purification of PHA from other Gram-negative heterotrophic bacteria. The most significant difference is in the generally lower PHB content in cyanobacteria. A creative and promising extraction may be the feeding of the bacterial biomass to mealworms, which can digest the cell components but not the PHB and release it with their faeces [75]. However, no generally new techniques were needed.

Impurities not found in PHA from heterotrophic bacteria are pigments (colour components), most probably residues and breakdown products from chlorophyll, carotenoids and phycobilins. Those need to be removed to meet the requirements and expectations of the PHA market [68]. Although only lab-scale experimental data are available, very good processing qualities can be expected from such a cleaned PHB (see Table 13.1). The separate marketing of carotenoids, phycocyanin and vitamins may contribute to the overall competitiveness of cyanobacterial PHB.

Practical experiences, such as long-term stability, seasonal influences and limits or maintenance intervals, are not available, as the photo-autotrophic PHB production is still in the research phase. The nutrient starvation requirement and the resulting stationary growth phase of the cells when they produce PHB makes the culture, generally speaking, highly susceptible to contamination (mainly green algae) [21] and predators (like ciliates) [112]. Therefore, closed systems are definitely preferred. Operating an open or another type of non-sterile photobioreactor, to save investment and some operation costs, is possible with some limitations: the cultivation systems needs to be auto-selective (by pH values above 10 [111] or by using halophilic cyanobacteria [81]); the operation alternates frequently between growth and sanitation phases [113] and genetically modified organisms cannot be used. With regard to productivity, it should be considered that biomass concentrations in photo-autotrophic systems are generally low. Typical concentrations are between 0.3 and 8 g L⁻¹ [85], which is much lower than in heterotrophic systems where biomass concentrations up to 250 g L⁻¹ can be achieved in fed-batch systems [40]. Biotechnological production systems with photo-autotrophic microorganisms always require the management of comparably high volumes, which increases all three costs: investment, operation and downstream processing, with the biggest factor being pumping cost [83]. Additionally, downstream processing is complex due to the removal of cyanobacterial pigments [67].

TABLE 13.1

Data of Phototrophic PHB from *Synechocystis* CCALA 192, Heterotrophic PHA from *Cupriavidus Necator* H16 and Commercially Available PHA of Different Origin. All Samples Were Analyzed in the Same Laboratory by Gel Permeation Chromatography (GPC), Differential Scanning Calorimetry (DSC) and Thermogravimetric Analysis (TGA)

Origin	Cell pre-treatment	Pigment removal	GPC			DSC				TGA	
			M_n [kg mol^{-1}]	M_w [kg mol^{-1}]	$Đ_M$ [–]	T_m [°C]	ΔH_m [J g^{-1}]	X_c [%]	T_g [°C]	T_d [°C]	T_{max} [°C]
Synechocystis salina CCALA 192 cultivated in mineral medium	drying	acetone/methanol[1]	420	888	2.1	181.6	72.5	49.7	7.5	259.6	272.1
	drying	acetone/ethanol[1]	368	800	2.2	177.6	71.1	48.7	4.9	255.8	267.6
	drying and mill	acetone/methanol[2]	413	943	2.3	180.9	58.3	39.9	7.5	248.7	259.0
	drying and mill	acetone/ethanol[1]	626	988	1.6	177.4	54.7	37.5	9.1	243.9	254.8
	ultrasound	–	314	569	1.8	176.2	50.8	34.8	4.9	239.4	248.6
	French press[3]	–	791	1390	1.8	183.7	94.9	65.0	6.4	275.1	286.9
Synechocystis salina CCALA 192 cultivated in digestate	–	ethanol/acetone[4]	1520	5820	3.8	179.3	129.4	88.6	2.9	283.3	292.5
	–	ethanol/acetone[4]	2660	7980	3.0	179.4	130.3	89.2	7.8	281.6	291.5
Cupriavidus necator H16	–	–	401	1136	2.8	178.1	87.3	59.8	5.7	284.2	296.4
Biomer P309[5]	–	–	120	800	6.5	174.6	105.0	71.9	0.45	286.7	294.1
ENMAT™ Y1000[5]	–	–	201	78	2.6	168.9	134.8	nd	nd	272.8	291.0
Mirel™ F1005[5]	–	–	149	72	2.1	166.7	63.3	nd	nd	283.3	302.4
Mirel™ F1006[5]	–	–	158	71	2.2	165.7	44.6	nd	nd	275.5	285.3

M_n: average molecular weight; M_w: average molecular weight; $Đ_M$: polydispersity index; T_m: melting temperature; ΔH_m: melting enthalpy; X_c: crystallinity; T_g: glass transition temperature; T_d: onset of degradation temperature; T_{max}: degradation temperature with the maximum decomposition rate; nd: not determined; [1] pigment extraction for 30 h at 4°C, solvent ratio of 70:30; [2] pigment extraction for 24 h at 4°C, solvent ratio of 70:30; [3] different batch, with the addition of sodium acetate (ca. 17 mmol); [4] pigment extraction without drying for 1 h, at room temperature; [5] commercially available polymers.

Data modified from [68]

13.10   RECOMBINANT CYANOBACTERIA FOR PHA PRODUCTION

Current research is aimed at producing high cellular amounts of PHA by recombinant bacterial cell factories [28]. The two main strategies are to either (1) magnify the synthesis capacity of natural PHA producers [46] or (2) transfer the PHA-synthesis genes into fast growing bacteria, such as *E. coli* [110]. All strategies result in genetically modified microorganisms (GMOs) of safety classes L1 or L2 harbouring highly active heterologous PHA biosynthetic genes. Such GMOs can accumulate sub-cellular PHA amounts up to 80% of bacterial cell dry mass [107]. Despite the advantage of high productivity, the process has elevated cost in the form of increased safety measures and raises ethical questions when converting edible carbohydrates into (single-use) plastic. Although the strain improvement methods for cyanobacteria are less deeply investigated, they are, most probably, the most promising way to go in the longer term [32].

In comparison with algae and plants, cyanobacteria are easier to genetically manipulate, either in cis, through chromosome modification, or in trans, through plasmid introduction [14]. Some cyanobacteria (as mentioned above) naturally possess genes for PHA biosynthesis and are inspiring starting points for the construction of an optimal recombinant PHA producer. Intense research on the heterotrophic bacterium *Cupriavidus necator*, a natural producer of high-molecular weight PHA, allowed the identification of the PHA biosynthesis operon [70].

Synechococcus PCC7942 represents the first example of a recombinant cyanobacterium for PHA production. The introduction of the *C. necator* PHA operon conferred to the cyanobacterium the ability to improve PHA accumulation within the cell from 3 to 25% of the cell dry mass [108]. After the entire genome of *Synechocystis* sp. PCC6803 was sequenced in 1996 [63], the four PHASyn genes were identified in two distinct loci, clustered two by two, forming the phaA-BSyn cluster and the phaE-CSyn cluster (Figure 13.2b). The phaA (slr1993) and the phaB (slr1994) genes are co-linear, putatively co-expressed and encode for the PHA-specific β-ketothiolase and the acetoacetyl-CoA reductase, respectively. The phaE-CSyn operon, harbouring the phaE (slr1829) and the phaC (slr1830) genes, encodes for the putative PHB synthase component and the PHA synthase. It is possible that phaE and phaC form a protein complex involved in the modulation of PHA polymers [99].

With the availability of the genome data, the first complete PHA biosynthesis pathway known in cyanobacteria was drawn (Figure 13.1). Additionally, it opened the possibility of genetically manipulating *Synechocystis* sp. PCC6803 for improved PHB synthesis and accumulation. The potential of *Synechocystis* as a model for PHA genetic engineering was demonstrated by the inactivation of PHA synthase through the disruption of the phaE-CSyn cluster using a PCR-based gene disruption method [116]. Considering that *Synechocystis* harbours from six to ten copies of its genome, the disruption procedure permits the replacement of all the genome target genes, increasing the possibility of genetically manipulating this organism, although, the mechanism of total replacement is still unknown [109].

The high-CO_2 response mechanisms described in microalgae like *Chlorella* sp., *Scenedesmus* sp., *Nannochloropsis* sp. and *Chlorococcum* sp. initiated increasing interest in the CO_2-fixing conserved enzyme RuBisCO (EC 4.1.1.39) [10] involved in the Calvin–Benson cycle. The RuBisCO promoter has been used successfully as a regulating sequence for pha-gene expression and production in *Synechocystis* sp. PCC6803 [72]. Future work will be aimed at optimizing as much as possible the polymer production potential of cyanobacteria factories by metabolic engineering and genetic modifications.

13.11 PHA ISOLATION FROM THE CELLS, PURIFICATION AND RESULTING QUALITIES

The cyanobacterial biomass contains the pigments chlorophyll a, chlorophyll b, carotenoids, phycocyanin and allophycocyanin. Their concentration depends on the available nutrients, especially nitrogen [4]. A remarkable concentration of phycocyanin up to 12% of the cell mass was analyzed in the PHB-producing *Synechocystis* CCALA192 [68]; other authors report up to 9% as high-producers [101]. In ecology, the phycocyanin content of aquatic habitats can be used for biomonitoring purposes, reflecting the nutrient status [71] or indicating heavy metal intoxication [43].

Isolated and purified pigments are high value commercial products for the food and feed industry, for cosmetics and in biomedical research [49]. As such, they may contribute to the overall economic situation of PHB production with cyanobacteria and should be removed from the PHB as they are contaminants in the expected colourless (or white) polymer. The conflict is in the natural metabolism of cyanobacteria; at elevated nutrient concentrations, they produce high amounts of pigments but low amounts of PHB and vice versa [4,68]. However, when *Synechocystis* was grown on a complex composed digestate (as a nutrient source), an acceptable level of PHB accumulation (6% compared with 8–11% in optimized synthetic medium) was observed in a well-growing culture without pigment bleaching [68]. This cannot be explained by the nutrient levels alone.

13.12 UTILIZATION OF RESIDUAL CYANOBACTERIA BIOMASS

The cyanobacterial cell residues still contain carbohydrates, lipids and proteins after PHB extraction and represent ca. 80–90% of the original biomass grown in the photobioreactor. Other than for PHA production with heterotrophic bacteria, the cyanobacterial biomass should be seen as a resource and not as waste. Utilization can follow one of two possible strategies: either anaerobic digestion for biogas (methane) [36] or as an valuable ingredient of animal feed [2]. The first contributes to the comparably high energy demand when operating a photobioreactor (during the comparably long growth phase) and produces an organic fertilizer as a secondary product [83]. The second supports the regional economy in reducing the need for soy or other protein imports for animal production [2].

A highly innovative idea combines PHA extraction with residual biomass digestion by feeding mealworms with the PHA-containing bacterial biomass, resulting in mealworms as animal feed and in significant PHA granule enrichment in the mealworms faeces [75]. Although currently done with heterotrophically grown bacteria containing high PHA content, this a promising concept to be tested for cyanobacteria as well. The procedure avoids the use of (maybe harmful) organic solvents and converts the bacterial biomass into an insect biomass with very high nutritional value [86]. However, using phototrophic biomass grown in open reactors needs careful investigation and certified conformity with food and feed regulations to avoid the introduction of contaminants or pathogens to the food chain.

13.13 COMPARING HETEROTROPHICALLY PRODUCED PHB WITH PHOTOTROPHICALLY PRODUCED PHB

Both heterotrophic bacteria and cyanobacteria can accumulate PHA under nutrient-limited conditions in the presence of excess carbon and an energy source. Cyanobacteria can metabolize CO_2 but may, additionally, use organic carbon sources (e.g. acetate) for growth and accumulation of storage products. For the cultivation of heterotrophic bacteria as well as a cyanobacteria alternative, nutrient sources can be used, but for this purpose, different pre-treatments are required. A major drawback of cyanobacteria is the significantly lower accumulation of PHA compared with heterotrophic bacteria. PHB accumulation can be increased to 14% in *Synechocystis salina* by optimizing the process control [113] and most probably also by supplementation with organic acids. Significantly more residual biomass after PHA extraction remains with comparably high concentrations of proteins and carbohydrates.

Mechanical and chemical properties of PHA and PHB from different sources are shown in Table 13.1. When comparing the characteristics of PHB derived from *Synechocystis salina* and PHA derived from *Cupriavidus necator* H16 with commercial PHA grades, the thermal history of the commercial grades needs to be considered [114]. Customization of PHA includes co-extrusion with additives for a commercial material with specified properties. But even when taking the sensitivity of PHA to thermal processing [80] into account, the molecular weight properties (M_n and M_w) and melting temperatures (T_m) of the cyanobacterial PHB were significantly better than those of commercial grades of PHA, leading to the assumption of overall improved processability and increased number of recycling runs before reaching the end-of-life stage.

13.14 CONCLUSIONS AND OUTLOOK

Industry is starting to show more and more interest in the use of CO_2 as a raw material for their processes. Societal and legal drivers are also going in this direction, as direct competition with food crops can be avoided. Nevertheless, industrial photo-autotrophic production of PHA has far to go as process optimization is still ongoing. There is an immense potential for the use of waste carbon dioxide either from flue gases or raw industrial CO_2 off-gases. Several challenges remain to establish an economically viable process:

- Elevated PHA content in the photo-autotrophic cells
- Increased cell density in the photobioreactor
- Scale-up of photobioreactors by innovative design
- Closed loop of nutrients and water – optimized downstream processing

- Definition of quality criteria for CO_2 containing off-gases
- Utilization of the surplus cyanobacteria biomass

Currently achieved PHA concentrations are very low, so that strain improvement of cyanobacteria is intensively discussed. Gene amplification and gene transfer are the most promising techniques to achieve this goal. Unfortunately, cell densities in photo-autotrophic cultivation are lower by at least a factor of ten than in heterotrophic cultivation. With the same factor, the volumetric productivity is lower, and the investment costs are higher. Directly proportional to the reactor volume is the effort for downstream processing. Volumes of open ponds cannot be increased infinitely; the same is true for tube length or tube diameter in closed photobioreactors. Carbon dioxide saturation and O_2 degassing is the limiting factor for all closed reactors. Reactor design needs to be optimized for energy efficiency, light use, flow management, mixing and distribution of cells and nutrients. The whole production plant needs to be designed for water and nutrient recycling and for biomass recovery, avoiding excessive need for finite resources. For example, coupling the photo-autotrophic growth with anaerobic digestion is a viable option. It seems to be obvious that acidic flue gas content, such as SO_2 or NO_X and toxic heavy metals, are unfavourable for photo-autotrophic microorganisms and need to be removed. Not so clear is the influence of solid particles (dust) or microbial contaminants (from biotechnological sources) on the performance and stability of a photo-autotrophic process.

Aiming towards a sustainable society is fundamental and cannot, for ethical reasons, be achieved by the production of single-use products from agricultural resources. Now PHA are successfully produced by heterotrophic fermentation and this will continue for the next decade, especially when residuals from food and feed production or when non-edible plant resources are used. Such residuals and waste sources (including CO_2) avoid the ethical conflict with food production (the "plate-versus-plastic conflict") but have other drawbacks, mostly on the quantitative aspects of mass production. Residuals, the input material of choice for second-generation biorefinery, are in high demand for materials and energy production. With seasonal and regional quality fluctuations, neither their price nor their quantitative availability can be predicted. CO_2 rich off-gas is currently rarely used as a carbon source but it has enormous potential in phototrophic biotechnological processes to regenerate (recover) organic biomass. Therefore, we believe in the success of PHA production by photo-autotrophic biotechnological processes in the long run (Figure 13.4).

Finally, the biotechnological production path of choice must consider all input and all production parameters and must combine them into the best option. It is never a matter of resources nor of investment nor of operation cost alone – the weight of the parameters may change over time. We see, for that reason, heterotrophic and phototrophic PHA production as complementing strategies. Independent from how production capacities and their dynamics will develop, a paradigm shift in consumer behaviour and market supply is necessary to support human wealth in the long term; biobased plastics, such as PHA, are an important part of it.

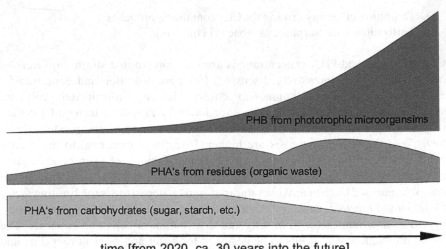

FIGURE 13.4 PHA production scenarios based on the three general paths according to first- and second-generation biorefinery and phototrophic production. The thickness of the bars represents estimated quantities (not to scale).

REFERENCES

1. Acién Fernández F G, Fernández Sevilla J M, Sánchez J A, et al. Airlift-driven external-loop tubular photobioreactors for outdoor production of microalgae: assessment of design and performance. *Chem Eng Sci* 2001; 56 (8): 2721–2732.

2. Adler A, Doppelreiter F, Galler A, et al. *Kontrollierte Futtermittel, Gesunde Tiere, Sichere Lebensmittel [Controlled Feed, Healthy Animals, Safe Food]*. AGES – Österreichische Agentur für Gesundheit und Ernährungssicherheit GmbH Institut für Tierernährung und Futtermittel, Wien, 2016.

3. Agler M T, Wrenn B A, Zinder S H, Angenent L T. Waste to bioproduct conversion with undefined mixed cultures: the carboxylate platform. *Trend Biotechnol* 2011; 29: 70–78.

4. Allen M M, Smith A J. Nitrogen chlorosis in blue-green algae. *Arch Mikrobiol* 1969; 69 (2): 114–120.

5. Andersen R A, Berges R A, Harrison P J, et al. Appendix A – recipes for freshwater and seawater media; enriched natural seawater media. In: Andersen R A (Ed.), *Algal Culturing Techniques*. Elsevier/Academic Press, 2005: 429–532.

6. Anonymous. Biotech company built a 230 kilometers vertical photobioreactor in just 10 months. *Algae World News 2018*. Available at: https://news.algaeworld.org/2018/10/biotech-company-built-a-230-kilometers-vertical-photobioreactor-in-just-10-months/.

7. Anonymous GRID-Arendal. *Global Plastic Production and Future Trends*. Internet reference 2019. http://www.grida.no/resources/6923.

8. Ansari S, Fatma T. Cyanobacterial polyhydroxybutyrate (PHB): screening, optimization and characterization. *Plos One* 2016; 11 (6): 1–20.

9. Asada Y, Miyake M, Miyake J, et al. Photosynthetic accumulation of poly-(hydroxybutyrate) by cyanobacteria—the metabolism and potential for CO_2 recycling. *Int J Biol Macromol* 1999; 25: 37–42.

10. Baba M, Shiraiwa Y. High-CO_2 response mechanisms in microalgae. In: Najafpour M M (Ed.), *Advances in Photosynthesis – Fundamental Aspects*. InTech, 2013.

11. Bazzanella A, Kramer D, Peters M. CO_2 als Rohstoff [CO_2 as resource]. *Nachr Chem* 2010; 58: 1226–1230.
12. Behrendt L, Schrameyer V, Qvortrup K, et al. Biofilm growth and near-infrared radiation-driven photosynthesis of the chlorophyll d-containing cyanobacterium *Acaryochloris marina*. *Appl Environ Microbiol* 2012; 78 (11): 3896–3904.
13. Behrens P W. Photobioreactors and fermentors: the light and dark sides of growing algae. In: Andersen R A. (Ed.), *Algal Culturing Techniques*. Elsevier, Oxford, 2005: 189–204.
14. Berla B M, Saha R, Immethu C M, et al. Synthetic biology of cyanobacteria: unique challenges and opportunities. *Front Microbiol* 2013; 4: Art 246.
15. Bolton J R, Hall D O. The maximum efficiency of photosynthesis. *Photochem Photobiol* 1991; 53 (4): 545–548.
16. Borowitzka M A. Commercial production of microalgae: ponds, tanks, tubes and fermenters. *J Biotechnol* 1999; 70 (1–3): 313–321.
17. Bothe H, Schmitz O, Yates M G, et al. Nitrogen fixation and hydrogen metabolism in cyanobacteria. *Microbiol Mol Biol Rev* 2010; 74 (4): 529–551.
18. Braunegg G, Lefebvre G, Genser K F. Polyhydroxyalkanoates, biopolyesters from renewable resources: physiological and engineering aspects. *J Biotechnol* 1998; 65 (2–3): 127–161.
19. Brennan L, Owende P. Biofuels from microalgae—a review of technologies for production, processing, and extractions of biofuels and co-products. *Renew Sust Energ Rev* 2010; 14 (2): 557–577.
20. Brock T D. Lower pH limit for the existence of blue-green algae: evolutionary and ecological implications. *Science* 1973; 179 (4072): 480–483.
21. Brunson S, Cheatham J, Clayton J, et al. Effects of algae growth in different light exposure. *J Introd Biol Invest* 2016; 4 (2): 17.
22. Carpine R. *Cyanobacteria for PHB Production*. PhD Thesis, Università degli Studi di Napoli Federico II, Italy, 2013.
23. Carvalho A P, Meireles L A, Malcata F X. Microalgal reactors: a review of enclosed system designs and performances. *Biotechnol Progress* 2006; 22 (6): 1490–1506.
24. Castenholz R W. *A Handbook on Habitats, Isolation, and Identification of Bacteria, Section B*. Springer, Berlin Heidelberg, 1981: 236–246.
25. Castilho L R, Mitchell D A, Freire M G. Production of polyhydroxyalkanoates (PHAs) from waste materials and by-products by submerged and solid-state fermentation. *Bioresour Technol* 2009; 100: 5996–6009.
26. Chen G-Q. A microbial polyhydroxyalkanoates (PHA) based bio- and materials industry. *Chem Soc Rev* 2009; 38: 2434–2446.
27. Chen G-Q, Wu Q, Jung Y K, Lee S Y. PHA/PHB. In: Moo-Young M (Ed.), *Comprehensive Biotechnology* (Second Edition). Elsevier, Amsterdam, 2011: 217–227.
28. Chee J-Y, Yoga S S, Lau N-S, et al. Bacterially produced polyhydroxyalkanoate (PHA): converting renewable resources into bioplastics. In: Méndez-Vilas A (Ed.), *Current Research, Technology and Education Topics in Applied Microbiology and Microbial Biotechnology*. Formatex Research Center, Spain, 2010: 1395–1404.
29. Damro R, Maldener I, Zilliges Y. The multiple functions of common microbial carbon polymers, glycogen and PHB, during stress responses in the non-diazotrophic cyanobacterium *Synechocystis sp.* PCC 6803. *Front Microbiol* 2016; 7: 966.
30. De Philippis R, Ena A, Guastini M, et al. Factors affecting poly-β-hydroxybutyrate accumulation in cyanobacteria and in purple non-sulfur bacteria. *FEMS Microbiol Rev* 1992; 103: 187–194.
31. De Vera J-P, Möhlmann D, Leya T. Photosynthesis activity of frozen cyanobacteria, snow alga and lichens as pre-tests for further on studies with simulation of Mars equatorial latitude temperatures. *Proceedings in European Planetary Science Congress 2009*, Potsdam, Germany, 2009: 355.

32. Drosg B, Fritz I, Gattermayr F, et al. Photo-autotrophic production of poly(hydroxyalkanoates) in cyanobacteria. *Chem Biochem Eng* Q 2015; 29 (2): 145–156.

33. Gattermayer F. *Design of a Production System for Synechocystis cf. salina and Investigation of its Growing Conditions*. Master Thesis at the University of Natural Resources and Life Sciences, Vienna, 2014.

34. Grima E M, Fernández F G A, Camacho F G, et al. Scale-up of tubular photobioreactors. *J Appl Phycol* 2000; 12 (3–5): 355–368.

35. Grima E M, Belarbi E-H, Fernández F G A, et al. Recovery of microalgal biomass and metabolites: process options and economics. *Biotechnol Adv* 2003; 20 (7–8): 491–515.

36. Gronald G. *CO2USE – Plastic from Bioreactors*. Energy Innovation Austria, 2017: 4. Available at: https://www.energy-innovation-austria.at/article/co2use-2/?lang=en.

37. Haas C, El-Najjar T, Virgolini N, et al. High cell-density production of poly(3-hydroxybutyrate) in a membrane bioreactor. *New Biotechnol* 2017; 37: 117–122.

38. Hai T, Hein S, Steinbüchel A. Multiple evidence for widespread and general occurrence of type-III PHA synthases in cyanobacteria and molecular characterization of the PHA synthases from two thermophilic cyanobacteria: *Chlorogloeopsis fritschii* PCC 6912 and *Synechococcus* sp. strain MA19. *Microbiology* 2001; 147 (11): 3047–3060.

39. Herero A, Muro-Pastor A M, Flores E. Nitrogen control in cyanobacteria. *J Bacteriol* 2001; 183 (2): 411–425.

40. Ienczak J L, Schmidell W, de Aragão G M F. High-cell-density culture strategies for polyhydroxyalkanoate production: a review. *J Ind Microbiol Biotechnol* 2013; 40: 275–286.

41. IPCC report. In: Stocker T F, Qin D, Plattner G-K, Tignor M, Allen S K, Boschung J, Nauels A, Xia Y, Bex V, Midgley P M (Eds.), *Climate Change 2013: The Physical Science Basis. Contribution of Working Group I to the Fifth Assessment Report of the Intergovernmental Panel on Climate Change*. Cambridge University Press, Cambridge, United Kingdom and New York, NY, USA.

42. Jander F. *Massenkultur von Mikroalgen mit pharmazeutisch nutzbaren Inhaltsstoffen unter Verwendung von CO_2 und $NaHCO_3$, gewonnen aus den Abgasen eines Blockheizkraftwerkes*. PhD Thesis, Christian-Albrechts-Universität Kiel, Germany, 2001.

43. Jusoh W N A W, Wong L S, Chai M K. Phycocyanin fluorescence in whole cyanobacterial cells as bioindicators for the screening of Cu^{2+} and Pb^{2+} in water. *Trans Sci Technol* 2017; 4 (1): 8–13.

44. Kallas T, Castenholz R W. Internal pH and ATP-ADP pools in the cyanobacterium *Synechococcus* sp. during exposure to growth-inhibiting low pH. *J Bacteriol* 1982; 149 (1): 229–236.

45. Kamravamanesh D, Pflügl S, Nischkauer W, et al. Photosynthetic poly-β-hydroxybutyrate accumulation in unicellular cyanobacterium *Synechocystis* sp. PCC 6714. *AMB Expr* 2017; 7: 143.

46. Khetkorn W, Incharoensakdi A, Lindblad P. Enhancement of poly-3-hydroxybutyrate production in *Synechocystis sp.* PCC 6803 by overexpression of its native biosynthetic genes. *Bioresour Technol* 2016; 214: 761–768.

47. Koch M, Doello S, Gutekunst K, et al. PHB is produced from glycogen turn-over during nitrogen starvation in *Synechocystis sp.* PCC 6803. *Int J Mol Sci* 2019; 20 (8): 1942.

48. Koller M, Atlic A, Dias M, et al. Microbial PHA production from waste materials. In: Chen G-Q (Ed.), *Plastics from Bacteria: Natural Functions and Applications*. Microbiology Monographs (Vol. 14). Springer Verlag, Heidelberg, Berlin, 2010: 85–119.

49. Koller M, Muhr A, Braunegg G. Microalgae as versatile cellular factories for valued products. *Algal Res* 2014; 6 (A): 52–63.

50. Koller M, Maršálek L. Cyanobacterial polyhydroxyalkanoate production: status quo and quo vadis? *Curr Biotechnol* 2015; 4 (3): 1–17.

51. Koller M, Chiellini E, Braunegg G. Study on the production and re-use of poly(3-hydroxybutyrate-*co*-3-hydroxyvalerate) and extracellular polysaccharide by the archaeon *Haloferax mediterranei* strain DSM 1411. *Chem Biochem Eng* Q 2015; 29 (2): 87–89.

52. Koller M, Maršálek L, Miranda de Sousa Dias M, et al. Producing microbial polyhydroxyalkanoate (PHA) biopolyester in a sustainable manner. *New Biotechnol* 2017; 37: 24–38.

53. Koller M. A review on established and emerging fermentation schemes for microbial production of polyhydroxyalkanoate (PHA) biopolyesters. *Fermentation* 2018; 4: 30.

54. Kovalcik A, Obruca S, Fritz I, et al. Polyhydroxyalkanoates. *BioResources* 2019; 14 (2): 2468–2471.

55. Kroon B M A. Variability of photosystem II quantum yield and related processes in *Chlorella pyrenoidosa* (chlorophyta) acclimated to an oscillating light regime simulating a mixed photic zone. *J Phycol* 1994; 30: 841–852.

56. Kucera D, Pernicová I, Kovalcik A, et al. Characterization of the promising poly(3-hydroxybutyrate) producing halophilic bacterium *Halomonas halophila*. *Bioresour Technol* 2018; 256: 552–556.

57. Kumar K, Dasgupta C N, Nayak B, et al. Development of suitable photobioreactors for CO_2 sequestration addressing global warming using green algae and cyanobacteria. *Bioresour Technol* 2011; 102 (8): 4945–4953.

58. Lee J, Lee S Y, Park S, et al. Control of fed-batch fermentations. *Biotechnol Adv* 1999; 17: 29–48.

59. Lee W S, Chua A S M, Yeoh H K, et al. A review of the production and applications of waste-derived volatile fatty acids. *Chem Eng J* 2014; 235: 83–99.

60. Lemoigne M. Produits de dehydration et de polymerisation de l'acide ß-oxobutyrique. *Bull Soc Chim Biol* 1926; 8: 770–782.

61. Liebergesell M, Rahalkar S, Steinbüchel A. Analysis of the *Thiocapsa pfennigii* polyhydroxyalkanoate synthase: subcloning, molecular characterization and generation of hybrid synthases with the corresponding *Chromatium vinosum* enzyme. *Appl Microbiol Biotechnol* 2000; 54 (2): 186–194.

62. Luimstra V M, Schuurmans J M, Verschoor A M, et al. Blue light reduces photosynthetic efficiency of cyanobacteria through an imbalance between photosystems I and II. *Photosynth Res* 2018; 138: 177–189.

63. Madison L L, Huisman G W. Metabolic engineering of poly(3-hydroxyalkanoates): from DNA to plastic. *Microbiol Mol Biol Rev* 1999; 63 (1): 21–53.

64. Mangan N M, Flamholz A, Hood R D, et al. pH determines the energetic efficiency of the cyanobacterial CO_2 concentrating mechanism. *PNAS* 2016; 113 (36): E5354–E5362.

65. Mannan R M, Pakrasi H B. Dark heterotrophic growth conditions result in an increase in the content of photosystem II units in the filamentous cyanobacterium *Anabaena variabilis* ATCC 29413. *Plant Physiol* 1993; 103: 971–977.

66. Masojídek J, Prášil O. The development of microalgal biotechnology in the Czech Republic. *J Ind Microbiol Biotechnol* 2010; 37 (12): 1307–1317.

67. Meixner K, Fritz I, Daffert C, et al. Processing recommendations for using low-solids digestate as nutrient solution for poly-ß-hydroxybutyrate production with *Synechocystis salina*. *J Biotechnol* 2016; 240: 61–67.

68. Meixner K. *Integrating Cyanobacterial Poly(3-Hydroxybutyrate) Production into Biorefinery Concepts*. PhD Thesis at the University of Natural Resources and Life Sciences, Vienna, 2018.

69. Metz B, Davidson O, De Coninck H, et al. *IPCC Special Report on Carbon Dioxide Capture and Storage*. Cambridge University Press, Cambridge, UK and New York, NY, USA.

70. Mifune J, Nakamura S, Toshiaki F. Engineering of pha operon on *Cupriavidus necator* chromosome for efficient biosynthesis of poly(3-hydroxybutyrate-co-3-hydroxyhexan oate) from vegetable oil. *Polym Degrad Stab* 2010; 95 (8): 1305–1312.

71. Mishra S, Mishra D R, Lee Z, et al. Quantifying cyanobacterial phycocyanin concentration in turbid productive waters: a quasi-analytical approach. *Remote Sens Environ* 2013; 133: 141–151.

72. Miyasaka H, Okuhata H, Tanaka S, et al. Polyhydroxyalkanoate (PHA) production from carbon dioxide by recombinant cyanobacteria. In: Petre M (Ed.), *Environmental Biotechnology – New Approaches and Prospective Applications*. InTech, 2013.

73. Mohsenpour S F, Willoughby N. Luminescent photobioreactor design for improved algal growth and photosynthetic pigment production through spectral conversion of light. *Bioresour Technol* 2013; 142: 147–153.

74. Mozejko-Ciesielska J, Szacherska K, Marciniak P. Pseudomonas species as producers of eco-friendly polyhydroxyalkanoates. *J Polym Environ* 2019; 27: 1151–1166.

75. Murugan P, Han L, Gan C-Y, et al. A new biological recovery approach for PHA using mealworm, *Tenebrio molitor*. *J Biotechnol* 2016; 239: 98–105.

76. Nikodinovic-Runic J, Guzik M, Kenny S T, et al. Carbon-rich wastes as feedstocks for biodegradable polymer (polyhydroxyalkanoate) production using bacteria. *Adv Appl Microbiol* 2013; 84: 139–200.

77. Oliver R L, Hamilton D P, Brookes J D, et al. Physiology, blooms and prediction of planktonic cyanobacteria. In: Whitton B A (Ed.), *Ecology of Cyanobacteria II*. Springer, Dordrecht, 2012: 155–194.

78. Oren A. Cyanobacteria in hypersaline environments: biodiversity and physiological properties. *Biodivers Conserv* 2015; 24 (4): 781–798.

79. Osanai T, Azumaa M, Tanaka K. Sugarcatabolism regulated by light- and nitrogenstatus in the cyanobacterium *Synechocystis sp.* PCC 6803. *Photochem Photobiol Sci* 2007; 6: 508–514.

80. Pachekoski W M, Dalmolin C, Agnelli J A M. The influence of the industrial processing on the degradation of poly(hidroxybutyrate) – PHB. *Mater Res* 2013; 16: 237–332.

81. Pade N, Hagemann M. Salt acclimation of cyanobacteria and their application in biotechnology. Life (Basel) 2015; 5 (1): 25–49.

82. Panda B, Sharma L, Mallick N. Poly-β-hydroxybutyrate accumulation in *Nostoc muscorum* and *Spirulina platensis* under phosphate limitation. *J Plant Physiol* 2005; 162 (12): 1376–1379.

83. Panuschka S, Drosg B, Ellersdorfer M, et al. Photoautotrophic production of polyhydroxybutyrate – first detailed cost estimations. *Algal Res* 2019; 41: 101558.

84. Pattanayak G K, Lambert G, Bernat K, et al. Controlling the cyanobacterial clock by synthetically rewiring metabolism. *Cell Rep* 2015; 13 (11): 2362–2367.

85. Pulz O. Photobioreactors: production systems for phototrophic microorganisms. *Appl Microbiol Biotechnol* 2001; 57 (3): 287–293.

86. Ravzanaadii N, Kim S-H, Choi W H. Nutritional value of mealworm, *Tenebrio molitor* as food source. *Int J Indust Entomol* 2012; 25 (1): 93–98.

87. Richmond A (Ed.). *Handbook of Microalgal Culture: Biotechnology and Applied Phycology*. John Wiley & Sons, 2008.

88. Riedel S L, Jahns S, Koenig S, et al. Polyhydroxyalkanoates production with *Ralstonia eutropha* from low quality waste animal fats. *J Biotechnol* 2015; 214: 119–127.

89. Rippka R, Deruelles J, Waterbury J B, et al. Generic assignments, strain histories and properties of pure cultures of cyanobacteria. *General Microbiol* 1979; 111 (1): 1–61.

90. Rodriguez-Perez S, Serrano A, Pantion A A, et al. Challenges of scaling-up PHA production from waste streams: a review. *J Environ Manage* 2018; 205: 215–230.

91. Rubio F C, Camacho F G, Sevilla J M F, et al. A mechanistic model of photosynthesis in microalgae. *Biotechnol Bioeng* 2003; 81: 459–473.

92. Sánchez J L G, Berenguel M, Rodríguez F, et al. Minimization of carbon losses in pilot-scale outdoor photobioreactors by model-based predictive control. *Biotechnol Bioeng* 2003; 84 (5): 533–543.

93. Schindler D W. Evolution of phosphorus limitation in lakes. *Science* 1977; 195 (4275): 260–262.

94. Schlegel H G. *Allgemeine Mikrobiologie* [*Common Microbiology*]. Georg Thieme, Stuttgart and New York, NY, 1992.

95. Schmid M T, Song H, Raschbauer M, et al. Utilization of desugarized sugar beet molasses for the production of poly(3-hydroxybutyrate) by halophilic *Bacillus megaterium* uyuni S29. *Proc Biochem* 2019; 86: 9–15.

96. Šejnohová L, Maršálek B. Microcystis. In: Whitton B A (Ed.), *Ecology of Cyanobacteria II*. Springer, Dordrecht, 2012.

97. Serafim L S, Lemos P C, Albuquerque M G E, et al. Strategies for PHA production by mixed cultures and renewable waste materials. *Appl Microbiol Biotechnol* 2008; 81: 615–628.

98. Shruthi M S, Rajashekhar M. Effect of salinity and pH on the growth and biomass production in the four species of estuarine cyanobacteria. *J Algal Biomass Utln* 2014; 5 (4): 29–36.

99. Silvestrini L, Drosg B, Fritz I. Identification of four polyhydroxyalkanoate structural genes in *Synechocystis cf. salina* PCC6909: in silico evidences. *J Proteom Bioinform* 2016; 9 (2): 28–37.

100. Smith A J. Modes of cyanobacterial carbon metabolism. *Ann Inst Pasteur/Microbiol* 1983; 134 (1): 93–113.

101. Sobiechowska-Sasim M, Stoń-Egiert J, Kosakowska A. Quantitative analysis of extracted phycobilin pigments in cyanobacteria—an assessment of spectrophotometric and spectrofluorometric methods. *J Appl Phycol* 2014; 26 (5): 2065–2074.

102. Sosik H M, Mitchell B G. Absorption, fluorescence, and quantum yield for growth in nitrogen-limited *Dunaliella tertiolecta*. *Limnol Oceanogr* 1991; 36 (5): 910–921.

103. Stal L J. Poly(hydroxyalkanoate) in cyanobacteria: an overview. *FEMS Microbiol Rev* 1992; 9 (2–4): 169–180.

104. Stal L J. Cyanobacterial mats and stromatolites. In: Whitton B A (Ed.), *Ecology of Cyanobacteria II*. Springer, Dordrecht, 2012: 65–125.

105. Stotz C E, Topp E M. Applications of model beta-hairpin peptides. *J Pharm Sci* 2004; 93 (12): 2881–2894.

106. Sudesh K, Taguchi K, Doi Y. Can cyanobacteria be a potential PHA producer? *RIKEN Rev* 2001; 42: 75–76.

107. Sudesh K, Taguchi K, Doi Y. Effect of increased PHA synthase activity on polyhydroxyalkanoates biosynthesis in *Synechocystis* sp. PCC6803. *Int J Biol Macromol* 2002; 30 (2): 97–104.

108. Takahashi H, Miyake M, Tokiwa Y, et al. Improved accumulation of poly-3-hydroxybutyrate by a recombinant cyanobacterium. *Biotechnol Lett* 1998; 20 (2): 183–186.

109. Taroncher-Oldenburg G, Nishina K, Stephanopoulos G. Identification and analysis of the polyhydroxyalkanoate-specific ß-ketothiolase and acetoacetyl coenzyme A reductase genes in the cyanobacterium *Synechocystis* sp. strain PCC6803. *Appl Environ Microbiol* 2000; 66 (10): 4440–4448.

110. Thirumala M, Reddy S V. Production of PHA by recombinant organisms. *Int J Life Sci Biotechnol Pharm Res* 2012; 1 (2): 40–62.

111. Touloupakis E, Cicchi B, Silva Benavides A M, et al. Effect of high pH on growth of *Synechocystis* sp. PCC 6803 cultures and their contamination by golden algae (*Poterioochromonas* sp.). *Appl Microbiol Biotechnol* 2016; 100: 1333–1341.

112. Troschl C, Meixnwer K, Drosg B. Cyanobacterial PHA production—review of recent advances and a summary of three years' working experience running a pilot plant. *Bioengineering (Basel)* 2017; 4(2): 26.

113. Troschl C, Meixner K, Fritz I, et al. Pilot-scale production of poly-β-hydroxybutyrate with the cyanobacterium *Synechocytis sp.* CCALA192 in a non-sterile tubular photobioreactor. *Algal Res* 2018; 34: 116–125.

114. Vaz Rossell C E, Mantelatto P E, Agnelli J A M, et al. Sugar-based biorefinery – technology for integrated production of poly(3-hydroxybutyrate), sugar, and ethanol. In: Kamm B et al. (Eds.), *Biorefineries-Industrial Processes and Products*. Wiley-VCH Verlag GmbH, 2005: 209–226.

115. Volova T, Demidenko A, Kiselev E, et al. Polyhydroxyalkanoate synthesis based on glycerol and implementation of the process under conditions of pilot production. *Appl Microbiol Biotechnol* 2019; 103: 225–237.

116. Wang B, Pugh S, Nielsen D R, et al. Engineering cyanobacteria for photosynthetic production of 3-hydroxybutyrate directly from CO_2. *Metabol Eng* 2013; 16: 68–77.

117. Whitton B A, Potts M. Introduction to the cyanobacteria. In: Whitton B A (Ed.), *Ecology of Cyanobacteria II*. Springer, Dordrecht, 2012.

118. Wu G F, Shen Z Y, Wu Q Y. Modification of carbon partitioning to enhance PHB production in *Synechocystis sp.* PCC6803. *Enzyme Microb Technol* 2002; 30: 710–715.

119. Wyman M, Fay P. Underwater light climate and the growth and pigmentation of planktonic blue-green algae (Cyanobacteria) I. The influence of light quantity. *Proc R Soc Lond Ser B Biol Sci* 1986; 227: 367–380.

120. Yamane T. Yield of poly-D(-)-3-hydroxybutyrate from various carbon sources: a theoretical study. *Biotechnol Bioeng* 1993; 41: 165–170.

121. Yu L-P, Wu F-Q, Chen G-Q. Next-generation industrial biotechnology – transforming the current industrial biotechnology into competitive processes. *Biotechnol J* 2019; 14: 1800437.

14 Coupling Biogas with PHA Biosynthesis

Yadira Rodríguez, Victor Pérez, Juan Carlos López, Sergio Bordel, Paulo Igor Firmino, Raquel Lebrero, and Raúl Muñoz

CONTENTS

14.1 INTRODUCTION

Massive research efforts have been devoted to the development and improvement of biodegradable and environmentally friendly alternative materials as a result of the severe environmental problems caused by petroleum-based plastics [1]. In this context, biopolymer-based plastics, such as the microbiologically synthesized polyhydroxyalkanoates (PHA), constitute a viable alternative for replacing those synthetic plastics [2]. PHA such as poly(3-hydroxybutyrate) [PHB, a.k.a. P(3HB)], poly(3-hydroxyvalerate) (PHV), and their copolymer poly(3-hydroxybutyrate-*co*-3-hydroxyvalerate) (PHBV) are biopolymers produced intracellularly as cytoplasmic inclusions under conditions of carbon excess and limitation in nutrients such as nitrogen, phosphorus, and even oxygen. PHA are used as carbon and energy storage compounds by several species of Gram-positive and Gram-negative bacteria, including methanotrophs [3].

PHA are considered as the most promising bioplastics because, besides being completely biodegradable, they possess outstanding mechanical properties similar to those of polyethylene (PE) and polypropylene (PP), and are biocompatible, piezoelectric, optically pure, thermoplastic, elastomeric, non-toxic, and more resistant to ultraviolet light degradation than PP [1,2,4]. Consequently, PHA hold untapped potential for multiple applications in medicine (e.g., tissue engineering), pharmacy (e.g., drug delivery), biofuel production (e.g., methyl esterification), agriculture (e.g., controlled release of pesticides), and manufacturing of disposable items (e.g., razors, packaging) [1,2].

However, despite their many advantages over petroleum-based plastics, PHA still have higher production costs (approximately 5–10 times) than those of conventional plastics [1]. In fact, up to 40–50% of the overall PHA production cost is attributed to the carbon source used in the biosynthesis process, of which glucose and fructose are the most common substrates [1,4]. Therefore, there is an urgent need to lower the costs of PHA production using residual materials (e.g., whey, wheat bran, molasses) as alternative feedstocks to make biopolymer-based plastics more competitive [1]. In this context, methane (CH_4), the main constituent of natural gas and biogas, has emerged as a more economical and environmentally friendly feedstock for PHA production due to its wide availability and abundance. This may reduce the carbon source cost by more than 50% [5,6] as well as the overall environmental impacts of the process, particularly global warming [7]. Consequently, methane-based bioplastic production has attracted significant investments over the past decade, where the companies Mango Materials and Newlight Technologies are pioneers in the development of biotechnologies for the production of such alternative materials [4,8]. The production of CH_4-based PHA relies on the assimilative metabolism of a specific group of methanotrophic Gram-negative bacteria belonging to the α-proteobacteria class, which are traditionally classified as type II methanotrophs. Under nutrient-limiting (usually nitrogen) conditions, type II methanotrophs can convert methane into PHA, mainly PHB. PHA production is governed by environmental factors such as pH, temperature, and concentration of methane, oxygen, carbon dioxide, macronutrients (e.g., nitrogen and magnesium), and micronutrients (e.g., copper and cobalt) [9]. Moreover, the low solubility of methane in water (dimensionless Henry's law constant of 30 at 25°C) still represents a key limitation to be addressed in PHA production from CH_4, especially at the industrial scale. Thus, novel high-performance off-gas bioreactors should be designed to enhance gas-liquid mass transfer and increase methane conversion yields [4,10].

Finally, it should be stressed that the use of methane, the second most abundant anthropogenic greenhouse gas (GHG) after carbon dioxide, as feedstock for PHA production, is not only an effective GHG-mitigating measure, but also a cost-competitive strategy of valorizing methane. This is particularly relevant in the field of anaerobic digestion, where biogas can be bioconverted into a high value-added product such as PHB instead of into electricity, and the rapid expansion of the global biogas market is expected to lower the price of this commodity [11]. In this chapter, PHA production from CH_4 will be critically reviewed from a holistic point of view, with a special focus on the use of biogas as a feedstock. The microbiological and engineering aspects of PHA biosynthesis from biogas will be described, and, finally, the process will be evaluated through a techno-economic analysis.

14.2   BIOGAS MARKET

Biogas can be generated from the anaerobic digestion of wastewater and biodegradable organic waste from the reduction of industrial CO_2 emissions with H_2 (produced from renewable energy) as an electron donor using a power-to-gas approach, or from the gasification of lignocellulosic biomass followed by CO reduction to CH_4 using renewable H_2 [12]. This chapter will focus on biogas generated from anaerobic digestion, which is the most common technology to generate biogas worldwide. Biogas from the anaerobic digestion of livestock waste, agro-industrial biomass, or municipal solid waste in closed digesters is composed of CH_4 (50–70%), CO_2 (45–30%), H_2S (0–2%), and other pollutants such as O_2, N_2, NH_3, volatile organic compounds, and siloxanes [13]. In the past few years, there has been a gradual change in the perception of anaerobic digestion from a simple waste reduction technology to a platform capable of generating renewable electricity. In contrast, in recent years, anaerobic digestion is regarded as the core of a multiproduct biorefinery. Indeed, anaerobic digestion is an effective carbon and nutrient recycling platform that can generate biofertilizers (from the nutrients contained in both the liquid and solid fraction of digestate), electricity and industrial heat (using internal combustion engines or turbines), biomethane for injection into natural gas grids or use as vehicle fuel, or used as a feedstock (biogas) for the production of fine and bulk chemicals [14,15]. Biogas quality requirements depend on the final use of biogas. Thus, internal combustion engines typically require siloxane concentrations lower than 9–44 ppm$_v$ and H_2S concentrations lower than 200–1000 ppm$_v$. Turbines and microturbines require very low levels of siloxanes but are very tolerant of H_2S (<10000–70000 ppm$_v$). Fuel cells require low levels of siloxanes and H_2S. In contrast, the quality of biomethane for injection into natural gas grids and use as a vehicle fuel is regulated in Europe by the Standard EN-16723 [16].

The number of biogas production plants in Europe has increased from 6000 in 2009 to almost 18000 by the end of 2017, while 2200 and 7000 full-scale biogas plants were in operation in 2015 in the United States and China, respectively [17,18]. Primary energy production from biogas in Europe has increased exponentially, although, in the past few years, its growth rate has decreased as a result of legal limitations in the use of energy crops and the lower tax incentives for electricity production from biogas. In 2016, 1.5 million tons of oil equivalent (Mtoe) of biogas were produced in wastewater treatment plants in Europe, 3 Mtoe in landfills, and 12 Mtoe from other sources such as livestock waste, urban solid waste, and energy crops [19]. The annual production of biogas in the EU will reach 41 Mtoe by 2030, according to the European Biogas Association, while the annual biogas production potential in the world has been estimated to 658 Mt CH_4. Similarly, the global electricity generation from biogas increased from 46108 GWh in 2010 up to 87500 GWh by the end of 2016 [18]. Today, the biogas sector employs more than 344000 people worldwide [18].

The number of biogas upgrading plants (converting biogas into biomethane) has increased exponentially in the past years, with 470 plants in operation by the end of 2016 in the member countries of the International Energy Agency and more than 700 plants worldwide [12,18]. Biomethane production in Europe rapidly increased from 750 GWh in 2011 to 19542 GWh by the end of 2017, which corresponds to an increase from 0.08 to 1.94 million Nm^3 [17]. Germany is by far the country with the largest number of biogas upgrading plants in the world, followed by the UK, Sweden, France, Switzerland, and the Netherlands. The price of biomethane decreased from

€0.12 kWh^{-1} at low upgrading capacities of 80 Nm^3 h^{-1} down to €0.08 kWh^{-1} at upgrading capacities of 500 Nm^3 h^{-1} [20]. Today, there is a large portfolio of CO_2 removal technologies available: water scrubbing, chemical scrubbing and organic solvent scrubbing, pressure swing adsorption, membrane separation, and cryogenic separation rank among the most popular physical-chemical technologies. In 2012, the biogas upgrading market was dominated by water scrubbing, followed by chemical scrubbing and pressure swing adsorption. Membrane separation accounted only for 10% of the market share. Today, water scrubbing is still the dominant technology closely followed by membrane separation, with 25% of the market share [12]. However, photosynthetic, hydrogenotrophic, and electromethanogenesis upgrading have emerged in the past decade as more sustainable technologies [21].

The reduction in tax incentives or feed-in tariffs for electricity production from biogas, as a result of the rapidly decreasing prices of solar or wind power electricity, requires innovative biogas valorization strategies. Thus, biomethane could be catalytically reformed into CO and used as a building block in the chemical industry for the synthesis of building blocks [15]. Biogas could also be bioconverted into biopolymers, ectoine, single-cell protein, exopolysaccharides, and chemical building blocks using methanotrophs. More specifically, the biogas produced during the anaerobic digestion of organic substrates can be bioconverted into polyhydroxyalkanoates using type II methanotrophs under nutrient-limiting conditions, which will be the core topic of this chapter.

14.3 METHANOTROPHS

Methanotrophs are methylotrophic microbes with the ability to use CH_4 as their sole carbon and energy source. These microorganisms are ubiquitous in nature—rice paddies, peat bogs, landfill cover soils, sewage sludge, or marine sediments among other locations—and can be both aerobic or anaerobic, the latter usually related to sulfate, nitrate, nitrite, manganese, and/or iron reduction processes [6]. Of them, aerobic methanotrophs represent those with higher potential as bacterial cell factories in CH_4 valorization platforms since they have been consistently proved to synthesize multiple bioproducts from a single fermentation process [11].

In this regard, aerobic methanotrophs mainly belong to two different taxonomical phyla: Proteobacteria (subdivisions γ- and α-proteobacteria corresponding to those traditionally classified as type I and II methanotrophs, respectively) and Verrucomicrobia (recently designated as type III methanotrophs and corresponding to few extremophilic genera). *Methylomonas, Methylobacter, Methylosarcina,* and *Methylococcus* are the most representative type I genera within γ-proteobacteria, while *Methylocystis, Methylosinus, Methylocella,* and *Methylocapsa* rank among the most representative type II genera within the α-proteobacteria phylum. Finally, *Methylacidiphilum* and *Methylacidimicrobium* genera are the most significant affiliates of Verrucomicrobia [22]. All these methanotrophs use the enzyme methane monooxygenase (MMO), either in its particulate (pMMO) or soluble (sMMO) form, to aerobically oxidize CH_4 sequentially to methanol and formaldehyde. Then, formaldehyde may be completely oxidized to CO_2 to generate reducing equivalents or assimilated into new biomass. In this context, it must be stressed that the different methanotroph types differ not only in their physiological and morphological features but also in their carbon assimilation pathways. Thus, type I and type II methanotrophs take up carbon (at the level of formaldehyde) via the ribulose monophosphate (RuMP) and the serine cycle, respectively. Type III methanotrophs do not assimilate carbon using these pathways, but they oxidize CH_4 to CO_2 and fix the latter via the Calvin–Benson–Bassham (CBB) cycle [23].

In terms of CH_4 bioconversion efficiencies through these metabolic pathways, the RuMP pathway of type I methanotrophs results in higher growth yields considering the lower quantity of adenosine triphosphate (ATP) molecules (1) required for the assimilation of each three formaldehyde molecules, compared with those of type II and III methanotrophs. These figures explain the fact that type I methanotrophs predominate in mixed methanotrophic cultures unless selective culture conditions favor the other types. In this regard, different selection strategies have been assessed in the last decade to enrich type II methanotrophs and subsequently stress them to produce high added-value products, such as short-chain-length polyhydroxyalkanoates (scl-PHA) [24].

The current key parameters selecting type II methanotrophs are the O_2:CH_4 ratio, temperature, pH, the nitrogen source and concentration, and micronutrient conditions. Specifically, the growth of type II methanotrophs can be promoted in mixed cultures under low O_2:CH_4 ratios (dissolved oxygen <30% saturation in air), which preferentially induce the expression of the sMMO enzyme (predominant in type II methanotrophs) [25]. This fact is of key relevance in scenarios where N_2 is used as the sole nitrogen source, since O_2 may inactivate the nitrogenase enzyme encoded in type II methanotrophs (and in a few type I strains) [26]. However, recent studies have also highlighted the key role of increasing temperatures over those typically employed (25°C) to successfully enrich high-rate PHA-producing type II methanotrophs from mixed cultures [27]. In addition, despite most methanotrophs preference to grow at neutral pH values, the selection of type II over type I methanotrophs in mixed cultures may be promoted at pH values of 3.6–6, possibly due to the CO_2 requirements in the serine cycle [28]. Finally, despite how nitrate typically results in the highest biomass production rates compared with N_2 or NH_4^+, unfortunately, nitrate does not represent the ideal nitrogen source to effectively select type II methanotrophs in mixed cultures [29]. In contrast, selective type II methanotrophic enrichments have been achieved using both N_2 (at limited dissolved oxygen levels) and ammonium [30,31] (see Table 14.1). In the particular case of ammonium, the most rapid enrichment and highest CH_4 biodegradation kinetics were obtained at low nitrogen concentrations (<10 mM NH_4^+), which might be attributed to the i) preferential growth of type II over type I methanotrophs at high C/N ratios and ii) lack of byproducts (i.e., hydroxylamine, nitrite) that might be produced at higher concentrations and ultimately inhibit the growth of type II methanotrophs. Finally, the type and concentration of micronutrients present in the cultivation medium (i.e., copper, carbonate) have been shown to influence the competition between type I and II methanotrophs. More specifically, copper concentration determines the synthesis of the enzyme sMMO in type II methanotrophs and should be minimized to promote the enrichment of PHA synthetizing methanotrophs. In this regard, Pieja et al. found that type II methanotrophs may be effectively selected from mixed cultures not only by combining low pH values and low carbonate concentration

TABLE 14.1

Enrichment Procedures Devoted to Preferentially Select Type II Methanotrophs in Mixed Microbial Cultures

Inoculum	Condition of Selection	Other Enrichment Conditions[a]	Predominant Genera (% abundance)	Ref.
M. trichosporium OB3b/*M. albus* BG8 co-culture	DO levels <30% saturation in air at constant CH$_4$ concentration (3.5% v/v)	Copper limitation	*M. trichosporium* OB3b (~100%)	[25]
Hot spring sediments	N$_2$ as nitrogen source	CH$_4$ excess ~1:2.5 O$_2$:CH$_4$	*Methylocystis/Met hylosinus*	[30]
Activated sludge	i) No Cu, diluted medium (10%) ii) pH 5, carbonate at 10 mM	–	*Methylocystis/Met hylosinus*	[28]
Mixed landfill sample	NH$_4^+$ (4 mM) as N source	5:1 O$_2$:CH$_4$	*Methylocystis* (>80%)	[31]
Sphagnum and activated sludge	37°C	–	*Methylocystis* (~30%)	[27]

[a] Unless otherwise specified, temperature and pH were adjusted at 25°C and 6.8–7.5, respectively.

(1mM) but also by combining a diluted mineral salts medium (10%) in the absence of copper [28] (see Table 14.1).

14.4 PHA BIOSYNTHESIS FROM METHANE

The process of accumulating PHA from methane corresponds to bacteria belonging to the α-proteobacteria class (also known as type II methanotrophs), which assimilate C1 compounds via the serine pathway [28] instead of the RuMP cycle, typically found in γ-proteobacterial methanotrophs. Methane conversion (Figure 14.1) is initiated by the enzyme MMO that mediates the production of methanol, which is further oxidized into formaldehyde by the enzyme methanol dehydrogenase (MeDH) [containing the pyrroloquinoline quinone (PQQ) as the catalytic center] [32]. In type II methanotrophs, formaldehyde can be metabolized into formate via two pathways involving the participation of tetrahydromethanopterin (H$_4$MPT) or tetrahydrofolate (H$_4$F) dependent enzymes, and whereby the substrate of the serine cycle (methylene–H$_4$F) is generated [22,33]. Methylene–H$_4$F reacts with glycine to form serine. The serine route ultimately yields malyl-CoA, which in turn produces glyoxylate and acetyl-CoA. The latter either enters the tricarboxylic acid (TCA) cycle in nutrient sufficient conditions or the PHA biosynthetic pathway in nutrient-limiting conditions [34]. More precisely, the presence of coenzyme A, which is released from the TCA cycle in nutrient sufficient conditions and acts as an inhibitor of one of the key enzymes (β-ketothiolase) of the PHA cycle, determines the active metabolic route [6]. In the PHA pathway, the β-ketothiolase enzyme (*phaA* gene) converts acetyl-CoA

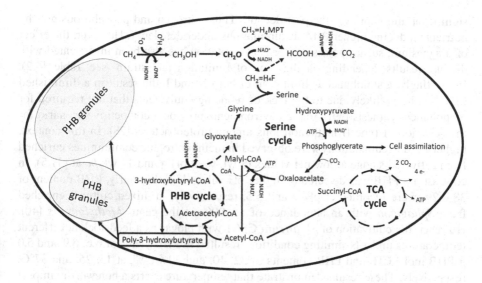

FIGURE 14.1 Schematic representation of methane oxidation and PHB synthesis in type II methanotrophs.

into acetoacetyl-CoA, followed by the production of hydroxyacyl-CoA thioester by the acetoacetyl-CoA reductase enzyme (*phaB* gene). Subsequently, the enzyme PHA synthase (*phaC* gene) enables PHA formation via ester linkages. It must be emphasized that type II methane-oxidizing bacteria produce only one type of PHA, poly(3-hydroxybutyrate) [PHB, a.k.a. P(3HB)], when grown in CH_4 as the sole energy and carbon source [8,30]. The role of the internally accumulated PHB and its mechanism of consumption by the methanotrophic bacteria is discussed in section 14.5. The theoretical yield for PHB accumulation reported in the literature varies from 0.54 [35] to 0.67 gPHB gCH_4^{-1} [36]. Higher yields result from not considering the regeneration of $NADP^+$ into NADPH needed for acetoacetyl-CoA production during PHB synthesis, which is presumably mediated by isocitrate dehydrogenase in the TCA cycle [35].

Experimental yields and PHB content strongly depend on the culture conditions enrichment strategy and the feast-famine strategy selected, which typically consists of two steps: i) period of cell growth and ii) a period of nutrient deprivation triggering PHA synthesis. The genus *Methylocystis*, which predominates in most of the enriched mixed cultures, and to a lesser extent, the genus *Methylosinus*, are the genera most intensively studied for PHA synthesis due to their potential to produce bioplastics [6,37]. In this regard, PHB content up to 60% [26], 54% [38], and 51% (on a dry weight basis) [39] have been reported for the species *Methylocystis parvus* OBBP, *Methylocystis hirsuta* CSC1, and *Methylosinus trichosporium* OB3b, respectively. Although the use of methane-utilizing mixed cultures seems to be preferred due to the potential benefits of a stable microbial community, under non-aseptic conditions, the use of pure strains is foreseen to be viable for continuous PHA production.

The influence of environmental factors (e.g., nutrient requirements and limitation, pH, temperature, and $O_2:CH_4$ ratio) on PHB cell synthesis from CH_4 has received significant attention over the past decade. Thus, nitrogen and phosphorous are the nutrients inducing higher PHA accumulations under deficiency. However, the effect of magnesium, sulfur, potassium, or iron limitation has also been investigated with distinct results depending on the type of limitation and strain (see Table 14.2). Interestingly, a simultaneous limitation of both N and P did result in a diminished PHB synthesis, likely due to the lack of reducing equivalents that are required for biopolymer synthesis [39]. Acidic environments provide a competitive scenario for the selection of type II methanotrophs among proteobacteria [28]. In this context, higher PHB accumulations were observed in methanotrophic communities enriched from activated sludge at low pH values (8 wt.-% at pH 4 and 14 wt.-% at pH 5). In contrast, no PHB was detected at higher pH values [28]. Similarly, PHB content of 38–42% were recorded at pH 7 and 5.5, respectively, in mixed cultures enriched from *Sphagnum* with an abundance of 85–90% in the genus *Methylocystis* [40]. However, the cultivation of *M. hirsuta* CSC1 with biogas as a feedstock at different temperatures under N-limiting conditions resulted in PHB yields of 6.6, 6.9, and 5.0 g PHB mol^{-1} CH$_4$ and PHB contents of 32, 40, and 30 wt.-%, at 15, 25, and 37°C, respectively. These results demonstrate that temperature exerts a noteworthy impact on the PHB synthesis capacity of methanotrophs [41].

A promising feature of these microorganisms is their ability to perform copolymerization of 3-hydroxybutyrate (3HB) with alternative hydroxyalkanoate (HA) monomers, thus allowing the biosynthesis of copolymers such as poly(3-hydroxybutyrate-*co*-3-hydroxyvalerate) with superior mechanical properties (i.e., lower melting and glass transition temperatures and higher elongation-to-break ratios) [8,34]. In this regard, process supplementation with valerate in nutrient-limiting conditions supported an hydroxyvalerate (HV) fraction in the copolymer of 60 mol-% by *M. parvus* OBBP and *Methylocystis* sp. WRRC1 [42,43], and 25 mol-% by *M. hirsuta* [38]. Recent approaches focused on the molecular design of the copolymer through the addition of ω-hydroxyalkanoate monomers as co-substrates. Thus, Myung et al. successfully tailored PHA consisting of monomer units other than 3HB and 3HV [poly(3-hydroxybutyrate-*co*-4-hydroxybutyrate), poly(3-hydroxyvalerate-*co*-5-hydroxyvalerate-*co*-3-hydroxyvalerate), and poly(3-hydroxybutyrate-*co*-6-hydroxyhexanoate-*co*-4-hydroxybutyrate)] using the strain *M. parvus* OBBP in nitrogen-limiting conditions [44]. Apart from the direct copolymer synthesis in methanotrophic cultures, other potential strategies for the production of versatile PHA using CH$_4$ as a primary feedstock include the use of a co-culture that promotes PHBV synthesis by accumulating species, such as *Ralstonia eutropha*, or downstream PHA modification, which would ultimately result in intensive and costly processes [34].

14.5   GENOME SCALE METABOLIC MODELS AS A TOOL FOR UNDERSTANDING THE METABOLISM OF PHB IN METHANOTROPHS

Genome scale metabolic models (GSMMs) are full compilations of all the metabolic reactions taking place in a particular microorganism. These models can be reconstructed based on genome annotations and experimental evidence such as outcomes

TABLE 14.2
PHA Yields and Content (wt.-%) in Type II Strains Using Different Sources of Methane and Nutrient Limitations

CH₄ source (Purity)	Type II Strain	System/O₂:CH₄ atmosphere % (v/v)	Limiting nutrient	PHA Content (wt.-%)	PHA yield (g_PHB g_CH₄⁻¹)	Ref.
Methane (>99.5%)	Methylocystis sp. GB25	Stirred tank reactor 80:20	N/P/Mg	51/47/28	0.52/0.55/0.37	[45]
Methane (>99.5%)	Methylocystis sp. GB25 (>86%)	Stirred tank reactor 75:25	P	46.2	–	[46]
Methane (>99.5%)	Methylocystis sp. GB25 (>86%)	Stirred tank reactor 75:25	K/S/Fe	34/33/10	0.45/0.40/0.22	[47]
Methane (NA)	Methylosinus trichosporium OB3b/ Methylocystis parvus OBBP	Serum bottles 50:50	N	45/60 (Grown in N₂ and NH₄⁺, respectively)	1.1/0.9	[26]
Methane (NA)	Methylocystis parvus OBBP	Microbioreactors 50:50	N (control)/N + Cu/N + Ca/N + K/N + P/N + Cu + Ca	18/35/39/28/31/49	–	[48]
Methane (NA)	Methylocystis trichosporium OB3b	Serum bottles 50:50	N	51/45/32 (grown in NO₃⁻/NH₄⁺/N₂, respectively)	–	[39]
Methane (>99.5%)	Methylocystis hirsuta CSC1	Serum bottles 66:33	Mn/K/N/(N + Fe excess)	8/13/28/19	–	[49]
Synthetic biogas (70% CH₄, 30% CO₂, 0.5 H₂S)	Methylocystis hirsuta CSC1	Serum bottles 65:35	N (NO₃⁻)	43.1	0.44	[38]
Synthetic biogas (70% CH₄, 30% CO₂)	Methylocystis hirsuta CSC1	Serum bottles 50:50/60:40/66:33	N (NO₃⁻)	16/15/45.3	0.14/0.16/0.43	[41]

NA: not available

of gene knockouts, the ability to grow on different substrates, or enzymatic tests. Metabolic modeling of methanotrophs is still in its infancy. Currently, only six GSMMs of methanotrophs have been published [50–53]. These models correspond to the species *Methylomicrobium buryatense*, *Methylomicrobium alcaliphilum*, and four *Methylocystis* species (*M. parvus*, *M. hirsuta*, *M.* sp. SC2, and *M.* sp. SB2). Four of these species are natural PHB producers.

GSMMs allow us to predict theoretical yields. The theoretical PHB yield on methane for *Methylocystis* species was found to be 0.10 mol PHB mol CH_4^{-1} or the equivalent to 0.53 g PHB g CH_4^{-1}. *Methylocystis hirsuta*, in nitrogen-starvation conditions and with an O_2:CH_4 ratio equal to 2:1, has been found to accumulate PHB with a yield of 0.088 ± 0.002 mol PHB mol CH_4^{-1}, which is very close to the predicted theoretical value. However, the main limitation to the efficient production of PHB is not its yield on methane but the capacity of the cells to accumulate large percentages of its dry biomass in the form of PHB.

The two initial steps of PHB synthesis are catalyzed by the enzymes acetyl-CoA, C-acetyltransferase, and acetoacetyl-CoA reductase. The genes coding these enzymes in *Methylocystis* (in all four species mentioned), phaA and phaB, respectively, form a gene cluster, flanked by a putative transcription repressor, phaR. The expression of this repressor is likely to be the reason for the activation or arrest of PHB synthesis. Therefore, its knockout or expression under the control of an inducible promoter could be a strategy to improve this process in the future.

Another factor to be considered to design efficient production processes, which involve cycles of nitrogen feeding and starvation, is the fact that PHB is used by the cells as a storage compound that can be used as an energy or carbon source for growth when nitrogen sources become available. The consumption of PHB in methanotrophs has been studied in *Methylocystis parvus* [54] (Figure 14.2), which revealed that when

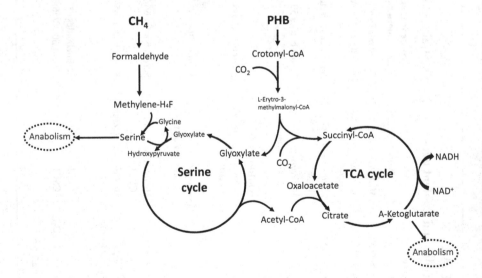

FIGURE 14.2 Schematic representation of methane and PHB co-consumption when both nitrogen and methane are supplied to PHB accumulating cells of *M. parvus*.

nitrogen supply is restored, PHB-containing cells use the biopolymer as a carbon and energy source for protein synthesis (evidenced by their increased protein content), but cells do not divide. However, when methane is available following N supply restoration, methanotrophic cells start dividing at a faster rate (compared with those without stored PHB), and methane and PHB are consumed simultaneously. Metabolic modeling shows that during this co-consumption, methane is used as an energy source, while PHB is used as a carbon source. This is because when methane is assimilated through the serine cycle (a mechanism common to all type II methanotrophs, including *Methylocystis*), it is transformed into acetyl-CoA, which itself is fed to the TCA cycle leading to the production of NADH and the subsequent production of ATP in the respiratory chain. For cells to grow, metabolic precursors must also be supplied for the synthesis of amino acids and other biomass building blocks. If metabolic precursors are to be drained from the serine or the TCA cycle, other reactions have to produce intermediates of these cycles to replenish the precursors drained for biosynthesis. These are called anaplerotic reactions and are essential for aerobic cell growth. In organisms that rely on glycolysis, the most common anaplerotic reaction is pyruvate carboxylase. For methanotrophs using the serine cycle, the hypothesis is that the anaplerotic function is played by glycine synthase [52], which produces glycine from methylene tetrahydrofolate and drains serine, from where the rest of biomass building blocks are synthesized. Simulations of methane and PHB co-consumption [53] revealed that in this scenario, PHB is degraded via crotonyl-CoA to glyoxylate and succinyl-CoA. Glyoxylate replenishes the serine cycle, and succinyl-CoA replenishes the TCA cycle, which allows cells to withdraw both serine and α-ketoglutarate for biosynthesis, leading to the observed decrease in duplication time for PHB-containing cells.

Table 14.3 shows the metabolic fluxes predicted by a GSMM constructed for *Methylocystis parvus* OBBP, which are consistent with the observed growth rates and methane and PHB consumption rates.

14.6 BIOREACTORS FOR BIOGAS BIOCONVERSION

Gas-phase packed bioreactors, such as biotrickling filters or biofilters, have been traditionally used for waste gas treatment. However, they are not suitable for bioproduct

---

TABLE 14.3

Rates of Key Metabolic Processes Active During PHB and Methane Co-Consumption by *Methylocystis Parvus* OBBP Compared with Growth on Methane [53]

		Growth on CH₄	Co-consumption of PHB and CH₄
μ	[1/h]	0.107	0.154
Methane consumption	[mmol h⁻¹ (g	14.9	13.2
PHB degradation	CDM)⁻¹]	0	0.608
Malyl-CoA lyase		12.3	11.8
Glycine synthase		1.3	0.41

recovery since biomass growth as biofilm hinders any subsequent biomass down-stream. Likewise, two-stage bioscrubbers exhibit a poor mass transfer for scarcely water-soluble pollutants such as CH_4 or O_2 (dimensionless Henry's law constants of 30 and 31.3 at 25°C, respectively) [55]. In this regard, suspended-growth bioreactors are the most suitable configuration for the bioconversion of CH_4-laden gas streams into added-value byproducts and their subsequent recovery.

An improved mass transfer in suspended-growth bioreactors can be achieved via an increase in the gas-liquid concentration gradient or the volumetric mass transfer coefficient ($K_la_{G/A}$). To date, the classical approach to enhance pollutant mass trans-port in so-called turbulent bioreactors was based on increasing the energy input to the bioreactor (to increase the turbulence in the liquid side) by intensive mixing and pumping. The most common turbulent bioreactors applied in industrial fermentative processes are mechanically stirred tanks and bubble column bioreactors with con-trolled nutrient feeding [10,56].

Stirred tank reactors (STRs) are vessels where the culture broth is mechanically agitated by an impeller, and the polluted gas is supplied via a sparger located at the bottom of the reactor. CH_4 removal efficiencies of 50–60% have been achieved at rela-tively low gas residence times of 4–10 min [57,58]. However, the high power to volume ratios, the difficulties of ensuring proper mixing and heat removal in large-scale bio-reactors, and the excessive shear stress are key drawbacks of this configuration [10].

In the case of bubble column bioreactors (BCBs), no mechanical agitation is provided, and the increase in the gas-liquid mass transfer is achieved either using micropore diffusers or installing a concentric draft-tube (riser) in airlift bioreactors (ALRs) [59]. Despite the lower power requirements compared to STRs, and therefore the higher cost-effectiveness of this configuration, non-homogeneous nutrient distri-bution due to poor liquid circulation, partial CH_4 utilization, and bubble coalescence are common shortcomings of BCBs [6,10]. The conventional approach to enhancing mass transport based on increasing the energy input to the bioreactor entails prohibi-tive operating costs during CH_4 bioconversion due to the intensive power consump-tion. In this context, novel reactor designs, such as Taylor flow, forced loop, and membrane diffusion bioreactors, and operating strategies, such as internal gas recy-cling or the addition of a non-aqueous phase, have been recently tested as suspended-growth platforms capable of supporting high CH_4 mass transfer rates compared with conventional biotechnologies at low-moderate energy demands [4,10].

14.6.1 Operating Strategies to Promote Mass Transfer

- Two-phase partitioning bioreactors (TPPBs) are characterized by the addi-tion of an immiscible, non-volatile, biocompatible, and non-biodegradable non-aqueous phase (NAP) with a high affinity for the target gas pollut-ant [60]. The NAP mediates an additional and more efficient pathway for the transport of CH_4 from the gas phase to the methanotrophic community (as a result of the increased concentration gradient) and an increase in the gas–water and gas–NAP interfacial areas (Figure 14.3a,b). For instance, the addition of 10% v/v silicone oil as NAP resulted in improved CH_4 removals of 30 and 47% in an STR and BCB, respectively [61,62].

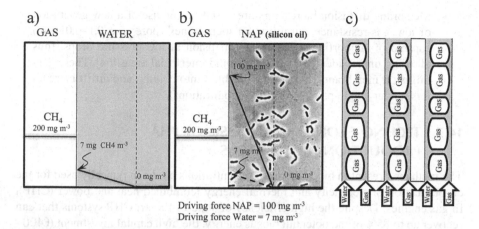

FIGURE 14.3 Concentration profiles and driving forces for CH_4 in (a) a conventional mass transfer limited bioreactor and (b) a two-phase partitioning bioreactor constructed with silicone oil as non-aqueous phase (NAP), and (c) segmented flow regime in Taylor flow bioreactors.

- The implementation of an internal gas recirculation allows the decoupling of the actual gas residence time and turbulence in the culture broth from the overall empty bed residence time. Studies evaluating this strategy for CH_4 bioconversion into PHB have achieved CH_4 removals over 70% at an empty bed residence time of 30 min and 0.50 m^3_{gas} $m^{-3}_{reactor}$ min^{-1} of the internal gas-recycling rate, recording a PHB content of up to 35% [49].

14.6.2 Novel Bioreactor Configurations

- Taylor flow bioreactors are capillary multi-channel units where the gas-liquid hydrodynamics consist of an alternating sequence of gas bubbles and liquid slugs. This segmented flow regime entails a high gas-liquid interfacial area along with a reduced liquid thickness and high turbulence at the liquid side (Figure 14.3c), which translates into mass transfer coefficients equivalent to those of turbulent reactors at the expense of one order of magnitude lower power consumptions. Nevertheless, this implementation for CH_4 abatement is limited to the study of Rocha-Rios et al. (2013) [63], where a 50% improvement in CH_4 removal was obtained compared with two-phase STRs at significantly smaller reactor sizes and pressure drops.
- Forced circulation loop bioreactors (FCLB) are a modification of ALRs that include internal and external recirculation pipes equipped with extra fittings to enhance gas-liquid mass transfer [6]. For instance, one of the most innovative industrial bioreactors designed to improve CH_4 abatement and the production of single-cell proteins is the U-Loop fermentor patented by Unibio A/S, which is capable of handling a large biomass concentration while providing a high gas to liquid mass transfer [4].

- Membrane diffusion bioreactors are based on the use of a new generation of low gas-resistance ultrafiltration membranes (pore size 0.1–0.01 μm) capable of supporting efficient gas diffusion at low-pressure drops, thus allowing unprecedentedly high gas-liquid interfacial areas [64]. These gas-diffusion membranes can be implemented into column and airlift bioreactors (under single-phase or TPPB configurations).

14.7 TECHNO-ECONOMIC ANALYSIS OF PHA PRODUCTION FROM BIOGAS

The methane present in biogas at high concentrations has been typically used for the co-production of electricity and thermal energy [combined heat and power (CHP)] in gas engines. Despite the high efficiency of state-of-the-art CHP systems that can recover up to 85% of the potential biogas energy, the high capital investment (€400–1100 kWh^{-1} installed), operating and maintenance costs (€0.01–0.02 kWh^{-1}), along with the limited lifespan (10 years) of CHP engines, constrain their economic viability [65]. Hence, large amounts of CH_4 (92 Mt y^{-1} of natural gas) produced in waste and wastewater treatment plants are being flared or vented to the atmosphere, substantially contributing to the global emission of greenhouse gases [66]. The current context of decreasing solar/wind energy prices and reduction of fiscal incentives for renewable energy production from biogas, together with the new strategies for waste and plastic management in the circular economy, have increased attention toward the valorization of the CH_4 contained in biogas as a precursor for the biopolymer industry [6]. This innovative biogas valorization can be particularly relevant for municipal and agro-industrial waste treatment facilities, which represents 80% of the total biogas plants installed in Europe [17].

Hence, the use of biogas generated from the anaerobic digestion of organic waste to produce PHA is emerging as an attractive alternative to conventional heat and power generation. Depending on the bacterial strain and the substrate used for PHA biosynthesis, the biopolymer production costs can range from €4–20 kg^{-1} PHA. At the same time, it is assumed that commercially viable selling prices should range between €3 and €5 kg^{-1} PHA [59,67]. In 1999, Choi and Lee identified the raw material cost (especially carbon source), energy requirements for sterilization, PHA yield on a substrate, and downstream processing as the major factors affecting PHA production costs [68]. A wide range of carbon feedstock such as glucose, sucrose, glycerol, and acetate have been investigated as candidates for bacterial PHA production. However, their acquisition costs represent 40–50% of the total biopolymer selling price. In this context, the use of biogas from landfills and waste treatment plants constitutes a reliable, low-cost, and globally available substrate for PHA production. Recent estimates revealed that biogas-to-PHA facilities conceived in a cradle-to-cradle design could potentially satisfy 20–30% of the global plastic demand [7]. In addition, a recent study has demonstrated the feasibility of using mixed methanotrophic cultures for the accumulation of PHA under nitrogen limitation using CH_4 as the only carbon and energy source, thus avoiding the need for sterilization and improving system robustness under long-term operation [27]. However, it has been estimated that biopolymer recovery and purification may account for up to 50% of the total polymer production costs, while a recent report estimated €1.5 kg^{-1} PHA as the maximum downstream cost for economically viable PHA production [59]. Therefore, the ideal extraction method should maintain polymer properties, achieve a high purity and recovery yield, and exhibit both low cost and low environmental impacts. In this context, cell digestion with chemicals (e.g., NaOH or sodium hypochlorite) or enzymatic cocktails has been regarded as the most applicable purification method [69].

Therefore, a techno-economic study was performed at the Institute of Sustainable Processes at the University of Valladolid, comparing the sustainability aspects of biogas combustion for CHP and biogas bioconversion into PHA. The scheme selected for the valorization of the biogas produced in a municipal waste processing plant (24,000 m^3 d^{-1}) into a commercially viable PHA included an anoxic desulfurization step prior to PHA biosynthesis in a bubble column bioreactor provided with internal gas recirculation. Subsequently, the PHA accumulated in the methanotrophic cells (40 wt.-%) was extracted by the addition of NaOH and subsequent washing with water and ethanol. This study demonstrated that PHA production from biogas constitutes is an economically viable, environmentally sustainable, and socially beneficial alternative to CHP in waste treatment plants. Despite the higher capital investment of PHA production from biogas (€5.7 million for CHP vs. €7.8 million for PHA production), this option did not present any additional financial risks in terms of net present value, internal rate of return, and payback period. Interestingly, it was demonstrated that PHA could be produced at a competitive price (€4.2 kg^{-1} PHA), regardless of the economy of scale, when a fraction of the biogas produced is used for internal energy provision. This value agrees with the current PHB resin price (€4.3 kg^{-1} PHA) and with those reported in a similar techno-economic study using pure CH$_4$ under pressurized conditions in a thermophilic reactor (€3.9 kg^{-1} PHA) [5,59]. Likewise, the values were below those first reported by Listewnik et al. [70] using natural gas (€11.5–14.0 kg^{-1} PHA) and below the median price found in the literature for all kinds of substrates (€6.8 kg^{-1} PHA) [5,70]. Lower prices were reported by Fernández-Dacosta et al. (€1.4–2.0 kg^{-1} PHA) [71] in a wastewater fermentation process in more favorable conditions for PHA accumulation (70% ww^{-1} vs. 40% ww^{-1} in our study) and oxygen transfer rate (500 vs. 60 g O$_2$ m^{-3} h^{-1} in our study). The price calculation included economic credits for the prevention of wastewater and biogenic carbon discharge [71]. Interestingly, Choi et al. reported a value of €1.5 kg^{-1} PHA during the co-production of H$_2$ and PHA from syngas [72].

Despite exhibiting higher land, water, energy, and chemical requirements, PHA production entails a significant reduction of atmospheric acidification and odor emissions compared with traditional biogas flaring and combustion in gas engines. Additionally, comprehensive lifecycle analyses have previously demonstrated a positive carbon footprint of biopolymer-from-waste schemes [7]. However, PHA production from biogas (and waste in general) is regarded as an opportunity for improving social and local acceptance of waste processing facilities. The increasing public demand for biobased products and greener processes as well as job creation associated with the production, commercialization, and product distribution of biopolymers were identified as key aspects for public acceptance within the new context of circular bioeconomy.

ACKNOWLEDGMENTS

The Ministry of Science, Innovation and Universities (BES-2016-077160 contract and CTM2015-70442-R project), the Regional Government of Castilla y León, and the EU-FEDER programme (CLU 2017-09, VA281P18 and UIC 71) are gratefully acknowledged.

REFERENCES

1. Raza Z, Abid S, Banat I. Polyhydroxyalkanoates: Characteristics, production, recent developments and applications. *Int Biodeterior Biodegrad* 2018; 126: 45–56.
2. Costa SS, Miranda AL, de Morais MG, et al. Microalgae as source of polyhydroxyal-kanoates (PHAs)—A review. *Int J Biol Macromol* 2019; 131: 536–547.
3. Laycock B, Halley P, Pratt S, et al. The chemomechanical properties of microbial poly-hydroxyalkanoates. *Prog Polym Sci* 2013; 38(3): 536–583.
4. Cantera S, Muñoz R, Lebrero R, et al. Technologies for the bioconversion of methane into more valuable products. *Curr Opin Biotechnol* 2018; 50: 128–135.
5. Levett I, Birkett G, Davies N, et al. Techno-economic assessment of poly-3-hydroxy-butyrate (PHB) production from methane—The case for thermophilic bioprocessing. *J Environ Chem Eng* 2016; 4: 3724–3733.
6. López JC, Rodríguez Y, Pérez V, et al. CH_4-Based polyhydroxyalkanoate production: A step further towards a sustainable bioeconomy. In: *Biotechnological Applications of Polyhydroxyalkanoates*, edn 1. Edited by: Kalia VC. Springer Singapore, 2019; pp. 283–321.
7. Rostkowski KH, Criddle CS, Lepech MD. Cradle-to-gate life cycle assessment for a cradle-to-cradle cycle: Biogas-to-bioplastic (and back). *Environ Sci Technol* 2012; 46(18): 9822–9829.
8. Pieja AJ, Morse MC, Cal AJ. Methane to bioproducts: The future of the bioeconomy? *Curr Opin Chem Biol* 2017; 41: 123–131.
9. Strong PJ, Xie S, Clarke WP. Methane as a resource: Can the methanotrophs add value? *Environ Sci Technol* 2015; 49(7): 4001–4018.
10. Stone KA, Hilliard MV, He QP, et al. A mini review on bioreactor configurations and gas transfer enhancements for biochemical methane conversion. *Biochem Eng J* 2017; 128: 83–92.
11. Strong PJ, Kalyuzhnaya M, Silverman J, et al. A methanotroph-based biorefinery: Potential scenarios for generating multiple products from a single fermentation. *Bioresour Technol* 2016; 215: 314–323.
12. IEA. *Green Gas. Facilitating a Future Green Gas Grid Through the Production of Renewable Gas.* IEA Bioenergy, 2018. http://task37.ieabioenergy.com/files/daten-re daktion/download/Technical Brochures/green_gas_web_end.pdf. Accessed: 20th August 2019.
13. Muñoz R, Meier L, Diaz I, et al. A review on the state-of-the-art of physical/chemical and biological technologies for biogas upgrading. *Rev Environ Sci Biotechnol* 2015; 14(4): 727–759.
14. Rodero MR, Ángeles R, Marín D, et al. Biogas purification and upgrading technolo-gies. In: *Biogas: Fundamentals, Process, and Operation*, edn 1. Edited by: Tabatabaei M, Ghanavati H. Springer International Publishing, 2018; pp. 239–276.
15. Verbeeck K, Buelens LC, Galvita VV, et al. Upgrading the value of anaerobic diges-tion via chemical production from grid injected biomethane. *Energy Environ Sci* 2018; 11(7): 1788–1802.

16. Bailón L, Hinge J. *Report: Biogas and Bio-Syngas Upgrading.* Danish Technological Institute, 2012. http://www.teknologisk.dk/_root/media/52679_Report-Biogas and syngas upgrading.pdf. Accessed: 20th August 2019.
17. EBA. *Statistical Report of the European Biogas Association 2018.* http://biogas.org .rs/wp-content/uploads/2018/12/EBA_Statistical-Report-2018_European-Overview-C hapter.pdf. Accessed: 20th August 2019.
18. WBA. *Global Potential of Biogas.* World Biogas Association, 2019. http://www.worl dbiogasassociation.org/wp-content/uploads/2019/09/WBA-globalreport-56ppa4_d igital-Sept-2019.pdf. Accessed: 20th August 2019.
19. EurObserv'ER. *Biogas Barometer 2017.* https://www.eurobserv-er.org/biogas-baro meter-2017/.
20. Biosurf. *D3.4 I Technical-Economic Analysis for Determining the Feasibility Threshold for Tradable Biomethane Certificates.* BIOSURF Fuelling Biomethane, 2016. https:// ec.europa.eu/research/participants/documents/downloadPublic?documentIds=08016 6e5aa5ba7aa&appId=PPGMS. Accessed: 20th August 2019.
21. Angelidaki I, Xie L, Luo G, et al. Chapter 33 – Biogas upgrading: current and emerging technologies. In: *Biofuels: Alternative Feedstocks and Conversion Processes for the Production of Liquid and Gaseous Biofuels*, edn 2. Edited by: Pandey A, Larroche C, Dussap C-G, Gnansounou E, Khanal SK, Ricke S. Academic Press, 2019; pp. 817–843.
22. Semrau JD, DiSpirito AA, Gu W, et al. Metals and methanotrophy. *Appl Environ Microbiol* 2018; 84(6): 1–17.
23. Kang CS, Dunfield PF, Semrau JD. The origin of aerobic methanotrophy within the proteobacteria. *FEMS Microbiol Lett* 2019; 366(9): 1–11.
24. Karthikeyan OP, Chidambarampadmavathy K, Cirés S, et al. Review of sustainable methane mitigation and biopolymer production. *Crit Rev Env Sci Technol* 2015; 45(15): 1579–1610.
25. Graham DW, Chaudhary JA, Hanson RS, et al. Factors affecting competition between type I and type II methanotrophs in two-organism, continuous-flow reactors. *Microb Ecol* 1993; 25(1): 1–17.
26. Rostkowski KH, Pfluger AR, Criddle CS. Stoichiometry and kinetics of the PHB-producing type II methanotrophs *Methylosinus trichosporium* OB3b and Methylocystis parvus OBBP. *Bioresour Technol* 2013; 132: 71–77.
27. Pérez R, Cantera S, Bordel S, et al. The effect of temperature during culture enrichment on methanotrophic polyhydroxyalkanoate production. *Int Biodeterior Biodegrad* 2019; 140: 144–151.
28. Pieja AJ, Rostkowski KH, Criddle CS. Distribution and selection of poly-3-hydroxybutyrate production capacity in methanotrophic proteobacteria. *Environ Microbiol* 2011; 62: 564–573.
29. López JC, Quijano G, Pérez R, et al. Assessing the influence of CH₄ concentration during culture enrichment on the biodegradation kinetics and population structure. *J Environ Manage* 2014; 146: 116–123.
30. Pfluger AR, Wu WM, Pieja AJ, et al. Selection of type I and type II methanotrophic proteobacteria in a fluidized bed reactor under non-sterile conditions. *Bioresour Technol* 2011; 102(21): 9919–9926.
31. López JC, Porca E, Collins G, et al. Ammonium influences kinetics and structure of methanotrophic consortia. *Waste Manag* 2019; 89: 345–353.
32. Khmelenina VN, Colin Murrell J, Smith TJ, et al. Physiology and biochemistry of the aerobic methanotrophs. In: *Aerobic Utilization of Hydrocarbons, Oils and Lipids.* Edited by: Rojo F. Springer International Publishing, 2018; pp. 1–25.
33. Vorholt JA. Cofactor-dependent pathways of formaldehyde oxidation in methylotrophic bacteria. *Arch Microbiol* 2002; 178: 239–249.

34. Strong JP, Laycock B, Mahamud NS, et al. The Opportunity for high-performance biomaterials from methane. *Microorganisms* 2016; 4(11): 1–20.
35. Yamane T. Yield of poly-D(-)-3-hydroxybutyrate from various carbon sources: A theoretical study. *Biotechnol Bioeng* 1993; 41(1): 165–170.
36. Asenjo JA, Suk JS. Microbial conversion of methane into poly-β-hydroxybutyrate (PHB): Growth and intracellular product accumulation in a type II methanotroph. *J Ferment Bioeng* 1986; 64(4): 271–278.
37. Cantera S, Bordel S, Lebrero R, et al. Bio-conversion of methane into high profit margin compounds: an innovative, environmentally friendly and cost-effective platform for methane abatement. *World J Microbiol Biotechnol* 2019; 35(16): 1–10.
38. López JC, Arnáiz E, Merchán L, et al. Biogas-based polyhydroxyalkanoates production by *Methylocystis hirsuta*: A step further in anaerobic digestion biorefineries. *Chem Eng J* 2018; 333: 529–536.
39. Zhang T, Zhou J, Wang X, et al. Coupled effects of methane monooxygenase and nitrogen source on growth and poly-β-hydroxybutyrate (PHB) production of *Methylosinus trichosporium* OB3b. *J Environ Sci* 2017; 52: 49–57.
40. Pérez R, Pérez V, Lebrero R, et al. Polyhydroxyalkanoates production from methane emissions in *Sphagnum mosses*: Assessing the effect of pH and nitrogen limitation. 2020. Manuscript in preparation.
41. Rodríguez Y, Firmino PIM, Arnáiz E, et al. Elucidating the influence of environmental factors on biogas-based polyhydroxybutyrate production by *Methylocystis hirsuta* CSC1. *Sci Total Environ* 2020. Manuscript under review.
42. Cal AJ, Sikkema WD, Ponce MI, et al. Methanotrophic production of polyhydroxybutyrate-co-hydroxyvalerate with high hydroxyvalerate content. *Int J Biol Macromol* 2016; 87: 302–307.
43. Myung J, Flanagan JCA, Waymouth RM, et al. Methane or methanol-oxidation dependent synthesis of poly(3-hydroxybutyrate-co-3-hydroxyvalerate) by obligate type II methanotrophs. *Process Biochem* 2016; 51(5): 561–567.
44. Myung J, Flanagan JCA, Waymouth RM, et al. Expanding the range of polyhydroxyalkanoates synthesized by methanotrophic bacteria through the utilization of omega-hydroxyalkanoate co-substrates. *AMB Express* 2017; 7(1): 118.
45. Wendlandt KD, Jechorek M, Helm J, et al. Producing poly-3-hydroxybutyrate with a high molecular mass from methane. *J Biotechnol* 2001; 86(2): 127–133.
46. Helm J, Wendlandt KD, Rogge G, et al. Characterizing a stable methane-utilizing mixed culture used in the synthesis of a high-quality biopolymer in an open system. *J Appl Microbiol* 2006; 101(2): 387–395.
47. Helm J, Wendlandt KD, Jechorek M, et al. Potassium deficiency results in accumulation of ultra-high molecular weight poly-β-hydroxybutyrate in a methane-utilizing mixed culture. *J Appl Microbiol* 2008; 105(4): 1054–1061.
48. Sundstrom ER, Criddle CS. Optimization of methanotrophic growth and production of poly(3-hydroxybutyrate) in a high-throughput microbioreactor system. *Appl Environ Microbiol* 2015; 81(14): 4767.
49. García-Pérez T, López JC, Passos F, et al. Simultaneous methane abatement and PHB production by Methylocystis hirsuta in a novel gas-recycling bubble column bioreactor. *Chem Eng J* 2018; 334: 691–697.
50. de la Torre A, Metivier A, Chu F, et al. Genome-scale metabolic reconstructions and theoretical investigation of methane conversion in *Methylomicrobium buryatense* strain 5G(B1). *Microb Cell Fact* 2015; 14(1): 188.
51. Akberdin IR, Thompson M, Hamilton R, et al. Methane utilization in *Methylomicrobium alcaliphilum* 20ZR: A systems approach. *Sci Rep* 2018; 8(1): 2512.

52. Bordel S, Rodríguez Y, Hakobyan A, et al. Genome scale metabolic modeling reveals the metabolic potential of three type II methanotrophs of the genus Methylocystis. *Metab Eng* 2019; 54: 191–199.

53. Bordel S, Rojas A, Muñoz R. Reconstruction of a genome scale metabolic model of the polyhydroxybutyrate producing methanotroph *Methylocystis parvus* OBBP. *Microb Cell Fact* 2019; 18(1): 104.

54. Pieja AJ, Sundstrom ER, Criddle CS. Poly-3-hydroxybutyrate metabolism in the type ii methanotroph *Methylocystis parvus* OBBP. *Appl Environ Microbiol* 2011; 77(17): 6012–6019.

55. López JC, Quijano G, Souza TSO, et al. Biotechnologies for greenhouse gases (CH_4, N_2O, and CO_2) abatement: State of the art and challenges. *Appl Microbiol Biotechnol* 2013; 97(6): 2277–2303.

56. Kraakman NJR, Rocha-Rios J, van Loosdrecht MCM. Review of mass transfer aspects for biological gas treatment. *Appl Microbiol Biotechnol* 2011; 91(4): 873–886.

57. Rocha-Rios J, Bordel S, Hernández S, et al. Methane degradation in two-phase partition bioreactors. *Chem Eng J* 2009; 152(1): 289–292.

58. Cantera S, Estrada JM, Lebrero R, et al. Comparative performance evaluation of conventional and two-phase hydrophobic stirred tank reactors for methane abatement: Mass transfer and biological considerations. *Biotechnol Bioeng* 2016; 113(6): 1203–1212.

59. CalRecycle. *Bioplastics in California – Economic Assessment of Market Conditions for PHA/PHB Bioplastics Produced from Waste Methane 2013*. https://www2.calrecycle.ca.gov/Publications/Download/1085. Accessed: 5th September 2019.

60. Pittman MJ, Bodley MW, Daugulis AJ. Mass transfer considerations in solid–liquid two-phase partitioning bioreactors: A polymer selection guide. *J Chem Technol Biotechnol* 2015; 90(8): 1391–1399.

61. Rocha-Rios J, Quijano G, Thalasso F, et al. Methane biodegradation in a two-phase partition internal loop airlift reactor with gas recirculation. *J Chem Technol Biotechnol* 2011; 86(3): 353–360.

62. Rocha-Rios J, Muñoz R, Revah S. Effect of silicone oil fraction and stirring rate on methane degradation in a stirred tank reactor. *J Chem Technol Biotechnol* 2010; 85(3): 314–319.

63. Rocha-Rios J, Kraakman NJR, Kleerebezem R, et al. A capillary bioreactor to increase methane transfer and oxidation through Taylor flow formation and transfer vector addition. *Chem Eng J* 2013; 217: 91–98.

64. Matsuura T, Rana D, Rasool Qtaishat M, et al. Recent advances in membrane science and technology in sea water desalination with technology development in the Middle East and Singapore. In: *Water and Wastewater Technologies*. Encyclopedia of Life Support Systems (EOLSS), 2011.

65. Wellinger A, Murphy J, Baxter D. *The Biogas Handbook: Science, Production and Applications*. Elsevier, 2013.

66. WorldBank. *Zero Routine Flaring by 2030*, 2015. http://www.worldbank.org/en/programs/zero-routine-flaring-by-2030. Accessed: 5th September 2019.

67. Castilho LR, Mitchell DA, Freire DMG. Production of polyhydroxyalkanoates (PHAs) from waste materials and by-products by submerged and solid-state fermentation. *Bioresour Technol* 2009; 100(23): 5996–6009.

68. Choi J, Lee SY. Factors affecting the economics of polyhydroxyalkanoate production by bacterial fermentation. *Appl Microbiol Biotechnol* 1999; 51(1): 13–21.

69. López-Abelairas M, García-Torreiro M, Lú-Chau T, et al. Comparison of several methods for the separation of poly(3-hydroxybutyrate) from *Cupriavidus necator* H16 cultures. *Biochem Eng J* 2015; 93: 250–259.

70. Listewnik HF, Wendlandt KD, Jechorek M, et al. Process design for the microbial synthesis of poly-β-hydroxybutyrate (PHB) from natural gas. *Eng Life Sci* 2007; 7(3): 278–282.
71. Fernández-Dacosta C, Posada JA, Kleerebezem R, et al. Microbial community-based polyhydroxyalkanoates (PHAs) production from wastewater: Techno-economic analysis and ex-ante environmental assessment. *Bioresour Technol* 2015; 185: 368–377.
72. Choi D, Chipman DC, Bents SC, et al. A techno-economic analysis of polyhydroxyalkanoate and hydrogen production from syngas fermentation of gasified biomass. *Appl Biochem Biotechnol* 2010; 160(4): 1032–1046.

15 Syngas as a Sustainable Carbon Source for PHA Production

Véronique Amstutz and Manfred Zinn

CONTENTS

15.1 INTRODUCTION

Several chemical and biotechnological processes are now available for replacing widely used, persistent, and petrol-based plastics with ideally biodegradable and renewable polymers. Materials, such as cellulose- or starch-based plastics, are obtained primarily from plant extractions and blended with biodegradable plasticizers. Poly(lactic acid) is synthesized by chemical polymerization of bio-based lactic acid monomers (fermentation). These materials are commercially available and take up a major part of the biodegradable bioplastic market [1].

Another emerging biopolymer is polyhydroxyalkanoate (PHA), a polyester that is biosynthesized by microorganisms, mostly by bacteria [2]. It is essential to focus on the nature of the feedstock serving as a substrate for the microorganisms for this technology to be considered renewable. Indeed, in this process, bacteria catalyze the transformation of organic or inorganic carbon-containing molecules into PHA materials using energy provided by an organic carbon source, light, or an inorganic chemical source (e.g., hydrogen).

The global production of plastics amounted to 348 million tons in 2017 [3] and 359 million tons in 2018 [4]; it is expected to exceed 500 million tons/year by 2050 [5]. It was calculated that 30000 tons of PHA were produced globally in 2018 [6]. Assuming that this amount was based entirely on a glucose substrate, that 1 kg of corn yields 0.67 kg [7] of glucose, and that the yield of PHA on glucose for *Cupriavidus necator* is shown to be around 0.3 g PHA/g glucose [8], the global production of PHA would require 149250 tons of corn. If one assumes that in 2050 the production of PHA would reach 5% of the global production of plastics, that is, 25 million tons of PHA, this would require 124.4 million tons of corn. For comparison, in 2018, the global production of corn was 43290 millions of bushels [9], which is 1100 million tons. On this basis, it becomes clear that the production of feedstock required for PHA synthesis will compete with food production and agricultural needs. Alternative feedstocks are required to avoid this.

If the organic carbon source is based on municipal sewage sludge, researchers calculated that using all the sewage sludge in the EU (10906610 t/a) would produce 879469 t of PHA/a [10]. It should be noted, however, that municipal organic waste, such as sewage sludge, contains a large variety of organic molecules and involves seasonal and geographical changes in its composition. This translates into a large variety of PHA produced in terms of molecular weight, composition of the copolymer, as well as the nature and concentration of the impurities. These variations can be associated with significant modifications of the mechanical properties, which could become troublesome for the further processing of the PHA material and its applications.

From this perspective, a way to obtain more constant PHA material compositions and characteristics is to use a single carbon substrate for production. Syngas, and especially its main component, carbon monoxide (CO), is a potentially game-changing substrate, as this gas mixture can be produced from renewable sources.

The objective of this chapter is to present the potential of syngas as a feedstock for sustainable PHA production. We will discuss in detail the nature of syngas and its production from organic waste. In addition, the metabolic pathways supporting syngas assimilation for growth and PHA accumulation will be presented, and some details on the bioprocess design will also be given.

15.2   SYNGAS

Syngas is a mixture of gas containing mostly hydrogen (H_2) and carbon monoxide. These two components primarily determine its specific heating value, even though the presence of methane (or other hydrocarbons) impurities may also significantly affect it. Syngas is not naturally present in high quantities. It is, therefore, not a natural resource and needs to be artificially synthesized. In conventional processes, this requires high temperatures (500–1600°C) and the presence of hydrocarbon or carbon and oxygen or water as reactants. These reactants may also necessitate specific catalysts and/or high pressures (50–80 atm). In principle, any carbon-based molecule can serve as a feedstock for syngas production. Typical feedstocks include coke, coal, petroleum residues, petroleum naphtha, natural gas, and biomass. The conversion of biogas was also considered [11]. The source of oxygen may be pure oxygen (partial oxidation or gasification), carbon dioxide (dry reforming), steam (steam reforming), or O_2-enriched steam (autothermal reforming). A summary chart of the feedstocks, the processes involved in syngas production, as well as the applications of syngas and their resulting final products is depicted in Figure 15.1. This shows not only the large variety of processes and process parameters but also the importance of syngas worldwide, as reflected by the large range of its applications.

15.2.1 Syngas Production Technologies

As previously reviewed from a historical perspective [12], the early syngas production process was based on the conversion of coke to syngas (Equation 1). It was then replaced by methane (Equation 2) from natural gas in the period 1920–1940, before being complemented by the use of light petroleum naphtha (Equation 3, with n-hexane as the representative hydrocarbon) in 1950–1960. At this point, the substrates were treated with steam (steam reforming), but another alternative was progressively developed: the gasification of low-quality petroleum residues, coal, or coke with oxygen at high temperatures (1200–1600°C), high pressure (up to 70 atm), and in the absence of any catalyst. The process of catalytic partial oxidation (Equations 4 and 5) also emerged during this period. Subsequently, the steam reforming process was significantly optimized by Haldor Topsøe in 1950–1960 to achieve so-called autothermal reforming [12]. The improvement was made at the engineering level by a combination of steam reforming with partial oxidation processes. In an autothermal reformer, the reaction of partial oxidation is exothermic and precedes the reaction of steam reforming in the reactor. Therefore, it provides the heat required for subsequent endothermic steam reforming. Consequently, the energy efficiency is greatly improved. A more recent approach to syngas production is dry reforming (Equation 6). In this last process, the challenge is to avoid the reaction of CO_2 with H_2, producing water and CO, while promoting the reaction of CO_2 with the organic carbon available.

Lately, interest has been directed toward catalytic partial oxidation using monolith precious metal catalysts, especially because the reaction is two orders of magnitude faster than in steam reforming and is mildly exothermic, therefore, not requiring constant heating [12]. Moreover, high selectivity (depending on the catalyst and conditions) and gas composition close to the theoretical equilibrium was observed. Finally, as will be discussed in Section 15.3, the use of biomass as feedstock for the production of syngas was intensively studied over the last few decades, as it can be considered as a renewable source of syngas.

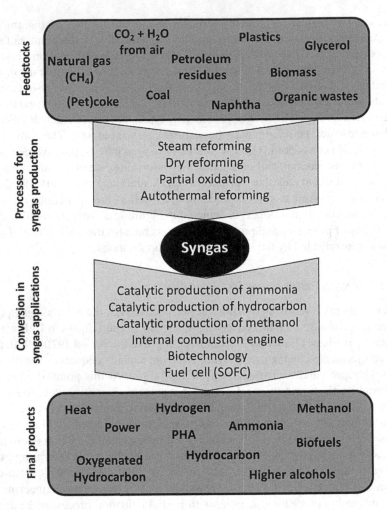

FIGURE 15.1   Feedstocks for syngas synthesis and products of its possible applications (SOFC: Solid oxide fuel cell).

$$C + H_2O \rightarrow CO + H_2 \tag{1}$$

$$CH_4 + H_2O \rightarrow CO + 3H_2 \tag{2}$$

$$C_6H_{14} + 6H_2O \rightarrow 6CO + 13H_2 \tag{3}$$

$$CH_4 + 0.5O_2 \rightarrow CO + 2H_2 \tag{4}$$

$$C_6H_{14} + 3O_2 \rightarrow 6CO + 7H_2 \tag{5}$$

$$CH_4 + CO_2 \rightarrow 2CO + 2H_2 \tag{6}$$

15.2.2 COMPOSITION OF SYNGAS

The exact composition of the gas depends on the design of the gasification process, the related conditions (temperature, pressure, catalysts, etc.), and is strongly related to the nature of the original feedstock. For instance, a partial oxidation process based on methane feedstock conducted above 850°C leads to a 2:1 H_2:CO ratio in syngas [12]. In comparison, the reforming of methane in the presence of steam as a co-reactant led to a higher H_2:CO ratio, since the water-gas shift reaction (Equation 7) was responsible for generating a higher amount of H_2 [12]. Similarly, as a consequence of this reaction, the moisture content of the biomass was shown to be of importance in determining the composition of syngas; lower CO concentration was observed with a higher moisture content [13,14]. The concentration of oxygen can also affect syngas composition. Finally, depending on the variety of hydrocarbon molecules present in the feedstock and their relative concentration, the composition of the gas can vary. Even seasonal or yearly changes can be observed using the petroleum of a single source. For the use of biomass, the variety is larger as more variability is observed in terms of the relative concentrations of hydrocarbon molecules in the feedstock.

$$CO + H_2O \rightarrow CO_2 + H_2 \tag{7}$$

Besides CO and H_2, syngas may also contain a substantial amount of carbon dioxide and methane, but also impurities of water, argon, and nitrogen (when air is used), hydrogen sulfur, and possibly other substances, such as ethane or ethylene [15]. Moreover, side products of syngas synthesis are carbon materials that were not totally oxidized into CO or CO_2. This includes small carbonaceous particles, ash, slag, tar, and char (also called oils). Their presence is highly related to the nature and particle shape of the feedstock. For instance, unsaturated hydrocarbon molecules can be precursors for coke formation during any high-temperature reforming or oxidation processes [16]. Coking in reforming is to be avoided as it involves poisoning the catalyst and lower feedstock availability for syngas conversion. Tar was defined by Qin et al. [17] as hydrocarbon molecules or oxygenated compounds with a molecular weight larger than benzene. Lignin, due to its aromatic structure, is considered a major precursor for tar compounds [18,19]. Their presence in the gas (viscous liquid at room temperature) results in subsequent fouling of the surfaces of heat exchangers and engines and catalysts. For example, a mean tar concentration of 78 g/m^3 (variations between 24 and 170 g/m^3) was reported by Oveisi et al. using woodchips as feedstock in an updraft gasifier using air [19].

15.2.3 PROPERTIES AND HANDLING OF SYNGAS

Due to the large variation in its composition, it is difficult to characterize syngas in terms of chemical and physical properties. One important parameter to be determined when syngas is considered as an energy carrier is its energy density. It depends on the syngas composition and is often calculated in regard to its application. The energy-containing components in syngas are hydrogen, carbon monoxide, methane, and other hydrocarbon impurities. Their associated lower heating values (LHV) are LHV(H_2) = 120 MJ/kg, LHV(CH_4) = 50 MJ/kg, and LHV(CO) = 10.1 MJ/kg. Therefore, the amount of H_2 is important for determining the energy density of syngas. Moreover, when syngas is used in a water-gas shift reaction, heat is provided so that the CO can transfer its energy to protons from water to form hydrogen. The typical volumetric energy densities of syngas are in the range of 4 to 15 MJ/Nm³ [20].

Another aspect of syngas, which is not often discussed but is essential when handling it, is the safety requirements associated with this gas mix. Indeed, the presence of hydrogen (and secondarily methane, when present in high amounts) is related to a risk of explosion if a sufficient amount of syngas contacts both air (O_2) and a source of ignition. Moreover, CO is highly toxic to humans at low levels due to its high affinity for hemoglobin. Therefore, safety measures need to be considered for handling syngas. In brief, this includes the use of proper pipes and connection materials to avoid gas leaks and corrosion. Besides the material, it is also the type of connection that is important, especially for avoiding any hydrogen leak. Furthermore, an ATEX environment (different levels can be defined depending on the risks and the potential hazard related to a leak) is established around the most critical zones. Finally, H_2 and CO detection systems need to be installed to trigger an alarm (different levels are possible) in case of a gas leak. The detection system may also interfere with the process, for example, close valves or shut down the installation according to an emergency procedure (e.g., flushing the system with N_2). These safety aspects are related to the production of syngas, but also to its transportation and finally to its use.

15.2.4 APPLICATION OF SYNGAS

The fact that syngas is a contraction of the words "synthesis" and "gas" reflects its original functions in essential synthetic processes for supporting humankind, such as 1) catalytic ammonia synthesis, 2) Fischer–Tropsch hydrocarbon synthesis for fuels, lubricants, and oxygenated hydrocarbons for further chemical synthesis, 3) catalytic methanol production, 4) synthesis of aldehydes by the "oxo" process (hydroformylation), and 5) the synthesis of higher alcohols using a direct catalytic synthesis process [12]. Furthermore, syngas is considered an energy carrier, which is harnessed in internal combustion engines for electricity, heat, or combined heat and power generation. Energy efficiency of more than 37% can be achieved, for instance, with Jenbacher gas engine technology [21]. It is worth noting, however, that the energy density of syngas is less than half that of natural gas. Similarly, within the frame of renewable energy conversion systems, syngas is fed to solid oxide fuel cell (SOFC) for cogeneration of power and heat. Finally, syngas is the source for most of the production of hydrogen worldwide [16], mostly by steam reforming of natural gas. The water-gas shift reaction under partial oxidation or steam reforming conditions facilitates the production of CO_2 and H_2. The production process is followed by purification to obtain pure H_2. Recently, the use of syngas to feed microorganisms in specific biotechnological processes has been shown and will be discussed in Section 15.4.

15.2.5 CURRENT STATUS OF SYNGAS PRODUCTION

Some statistics regarding the current status of gasification worldwide follows, referring to the website of the Global Syngas Technologies Council [22]. It seems that these numbers are a cumulative calculation of the data available in 2014–2015 and the projects planned until the end of 2019 (it can be noted, however, that based on the actual data of 2015, the trends are similar to those described below for 2019). All the following numbers are, therefore, only estimates and should be confirmed based on a new set of data. Numbers were gathered from graphs available on the above-cited website; they remain the most precise data found by the authors. In 2019, there should be 272 operating gasification plants worldwide for a total of 686 gasifiers. In addition, 74 supplementary plants were under construction until the end of 2019 (adding 238 gasifiers), corresponding to an additional 83 MW_{th} (megawatt thermal). As a result, by the end of 2019, an overall production capacity of about 300000 MW_{th} synthesis gas should be reached.

In terms of location, this capacity is strong in Asia/Australia, accounting for 240000 MW_{th}, compared with about 30000 MW_{th} for Africa and less than 20 MW_{th} for Europe and North America, individually. The analysis goes further and distributes the capacity of syngas production as a function of its main four applications: a total of 140000 MW_{th} of syngas is dedicated to chemical synthesis (mostly ammonia and methanol), while 55000 MW_{th} are transformed into liquid fuels, 25000 MW_{th} are used for power generation, and 85000 MW_{th} for gaseous fuel generation.

Regarding the classification of gasifiers as a function of the nature of the feedstocks, a large majority of gasifiers have been built for coal gasification and petroleum feedstocks (88.5%), and only a small number were designed for biomass and waste gasification (6%) while the remaining part was dedicated to gas or petcoke gasification (5.5%). In a general perspective, gasification projects based on coal or petroleum feedstocks tend to become larger, while those for biomass and waste gasification have to be flexible because they need to adapt to the nature of the feedstock. Consequently, these installations are smaller and ideally may also move to locations where feedstock is available.

15.3 PRODUCTION OF SYNGAS FROM ORGANIC WASTE AND BIOMASS

Renewable carbonaceous feedstocks, such as biomass and carbonaceous waste, are considered in this section to convert the production of syngas from limited and polluting fossil fuel feedstocks into a sustainable production of syngas. The expected effects of using such alternatives are reduced greenhouse gas emissions and long-term availability of feedstocks. Indeed, biomass materials are considered a CO_2-neutral fuel since the released CO_2 is reused for its production (photosynthesis). It is estimated that the global biomass produced per year is 220 billion dry ton/a [23] (this includes all plant-derived matter, including all products from agriculture and forestry).

15.3.1 BIOMASS AND ORGANIC WASTE

The same approach is followed for the production of biofuels (bioethanol and biodiesel) mostly using biomass from agricultural origin. It includes sugar cane, cereals, beets, vegetable oil, and corn (first generation biofuels). As a consequence of increasing biofuel production from these feedstocks, competition with food production may occur and lead to ethical and societal conflicts. Therefore, other feedstocks were investigated, the so-called lignocellulosic biomass and especially organic waste. Using waste, which would otherwise be lost [24], especially in landfills or transformed to CO_2 in waste incineration plants, has the advantages of recycling carbon, thus lowering CO_2 emissions and indirectly decreasing environmental pollution. More specifically, waste from forestry (forest clearing, wood processing), agricultural activities, and food industry processes, as well as urban waste (mostly solid waste, wastewater sludge, and composting), are targeted. Furthermore, waste from industry (packaging, pallets, paper, cardboard, wood) have a high heating value and, therefore, a high potential as feedstock. Finally, other alternative feedstock, such as plastics waste, glycerol, and CO_2 from the air (or richer sources), are also being investigated as carbon sources for the production of syngas. Glycerol is a co-product of biodiesel production and is now available in large amounts at a low cost. Glycerol reforming has, therefore, been studied over the last few years. The catalysts, conditions of reaction, and yields were thoroughly reviewed by Abatzoglou and Fauteux-Lefebvre [16].

Targeting waste as feedstock for the production of syngas, which is further used for fuel, power production, or chemical synthesis, offers economic advantages, especially as it is inexpensive feedstock. However, this may be counterbalanced for specific materials with the requirement for pretreatment before their use as a feedstock or posttreatment by purification of the produced syngas. Biomass gasification enables the reuse of the carbon and the energy content of the otherwise useless waste and allows onsite power production. Moreover, it is suitable for all types of waste, as, for example, complex municipal waste and wastewater sewage sludge.

15.3.2 BIOMASS AND ORGANIC WASTE GASIFICATION

The conversion of biomass into syngas is mostly based on gasification, a thermochemical process in which biomass is transformed into a gas mixture by a gasifying agent, similar to the processes discussed in Section 15.2. The advantages of this process are 1) it provides high flexibility in terms of variability of the feedstocks, 2) it is more efficient and is associated with a lower environmental impact compared with combustion-based technologies, 3) the process can be further improved to integrate CO_2 separation and concentration at affordable costs, 4) the products of gasification are easily treated (H_2S, NO_x) compared with combustion products, at lower costs and with better efficiency; and 5) ash and slag are not toxic and can be deposited in landfills or used for higher-value products without further treatments [15]. The use of biomass as a feedstock for gasification also presents some drawbacks and challenges, such as moisture content, low initial energy density, and the need to reduce the materials prior to its gasification.

The process can be divided into four consecutive steps associated with the chemical reactions of Equations 8–17 [25]:

1. Drying (endothermic): Evaporation of water from the biomass.
2. Pyrolysis (endothermic):

$$\text{Biomass} \rightarrow CO + H_2 + CO_2 + CH_4 + H_2O + \text{Tar} + \text{Char} \qquad (8)$$

3. Oxidation (exothermic):

$$\text{Char oxidation}: \text{Char} + O_2 \rightarrow CO_2 \qquad (9)$$

$$\text{Partial oxidation}: C + \frac{1}{2}O_2 \rightarrow CO \qquad (10)$$

$$\text{Hydrogen oxidation}: H_2 + \frac{1}{2}O_2 \rightarrow H_2O \qquad (11)$$

4. Reduction (endothermic):

$$\text{Reforming of char}: \text{Char} + H_2O \leftrightarrow CO + H_2 \qquad (12)$$

$$\text{Boudouard reaction}: C + CO_2 \rightarrow 2CO \qquad (13)$$

$$\text{Water} - \text{gas shift reaction}: CO + H_2O \leftrightarrow CO + H_2 \qquad (14)$$

$$\text{Methanation reaction}: C + 2H_2 \rightarrow CH_4 \qquad (15)$$

$$\text{Steam reforming of methane}: CH_4 + H_2O \rightarrow CO + 3H_2 \qquad (16)$$

$$\text{Dry reforming of methane}: CH_4 + CO_2 \rightarrow 2CO + 2H_2 \qquad (17)$$

The gasification can be extended to a fifth step consisting of tar reforming into light hydrocarbons (from large molecular weight tar compounds) [25], according to Equation 18.

$$\text{Steam reforming of Tar}: \text{Tar} + H_2O \rightarrow H_2 + CO_2 + CO + C_xH_y \qquad (18)$$

Regarding the initial drying process, as alternatives to conventional mechanical (filter presses, centrifuges) or thermal (direct/indirect fired rotary dryers, conveyor dryers, cascade dryers, superheated dryers, microwave dryers, etc.) dewatering processes, novel strategies for pretreatment include pelletization (reduction to a suitable size) and torrefaction (reduction of moisture content) [26].

In terms of energy requirement for biomass gasification, the heat demand for the endothermic steps can be provided by external sources by recycling the heat generated by the exothermic oxidation step or through the use of a multicomponent reactor as in autothermal reforming. Both the heating rate and temperature are important in gasification [25] as they influence the CO, H_2, and tar content. Another parameter, the pressure in the reactor during gasification, affects the reactivity of the char, and higher pressure leads to a more complete carbon conversion and a lower yield of light hydrocarbons and tar. Finally, when pressurized syngas is required, it is usually also produced under pressure.

15.3.3 TUNING PARAMETERS IN BIOMASS AND WASTE GASIFICATION AND DIFFERENT REACTOR SYSTEMS

An essential parameter of the gasification process is the gasifying agent. It influences the composition and, therefore, the final heating value of the produced syngas and acts on the reactivity of the char. Air, oxygen, steam, or CO_2 can be used as gasifying agents. A comparison between them is provided by Molino et al. [25]. Briefly, the use of air involves the production of nitrogen in the gasifier, thus reducing the CO and H_2 content and the heating value of the produced syngas. Moreover, more CO_2 is present due to the combustion of CO and H_2. Steam, as a gasifying agent, leads to a higher H_2 content due to the presence of water, which sustains the water-gas shift reaction. However, this combination is associated with a higher energy need because this is an endothermic reaction. Pure oxygen, as a gasifying agent, leads to high heating value syngas with higher concentrations of CO, H_2, and low formation of tar, but it is expensive. CO_2-based gasification results in a CO-rich syngas with a high heating value but requires external heat. The equivalence ratio (ER), the air to biomass ratio required for gasification, mostly controls the combustion of H_2 and CO; when it is too high, it will lead to an increased concentration of CO_2. The ER is usually between 0.2 and 0.4. Similarly, the steam-to-biomass ratio (on a flow rate basis) influences the composition, and it was observed that an increased ratio led to higher H_2 and CO_2 concentrations, but lower CO and tar concentrations (water-gas shift, reforming, and cracking reactions), and thus to a higher heating value.

Gasification processes are commonly performed in fixed bed, fluidized bed, or entrained flow gasifiers [25]. Fixed-bed reactors are mostly used for small scale plants (power generation of maximum 10 MW), while fluidized bed gasifiers are selected for larger-scale plants due to their easy scale-up. When the biomass is fed to the reactor from the top, and the gasifying agent is supplied from the bottom of a fixed-bed reactor at the same time, it is called an updraft gasifier (counter-current), while a downdraft gasifier is when both the biomass and agent are supplied from the top of the reactor (co-current). This difference influences the sequence of the reaction. In an updraft reactor, the biomass is subjected to drying, followed by pyrolysis, reduction, and is only then oxidized and collected from the top of the reactor. In the downdraft reactor, the biomass is dried, pyrolyzed, oxidized, and finally reduced before being collected from the bottom of the reactor. Updraft gasifiers provide a more efficient heat transfer but result in a high tar content. In contrast, downdraft gasifiers lead to low tar and low particulate syngas, but have a complicated temperature control and require low moisture, low initial ash content, and homogeneous biomass.

Fluidized bed reactors are associated with high rates of mass and heat transfer and are therefore more tolerant of feedstock variability. There are two types of fluidized bed reactors: 1) the bubbling fluidized bed gasifier where the gasification agent is supplied from the bottom and contributes to the fluidization of the solid bed and, therefore, the gasification process occurs in the fluidized bed and 2) the dual bed gasifier with two separated chambers in which the process is based on two steps. In the first step, the combustion occurs in the so-called combustion chamber and generates heat. It is followed by the second step, in a second chamber, by the pyrolysis and gasification reactions with a high-speed gas in a bubbling fluidized bed, where the gas is then separated with a cyclone separator. The addition of a catalyst enables the higher oxidation of tar into syngas.

Entrained gas flow gasifiers are mostly used for large-scale plants; they allow high operating temperatures and the use of oxygen as the gasification agent. They are usually operated with a pressurized reaction chamber. As a consequence, only low amounts of tar and higher conversion efficiencies are obtained. However, only fine powder feedstock can be used in this type of reactor, therefore, requiring a size reduction pretreatment step. Entrained flow gasifiers are divided into two categories: 1) top-fed gasifiers where fine particles are pulverized concurrently to the gasification agent to form a jet from the top of a vertical cylinder; the jet is subjected to a burner and the syngas is collected below; 2) side-fed gasifiers in which the powdered biomass and gasification agents are fed co-currently by nozzles placed in the lower part of the reactor (burning chamber), where it is directly reacted. The resulting syngas is collected at the top of the reactor (cylinder on top of the burning chamber).

The configuration of the gasifier strongly influences the performance of the reactor (see a summary of performances by Molino et al. [25]), while the bed material can be inert or act as a heat transfer medium, catalyst carrier, or catalyst (for reforming, cracking reactions). Mostly silica, limestone, dolomite, olivine, and alkaline metal oxides and Ni- and K-based catalysts are used in these reactions [25].

15.3.4 ADVANCED APPROACHES TO BIOMASS AND WASTE GASIFICATION

Further optimization of the conventional biomass gasification process is still ongoing and key strategies were reviewed by Heidenreich et al. [27]. In particular, these strategies include polygeneration, which is the production of more than one product in a single process, and, aims at a more careful use of the generated heat. Moreover, in the last few decades, supercritical water gasification has been applied to high moisture content biomass feedstock, mostly for the application of hydrogen production (see review articles [28,29]). In this process, the biomass does not need to be initially dried because the presence of water in a supercritical state implies specific pathways for biomass degradation, in particular in spots where its chemical composition is suitable (e.g., high salt and protein content). Moreover, a lower temperature (ca. 600°C) is required in comparison to conventional gasification processes. However, it still requires a large amount of energy for heating the water and, therefore, a system for heat recovery is needed to improve the energy balance of the system.

Another advanced technology for syngas production is plasma gasification of biomass. It relies on the high temperature (up to 5000 K) generated under the application of a direct or alternate current discharge, radio frequency induction discharge, or microwave discharge [27,30]. Often an AC or DC arc plasma torch generator is used. Cold or thermal plasma is generated under vacuum or at atmospheric pressure, respectively. The gasification process is very fast and takes place without intermediary reaction. It is tolerant of biomass with high water content (e.g., sewage sludge) and is insensitive to the particle size and structure of the biomass. It leads to high H_2 and CO content and, therefore, high heating value and low CO_2 and tar content. However, it requires high electricity input and presents relatively low energy efficiency while requiring high investment costs [27]. It is worth mentioning the fact that this system has already been created on a large scale in Japan, where a plasma gasification plant was put in operation in 2002: 300 t/d of municipal solid waste and automobile shredder residues were treated, producing 7.9 MWh electricity, from which 4.3 MWh were fed to the local grid [30].

An alternative emerging process enabling the conversion of biomass to syngas is microwave-induced pyrolysis. It is highly effective because the microwaves give fast and volumetric heating [31] of the biomass and, therefore, lead to a significantly improved heat transfer compared with the classic technologies previously discussed. It was applied, for instance, to the pyrolysis of wood sawdust and corn stover [32], coffee hull [33], rice straw [34], wood blocks [35], and microalgae [36]. This technology uses the fact that carbon materials and metal oxides are recognized to be good adsorbents for microwaves (microwave receptors) and, therefore, transform the microwave energy into heat [37] that can be further transferred to the surrounding materials. Therefore, when carbon particles are added to biomass and the mixture is subjected to microwaves, the temperature is increased and the steps of drying, pyrolysis, and gasification occur simultaneously. The method leads to a high purity syngas (mostly formed of CO and H_2) with a high conversion efficiency [11,37]. The char produced in the process acts itself as a microwave receptor, increasing the process' efficiency, and is partially converted to syngas.

15.3.5 PURIFICATION OF SYNGAS

After its production, syngas may contain impurities that need to be removed, either to avoid corrosion of the associated equipment or flue gas emission of toxic gas in the atmosphere (e.g., in combustion applications). Moreover, in some applications, for example, for hydrogen production, one or more components are isolated from the syngas mixture. The purification procedure is, therefore, designed specifically according to the syngas application (and the associated temperature), which also determines the choice of syngas production process and the selection of a suitable feedstock. The purification process is usually done in two steps: 1) primary cleaning methods performed during gasification focus on the removal of solid contaminants, while 2) secondary cleaning methods, occurring subsequently to gasification, target the removal of fluid pollutants [38]. Moreover, the cleaning procedures also rely on the gas temperature. In principle, they are classified into wet and dry and hot and cold cleaning methods [38].

The substances to be removed in a purification process can be acidic and alkaline species, CO_2, oxygen, sulfur molecules [mostly carbonyl sulfur (COS) and hydrogen sulfide (H_2S)], heavy metals, halogens, cyanides, nitrogen, nitrogen oxides, and phosphorus. Moreover, often, dust particles (char residues, ash, etc.), char, and tar need to be removed before further processing of syngas.

A thorough discussion of the main purification methods is provided by Pérez-Fortes et al. [38] and is summarized here. Mechanical methods are used to separate the solids from gas mixtures and include the use of a cyclone placed in the reactor itself (only for larger particles), sintered filters, wet scrubbers, and ceramic candles. Wet scrubbing is also used for nitrogen, halides, and cyanide removal. Chemical and physical absorption methods separate acidic and alkaline species from gas mixtures. They involve bubbling the produced syngas in a liquid solvent (methanol, dimethyl ethers of polyethylene glycol, amines solutions, etc.) to enable the selective solubilization of the undesired compounds. Once saturated, the solvent is usually treated (desorption) in order to be reused. Finally, in the process of adsorption, syngas is flowed through a fixed bed, whose material specifically adsorbs the unwanted component of the gas. Again, the material can be regenerated by applying a higher temperature and/or a lower pressure.

In Table 15.1, typical syngas compositions are provided for various biomass and waste feedstock under various gasification conditions and processes. Data for a larger variety of feedstock are provided elsewhere [20,25,39].

15.3.6 Variety and Challenges in Biomass and Waste Gasification

Designing a process for biomass or waste gasification relies on interconnections between a number of parameters. Especially important are the feedstock's composition, the presence and nature of a catalyst, the nature of the gasifying agent, and the selection of the reactor configuration. Consequently, the applications of the syngas need to be defined at a very early stage so that an appropriate purification system can be applied. In view of these aspects, it appears that it is not possible to standardize a biomass or waste gasifying plant since it needs to be tailored to the feedstock available and the expected syngas applications to satisfy the associated quality requirements.

For illustration, in terms of the feedstock characteristics that influence the final syngas quality, the type of biomass, moisture content, particle size, and ash content are important [25]. In brief, a higher cellulose and hemicellulose to lignin ratio leads to higher syngas yields, while the presence of lignin results in the increased formation of residues (e.g., tar). The moisture content will affect the higher heating value, tar content, and energy efficiency of the process (in particular due to the presence of the water-gas shift reaction). The particle size is directly related to the surface area available for the reaction; the lower it is, the lower the heat and mass transfer resistance is, therefore, leading to both a higher conversion yield to syngas and a higher reaction rate. As a result, less tar and char are produced. However, these effects are reduced at high temperatures. Particle sizes between 0.15 and 51 mm may be applied, depending on the type of reactor used (smaller size for entrained flow gasifiers and larger ones for fixed-bed reactors). Nevertheless, biomass pretreatment for size reduction may increase the process cost. The initial ash content of the biomass is critical as it leads to slag formation. Therefore, feedstocks with low ash content (<2 wt.-%) are preferred over those with high ash content (>10 wt.-%, e.g., cereal, oil seed, root crops, or grasses and flowers) or very high ash content (>20 wt.-%, e.g., rice husk).

The commercialization of units for syngas generation from lignocellulosic biomass, especially from agricultural, forestry, and food industry waste, is underway.

TABLE 15.1
Syngas Composition for a Variety of Biomass and Waste Feedstocks, as well as Gasification Process Configurations and Conditions

Feedstock	Process	Syngas composition (unless specified mol-%)	Comments	Ref.
Glycerol	Dry reforming	$1 < H_2:CO < 4.5$ (850 K)	- Ratio depends on CO_2-to-glycerol ratio and on temperature - Based on thermodynamic analysis	[40]
Glycerol	Steam gasification	$H_2:CO = 2$ (pure glycerol) $H_2:CO = 3$ (crude glycerol)	- Fixed-bed gasifier with quartz or SiC particles - 800°C - Ni/Al_2O_3 catalyst	[41]
Wood chips and scraps from construction	Fixed-bed updraft steam gasification	CO: 20–30% H_2: 8–10% N_2: 55–60% CH_4: 0–5% CO_2: 5–12% ($H_2:CO = 0.26$–0.54)	- In commercial gasifiers - Effects of chips size and biomass moisture content - Air as gasifying agent	[19]
$H_2O + CO_2$ extracted from air	Co-thermolysis using energy from a solar concentrator	$CO:H_2 = 2$–5.5	- Research scale	[42]
Palm oil waste	Steam gasification with a fixed-bed reactor	15–25% CO 48–60% H_2 4–5% CH_4 20–25% CO_2	- LHV: 9.1–11.2 MJ/Nm^3 - Yield: 1.79–2.48 Nm^3/kg biomass - Trimetallic catalyst - Effects of temperature, steam-to-biomass ratio, biomass particle size - 750–900°C	[43]
α-cellulose, bagasse, mushroom waste	Air–steam gasification with a fluidized bed reactor	13.6–23.6% CO 56.9–33.3% N_2 6.0–29.5% H_2 1.9–2.7% CH_4 21.6–10.9% CO_2	- LHV: 7.0–10.7 MJ/Nm^3 - 600–1000°C, ER 0.20–0.34 - Various steam-to-biomass ratio	[44]
Bamboo	Steam gasification in fluidized bed	23.5–30.6% CO 6.6–8.1% H_2 4–5% CH_4 59–63% CO_2	- LHV: 1.6–1.9 MJ/Nm^3 - 400–600°C, ER 0.4 - Yield: 1.9–2.0 Nm^3/kg biomass - Calcined dolomite as catalyst - Catalyst promotes tar reforming	[45]

(Continued)

TABLE 15.1 (CONTINUED)

Syngas Composition for a Variety of Biomass and Waste Feedstocks, as well as Gasification Process Configurations and Conditions

Feedstock	Process	Syngas composition (unless specified mol-%)	Comments	Ref.
Pine sawdust	Air–steam in fluidized bed reactor	35–43% CO 21–39% H_2 6–10% CH_4 18–20% CO_2	- LHV: 6.7–9.1 MJ/Nm³ - Yield: 1.43–2.57 Nm³/kg biomass - Effects of temperature, steam-to-biomass ratio, biomass particle size, ER - 700–900°C	[46]
Empty fruit bunch	Air in fluidized bed reactor	21–36% CO 10–38% H_2 5–14% CH_4 10–65% CO_2	- LHV: 7.5–15.5 MJ/Nm³ - 700–1000°C	[47]
Municipal solid waste (MSW)	Pyrolysis	23–30.1% CO 24.4–36.2% H_2 18.9–16.2% CH_4 10.8–20.4% CO_2 (H_2:CO = 1.06–1.2)	- LHV: 13.4–14.0 MJ/Nm³ - 750–900°C - Dolomite catalyst - Carbon conversion efficiency: 55.4–90.4 wt.-% - 5.3–8.1 mol-% C_2H_4 and 1.3–4.2 mol-% C_2H_6 - Effects of temperature, biomass-to-catalyst ratio, presence of catalyst	[48]
MSW	Steam gasification in fixed-bed reactor	11.3–22.7% CO 34–54.2 %l H_2 1.3–10.3% CH_4 20.6–38.2% CO_2 (H_2:CO = 2.4–3.0)	- MSW composed of kitchen garbage (69 wt.-%), paper (10%), textile (2%), wood (7.5%), plastics (11.5%) -Yield: 0.88–1.75 Nm³/kg MSW - Tar content: 0 (900°C)–8.3% - 0.1–3.3 mol-% C_2H_4 and 0.5–2.8 mol-% C_2H_6 - 700–900°C - Catalysts: NiO/$_\gamma$-Al_2O_3 or calcined dolomite - Effects of temperature, steam-to-carbon ratio, and catalyst type	[49]

(Continued)

TABLE 15.1 (CONTINUED)

Syngas Composition for a Variety of Biomass and Waste Feedstocks, as well as Gasification Process Configurations and Conditions

Feedstock	Process	Syngas composition (unless specified mol-%)	Comments	Ref.
MSW	Steam or air plasma gasification melting process (updraft moving bed)	12–15% CO 20–25% H_2 52–60% CO_2 + N_2 (H_2:CO = 1.4–2.1)	- Pilot plant: 300 kg/h MSW - MSW composed mostly of paper (50%), textile (5.5%), wood (11%), plastics (10%) - LHV: 6 kg/h 7 MJ/Nm³ - Yield: 1 kg/h 1.4 Nm³/ kg MSW - 6000°C, plasma power 240 kW - Max energy efficiency: 58%	[50]
Dried Sewage sludge	Steam gasification	10–18% CO 47–56% H_2 8–10% CH_4 20–28% CO_2	- 700–1000°C - Yield: 0.66–1.14 g_{gas}/ g_{solid}. - Same reference compared with pyrolysis: lower yields obtained	[51]
Algal biomass	Steam gasification	21–56% CO 25.1–43.9% H_2 0.8–6.0% CH_4 18.0–30.6% CO_2	- Catalyst: alkali and alkaline-earth metal - 700–900°C, ER 0.4 - Effect of catalyst on yield and char and tar content, biomass particle size, steam-to-biomass ratio, temperature	[52]

As discussed by Molino et al. [25] in 2018, only a very few plants (2) were identified as commercially viable among all biomass gasification projects. As previously stated, the installed biomass gasifiers represent only a small minority (6%) of the total number of gasifiers installed worldwide.

15.4 CONCEPT OF BACTERIAL PHA SYNTHESIS FROM SYNGAS

An alternative to the conversion of syngas by Fischer–Tropsch processes into hydrocarbon molecules for fuels or building blocks for the chemical industry is its bioconversion to the same or similar hydrocarbon molecules. A larger range of molecules is available with this approach due to the variety of microorganisms, their diverse metabolic pathways, and the countless possibilities offered by genetic engineering. Bioconversion by syngas bacterial fermentation has the following advantages [53]:

- Higher tolerance to impurities such as sulfur compounds (this may, however, depend on the microorganism involved).
- Larger range of gas composition (CO, H_2, CO_2 partial pressures) can be fed to the process.
- Lower temperature and pressure of operation.
- Higher product yields.
- Higher selectivity (especially for chiral compounds).

However, the critical aspects or limitations of the implementation of this technology can be summarized as [53]:

- Low product titers (especially on a volumetric basis).
- Limited mass transfer from gas to liquid due to low solubility of the gas components in water.
- Product inhibition due to toxicity for the microorganisms at concentrations below the thermodynamic limitation.
- Lack of understanding of the metabolic pathways and energy conservation processes of microorganisms.

A focus on the microorganisms able to grow on syngas and to generate hydrocarbon molecules shows that most bacteria and a few *Archaea* genera are involved and that they apply to a small variety of metabolic pathways, especially in relation to the product(s) synthesized. Except for a few methanogens, most of the bacteria and *Archaea*-metabolizing syngas are acetogens [53]. The part of their metabolism supporting this ability will, therefore, be discussed in Section 15.5.

More specifically, using renewable syngas for the production of polyhydroxyalkanoate (PHA) compounds through microbial fermentation has been investigated to produce biodegradable and bio-sourced plastics as a replacement for conventional petrol-based plastics. In this process, bacteria are incubated in a bioreactor containing a medium with the necessary nutrients and a carbon source in ample amounts (CO, CO_2, CH_4, etc.). In case this carbon source is not also an energy source to the bacteria, a supplementary energy source (hydrogen, CO, light, etc.) must be provided. Under optimal conditions (pH, temperature, osmotic pressure, etc.), the bacteria grow and increase the biomass concentration in the bioreactor. Moreover, in specific bacteria and depending on the growth conditions, for example, nutrient limitations, metabolic products are favorably synthesized. These products may be stored intracellularly or are secreted into the culture broth. Their quantity depends on their metabolic role(s), cell toxicity, and thermodynamic considerations. PHA is an intracellular storage product that is naturally produced by some bacteria with the objective of constituting an energy and carbon storage compound that can be recycled later during conditions of starvation. PHA is accumulated in the cytoplasm of bacteria in the form of granules, called carbonosomes. In optimized conditions, they can reach more than 80 wt.-% of the cell dry mass (CDM, i.e., the total weight of biomass, including PHA after being dried).

The term "PHA" describes a class of polyester compounds that are usually divided into two subclasses: short-chain-length PHA (scl-PHA, with 3–5 carbon atoms in the monomeric unit) and medium-chain-length PHA (mcl-PHA, with 6–14 carbon atoms in the monomeric unit). Most of PHA polyesters are based on chiral (R)-3-hydroxy ester bonds. The length of the side chains affects the polymer crystallinity and, therefore, most of their mechanical properties (hardness, elasticity, toughness), thermal properties (fusion temperature, glass transition temperature), and some of their biodegradability properties (biocompatibility, biodegradability). Furthermore, the polymer can be naturally produced in the form of homopolymers or copolymers, depending on the substrate(s) available, the strain, and the conditions of fermentation. Therefore, depending on the application of the PHA material, these parameters can be tailored during biosynthesis.

Organic substrates for bacteria, such as hexoses, mostly glucose and fructose from various sources, are commonly fed to bacteria for high yield production of PHA, especially scl-PHA. However, for large bioprocesses targeting PHA, the cost of this substrate is high and is estimated to account for 30% of the production cost [54]. Therefore, focusing on inexpensive carbon substrates, especially those recycled from organic waste, is beneficial for PHA commercialization. Some of the waste could be used as "received," such as some vegetable oils or glycerol for the synthesis of mcl-PHA. However, in most cases, these waste streams contain contaminants that affect bacterial growth and accumulation of PHA. Therefore, an approach consisting of the preliminary conversion of any carbon-based waste into syngas, which may be purified when necessary, and fed to bacteria is of great interest for the sustainable production of PHA. It allows a wider range of initial substrates to be used in a single bioprocess. The conversion of biomass and waste into syngas is discussed in section 15.3 and is the subject of a large body of research; in contrast, the bacterial conversion of syngas into PHA is only sparsely discussed in the literature. It appears that it was only investigated about one decade ago in the field of syngas fermentation [55].

One of the most studied microorganisms for the conversion of syngas to PHA is not an acetogen, but an alphaproteobacterium from the family of the Rhodospirillaceae, *Rhodospirillum rubrum*. Its metabolism of growth on syngas and its ability to produce PHA will be examined in section 15.6. Another alphaproteobacterium, *Pseudomonas carboxyhydrogena*, was shown to grow on syngas and produce PHA and will be discussed in section 15.7. Finally, hydrogen-oxidizing bacteria able to grow in syngas will be presented with some examples in section 15.8. Several attempts to genetically modify bacteria, either to enable them to produce PHA or to enhance their productivity, are also reviewed in the respective sections.

15.5 PRODUCTION OF PHA BY ACETOGENS BASED ON SYNGAS AS SUBSTRATE

Acetogens are associated with the ability to perform simultaneous carbon fixation and anaerobic respiration through the acetyl-coenzyme A (acetyl-CoA) pathway. It is also called the Wood–Ljungdahl (WL) pathway after the names of both researchers involved in the discovery. This well-known metabolism is believed to be the most ancient carbon fixation pathway and is composed of two branches that will be described [56]. Since the metabolism of acetogens has been reviewed elsewhere [57–59], we will focus only on the assimilation of CO and CO_2 and the use of H_2.

Under autotrophic conditions (i.e., the absence of an organic carbon source), inorganic carbon serves as feedstock for synthesizing building blocks and complex molecules in the bacteria. The task of the WL pathway machinery is to transform inorganic carbon into the specific building block, acetyl-CoA, which also leads acetate secretion in specific conditions. However, the WL pathway is not self-sufficient, and an energy source is also necessary for the bacteria to drive it and proliferate. Energy is either provided by high-energy electrons (reducing equivalents) or by the breaking of a high-energy chemical bond [e.g., adenosine triphosphate (ATP)]. In the second case, the bond was previously formed during the transfer of energy in a redox reaction involving high-energy electrons. External energy is either obtained from light absorption (photoautotroph), which will result in the promotion of an electron to a higher energy state in a photosensitizer molecule, or by the transfer of a high-energy electron from a chemically reducing species (lithoautotroph or chemoautotroph). The available high-energy electrons are then transported into the cytoplasm of the bacteria by mediator molecules (NADPH, NADH, ferredoxin, etc.) and are finally used for the reduction of inorganic carbon molecules (CO or CO_2) into reduced carbon species that constitute building blocks in the metabolism of growing bacteria.

The WL pathway represents an essential interconnection between carbon and energy flux in bacteria. Based on syngas with various contents of CO, H_2, and CO_2, three alternatives for carbon fixation can take place in acetogens, with a possible co-existence, depending on the bacteria (when CH_4 is also present, other metabolic pathways may also be involved):

1. CO_2 is the carbon source and its assimilation is enabled by H_2, which acts as an energy (or electron) source;
2. Similarly, CO is the carbon source, while H_2 is the energy source;
3. CO reacts with water in an enzymatically catalyzed water-gas shift reaction to provide CO_2, two high-energy electrons, and two protons (that can combine into a hydrogen molecule). CO_2 is further used as the carbon source and H_2 or the generated high-energy electrons as the energy source, as in 1.

First, the energy conservation pathways and the processes of electron transfer from hydrogen are not fully elucidated for any acetogen bacteria. In a proposed initial model based on joint knowledge gained on several bacteria, the energy of hydrogen electrons is harnessed by the membrane and cytoplasmic hydrogenases [60], which reduces solubilized hydrogen molecules into two protons, while collecting both high-energy electrons. For membrane hydrogenases, these reducing equivalents are further transferred through the membrane via cytochromes and quinone intermediates. At the same time, the transfer of protons through the membrane occurs to finally drive the ATP synthase machinery responsible for producing energy in the form of ATP for the bacteria. The electrons collected by cytoplasmic hydrogenases are transferred to a cytoplasmic electron mediator, such as NAD^+ or $NADP^+$, which, once reduced, can further transfer the electrons to ferredoxin proteins or other intracellular redox mediators. The high-energy-containing molecules of ATP, NADH, NADPH, and ferredoxin are used in the WL pathway, but also any other metabolic transformation requiring energy.

It was observed that many acetogens do not exhibit the proteins of the electron transport chain sustaining the action of ATP synthase by forming a proton gradient. New models were therefore suggested based on observations made of several different acetogens (i.e., none of these observations were made in a single bacterium species so far) [53]. In fact, several so-called flavin-based electron bifurcation (FBEB) complexes were discovered. First, an Rnf (NADH:quinone oxidoreductase membrane complex) membrane-associated complex acts simultaneously on the pumping of protons or sodium ions out of the cell and on the promotion of an electron transfer from reduced ferredoxin to NAD^+. Second, a cytoplasmic [FeFe]-hydrogenase able to reduce ferredoxin and NAD^+ by oxidizing hydrogen was identified. Another cytoplasmic [FeFe]-hydrogenase specific to NADP was also identified. Third, an Nfn (NADP oxidoreductase) complex was found, which was able to transfer an electron from NADH or ferredoxin to $NADP^+$. Finally, it is believed that the methylenetetrahydrofolate reductase, involved in the WL pathway, also acts as an FBEB. In summary, FBEBs are membrane and cytoplasmic entities that transfer electrons from one mediator to another and therefore act similarly to the electron transport chain to provide high-energy electrons to the bacteria (NADH, NADPH, reduced ferredoxins, etc.) and thus participate in the chemiosmotic gradient and ATP synthesis.

Regarding CO_2 assimilation (point 1 above), it first diffuses across the membrane and is then subjected to a six-electron reduction chain leading to a methyl group, which is transferred on a corrinoid iron-sulfur protein (CoFeSP) for participation in the formation of acetyl-CoA. This requires one ATP molecule and six electrons (provided by one NADPH and two NADH molecules) and corresponds to the methyl branch of the WL pathway. Alternatively, CO_2 can be simultaneously reduced to CO by the carbon monoxide dehydrogenase/acetyl-CoA synthase (CODH/ACS) complex using one ferredoxin molecule (i.e., two electrons). This enzyme-bound CO and a methyl group can undergo a reaction of condensation and be transferred on a CoA unit to finally form acetyl-CoA. The transformation of CO_2 into a bound CO constitutes the carbonyl branch of the WL pathway toward the formation of acetyl-CoA.

Finally, CO is either directly bound to the CODH/ACS complex (carbonyl branch), or its oxidation to CO_2 is catalyzed by the CODH (providing two protons and two electrons to ferredoxin, which can be used later in formate fixation). The generated molecule of CO_2 undergoes the whole reductive chain until the methyl group is transferred on a CoA unit to participate in acetyl-CoA production (methyl branch). Finally, similar to CO_2 assimilation, to form an acetyl-CoA unit, one methyl group provided by the methyl branch (i.e., methyl-COFeSP) is condensed with the CODH/ACS complex with one bound carbonyl group to form an acetyl-CoA unit while regenerating one CODH/ACS and one COFeSP complex. In purely autotrophic conditions, four electrons (i.e., two hydrogen molecules) are required to fix one carbon atom, and therefore eight electrons will be needed for one acetyl-CoA unit [53].

The fate of acetyl-CoA depends on the environmental growth conditions of the bacterium. It can be directed either in the formation of biomass via other metabolic pathways or toward the formation of acetate. In the latter transformation, one ATP molecule is generated. Therefore, when acetate is synthesized, no net ATP molecules are generated, as one molecule of ATP was used in the reductive methyl branch of the WL pathway. Therefore, the energy needed for further conversion processes can only be provided by high-energy electrons from hydrogen and/or CO.

The use of syngas, and especially CO by acetogens, is often assessed by comparing the bioreactor inlet and outlet gases composition. As a consequence of CODH and the promoted water-gas shift reaction, H_2 and CO_2 are produced, and their increased concentration can be measured when syngas is continuously flowing through the reactor, while the CO concentration decreases (e.g., in [55]). Moreover, from the perspective of these explanations, it becomes clear that, due to the catalysis of the water-gas shift reaction by CODH, the ratio of H_2:CO in the fed syngas gas mix can be wide (gas flexibility). Even if an optimal composition could be determined on the basis of an energy balance, it will not necessarily fit reality because of possible kinetics limitations.

15.5.1 PRODUCTS OF SYNGAS FERMENTATION WITH ACETOGENS

The main products that can be naturally synthesized by acetogens in autotrophic conditions are acetate, ethanol, 2,3-butandiol, butyrate, formate, H_2, H_2S, and traces of methane. Moreover, other products can be targeted, but with a lower selectivity: isopropanol, acetone, 1-butanol, 1,3-butanediol, citramalate, lactate, succinate, branched-chain amino acids, methyl ethyl ketone, acetoin, 2-butanol, 1,2-propanediol, 1-propanol, aromatics, isoprene, mevalonate, and 3-hydroxypropionate. In total, LanzaTech proposes that more than 50 products can be obtained from their proprietary strain [61,62].

Acetate is produced from acetyl-CoA via an acetyl-phosphate intermediate on the basis of catalysis by phosphotransacetylase and acetate kinase as the first and the second step, respectively. This second enzyme involves substrate-level phosphorylation and therefore yields an ATP molecule, the source of energy in the bacteria. This conversion is thus expected to be growth related [63]. In autotrophic conditions (H_2 + CO_2), cultures of *Acetobacterium woodii* and two genetically modified strains for overexpression of WL pathway enzymes in a continuous stirred tank reactor (CSTR) yielded 8.2–9.6 g/L acetate, with a specific production rate of 20.5–21.8 g g^{-1} d^{-1} [64]. A yield of 44 g/L (0.8 M) of acetate could be obtained with a maximum specific production rate of 6.9 g g^{-1} d^{-1} using the same microorganism and a CO_2 and H_2 substrate [65].

The synthesis of ethanol is based on acetyl-CoA and is based on a two-step process. First, the enzyme acetaldehyde dehydrogenase catalyzes the conversion of acetyl-CoA to acetaldehyde, which is further reduced to ethanol with the help of ethanol dehydrogenase. This production process is supposed to be not growth related, and an increased ethanol accumulation (9.43 mM) was observed when growth was restricted by a nitrogen source limitation compared with a growth medium: the ratio of ethanol:acetate changed from 1:7.8 mol/mol (in growth medium) to 1:4.5 mol/mol (in the absence of nitrogen nutrients) [66]. Besides the effect of nitrogen limitation, other strategies for increasing the ethanol:acetate ratio were suggested: to apply heat shocks to bacteria, to add a reducing agent, pH shift, to add H_2, and medium supplementations [63]. Moreover, the conversion of acetate to H_2 and CO_2 (microbial oxidation) was also applied.

Acetone, butanol, and ethanol (ABE) co-synthesis, which is also called ABE fermentation was investigated in the early 1980s for several clostridial species [67–69]. Using *Clostridium acetobutylicum* grown on molasses, up to 18–20 g/L of butanol and acetone could be produced with a mass ratio of butanol:acetone of 2:1 to 3:1 [67]. The same strain, immobilized in the form of spores on a calcium alginate gel, yielded a 4.5–5.1:1 butanol:acetone mass ratio on glucose and butyric acid substrates, and the yield of glucose to butanol was calculated to comprise between 12.3 and 20.1% in these conditions [68]. Over a whey filtrate substrate, this strain was reported to have lower productivity due to a slower metabolism compared with glucose substrates, and produced a 1.5 wt/vol% butanol, 0.15 wt/vol% of ethanol and 0.15 wt/vol% acetone, yielding, therefore, a mass ratio of butanol:acetone:ethanol of 10:1:1 [69]. These promising yields were, however, not sufficient for bringing the bioprocess to a commercial application, probably because of the high cost of the substrates, the low achieved titers due to solvent toxicity, the low productivity, and the costs related to solvents recovery and purification [70].

15.5.2 PHA SYNTHESIS BY ACETOGENS GROWING ON SYNGAS

The fact that acetogens can grow on CO_2 and H_2 while being resistant to the presence of CO, or even growing on CO, makes them very interesting for the production of PHA from syngas. Indeed, as CO is usually toxic (i.e., growth inhibitory), it may poison hydrogenases and other iron-based enzymatic centers in the cell. However, so far, despite this ability to use CO, no wild-type acetogenic microorganisms accumulating PHA on syngas as the sole substrate have been reported in the literature to the best of the authors' knowledge. However, it should be noted that PHA can be produced by some acetogenic bacteria, but in heterotrophic conditions [71].

Nevertheless, a recent report focused on the genetic engineering of two recombinant clostridial acetogens for 3-HB monomers and poly(3-hydroxybutyrate) (PHB) synthesis [71]. Genes coding for enzymes required for PHB production were integrated into *Clostridium ljungdahlii* and *Clostridium coskatii*, which are known to grow on the syngas substrate. The clones only grew when these genes originated from phylogenetically closely related microorganisms. The study focused on genetically redirecting the acetyl-CoA into PHB production with the addition of the type III PHA synthase and the *R*-enoyl-CoA enzymes. This approach was successful as transmission electron microscopy analysis showed that part of the cells clearly produced PHB, which could also be detected by extraction and gas chromatography analysis. The conditions for aeration and stimulating PHB formation could, however, be subjected to optimization for more efficient accumulation of PHB.

15.6 PHA PRODUCTION BY *RHODOSPIRILLUM RUBRUM* GROWN ON SYNGAS

Most of the studies in the field of PHA biosynthesis on syngas substrate were conducted with the purple, non-sulfur, Gram-negative bacterium *Rhodospirillum rubrum* [72]. It is a facultative photosynthetic bacterium that can either grow heterotrophically or autotrophically and in both aerobic and anaerobic environments [24]. Various organic compounds and the associated conditions were reviewed by Fuller [73] and will, therefore, not be discussed in more detail here. Other properties of *R. rubrum* were also discussed in the literature, such as its different metabolisms, nitrogen fixation, and hydrogen production. [74–81]. In terms of PHA accumulation, Brandl et al. achieved 20 and 46 wt.-% PHB under anaerobic, photoheterotrophic conditions, using either 30 mM acetate or 3-hydroxybutyrate as the carbon source, respectively, and applying nitrogen limitation [82].

A general scheme of the metabolic pathways involved in the growth of *R. rubrum* on syngas is presented in Figure 15.2. In this microorganism, under anaerobic conditions and in the presence or absence of light, CO is enzymatically converted to CO_2 by a CODH (EC 1.2.7.4) coupled with a CO-insensitive hydrogenase. This CO metabolism is regulated by CooA, a CO-sensitive transcription activator, and therefore the concentration of *cooA* transcripts can be considered as a probe of an active CO metabolism. It was demonstrated based on ^{13}C syngas labeling, that each CO_2 resulting from the water-gas shift reaction was further assimilated in biomass and PHA [31]. On the basis of a six-fold higher transcript level in the presence of syngas, the same authors proposed dual assimilation of CO_2 through the action of two carboxylases: the crotonyl-CoA reductase (*Ccr*, encoded by the gene *Rru_A3063*) and the ferredoxin-dependent pyruvate synthase (PFOR, encoded by the gene *Rru_A2398*).

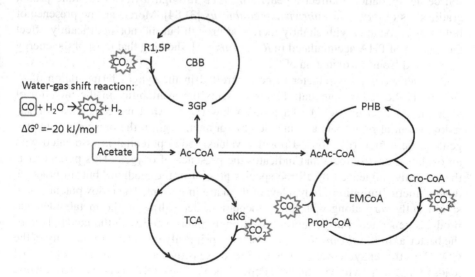

FIGURE 15.2 Simplified metabolic pathways involved in syngas assimilation and PHA synthesis in *Rhodospirillum rubrum* [24].

No increase in the transcript number was observed for other carboxylases. Moreover, despite a high level of Rubisco expression, no flux in the Calvin–Benson–Bassham cycle was measured in the presence of light, compared to darkness, in the presence of syngas. Therefore, the assimilation of CO_2 seems to be mostly occurring through the reductive tricarboxylic acid cycle and the ethylmalonyl-CoA pathway. It is interesting to note that other authors measured increased CO_2 and H_2 concentrations in the headspace of the reactor in dark growth conditions [24], which indicates that part of the oxidized carbon is not assimilated and that CO also acts as an energy source in this microorganism. This contradiction may be attributed to differences in growth conditions, but this would need to be confirmed experimentally for full elucidation.

It is important to note that the assimilation of CO by *R. rubrum* was shown to be dependent on the presence (even in traces) of an organic co-substrate, such as yeast extract, acetate, or malate [24,31,55,83]. For instance, the removal of acetate from the medium led to a 50% decrease in PHA production [55]. Revelles et al. observed that, on the one hand, cell growth increased by a factor of four when 5–10 mM acetate was added and, on the other hand, that PHA accumulation correlated with the presence of acetate in the medium [31]. The same study also showed that the presence of acetate correlated with an increased level of *cooA* transcripts. It was hypothesized that acetate promoted CO uptake so that energy for growth and PHA synthesis is generated.

Syngas fermentation of *R. rubrum* usually occurs in anaerobic conditions. The growth on non-fermentable acids, such as acetate, cannot occur under these conditions due to the lack of ATP produced via substrate-level phosphorylation. Therefore, another source of energy (specifically of ATP) is necessary in anaerobic conditions, especially with the absence of light for the assimilation of acetate. As growth occurred only in the presence of syngas, it appeared that CO also acted as an energy source, its oxidation indirectly generating ATP through a transmembrane proton gradient, as observed in *Rubrivivax gelatinosus* [31,84]. Moreover, the presence of light was associated with slightly increased growth but did not significantly affect the amount of PHA accumulated in *R. rubrum* [31], showing that most of the energy is collected from the oxidation of CO.

Another essential parameter to be controlled in the syngas fermentation of *R. rubrum* is the redox potential. This represents the availability in the medium of high-energy electrons for the bacteria. A low (i.e., toward more negative values) redox potential represents a reductive environment (e.g., in the presence of hydrogen), while a high (i.e., toward positive values) redox potential is associated with an oxidative environment and indicates the presence of oxygen. This parameter is theoretically influenced by all the species present in the medium, but, in practice, due to kinetic limitations, only a few of them are important. The redox potential can significantly vary along with the fermentation, depending on the metabolites and products generated. Furthermore, it is strongly interrelated with the metabolism of the bacteria. In *R. rubrum*, the redox potential primarily influences the activity of the CODH, as this enzyme was shown to be redox sensitive: a decreased redox potential leads to higher activity [55] and the dependence of the CODH activity as a function of the redox potential follows a Nernst equation shape. The standard redox potential of the enzyme is still under discussion as rather wide ranges of redox potentials were obtained (-316 vs. -418 mV). The enzymatic reaction seems to involve more than one conformation of the enzyme-substrate complex, and therefore the overall mechanism and the determination of the standard redox potential for the enzymatic reaction are also still under investigation [85,86]. The effect of the oxidation state of the substrate on the medium redox state and the metabolism for redox regulation needs to be considered.

In *R. rubrum*, the PHA produced is mostly composed of poly(3-hydroxybutyrate) containing a low weight percent of poly(3-hydroxyvalerate) (PHV). Independent of the energy and carbon sources in *R. rubrum*, PHB is produced in three steps from the acetyl-CoA central intermediate:

1. Condensation of two acetyl-CoA molecules to form acetoacetyl-CoA. This is catalyzed by 3-ketothiolase (PhaA);
2. Reduction of acetoacetyl-CoA by NADH to form 3-hydroxybutyryl-CoA. This reaction is catalyzed by acetoacetyl-CoA reductase (PhaB);
3. Polymerization of the 3-hydroxybutyryl-CoA monomers to form PHB, catalyzed by the PHB synthase (PhaC).

When carbon and/or energy is lacking in the medium, depolymerization of PHB can occur in the exact opposite reactions, and the whole reaction chain is reversed to produce the 3-hydroxy fatty acids that can be linked to an acetyl-CoA unit. This intermediate will be further directed toward other metabolic pathways.

It is interesting to note that PHA has often been shown to be produced more efficiently under nutrient limitation. For *R. rubrum*, significantly higher PHB content could be achieved with a dual carbon (acetate and CO) and phosphorus limitation, compared with only a carbon limitation and a dual carbon and nitrogen limitation. For both dual nutrient limitation situations, the growth was set to be acetate limited (i.e., with an acetate feed adapted to the growth rate). The results were explained by an increased accumulation (30 wt.-% CDM) compared with the sole carbon limitation (11.9 wt.-% CDM) and the dual carbon nitrogen limitation (11.1 wt.-% CDM) [24].

From the above-described reports, it appears that PHA can be produced from syngas by *R. rubrum*, especially in the presence of the co-substrate acetate. However, the specific and volumetric production rates remain low, despite efforts made to optimize growth conditions. Based on a genetic engineering study, this low productivity was at least partially imputed to the low activity of the PHA synthase in wild-type *R. rubrum* [87]. In this work, the gene coding for the PHA synthase was replaced by heterologous PHA synthase genes, and all strains showed were active for PHA synthesis on syngas. A follow-up study suggested that an *R. rubrum* strain, which was modified with several enzymes involved in the synthesis of PHA in *Pseudomonas putida* KT2440, was able to produce *mcl*-PHA from syngas [88]. Accumulation of a copolymer, poly(3-hydroxydecanoate-*co*-3-hydroxyoctanoate) up to 7.1 wt.-% CDM exclusively on syngas containing both CO and CO_2 carbon source and hydrogen. Even though the *mcl*-PHA content remained relatively low, this result showed that it is possible to extend the number of PHA materials that can be synthesized from syngas by *R. rubrum*. More importantly, these two last reports illustrate the importance of genetic engineering for enlarging the abilities of *R. rubrum* grown on syngas and provide directions for future enhanced PHA productivities.

15.7 SYNTHESIS OF PHA BY CARBOXYDOBACTERIA GROWN ON SYNGAS

A small number of bacterial strains were identified as so-called carboxydobacteria. They are characterized by their ability to perform, under aerobic conditions, the reaction of CO oxidation as a source of energy, and to assimilate the resulting CO_2 as a source of carbon (CO as sole energy and carbon source). In these microorganisms, CO_2 is further assimilated via the reductive pentose phosphate cycle [89]. Moreover, they can use hydrogen as alternative energy source using a hydrogenase, just like the hydrogen-oxidizing bacteria (see below).

Three examples of this category of bacteria are *Comamonas compransoris* Z-1155, *Pseudomonas gazotropha* Z-1156, and the originally called *Seliberia carboxydohydrogena* Z-1062 [89]. The *S. carboxydohydrogena* bacterium was originally studied by Sanjieva and Zavarzin and was reclassified and called *Pseudomonas carboxydohydrogena* (DSM 1083) by Meyer et al. [90]. As described by the authors, this Gram-negative bacterium is an alphaproteobacterium; it is motile (monotrichious) and has a rod shape. It is strictly aerobic and can grow heterotrophically or autotrophically. Heterotrophically, it grows only on a limited number of substrates (acetate, pyruvate, lactate, citrate, fumarate, malate, aspartate, fructose, sucrose, and aspartate). Autotrophically, it can either grow on $CO + O_2 + N_2$ gas mix and $H_2 + CO_2 + O_2$ mix. The enzyme for CO oxidation was shown to be soluble and only expressed in the presence of CO, while the membrane-bound hydrogenase was always present in autotrophic growth. The bacteria is robust as its metabolism and PHA accumulation was reported to be insensitive to unfavorable growth conditions (non-optimal pH, temperature, O_2, CO_2, or H_2 excess or deficiency) [89]. In the same work, it was observed that the consumption of CO_2, O_2, and H_2 remained unchanged while the CO concentration was increased from 10 to 20 vol% CO, but increased consumption of O_2 and H_2 was measured at 30 vol% CO.

Interestingly, Volova et al. reported that when a gas mixture of CO, CO_2, H_2, and O_2 is provided, the concentrations of CO_2, H_2, and O_2 decrease faster than that of CO. This result suggests that assimilation of carbon from CO_2 coupled to the harnessing of hydrogen energy is faster than the CO oxidation reaction [89]. Moreover, the inhibitory effect of CO is not avoided as an increasing CO concentration was related to a decreasing specific growth rate. However, this effect did not influence the synthesis of nitrogen-based compounds and was increasing the activity of the hydrogenase. For instance, when the concentration of CO in the gas mix moved from 20 to 30 vol%, the biomass growth was slower, while no decrease was observed when moving from 10 to 20 vol% CO [89]. While increasing the CO concentration, a change of the membrane was observed too.

As discussed by the same author, *P. carboxydohydrogena* was also shown to be able to produce PHA, particularly when subjected to a sulfur, nitrogen, or phosphorus limitation. An accumulation of PHB of 28 wt.-% could be achieved in these conditions in the presence of CO, and 40–45 wt.-% without CO (only in the presence of CO_2, H_2, and O_2). A medium supplemented with hexanoic and valeric acids allowed the bacteria to synthesize a terpolymer including 3-hydroxyhexanoate (15–25 mol-% and 3-hydroxyvalerate) (1–3.2 mol-%) together with 3-hydroxybutyrate (74.83 wt.-%), but with a low PHA content (3.4–8.7 wt.-%) [89]. Overall, in terms of accumulation, with a limitation of nitrogen or phosphorus, a maximum PHB content of 63 wt.-% could be achieved associated with polymer productivity of 0.22 g L^{-1} h^{-1} in 56-h cultivation [89].

From this study, it appears that even if CO can be used as an energy source and potentially as a carbon source for the bacteria, its oxidation rate is rather slow, and without CO_2 and H_2 in the gas mixture, the growth is speculated to be slower (the experimental evidence is not provided in the literature). The reactions of these bacteria in the presence of CO seem to be protective (increased CODH activity, change in the membrane, etc.) [89] to avoid a high cytoplasmic concentration of CO. This makes this metabolic pathway less attractive in terms of PHA productivities for syngas with high CO content or low content of CO_2 and/or H_2.

15.8 PHA PRODUCTION BY CO-TOLERANT HYDROGEN-OXIDIZING STRAINS ON SYNGAS

Another category of bacteria that could be considered as potential candidates for syngas bioconversion to PHA is hydrogen-oxidizing bacteria. They can facultatively grow on a CO_2 and H_2 gas mix, in the presence of O_2. The concentration of O_2 needs to remain relatively low (micro-aerobic conditions) to avoid the inhibition of hydrogenase activity. Similarly, CO can affect cell energy intake by strongly binding to the active centers of hydrogenases. However, some hydrogen-oxidizing bacteria were reported to be carbon monoxide tolerant. This includes *Ralstonia eutropha* B5786 [91]. Bacteria of the *Cupriavidus necator* (this includes *R. eutropha* B5786) genera are generally recognized to be efficient PHA producers. They produce PHA from the acetyl-CoA intermediate using the same three enzymes as in *R. rubrum* (3-ketothiolase, acetoacetyl-CoA reductase, and PHB synthase). For *R. eutropha* B5786, it was observed that in the presence of CO, both growth metabolism and PHB accumulation capacities remain unchanged. Indeed, increasing the concentration of CO from 0 to 20 vol% only slightly decreased the final biomass concentration (from 20 to 17.5 g/L) and PHA content (from 76.4 to 72.8 wt.-%) in batch cultures [91]. It is interesting to note that the respective activity of hydrogenase and cytochrome oxidase increased when the CO concentration rose. Finally, it was observed that this bacterium was able to use CO as a substrate but only at low rates (30–50 µmol CO min^{-1} mg$_{protein}$$^{-1}$).

Another hydrogen-oxidizing bacterial strain, *Ideonella* sp. O-1 (JCM17105), was reported to be CO tolerant, grow on syngas, and concomitantly synthesize PHB [92]. This strain was also shown to be tolerant to high O_2 content in the gas mix as its specific growth rate only decreased by a factor two (without considering the dilution effect of the other gas components) in the presence of 30 vol% O_2, while two *Ralstonia eutropha* and one *Alcaligenes latus* strains showed no growth under such conditions. A similar trend was observed for the tolerance to CO; the specific growth rate of the *Ideonella* strain exhibited no significant decay even at concentrations as high as 70 vol% CO, whereas both other strains were already strongly affected at a concentration of 10 vol% CO. However, it should be noted that CO was not consumed by the bacterium and remained in the headspace and in the culture broth. The amount of PHA accumulated by *Ideonella* was slightly lower in the presence of CO as it reached 66 wt.-% at 20 vol% CO while being 78 wt.-% in the absence of CO ($H_2:CO_2:O_2 = 7:1:1$).

This class of bacteria is promising for the use of syngas with low CO content, especially due to its ability to accumulate high amounts of PHA. However, they will not assimilate CO or do so only very slowly, implying that its concentration will grow in the bioreactor, and it will become progressively more toxic to the bacteria. From this perspective, these bacteria are more interesting for growth on crude CO_2 and H_2 gas mixtures, which have CO impurities. It has to be noted that this bacterium requires the presence of oxygen, which is usually absent in syngas and therefore would need to be supplemented.

15.9 BIOPROCESSES FOR PHA PRODUCTION ON SYNGAS

Given the above discussions on the diversity of syngas composition, sensitivity of bacteria to this composition, the four syngas-assimilation metabolic schemes presented, and the ability of the studied bacteria to accumulate PHA, a few notes on the design of bioprocesses will be discussed in this section.

One important aspect is the gas mix composition and its effect on bioprocess performance, especially on specific growth and PHA accumulation rates. Moreover, it may affect the composition and molecular weight of the synthesized polymer, which are important for industrial applications. Indeed, as discussed above, syngas needs to simultaneously provide energy and carbon to bacteria. Moreover, the presence of oxygen is also required for carboxydobacteria and hydrogen-oxidizing bacteria. Therefore, the ideal relative proportions of the gas mix components are theoretically dictated by the stoichiometry involved in the biomass and PHA synthesis reactions. For alternative gas compositions, one of the essential elements may limit the growth and/or PHA accumulation processes. It should be noted, however, that the kinetics of mass transfer combined with enzymatic kinetics makes it more difficult to model the exact proportions needed for each of the gas components.

Furthermore, the toxicity effect of syngas, in particular, has to be considered. In this regard, it is worth mentioning the study by Karmann et al. [24], which showed the positive effect of diluting syngas with nitrogen. In this work, the initial syngas composition was 40 mol-% CO, 40 mol-% H_2, 10 mol-% CO_2, and 10 mol-% N_2. An optimal dilution with nitrogen to 60 vol% resulted in the highest specific growth rate in batch cultures. The effect of syngas dilution on the PHA content was, however, not further evaluated. The obtained results showed that the composition of the gas is of importance, as a growth rate four times lower was observed for the non-diluted gas under otherwise identical conditions. Based on sole dilution by nitrogen, it is not possible to explain the observed effect, but one of the hypotheses is the possible toxicity of CO to the concerned bacteria. Kerby et al. [93] reported that the doubling time of *R. rubrum* increased (4.8, 5.7, and 8.4 h) when the CO concentration was increased (25, 50, and 100 vol% CO). It is worth noting that only a limited CO toxicity effect was observed for the strain *P. carboxydohydrogena*, studied by the Volova group [89]. The toxicity of CO is, however, strain-dependent and has not been exactly quantified and understood for any of the microorganisms presented herein. Moreover, it can be expected that bacteria may require an adaptation period with a progressively increasing CO concentration. Other factors may also be involved, such as the relative concentration of the gas mix components and the nature of the enzyme's active site. Indeed, the toxicity of CO in the gas mix will at least partially depend on the relative affinity of the substrates to the CODH enzyme, besides its effects on other enzymatic systems in the microorganism.

Finally, the requirement for oxygen in some syngas fermentation (depending on the type of bacteria involved) implies the implementation of an accurate gas mixing system. The presence of oxygen in syngas is also associated with a risk of explosion in the presence of an ignition source when its concentration is not maintained below the lower explosivity limit 84 mol-% O_2 in a hydrogen and oxygen gas mix. Therefore, the gas supply may become limited in oxygen under these conditions, in particular, at high cell densities.

The presence of impurities when using crude syngas needs to be considered, as they may be inhibitors of growth or PHA accumulation or may present a catalytic effect. For instance, NO_x impurities may inhibit the activity of CODH. Similarly, for some cultivations, the presence of oxygen is considered a contaminant as it also negatively affects the CODH activity. However, to illustrate the positive effect of contaminants, Do et al. grew R. rubrum using an industrial synthesis gas and observed that tar, ash, and char accumulation in the gas input pipe might have the beneficial effect of purifying the gas of toxic impurities, especially high molecular weight, incompletely gasified compounds [55]. The same study also showed that the presence of tar and char and the use of crude synthesis gas compared with an artificial gas mix had an immediate positive effect on the activity of the CODH. It was proposed to be related to the presence of H_2S, which had the double effect of reducing the redox potential and activating the CODH. Indeed, in some bacteria, H_2S can be used as the sole energy source, in which case, elemental sulfur is observed in the form of an extracellular deposition.

However, the biggest challenge of syngas fermentation is increasing the mass transfer of the gas components into the growth medium. It is possible to act both on the limited concentration of the dissolved gas and the low rate of transfer of gas from the bubbles to the liquid. When considering Henry's constants of the involved gas, it appears that CO and H_2 have very low solubility in water and, therefore, will quickly limit the growth or product formation in the absence of an efficient supply, in particular, at high cell density. Indeed, in a bioreactor with growing cells and a constant, low supply of syngas in the medium, the concentration of dissolved gas is very close to zero and can be considered constant. In this steady-state situation, the rate of dissolved gas consumption is equal to the rate of the dissolution of gas. Therefore, the growth is gas supply limited and especially limited by the gas component that has the least efficient mass transfer. In practice, batch cultures will show a linear relation to the growth of biomass. This was observed in the bioreactor studies of Karmann et al. [24] and Do et al. [55]. The CO concentration was reported to be below the detection limit, especially during the growth phase. As a consequence, the growth was limited by the mass transfer of CO [55]. It is indeed expected that CO is the limiting gas in most syngas fermentation, but this depends on the syngas composition and the bacterium of concern.

Agitation and sparging systems, as well as the selection of a suitable type of reactor, need to be considered to improve mass transfer. Mass transfer in one specific reactor and medium is usually assessed by the measure of the oxygen k_La. In the literature, the system exhibiting high performance in PHA production based on gas fermentation presented k_La of 460 h^{-1} [89] and 2970 h^{-1} [94]. Finally, in terms of reactor design, the objective is to improve the mass transfer of gas, while maintaining suitable conditions for cell growth and high productivities. There are four main types of reactors that can be used for gas fermentations: stirred tank (STR), CSTR, bubble column, and packed bed column [63]. CSTRs are continuously fed with gas flow and keep a constant liquid volume while being agitated. Even though CSTRs are associated with high mass transfer coefficients, they involve strong agitation of the medium, which can affect cell growth and PHA production (shear stress). A trade-off between the high dispersion of the gas by the stirrer and low shear force on the microorganisms is necessary. Bubble columns are interesting when a long retention time for the liquid is necessary (e.g., for higher titer of the product(s) before separation). Finally, packed column reactors are specifically used for corrosive products, but also for low liquid hold-up time and small pressure differences [63].

15.10   CONCLUSIONS AND OUTLOOK

The technology of biomass and waste gasification has been the subject of a large body of research and is now on the verge of being commercialized. Compared with classic fossil fuel gasification processes, it presents several specificities, such as high moisture content of the feedstocks, diversity of feedstocks, and a resulting variable syngas composition. As a consequence, two approaches will be compared when designing such gasifiers for biomass or waste: 1) an unspecific feedstock gasifier, which will lead to lower reproducibility of the syngas composition but a larger-scale capacity; 2) a smaller scale gasifier, which can be tuned to the specific feedstock and produce a syngas of a more controllable composition. The technology of gasification is selected in regard to the application of the syngas produced, particularly in terms of the purification unit(s). As such, the whole material chain is considered when designing a syngas producing unit.

The use of syngas in bioprocesses for the production of hydrocarbon compounds, such as acetate and ethanol, was developed based on acetogenic microorganisms and is supported by a significant body of literature. However, so far, no wild-type acetogen growing on syngas has been reported to synthesize PHA. Other bacteria, such as *Rhodospirillum rubrum*, carboxydobacteria, and CO-tolerant hydrogen-oxidizing bacteria, can grow on specific syngas compositions for the production of PHA (mostly PHB). Two approaches based on recombinant bacteria were also proposed. The first approach consists of the modification of *R. rubrum* with *mcl*-PHA enzymatic apparatus genes leading to the accumulation of 7 wt.-% P(3HD-*co*-3HO) on the syngas substrate. The second approach focuses on providing PHA synthesis ability to *Clostridia acetogens* and produced about 1 wt.-% of PHA; it also produced 3-HB monomers using syngas as carbon and energy sources. These two studies pave the way toward using genetic engineering to creating bacteria able to grow on syngas and efficiently produce PHA. To summarize the performance of the different strains in terms of the yield of PHA accumulated on syngas, the results of the identified studies in the field of PHA production on syngas available in the literature are presented in Table 15.2.

TABLE 15.2

Performances of Microbial PHA Synthesis on Syngas or Similar Substrates. Only Wild-Type Bacteria Were Considered here as Recombinant Bacteria are Only at the Very Beginning of Their Development

Strain	Medium and gas composition (unless specified: mol-%)	PHA content [wt.-% in CDM]	Growth rate [h⁻¹] / productivity [mg PHA L⁻¹ day⁻¹]	Remarks	Ref.
R. rubrum (ATCC 11170)	RRNCO medium + 15 mM fructose 10 mM acetate 40% CO + 40% H_2 + 10% CO_2 + 10% N_2 Darkness, anaerobic	28 ± 10	0.021–0.029/ –	- Light improves slightly μ, but no effect on PHA - Acetate needed for PHA accumulation	[31]
R. rubrum	RRNCO medium with acetate 10% H_2 + 55% N_2 + 17% CO + 16% CO_2 Anaerobic (darkness?)	38	–/59.2	- PHBHV (86:14)	[55]
R. rubrum S1 (ATCC 11170)	Modified RRNC medium 0.59 g/L (= 10 mM) acetate Darkness, anaeronic	30.0	–/324	- Dual C, P limitation better compared with C or C, N dual limitation	[24]
P. carboxydohydrogena	CO: 10, 20, 30 vol% CO_2: 10 vol% H_2: 50, 60, 70 vol% O_2: 10vol%	63	–/5520	- High $k_L a$ (460 h⁻¹) - S, N and S + N limitation - Inhibitory effect of CO (CO > 20%)	[89]

(Continued)

TABLE 15.2 (CONTINUED)
Performances of Microbial PHA Synthesis on Syngas or Similar Substrates. Only Wild-Type Bacteria Were Considered here as Recombinant Bacteria are Only at the Very Beginning of Their Development

Strain	Medium and gas composition (unless specified: mol-%)	PHA content [wt.-% in CDM]	Growth rate [h^{-1}] / productivity [mg PHA L^{-1} day^{-1}]	Remarks	Ref.
R. eutropha B5786	Schlegel mineral salts medium CO: 0–20 vol% CO$_2$: H$_2$: O$_2$: unknown	72.8–76.4	–	- No effect of CO on accumulation - Nitrogen limited or nitrogen-free - Slight consumption of CO	[91]
Ideonella sp. O-1 (JCM17105)	Mineral salts medium H$_2$/O$_2$/CO$_2$ = 7:1:1 CO: 0–70 vol%	65.7–78% w/vol	0.17–0.26 h^{-1}/–	- Tolerant to CO (>70 vol%) but growth decreases if CO>20 vol% - No consumption of CO	[92]

Regarding the further development of the bioprocess, it was estimated that 1 kg of PHA produced from syngas would cost less than 1 USD [54], but it should be noted that this study is more than 20 years old. It would be valuable to conduct such a study again to evaluate the industrial and renewable syngas-based production of PHA as a biodegradable material. Three companies have been identified to be working on the biological conversion of syngas into useful products:

1. LanzaTech, founded more than 10 years ago, is developing biomass to bio-fuel process chain and has built several large-scale systems with several 500000-L vessels. Their larger plants are designed for producing 300 million tons of ethanol per year. From demonstration plants installed in the USA, China, Taiwan, Belgium, Japan, and New Zealand, the strategy is now ready to move to commercial installations [62]. The company has its own proprietary strain, *Clostridium autoethanogenum*, originally able to produce ethanol on a syngas substrate, that has been further optimized for the production of other products (acetone, isopropanol, isobutylene, and higher alcohols) by means of designing high-throughput pathways. The production of PHA was also considered but is not the main focus. However, another biopolymer, 2-hydroxybutyric acid, was developed in collaboration with Evonik.
2. Synata Bio is an American company that inherited the technology from another company, Coskata, Inc. It is based on the microorganism *Clostridium carboxidivorans*, initially for the production of ethanol.
3. In a joint venture with New Planet Energy, INEOS Bio designed a commercial factory based on *Clostridium ljungdahlii*, for syngas fermentation. Its capacity is meant to be 8 million gallons of ethanol per year.

Even though the fabrication of bioplastics was not the first priority for the last few decades, research in this direction has been conducted over the world. A new consciousness of the ever-increasing demand for fossil fuel-based plastics and the associated alarming environmental impact of these materials is increasing the interest in processes that enable the sustainable fabrication of bioplastics. The use of heterogeneous organic waste transformed into syngas, and especially the step of its subsequent transformation into useful products, such as bioplastics, has been scarcely studied. More efforts are required, in particular, for the screening and genetic engineering of robust strains, which are, ideally, syngas composition insensitive (especially toward CO), impurities tolerant, and present fast growth and a high content of PHA in optimized conditions on a sole syngas substrate. Moreover, design of efficient bioreactors for syngas fermentation, especially focusing on maximizing the mass transfer coefficient of the gas components, is required. Cost-efficient syngas purification technologies and PHA downstream processing methods also need to be the object of further research.

Another approach for harnessing the carbon and energy content of complex and variable substrates, such as syngas, would be to develop a consortium of syntrophic bacteria (or mixed culture or co-culture) [95]. This approach has been reported to be successful for more efficient biogas production (higher methane content) [96].

It could be imagined, for instance, that a microorganism is slowly growing but is efficient for CO oxidation (efficient CODH) while being insensitive to low amounts of oxygen, and is complemented by a hydrogen-oxidizing bacterium, which uses the CO_2 and H_2 provided for growth and PHA accumulation. Alternatively, a protected form of an O_2-tolerant CODH could be added to the bioreactor together with hydrogen-oxidizing bacteria, as these bacteria present a high accumulation of PHA. Similarly, a classic CODH coupled with carboxydobacteria, would also be possible. Finally, the same approach could be obtained by genetic engineering through recombinant strains containing the genes for the CODH.

Based on the discussion developed in this chapter, the use of the wide range and large amounts of available waste and biomass for the production of bioplastics through a syngas intermediate is of interest for the future sustainable supply of plastics. This requires, however, more effort in terms of microorganism development before becoming a reality.

REFERENCES

1. European Bioplastics. *Bioplastic Materials* [Internet]. Accessed 2019 Oct 30. Available from: https://www.european-bioplastics.org/bioplastics/materials/.
2. Ragaert P, Buntinx M, Maes C, et al. Polyhydroxyalkanoates for food packaging applications. In: *Reference Module in Food Science*. Elsevier, 2019.
3. Plastics Europe. *Plastics – The Facts 2018* [Internet]. Accessed 2019 Oct 30. Available from: https://www.plasticseurope.org/application/files/6315/4510/9658/Plastics_the _facts_2018_AF_web.pdf.
4. Senet S. *Plastic Production on the Rise Worldwide but Slowing in Europe* [Internet]. Euractiv. Accessed 2019 Oct 30. Available from: https://www.euractiv.com/section/en ergy-environment/news/while-global-plastic-production-is-increasing-worldwide-it-is -slowin-down-in-europe/.
5. Sardon H, Dove AP. Plastics recycling with a difference. *Science* 2018; 360(6387): 380–381.
6. European Bioplastics. *Bioplastics Market Data* [Internet]. Accessed 2019 Oct 30. Available from: https://www.european-bioplastics.org/market/.
7. Jiang G, Hill DJ, Kowalczuk M, et al. Carbon sources for polyhydroxyalkanoates and an integrated biorefinery. *Int J Mol Sci* 2016; 17(7): 1157–1178.
8. Kim BS, Lee SC, Lee SY, et al. Production of poly(3-hydroxybutyric acid) by fed-batch culture of *Alcaligenes eutrophus* with glucose concentration control. *Biotechnol Bioeng* 1994; 43(9): 892–898.
9. *World Corn Production 2018–2019* [Internet]. National Corn Growers Association. Accessed 2019 Oct 30. Available from: http://www.worldofcorn.com/#world-corn -production.
10. Pittmann T, Steinmetz H. Potential for polyhydroxyalkanoate production on German or European municipal waste water treatment plants. *Bioresour Technol* 2016; 214: 9–15.
11. Domínguez A, Fernández Y, Fidalgo B, et al. Biogas to syngas by microwave-assisted dry reforming in the presence of char. *Energ Fuel* 2007; 21(4): 2066–2071.
12. Reyes SC, Sinfelt JH, Feeley JS. Evolution of processes for synthesis gas production: Recent developments in an old technology. *Ind Eng Chem Res* 2003; 42(8): 1588–1597.
13. Atnaw SM, Sulaiman SA, Yusup S. Influence of fuel moisture content and reactor temperature on the calorific value of syngas resulted from gasification of oil palm fronds. *Sci World J* 2014; 2014: 1–9.

14. James AM, Yuan W, Boyette MD. The effect of biomass physical properties on top-lit updraft gasification of woodchips. *Energies* 2016; 9(4): 283–296.
15. Puigjaner L. Introduction. In: Puigjaner L (ed.): *Syngas from Waste Emerging Technologies*. London: Springer, 2011; pp. 1–10.
16. Abatzoglou N, Fauteux-Lefebvre C. Review of catalytic syngas production through steam or dry reforming and partial oxidation of studied liquid compounds. *WIREs Energ Environ* 2015; 5(2): 169–187.
17. Qin Y-H, Feng J, Li W-Y. Formation of tar and its characterization during air–steam gasification of sawdust in a fluidized bed reactor. *Fuel* 2010; 89(7): 1344–1347.
18. Font Palma C. Model for biomass gasification including tar formation and evolution. *Energ Fuel* 2013; 27(5): 2693–2702.
19. Oveisi E, Sokhansanj S, Lau A, et al. Characterization of recycled wood chips, syngas yield, and tar formation in an industrial updraft gasifier. *Environments* 2018; 5: 84–97.
20. Solarte-Toro JC, Chacón-Pérez Y, Cardona-Alzate CA. Evaluation of biogas and syngas as energy vectors for heat and power generation using lignocellulosic biomass as raw material. *Electron J Biotechnol* 2018; 33: 52–62.
21. *Syngas Cogeneration Combined Heat & Power* [Internet]. Clarke Energy, a Kohler Company. Accessed 2019 Oct 30. Available from: https://www.clarke-energy.com/synthesis-gas-syngas/.
22. *The Gasification Industry* [Internet]. Global Syngas Technologies Council GSTC. Accessed 2019 Oct 30. Available from: https://www.globalsyngas.org/resources/the-gasification-industry/.
23. Hislop D, Hall DO. *Biomass Resources for Gasification Power Plant* [Internet], 1996. Accessed 2019 Oct 30. Available from: https://www.google.com/url?sa=t&rct=j&q=&esrc=s&source=web&cd=4&ved=2ahUKEwip9YydgNPlAhUisKQKHakrAIQQFjADegQIBBAC&url=http%3A%2F%2Fwww.ieabioenergytask33.org%2Fapp%2Fwebroot%2Ffiles%2Ffile%2Fpublications%2FIEABMFeedHall.pdf&usg=AOvVaw17oJjkoSLCy2JhzFXUQp7d.
24. Karmann S, Panke S, Zinn M. Fed-batch cultivations of rhodospirillum rubrum under multiple nutrient-limited growth conditions on syngas as a novel option to produce poly(3-hydroxybutyrate) (PHB). *Front Bioeng Biotechnol* 2019; 7: 59–70.
25. Molino A, Larocca V, Chianese S, et al. Biofuels production by biomass gasification: A review. *Energies* 2018; 11(4): 811–842.
26. Rubiera F, Pis JJ, Covadonga P. Raw materials, selection, preparation and characterization. In: Puigjaner L (ed.): *Syngas from Waste Emerging Technologies*. London: Springer, 2011; pp 11–22.
27. Heidenreich S, Foscolo PU. New concepts in biomass gasification. *Prog Energ Combust* 2015; 46: 72–95.
28. Kruse A. Supercritical water gasification. *Biofuel Bioprod Biorefin* 2008; 2(5): 415–437.
29. Guo Y, Wang SZ, Xu DH, et al. Review of catalytic supercritical water gasification for hydrogen production from biomass. *Renew Sust Energ Rev* 2010; 14(1): 334–343.
30. Tang L, Huang H, Hao H, et al. Development of plasma pyrolysis/gasification systems for energy efficient and environmentally sound waste disposal. *J Electrostat* 2013; 71(5): 839–847.
31. Revelles O, Tarazona N, García JL, et al. Carbon roadmap from syngas to polyhydroxyalkanoates in *Rhodospirillum rubrum*. *Environ Microbiol* 2015; 18(2): 708–720.
32. Borges FC, Du Z, Xie Q, et al. Fast microwave assisted pyrolysis of biomass using microwave absorbent. *Bioresour Technol* 2014; 156: 267–274.
33. Domínguez A, Menéndez JA, Fernández Y, et al. Conventional and microwave induced pyrolysis of coffee hulls for the production of a hydrogen rich fuel gas. *J Anal Appl Pyrol* 2007; 79: 128–135.

34. Huang YF, Kuan WH, Lo SL, et al. Hydrogen-rich fuel gas from rice straw via micro-wave-induced pyrolysis. *Bioresour Technol* 2010; 101(6): 1968–1973.

35. Miura M, Kaga H, Sakurai A, et al. Rapid pyrolysis of wood block by microwave heating. *J Anal Appl Pyrol* 2004; 71(1): 187–199.

36. Beneroso D, Bermúdez JM, Arenillas A, et al. Microwave pyrolysis of microalgae for high syngas production. *Bioresour Technol* 2013; 144: 240–246.

37. Menéndez JA, Arenillas A, Fidalgo B, et al. Microwave heating processes involving carbon materials. *Fuel Process Technol* 2010; 91(1): 1–8.

38. Pérez-Fortes M, Bojarski AD. Main purification operations. In: Puigjaner L (ed.): *Syngas from Waste Emerging Technologies*. London: Springer, 2011; pp 89–120.

39. Ahmad AA, Zawawi NA, Kasim FH, et al. Assessing the gasification performance of biomass: A review on biomass gasification process conditions, optimization and economic evaluation. *Renew Sust Energ Rev* 2016; 53: 1333–1347.

40. Wang X, Li M, Wang M, et al. Thermodynamic analysis of glycerol dry reforming for hydrogen and synthesis gas production. *Fuel* 2009; 88(11): 2148–2153.

41. Valliyappan T, Ferdous D, Bakhshi NN, et al. Production of hydrogen and syngas via steam gasification of glycerol in a fixed-bed reactor. *Top Catal* 2008; 49(1–2): 59–67.

42. Tou M, Jin J, Hao Y, et al. Solar-driven co-thermolysis of CO_2 and H_2O promoted by in situ oxygen removal across a non-stoichiometric ceria membrane. *React Chem Eng* 2019; 4(8): 1431–1438.

43. Li J, Yin Y, Zhang X, et al. Hydrogen-rich gas production by steam gasification of palm oil wastes over supported tri-metallic catalyst. *Int J Hydrogen Energ* 2009; 34(22): 9108–9115.

44. Chang ACC, Chang H-F, Lin F-J, et al. Biomass gasification for hydrogen production. *Int J Hydrogen Energ* 2011; 36(21): 14252–14260.

45. Wongsiriamnuay T, Kannang N, Tippayawong N. Effect of operating conditions on catalytic gasification of bamboo in a fluidized bed. *Int J Chem Eng* 2013; 2013: 1–9.

46. Lv PM, Xiong ZH, Chang J, et al. An experimental study on biomass air–steam gasification in a fluidized bed. *Bioresour Technol* 2004; 95(1): 95–101.

47. Mohammed MAA, Salmiaton A, Wan Azlina WAKG, et al. Air gasification of empty fruit bunch for hydrogen-rich gas production in a fluidized-bed reactor. *Energ Convers Manage* 2011; 52(2): 1555–1561.

48. He M, Xiao B, Liu S, et al. Syngas production from pyrolysis of municipal solid waste (MSW) with dolomite as downstream catalysts. *J Anal Appl Pyrol* 2010; 87(2): 181–187.

49. Luo S, Zhou Y, Yi C. Syngas production by catalytic steam gasification of municipal solid waste in fixed-bed reactor. Integration and Energy System Engineering, European Symposium on Computer-Aided Process Engineering 2011. *Energy* 2012; 44(1): 391–395.

50. Zhang Q, Dor L, Fenigshtein D, et al. Gasification of municipal solid waste in the plasma gasification melting process. Energy Solutions for a Sustainable World, Special Issue of International Conference of Applied Energy, ICA2010, April 21–23, 2010, Singapore. *Appl Energ* 2012; 90(1): 106–112.

51. Nipattummakul N, Ahmed II, Kerdsuwan S, et al. Hydrogen and syngas production from sewage sludge via steam gasification. *Int J Hydrogen Energ* 2010; 35: 11738–11745.

52. Ebadi AG, Hisoriev H, Zarnegar M, et al. Hydrogen and syngas production by catalytic gasification of algal biomass (*Cladophora glomerata* L.) using alkali and alkaline-earth metals compounds. *Environ Technol* 2019; 40(9): 1178–1184.

53. Latif H, Zeidan AA, Nielsen AT, et al. Trash to treasure: Production of biofuels and commodity chemicals via syngas fermenting microorganisms. *Curr Opin Biotechnol* 2014; 27: 79–87.

54. Choi J-I, Lee SY. Process analysis and economic evaluation for poly(3-hydroxybutyrate) production by fermentation. *Bioprocess Eng* 1997; 17(6): 335–342.

55. Do YS, Smeenk J, Broer KM, et al. Growth of *Rhodospirillum rubrum* on synthesis gas: Conversion of CO to H2 and poly-β-hydroxyalkanoate. *Biotechnol Bioeng* 2007; 97(2): 279–286.

56. Martin WF. Hydrogen, metals, bifurcating electrons, and proton gradients: The early evolution of biological energy conservation. *FEBS Lett* 2012; 586(5): 485–493.

57. Ragsdale SW. Enzymology of the Wood-Ljungdahl pathway of acetogenesis. *Ann NY Acad Sci* 2008; 1125: 129–136.

58. Drake HL, Gößner AS, Daniel SL. Old acetogens, new light. *Ann NY Acad Sci* 2008; 1125: 100–128.

59. Balch WE, Schoberth S, Tanner RS, et al. Acetobacterium, a new genus of hydrogen-oxidizing, carbon dioxide-reducing, anaerobic bacteria. *Int J Syst Evol Microbiol* 1977; 27(4): 355–361.

60. Madigan MT, Martinko JM, Stahl DA, et al. *Brock Biology of Microorganisms*. Ed 13. Benjamin Cummings, 2010.

61. *Presentation by LanzaTech* [Internet]. LanzaTech. Accessed 2019 Oct 30. Available from: https://www.energy.gov/sites/prod/files/2017/07/f35/BETO_2017WTE-Works hop_SeanSimpson-LanzaTech.pdf.

62. Presentation by Lanzatech. *CCU-Now: Fuels and Chemicals from Waste* [Internet]. LanzaTech. Accessed 2019 Oct 30. Available from: https://ec.europa.eu/energy/sites/e ner/files/documents/25_sean_simpson-lanzatech.pdf.

63. Slivka RM, Chinn MS, Grunden AM. Gasification and synthesis gas fermentation: An alternative route to biofuel production. *Biofuels* 2014; 2(4): 405–419.

64. Straub M, Demler M, Weuster-Botz D, et al. Selective enhancement of autotrophic acetate production with genetically modified *Acetobacterium woodii*. *J Biotechnol* 2014; 178: 67–72.

65. Demler M, Weuster-Botz D. Reaction engineering analysis of hydrogenotrophic production of acetic acid by *Acetobacterium woodii*. *Biotechnol Bioeng* 2010; 108(2): 470–474.

66. Cotter JL, Chinn MS, Grunden AM. Ethanol and acetate production by *Clostridium ljungdahlii* and *Clostridium autoethanogenum* using resting cells. *Bioprocess Biosyst Eng* 2009; 32(3): 369–380.

67. Calam CT. Isolation of *Clostridium acetobutylicum* strains producing butanol and acetone. *Biotechnol Lett* 1980; 2(3): 111–116.

68. Häggström L, Molin N. Calcium alginate immobilized cells of *Clostridium acetobutylicum* for solvent production. *Biotechnol Lett* 1980; 2(5): 241–246.

69. Maddox IS. Production of n-butanol from whey filtrate using *Clostridium acetobutylicum* N.C.I.B. 2951. *Biotechnol Lett* 1980; 2(11): 493–498.

70. Gapes JR. The economics of acetone-butanol fermentation: Theoretical and market considerations. *J Mol Microb Biotechnol* 2000; 2(1): 27–32.

71. Flüchter S, Follonier S, Schiel-Bengelsdorf B, et al. Anaerobic production of poly(3-hydroxybutyrate) and its precursor 3-hydroxybutyrate from synthesis gas by autotrophic Clostridia. *Biomacromolecules* 2019; 20(9): 3271–3282.

72. Koller M. A review on established and emerging fermentation schemes for microbial production of polyhydroxyalkanoate (PHA) biopolyesters. *Fermentation* 2018; 4(2): 30.

73. Fuller RC. Polyesters and photosynthetic bacteria. In: Blankenship RE, Madigan MT, Bauer CE (eds.): *Anoxygenic Photosynthetic Bacteria*. Dordrecht: Springer Netherlands, 1995; pp 1245–1256.

74. Schultz JE, Weaver PF. Fermentation and anaerobic respiration by *Rhodospirillum rubrum*; and *Rhodopseudomonas capsulata*. *J Bacteriol* 1982; 149(1): 181–190.

75. Lehman LJ, Roberts GP. Identification of an alternative nitrogenase system in *Rhodospirillum rubrum*. *J Bacteriol* 1991; 173(18): 5705–5711.

76. Grammel H, Gilles E-D, Ghosh R. Microaerophilic cooperation of reductive and oxidative pathways allows maximal photosynthetic membrane biosynthesis in *Rhodospirillum rubrum*. *Appl Environ Microbiol* 2003; 69(11): 6577–6586.
77. Najafpour G, Younesi H, Mohamed AR. Effect of organic substrate on hydrogen production from synthesis gas using *Rhodospirillum rubrum*, in batch culture. *Biochem Eng J* 2004; 21(2): 123–130.
78. Najafpour GD, Younesi H. Bioconversion of synthesis gas to hydrogen using a light-dependent photosynthetic bacterium, *Rhodospirillum rubrum*. *World J Microbiol Biotechnol* 2007; 23: 275–284.
79. Younesi H, Najafpour G, Ku Ismail KS, et al. Biohydrogen production in a continuous stirred tank bioreactor from synthesis gas by anaerobic photosynthetic bacterium: *Rhodopirillum rubrum*. *Bioresour Technol* 2008; 99(7): 2612–2619.
80. Rudolf C, Grammel H. Fructose metabolism of the purple non-sulfur bacterium *Rhodospirillum rubrum*: Effect of carbon dioxide on growth, and production of bacteriochlorophyll and organic acids. *Enzyme Microb Technol* 2012; 50(4–5): 238–246.
81. Narancic T, Scollica E, Kenny ST, et al. Understanding the physiological roles of polyhydroxybutyrate (PHB) in *Rhodospirillum rubrum* S1 under aerobic chemoheterotrophic conditions. *Appl Microbiol Biotechnol* 2016; 100(20): 8901–8912.
82. Brandl H, Knee EJ, Fuller RC, et al. Ability of the phototrophic bacterium *Rhodospirillum rubrum* to produce various poly (β-hydroxyalkanoates): Potential sources for biodegradable polyesters. *Int J Biol Macromol* 1989; 11(1): 49–55.
83. Karmann S, Follonier S, Egger D, et al. Tailor-made PAT platform for safe syngas fermentations in batch, fed-batch and chemostat mode with *Rhodospirillum rubrum*. *Microb Biotechnol* 2017; 10(6): 1365–1375.
84. Maness PC, Huang J, Smolinski S, et al. Energy generation from the CO oxidation-hydrogen production pathway in *Rubrivivax gelatinosus*. *Appl Environ Microbiol* 2005; 71(6): 2870–2874.
85. Heo J, Halbleib CM, Ludden PW. Redox-dependent activation of CO dehydrogenase from *Rhodospirillum rubrum*. *Proc Natl Acad Sci USA* 2001; 98(14): 7690–7693.
86. Feng J, Lindahl PA. Carbon monoxide dehydrogenase from *Rhodospirillum rubrum*: Effect of redox potential on catalysis. *Biochemistry* 2004; 43(6): 1552–1559.
87. Klask C. Heterologous expression of various PHA synthase genes in *Rhodospirillum rubrum*. *Chem Biochem Eng Q* 2015; 29(2): 75–85.
88. Heinrich D, Raberg M, Fricke P, et al. Synthesis gas (syngas)-derived medium-chain-length polyhydroxyalkanoate synthesis in engineered *Rhodospirillum rubrum*. *Appl Environ Microbiol* 2016; 82(20): 6132–6140.
89. Volova T, Zhila N, Shishatskaya E. Synthesis of poly(3-hydroxybutyrate) by the autotrophic CO-oxidizing bacterium *Seliberia carboxydohydrogena* Z-1062. *J Ind Microbiol Biotechnol* 2015; 42(10): 1377–1387.
90. Meyer O, Lalucat J, Schlegel HG. *Pseudomonas carboxydohydrogena* (Sanjieva and Zavarzin) comb. nov., a monotrichous, nonbudding, strictly aerobic, carbon monoxide-utilizing hydrogen bacterium previously assigned to seliberia. *Int J Syst Evol Microbiol* 1980; 30(1): 189–195.
91. Volova T, Kalacheva G, Altukhova O. Autotrophic synthesis of polyhydroxyalkanoates by the bacteria *Ralstonia eutropha* in the presence of carbon monoxide. *Appl Microbiol Biotechnol* 2002; 58(5): 675–678.
92. Tanaka K, Miyawaki K, Yamaguchi A, et al. Cell growth and P(3HB) accumulation from CO_2 of a carbon monoxide-tolerant hydrogen-oxidizing bacterium, Ideonella sp. O-1. *Appl Microbiol Biotechnol* 2011; 92(6): 1161–1169.
93. Kerby RL, Ludden PW, Roberts GP. Carbon monoxide-dependent growth of *Rhodospirillum rubrum*. *J Bacteriol* 1995; 177(8): 2241–2244.

94. Tanaka K, Ishizaki A, Kanamaru T, et al. Production of poly(D-3-hydroxybutyrate) from CO_2, H_2, and O_2 by high cell density autotrophic cultivation of *Alcaligenes eutrophus*. *Biotechnol Bioeng* 1995; 45(3): 268–275.
95. Peng XN, Gilmore SP, O'Malley MA. Microbial communities for bioprocessing: Lessons learned from nature. *Curr Opin Chem Eng* 2016; 14: 103–109.
96. Müller B, Sun L, Schnürer A. First insights into the syntrophic acetate-oxidizing bacteria-a genetic study. *MicrobiologyOpen* 2013; 2: 35–53.

Index

Printed in the United States
by Baker & Taylor Publisher Services